15

W9-AQM-258

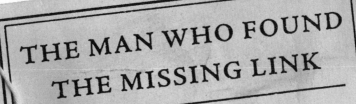

THE MAN WHO FOUND
THE MISSING LINK

Eugène Dubois and His Lifelong
Quest to Prove Darwin Right

PAT SHIPMAN

SIMON & SCHUSTER   New York · London · Toronto · Sydney · Singapore

Also by Pat Shipman

Taking Wing

The Evolution of Racism

The Wisdom of the Bones *with Alan Walker*

The Neandertals *with Erik Trinkaus*

 SIMON & SCHUSTER
Rockefeller Center
1230 Avenue of the Americas
New York, NY 10020

Copyright © 2001 by Pat Shipman
All rights reserved,
including the right of reproduction
in whole or in part in any form.

SIMON & SCHUSTER and colophon are registered trademarks
of Simon & Schuster, Inc.

Book design by Christopher Kuntze and Claire Van Vliet

Manufactured in the United States of America

10 9 8 7 6 5 4 3 2 1

Library of Congress Cataloging-in-Publication Data
Shipman, Pat, date.
    The man who found the missing link : Eugène Dubois and his lifelong quest to
prove Darwin right / Pat Shipman.
      p.   cm.
    1. Java man. 2. Dubois, Eugène, 1858–1940. 3. Physical anthropologists—
Netherlands—Biography. 4. Physical anthropologists—Indonesia—Java—Biography.
I. Title.
GN284.6 .S55 2001
569.9—dc21   00-044049

ISBN 0-684-85581-X

For M. E. F. T. D.,
of course

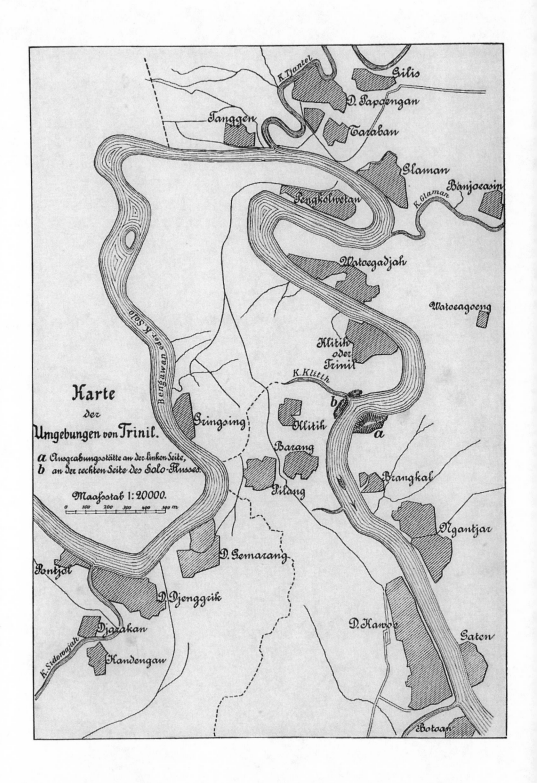

Karte
der
Umgebungen von Trinil.

a Ausgrabungsstätte an der linken Seite,
b an der rechten Seite des Solo-Flusses.

Maaßstab 1 : 20000.

0   100   200   300   400   500 m

CONTENTS

Author's Note — ix

Chapter 1 An Echo of the Past — 1

Chapter 2 The Beginning — 9

Chapter 3 The Game — 21

Chapter 4 Ambition — 25

Chapter 5 Lightning Rod — 31

Chapter 6 Love and Conflict — 41

Chapter 7 Turning Point — 49

Chapter 8 To Find the Missing Link — 59

Chapter 9 Logistics — 66

Chapter 10 Padang — 80

Chapter 11 Pajakombo — 95

Chapter 12 Fossils — 99

Chapter 13 Garuda — 107

Chapter 14 Fevers and Spells — 110

Chapter 15 To Java — 118

Chapter 16 Java Fossils — 128

Chapter 17 Coolies — 132

Chapter 18 Discoveries at Trinil — 138

Chapter 19 Gathering Resources — 145

Chapter 20 Friendship — 153

Chapter 21 Trinil — 159

Chapter 22 The Birth of *Pithecanthropus* — 165

Chapter 23 1893 — 171

Chapter 24 Disaster — 182

Chapter 25 Letters from a Friend — 186

Chapter 26 Aftermath — 193

Chapter 27 Perseverance — 196

Chapter 28 The Monograph — 202

Chapter 29 Writing Up — 207

Chapter 30 Separation and Loss — 211

Chapter 31 Intermission — 218

Chapter 32 To India   223

Chapter 33 Calcutta   227

Chapter 34 Sirmoor State   243

Chapter 35 Siwalik Adventures   249

Chapter 36 Leaving India   256

Chapter 37 Toeloeng Agoeng   261

Chapter 38 Departure   267

Chapter 39 Europe   271

Chapter 40 The Battlefield   277

Chapter 41 More Skirmishes   297

Chapter 42 Using His Brains   305

Chapter 43 Betrayal and Resurrection   316

Chapter 44 Family   322

Chapter 45 The New Century   327

Chapter 46 Diversions   335

Chapter 47 Tragedy   346

Chapter 48 Dangerous Times   354

Chapter 49 A New Skull   366

Chapter 50 Rumors and Isolation   373

Chapter 51 Brain Work   378

Chapter 52 The Diligent Assistant   386

Chapter 53 New Skulls from Java   408

Chapter 54 A Worthy Opponent   415

Chapter 55 To the Battlefront   422

Chapter 56 The Letter   428

Chapter 57 Pretender to the Throne   431

Chapter 58 Old Friends   441

Chapter 59 The Final Conflict   445

Epilogue   452

Notes   455

Glossary   476

Bibliography   479

Acknowledgments   497

Index   499

AUTHOR'S NOTE

After Dubois' death, his daughter Eugénie spent days burning materials that she did not want to go to the Dubois Archives (letter, Eugénie to Brongersma, January 21, 1941). Extensive documentary evidence remains intact and is cited in the endnotes; sources have been edited lightly for clarity, if at all. Translations from the Dutch and German are almost all the work of my invaluable research assistant, Dr. Paul Storm. The endnotes also indicate where I have filled in intriguing omissions resulting from Eugénie's actions. I have followed the Dutch conventions in capitalization of names.

We live by admiration, love, and hope.

—Wordsworth

Words copied by Dubois onto the frontispiece of the
field notebook he used starting September 1893.

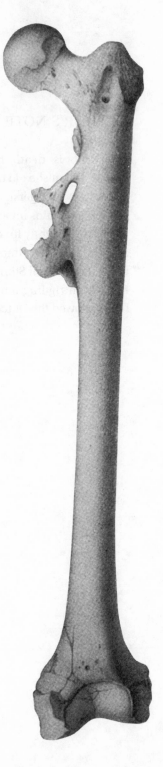

THE MAN

WHO FOUND THE

MISSING LINK

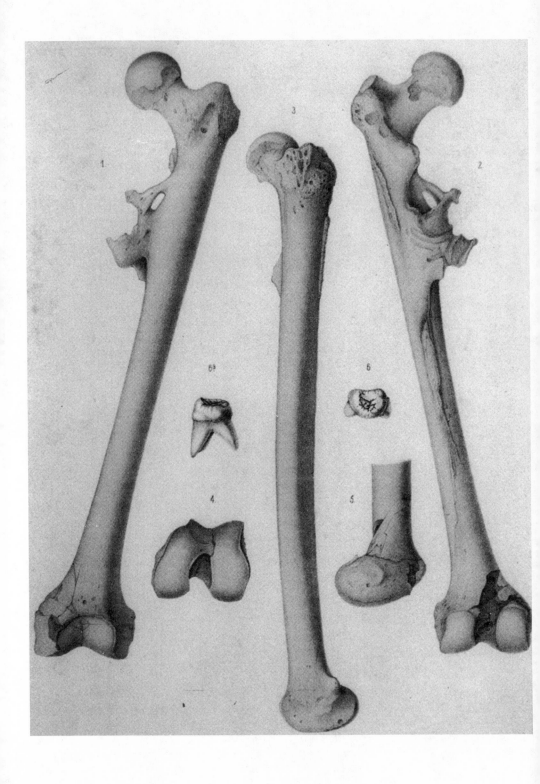

CHAPTER 1 AN ECHO OF THE PAST

The letter comes by the last post on a weakly sunny afternoon in February of 1937. Looking out the window, Dubois searches for the slightest hint of green that he knows will come first to the willow trees at De Bedelaar. It is still very brisk out, not yet warm. The promise of renewal seems cruel when any real hope of it is still far away.

He is slow to realize what has come, it is so unexpected. His mind is not so quick as it once was, now that he is in his eightieth year, although he has not become so vague as his wife, Anna. The servant girl brings the letters in as usual; he sits down at the desk, puts on his glasses, and takes up his letter opener. He carefully inserts the blade into the corner of the flap and slits each envelope neatly. It is his habit to read his letters in order, placing them in a tidy pile before pausing to compose his answers. While he skims the letters, Anna prattles on, unaware that anything of significance is happening. Most of the post is ordinary—bills, a few letters of inquiry from colleagues or students. When he picks up the last letter, unsuspecting, he is momentarily confused by the two handwritings on the envelope. The hand that wrote his name is crabbed and somehow familiar, but he cannot place it immediately; the other, which wrote the address, is completely unknown to him. When he opens that envelope and sees the tissue-thin paper inside, something stirs in his memory. As soon as he reads the salutation, he *knows,* as if he has been expecting this letter for years.

He decides that he cannot read it in front of Anna and rings for the servant. "Take your mistress to sit in the back garden for a while," he says to her. He waits until they have left and the room is quiet to unfold the translucent, crinkling pages. He doesn't need to turn over the last page to look for the signature. There is only one person from whom it can have come, only one person in Kediri, in faraway Java, who would be writing him.

"My dear doctor," the letter begins. He always addressed Dubois so, even when they saw each other daily. How long has it been since Dubois heard from him? It must have been *forty* years. Forty years since his friend called him "My dear doctor." From anyone else, these words might be only a courtesy, an acknowledgment of his professorial status and medical degree. From Prentice they were a term of endearment, an evocation of the intimacy and friendship they shared so long ago.

Kediri   February 7, 1936

> *An echo of the Past!*
> *"Dost thou recall?"*

My dear doctor,

You will hardly expect a letter from me! It is long, so very long since last we saw each other.

The philosopher, Renan, in addressing the shade of his departed sister who while in life had accompanied him in his sojourn in the Holy Land, said:—

> *"Dost thou recall from the bosom of God where thou reposest"—*

and I might say now:—

Dost thou recall from the quietness of your peaceful study in the homeland—the days now long, long flown which we passed together in the peaceful atmosphere of dear old Mr. Boyd's Koffeeland Mringin,—the good old man's dwelling Ngrodjo, Willisea the block house he put up for you at Jonojang, my own quiet abode at remote Tempoersarie?

Do you remember the many pleasant meetings we had at Ngrodjo when the old gentleman & I listened with so much interest to your enlightening & informative conversation? Indeed we learned much from you, and our minds ever reverted with satisfaction to the many agreeable meetings we three had together. Do you recall our excursion to Trinil the scene of your labors (where the famous Pithecanthropus erectus *was found), when contrary to your wont you regaled us at dinner in the evening with a bottle of wine saying it "aided digestion." Do you remember our bathing next day in the river, our pleasant walk in the afternoon to the station along the country road where a snake swallowed a frog and you at once ran to the rescue forcing the snake to disgorge the frog which, still quite alive, first looked to the right & to the left, and then lightheartedly plunged into the stream by the roadside? Do you remember the beautiful flowers at the station which we looked at while waiting for the train? One had a delicate light blue tint and you said that was well nigh your favorite color!*

Do you remember the long walks you & I had through the widespread coffee gardens at Mringin? . . .

Do you remember the two corporals of the engineers who looked after your team of convicts, at the excavation work? Their mode of life ever amused you—living like kings at the beginning of each month when *money was plentiful, and ever on <u>very</u> short rations towards the <u>end</u> of the month when the money was <u>all spent!</u> Through your favorable report they got promoted in time to the rank of Sergeant. Then you photographed them & noticed how they were maneuvering to bring full into the picture their arm shewing the new sergeant's <u>stripes!</u> . . . And Mr. Mulder, P. T. Sanvraar, & Mr. Turner, controller, at Toeloeng Agoeng. Do you remember our age—you, Mr. Mulder & I—was 34 years. Ah, yes, the golden days of youth! Perhaps we had our troubles, too, but we had <u>youth, health, home,</u> and <u>length of days</u> before us! Dost thou recall?*

As oft as I look back, the recollection of that happy time is a green spot in my memory, and will endure as long as life lasts!

Good old Mr. Boyd died in 1902 at Kediri under Van Buren's care from <u>cancer of the throat,</u> aged 74, & was buried at Toeloeng Agoeng. We were all present, and the Asst. Resident, Regent & etc & etc attended also. It may be the kind old man smoked too heavily?

While he still lived we often spoke of you after your departure from Java, very, very, often, & always with esteem & affection. Yes, we both loved you, and never could forget you! Like a sun that had come into our orbit you brought us light and happiness—it was just a chance in life never likely to recur, for <u>when</u> does it happen that a man of learning ever comes to live on a coffee estate <u>for any length of time?</u>

Later on coffee prices fell by 50% and all profit was gone. After Mr. Boyd's death Mringin was sold. Eventually it was given back to government & is now <u>Bosch Reserve,</u> no Ngrodjo, Willisea, Tempoersarie are all <u>forest</u> now, with not a soul living there any more, and only you & I remain today to muse over the past! Who would have foretold <u>that,</u> 40 years ago, when you & I were young!

My own coffee place near Mringin (Djaean was the name) suffered from the crisis in coffee & after 5 years I left it to be manager of the tapioca fabriek—the plantation called Brangganan at Ngadiloewek, 8 miles from Kediri on the high road to Kras, which an intimate friend of mine had taken over. Djaean we kept on, & sold later to rubber tree people in Lombok, who still work the estate. So it is <u>not</u> closed like good old Mringin. There I remained about 20 years. Much money was made & eventually my friend sold it for a fancy price & my work there ended. I was then 60 years of age. Since that time I have been interested in other things. I made enough money, but owing to the terrible slump of the past 6 years (malaise) nearly everybody in Java is about bankrupt! Like the majority I *have lost cruelly. At least ⅔ of what I had are gone. Today I have to live fru-gally to manage. I am still dependent on coffee for part of my income & coffee alas is down to f8–9.70 per picol—(formerly f55 to f60), simply ruinous; but there it is, & nothing can be done for no one is to blame—it is just a long spell of <u>bad times</u> with all values desperately <u>low</u>—no profit possible, and—*

"What can't be cured
Must be endured!"

One has just to make the best of things. The depression has been world-wide, & all have suffered. In Java those who had anything, have lost ½, or ¾, or all. . . . Last year (1935) only ±16% of the sugar mills of Java worked. The rest were closed & the staff discharged. A terrible loss to the Treasury, & to the country. . . .

However nothing can be done save to live on quietly in hope, or wait on better days coming soon. So I shall leave this gloomy theme & not depress you with a tale of woe!

I saw in the papers that you retired when 70 years old in 1928. Of course Mr. Boyd & I <u>fully expected</u> you to become <u>Professor;</u> it could not have been otherwise for your whole mind was ever bent on the acquisition of knowl-edge. I hope your life at home has been agreeable & satisfactory—that you have but <u>few</u> regrets & have experienced <u>no</u> heavy losses of whatever sort.

I must now say goodbye. I trust you are in good health. With all good wishes to you & yours,

Believe me as always,
Sincerely yours,
Adam Prentice

P.S. My own health is fairly good. I never had any very serious sick-ness to speak about—a little dysentery once or twice, years ago; and I have still pleasure in existence. But we are getting up in years & haven't the vigor of former days. I will be 78 years old in a few months. You will be about the same, and I fear not many of our friends remain today!

N.B. This letter will be forwarded to you by Mr. C. Van den Koppel, a state official of Batavia now traveling to Holland via Australia & America. He will find out your address at present in Holland.

Again goodbye—"Fare thee well"!

Dubois folds the letter back up and places it in his lap for a moment. When was the last time someone told him that they loved him? When was the last time that he brought light and happiness into someone's life? He is close to weeping as the memories flood his mind.

To know that Prentice remembers, too—that the past echoes for him, too—is sweet, but the pleasure is tinged with sad mockery. He and Prentice are no longer the handsome young men they once were, when they were together in Java. Thirty-four years old! What an age: so young, so hopeful, so naïve. Now Dubois feels old and fat and cold and tired.

Apparently Prentice never made enough money to return to Europe, as he once wished. But maybe he got used to the way of living there in Java, with the warmth and the sunshine and the servants. He had a beautiful nyai, a sweet and quiet native mistress, who looked after him in those years after his wife died. Is she still there, aged but graceful, looking after him even now? Those who come back can't live like that, with servants and large houses with fine gardens. The trouble is that Java could never have been Dubois' home. Even if he had adapted enough to be considered Indische, he would always have been an outsider, born in Europe. As for himself, Dubois couldn't have stayed; his work was here. He is glad to be here at his beautiful De Bedelaar, his own Dutch corner of nature, near Haelen.

Dubois catalogues his possessions in his mind, systematically. He has his house and his garden, his lake, his woods, his birds, his library, his specimens. He does not have youth or hope, or even much ambition, anymore. He does not have Prentice, or anyone who loves him. Prentice was a true companion. They were young together, and so sure, so certain—and now, no one else remembers but Prentice.

He hears the outer door open and close and knows that Anna has come back into the house. He isn't ready to see her, or anyone else. It is impossible, with so much in his heart and on his face, too, no doubt. How can he explain why the letter has affected him so? He could not let her read it; it would be a desecration. He takes off his glasses and carefully places them in his top pocket; the letter, too, goes there for safekeeping. Then he rises, goes into the hallway, and puts on his felt hat and a loose coat, not bothering to button it properly. He leaves by the front door, walking out slowly toward the bench by the lake. He often sits there in the afternoons; it is a good time to watch the birds and note which ones have returned from their winter farther south. No one will think it strange of him to sit alone there, thinking, for some hours. He lives a solitary life anyway, even though Anna still visits occasionally.

The earth along the path through the beechwood is springy and soft, from centuries of accumulation of fallen leaves and moss. The ground is no longer hard and bare; with his scientist's eye, he notices the few brave plants that are poking their green noses up through the leaf litter. But there are not many yet; it is too early, too cold still for them, he thinks. Nature always has time; she does not hurry. The woods know that life returns, that nothing is ever gone forever. It is a luxury of the young and of plants, to be so certain.

He reaches the bench and uses his sleeve to wipe off a faint skim of frost, leaving a shadow of moisture and dirt on the wool. He sits heavily and removes the letter from his pocket. He unfolds it carefully, but he doesn't read it again right away. He simply holds the pages in his large, soft, liver-spotted hand. His hands were once so skilled; they dissected the finest anatomical structures, drew close likenesses, and even sculpted a figure once or twice; these hands painstakingly removed rock matrix from price-less fossil specimens. Once he was envied for his fine manual skills as well as for his brains; too many scholars, even physicians, were ham-handed and clumsy, needing others to carry out the detailed work. Not him, not then. And now his hands lie gnarled and crooked-fingered in his lap, hold-ing a letter from his past.

To have Prentice suddenly reappear like that, sounding as full of joy and life as always, understanding Dubois' mind better than he ever did him-self . . . it is almost too much to bear. Dubois cannot think clearly. Indeed, he can hardly breathe for the shock of it. It is like having an attack of asthma; his chest is tight. He takes off his hat, crumpling the letter a little in the process. With the other hand, he rubs his bald head, disarranging the fringe of snow-white hair over his ears. It is a gesture he makes often when he is thinking now; it gives him an oddly wild look, like a merganser chick: untamed, startled, perhaps about to try to take flight. After a moment, he puts his hat back on his head for warmth, but carelessly, not setting it straight. He is an old man. He is not concerned with his appearance.

He feels the texture of the letter in his hand and sits very still, looking at the lake, watching for coots and ducks. He is remembering those years in Java, and that companionship. There was never another time like it.

He has achieved everything he set out to, even though all about him scoffed. It was in Java that he became the man who found the missing link: him, Marie Eugène François Thomas Dubois. In Java, he met his true des-tiny and began his true life. The missing link has been the most important thing in his life, as he knew it would be. It was . . . everything.

But now he realizes what he left behind in Java, something he did not know was of such value. For there has never been another friend like Prentice, never another companion of his heart and mind like him. It was a little improbable, that friendship between the Dutch scientist and the Scottish planter. Their backgrounds were different, their training had little overlap, but they both shared a few important things: burning curiosity, physical vigor, and an urgent need to show what they could do when they were freed from the petty restrictions of small-minded European society.

He turns the facts over and over in his mind, musing. How rarely such a friendship comes into a man's life; he didn't know then that it was his only chance. Then, he only thought of finding the missing link and achieving something important in science. He knew Prentice's companionship was a great comfort, for the Scotsman understood the significance of what Dubois was doing. They had in common that need to do something grand. How could a man bear to live without trying to make his mark in the world? In the end, Dubois accomplished something important, very important. He should have valued Prentice more highly, though; he was too preoccupied with fossils. He didn't know how cold and lonely the years after the discovery would be. The years of his greatest professional accomplishments, the years in which his name and discoveries have become famous around the world, have left him living here, in solitude. He has only the trees and the birds and the lake for company.

He stares at the thin light playing on the water. Sometimes a ripple catches the sunshine, reflecting it brightly enough to make his pale blue eyes water in sympathy.

The light was different in Java: hotter, more merciless, sometimes inca-pacitating. Sometimes there was so much light that he couldn't bear to be outside, and then he would seek the shade, any meager, pale scrap of shade.

In the forests, the light was dappled, filtered through a thousand tones of green and yellow, and the air was tangible with humidity. It was like a wall that pushed against him at every step. Sometimes it was so quiet that the song of an unseen bird would split the air, like a gunshot, making him jump. And there were those interminable insects, which creaked and popped and whirred like a madman's confounded invention, deep in the forest where no inventors ever strayed, where no books were ever read. There was just forest and green and leaves and more green. Those locusts would carry on and on and on with their rasping noise until that became embedded in his brain, echoing in the rhythm of his breathing. And then they'd stop, leaving a silence so profound that it could wake him out of a sound sleep. Or it might be a pair of long-armed gibbons sitting on a branch bellowing their gurgling, echoing call through the forest, the call that he thought must be heard all the way to Sumatra.

Forest! What a word for it. It was no forest like this beechwood at De Bedelaar. It was another thing entirely, a jungle, a creature in and of itself. In some places, he walked in a cathedral of trees, forest giants that soared above his head blocking out most of the light. In other places, he could not walk because there was so much vegetation. He'd fight his way up the hillsides, sweating and straining up the steep slopes, leading his weary horse, urging the men on, slashing at the alang-alang grass that shredded their clothes and skin. Tiny pieces of alang-alang work their way into the skin, leaving itching, red rashes that take days to heal.

Every night, the routine was the same; Dubois saw to that. Stop, make camp, talk to each man—not just the engineers—to find out if any of them had seen anything useful. Then clean the new wounds and blisters and lesions, bandage the worst of them, hope no infections would grow tomorrow. Salve the rashes and insect bites, clean and treat yesterday's sores that were still swollen and red, dose those men with fever. While he attended to the health of the coolies, he'd have one of the engineers assess the supplies while the other sent sound men to get water and firewood. The cook, the kokkie, would be setting up the kitchen and he'd remind her, every night, to boil the drinking water thoroughly before using it. Then Dubois would have a long drink of water, or juice if there was any, and a bath. While he waited for dinner, he'd make notes and consult the map and plan the next day's survey.

Nothing was easy, even with the engineers and all the coolies to do the carrying and the heavy labor. Sometimes he thought they only made more people to look after, until he remembered what it was like to work on his own, packing and unpacking his horse, making the fire, buying the food and cooking it too. Still, he had to do so much himself, all the medical work, all the thinking, all the decisions. No one else knew anything about fossils, and even the engineers didn't know much useful geology. But they went on, they always went on—looking for caves, looking for fossils, looking for fame and glory, not knowing what he had already found.

He sits on the bench remembering until the sun starts to go down and it grows cold. It is time to go back to the large white house and return to the present. He reads the letter one more time and returns it to his top pocket next to his glasses. As he rises and turns to go, the breeze plays across his face and he realizes that his face is wet. Tears have plunged down the deep crevasses beside his nose and mouth, leaving damp streaks in their wake. He can't go into the house like that; he can't let anyone see. He rummages impatiently in his pockets for a handkerchief, turning out rocks and feathers, a few small bones, a dried bit of fern . . . He finds only one crumpled, soiled square of linen. Why doesn't he have a fresh one? Why . . . Oh, it doesn't matter, not really. He wipes his face as well as he can. He last used the handkerchief a few days ago, to wrap up some tiny seedlings he was transplanting, and he isn't certain that the effect will be all that could be desired. It will have to do. Anna will not notice a few fragments of dried leaf or soil caught in the white stubble on his chin.

She never really looks at him these days, anyway. Most of what she sees is long ago and far away. For tonight, perhaps, maybe the two of them will be the same. All that matters to him, too, is long ago and far away now.

And he smiles a little and stumps back toward the house, a solitary, solid man, a little less alone in the last light of the evening than he was the day before.

CHAPTER 2 THE BEGINNING

In the days that follow, Dubois is held hostage by the events of his past. He is absent-minded, searching for the pattern of his life: How has he come to this in old age? What signs marked his path, for good or ill? Where were the turning points?

The beginning of the story is long ago, perhaps even before his birth, for no scientist is ever born in isolation. He was born on January 28, 1858, an interesting between-time in science. It was some eighteen months after the first Neanderthal skeleton was found in Germany and a little more than a year before Charles Darwin published *The Origin of Species* in England. The one was the first tangible proof of human evolution; the other was the theory that would make the find comprehensible by placing it in a context. He sees now that those momentous events stand like gateposts through which he passed as he started down the path to his future.

Eijsden, where Dubois grows up, is a small village in south Limburg, the little piece of Holland that stretches southward between Belgium and Germany. Limburgers, with their strong regional accent, are often carica-

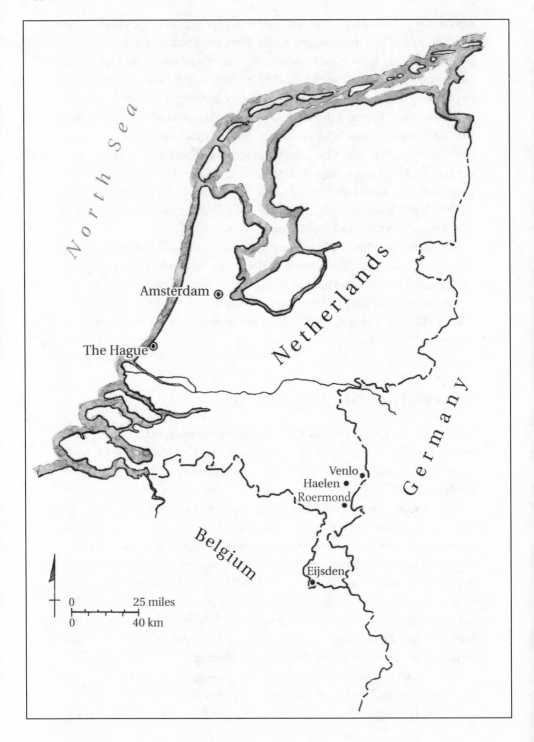

tured as provincial or countrified by other Hollanders. Many families in Limburg are Catholic, as the Dubois family is, and conservative. There are few large cities in Limburg, only the river and the beautiful, peaceful countryside. There Dubois returns in his old age, seeking that peace and calm at De Bedelaar, where he can rest and reflect.

His father is Jean Joseph Balthasar Dubois, an apothecary and sometime mayor of the village, one of its most educated men. Mayor Dubois is a large, sturdy man, not tall but strongly built, with a certain portliness and a wide, florid face that bespeak a good life and few anxieties. He is blond and blue-eyed and projects an air of authority, of certainty, and of great dignity. He is by nature deliberate, unexcitable, and thorough. The Dubois family consists of Jean Joseph; his wife, Trinette (née Marie Catherine Floriberta Agnes Roebroeck); and their four children. Trinette is a small, plump, good-hearted woman, who suffers from a certain amount of social ambition. They are an important family in Eijsden, highly respectable, and she intends that her children shall do them proud. She is well content with her life, her house—one of the grandest in the town, a good solid brick house that takes up a full block of the village—and her handsome family.

The family motto is "Recte et fortiter," "Straight and strong," and young Eugène takes seriously the charge of maintaining the family honor. Indeed, his habitual posture is so square-shouldered and erect that people often remark on it. Combined with his energy and his habit of looking at people directly with his pale, blue eyes, this posture gives Dubois a striking appearance, even as a youth. Eugène is sturdy like his father, but much quicker and far more impatient. He will grow up to be a handsome, irrepressibly curious man, taller than Jean Joseph and with a temperament so restless that sometimes it is hard for him to sit still. It is impossible for him not to think, impossible to be still in his mind.

His brother Victor, younger by one year, is a different sort of being entirely. Victor is broad and solid of build, like their father, with their mother's coloring. He is slow in speech and lax in thought, because he is unable to fix his attention on anything difficult for long. Eugène is fair, tall, and relatively slender, all ambition and fire; his brother dark, broad-faced, phlegmatic, and slow. The girls, Marie Antoinette and Jeannette Gérardine,

OPPOSITE: *Eugène Dubois was born in Eijsden and educated in Roermond and Amsterdam; he lived in The Hague, in Amsterdam, and finally at De Bedelaar near Haelen. He is buried in Venlo.*

The family Dubois, photographed in 1886. Front row, seated, left to right: Eugène's sister Marie (Mère Marie-Angélique) and his mother, Trinette. Back row, standing, left to right: Eugène Dubois; his sister Gérardine; his brother, Victor; and his father, Jean Joseph.

repeat the dichotomy. Marie, two years younger than Victor, has Dubois' coloring and is full of curiosity. She and Eugène are natural allies in a family dominated by slower, more cautious people. The baby, Gerardine, is five years younger than Dubois, only seven years old when he goes away to school. She has her mother's dark eyes, dark hair, and sallow complexion, like Victor, and much of the complacency, too.

In school, Dubois works hard. Learning and stimulation of his mind please him more than almost anything else, although he is also a powerful and expert swimmer. The truth is that he is smarter than the other pupils; he knows this without being told. Still, he does not let himself drift along, for he burns to know more than anyone else; he is hungry for knowledge. He wants to ensure that he will be at the top of his classes; he needs, for his own satisfaction, to be the best. He never believes his quick wits alone will suffice. He needs knowledge and facts with an urgency that the others do not feel. He sees that many of his fellows are stupid or lazy or both. It makes him a little self-satisfied, but his assessment is valid. He is a brilliant student, especially in the natural sciences. He vows to discover a truth, a fact as solid and important as a brick, one that will last forever. The name of Eugène Dubois will be known in science.

The first time anything really happens to Dubois is in 1868, when he is only ten years old. There are notices in the newspapers that the renowned German biologist Karl Vogt will come to Limburg to lecture to the public on evolutionary theory. Dubois' science master has spoken of the man. Dubois begs to attend the event, but his father does not think it appropriate fare for a youngster, even a precocious child interested in science. All Dubois can do is follow the reports in the newspaper of the event and its aftermath. They are exciting enough; it is the first step in his awakening.

The morning after the lecture, Dubois waits until his father has read and discarded the newspaper and then snatches it up to take into the large walled garden behind the house to read, looking for an account of Vogt's lecture. Dubois' favorite tree is huge and full, its trunk surrounded by a circular bench, and he takes the paper there to study it. He often sits here to read his precious books, or to watch and draw the birds, insects, and plants of the garden; it is where he ponders discoveries and theories and experiments. Under his tree, he feels protected, safe, and invisible, even though his mother can glance through the sheer curtains and see him if she chooses. As long as she sees his blond hair shining in the sun, his eyes fixed on some book or creature, she does not worry. It is only that he has a certain tendency to sneak off for a swim in the river when he ought to be at church or doing his schoolwork, so he needs to be watched.

The newspaper account seems to suggest that Vogt's lecture was almost a call to arms, the beginning of a tremendous battle of beliefs. Vogt's presentation of the evidence for evolutionary theory touches everyone: students, teachers, doctors, lawyers, bankers, newspaper editors, clergymen, educated people from many walks of life, and even some of the poorly educated but inquisitive working class. Dubois knows something of the principles of evolutionary theory—natural selection and survival of the fittest—from his science master and his reading. In his science class, he was captivated by the way these few simple ideas explain the entire natural world and the organisms in it. What he did not know beforehand, for he is still very young and naïve about the ways of the world, is that Vogt's well-delivered speech could turn the normally placid Dutch audience into combatants in a war over the truth.

What is at stake is a view of the world, of the very essence of reality. Most people in Europe believe life—particularly human life—is largely pre-ordained and that the order of society is static. Those with power and wealth guide the workings of the world because they are inherently superior. That inherent superiority, the privilege of the privileged classes, derives from God, who has created all creatures in their appropriate and perfect places. As snakes and worms are meant to crawl on the ground, so the poor are best suited to menial tasks and are most vulnerable to disease and vice. As the birds and higher beasts of the land rise above those lowly, crawling creatures, so, too, are the wealthy more able, more wise, and more suited to rule over other creatures and over other, less able men. Though individual effort and abilities count for something, dramatic changes in status or life are not to be expected. That is the stuff of fairy tales, in which the prince is mistaken for a pauper or the ash-sweeping stepchild marries a prince. In the real world, those who live proper, God-fearing lives succeed, and those who do not suffer and perish.

But this smug, orderly world is not the one pictured by evolutionists. The watchword of evolution is change: descent with modification over time. Evolution thus re-creates the world as a different sort of place, one of process and flux, of struggle marked by the success (in terms of survival or number of offspring) of those who are most fit. Evolution is revolution, asserting that positions in the grand scheme of things change, and always will change, because it is in their nature to change. The history of life is dynamic, a story marked by struggle and competition, not by stodgy pre-existing sameness. Many find this a deeply menacing view of the world.

Vogt's lecture demonstrates the enormous body of evidence concerning the likenesses between humans and other organisms. He argues that all creatures are constructed to such a detailed, common plan that they must surely share one ancestry. The creatures of the present have evolved by imperceptible changes from that common ancestor, as has man. Humans are a part of the natural world, the animal world, he asserts, not utterly apart from it. He gives special attention to delineating the links between the apes and man, for if man is but a cousin to the savage and bestial apes, then all creatures on the earth are united in a common existence, a common descent, and a common history of upheaval, struggle, and change.

For the first time, reading about the reactions to Vogt's lectures, Dubois begins to intuit the intimate link between ideas and society, between science and the people. What becomes evident is that acceptance of a theory is not a matter of science alone but also of human nature and human emotions. The lecture in Limburg provokes a scandalous mêlée. The newspaper accounts are guarded, but it is clear to Dubois' sharp mind that the occasion was much more sensational than anyone dares admit directly.

There, sitting on that bench, *his* bench, poring over the newspaper accounts of Vogt's lecture, Dubois longs to be a man of science, a man to whom everyone listens, a man with great ideas who will discover the truth. He opens the back door of the house to return the newspaper to its place, in case his father wants to see it again. The sunlight streams across the black and white marble floor, reaching toward the large front door. The house is lovely and welcoming, with generous-sized rooms with high ceilings and elegant woodwork. In the hallway is a large spiral staircase with a fine wooden banister that curves up to the first floor.

It is afternoon now and only a few customers have stepped past the large windows and entered the apothecary shop through its tall, double wooden door with a clear transom above. The people in the village approve of Mayor Dubois' house; there is no ostentation about it, except for its sheer size and the separate, arched entrance where carriages can drive directly from the street into the courtyard. The public face of the house is austere and unadorned, utterly respectable, and inexpressibly Dutch.

The apothecary is in one of the large front rooms looking out over the street. Dubois enters the shop through a door from the interior of the house, quietly so as not to attract attention. It is a grand and very somber room, decorated in dark green and deep red and black, with dark woodwork. It is one of Dubois' favorite places in the entire house. The lace curtains draping the large front windows strain the light—it is like the pale green sea passing through a fisherman's net—and ensure a certain privacy for the customers. Important business is conducted in that room: his father listens to ailments and complaints and issues prescriptions to make people well again. In the whole village, only he and the doctor know the right mixtures and the right doses to heal the sick.

Dubois especially likes the rows and rows of jars and bottles full of medicines that glint when the sunlight strikes them. He stands in front of them, reading the names to himself like an incantation: alum and antimony, ammonia, arsenic, benzoin, bismuth, calomel, camphor, carbonate of soda, castor oil, chamomile, compound of chalk, henbane, ipecacuanha, laudanum, lime water, madder, magnesia, morphia, niter, opium, peppermint, prussic acid, quinine, rhubarb, sal volatile, squill, tincture of iodine, zinc oxide. The names are like a secret scientific spell to banish sickness and evil. Better even than the jars of mysterious mixtures are the shelves full of thick-backed leather books—solid books, stuffed full of long words and important facts. It seems like a church or a temple to Dubois. Nothing is hurried in the pharmacy, nothing is urgent; all is known and all will be cured in time. His father is a precise and knowledgeable man.

Standing there quietly, Dubois overhears some of the customers discussing Vogt's lecture. He is amazed to learn that the audience challenged Vogt's ideas, his interpretations, his very evidence, despite his reputation as a great scholar. Dubois has been taught all his life to treat authorities with due respect for their position and wisdom. How could anyone doubt the word and knowledge of such a renowned man of science? Could Vogt be mistaken, even deceptive? How could the people of Limburg be so rude, so insulting to a famous visitor?

And yet, apparently, it is true. An elderly lady, Mevrouw Schilling, is waiting with her companion for a tonic. Thin and sharp, she is still quick-tongued and lively despite her white hair and wrinkled face. Apparently she attended Vogt's lecture. She describes how a member of the audience stood up to ask a question in German, interrupting Vogt's lecture. That small fact puzzles Dubois: what was the purpose of asking the question in German, not Dutch? Not everyone in the audience would speak German, although many might. Was the speaker attempting to demonstrate his own high level of education? Or was he perhaps trying to ensure that the German-speaking Vogt understood him perfectly? Whatever the point of the choice of language, the query itself strikes Dubois as very peculiar.

"What the man said was this," repeats Mevrouw Schilling, now curling her mouth oddly to mimic a pedantic expression and tone of voice, "'Dr.

Vogt, tell me this: do apes have churches? Do they have libraries?'" She cocks her head in a questioning attitude.

The boy wonders what the man meant by that. Is building churches and libraries, then, the test of humanness? Where is the logic in that? He does not want to eavesdrop, which is rude, but he thinks it acceptable since Mevrouw Schilling's bell-like tones can be heard throughout the apothecary.

"After the question," she declares, "I heard some of the students laugh. Then one of them said that neither the illiterate peasants of Russia nor the savages of darkest Africa have libraries or churches, yet they are still human. Well, they may be human, I suppose, but they are certainly benighted and unenlightened, that's all I have to say!" She shakes her head so firmly that some of her silver locks, until now confined to an orderly and well-behaved bun at the nape of her neck, quiver and fall free.

"They didn't say such a thing!" exclaims her companion. Dubois recognizes her, too, vaguely; she is, he thinks, a married daughter, Mevrouw . . . no, he cannot recall her name. She is a middle-aged, maternal woman. Her buxom figure is encased in a dark brown bombazine dress with matching cape that rustles majestically as she walks. She is the personification of conservative rectitude, from the fine leather on her buttoned shoes to the discreet feather on her modest hat.

"They did. But what did he mean?" Mevrouw Schilling persists, her voice growing querulous with confusion. "I don't understand these clever young people anymore. I mean, of course apes don't have libraries or churches. That is an utterly ridiculous idea. So what does it all have to do with this new theory everyone is so angry about?"

"I think, my dear," her daughter says comfortingly, her ample bosom jiggling slightly as she pats her mother's arm, "the students thought that a lack of civilized institutions did not disqualify apes from being closely related to man."

"No!" replies the other, snorting slightly and peering through her glasses. She taps the tip of her furled parasol in indignation. "Related to apes? Is that what that dreadful German was speaking of? Apes? I most certainly am not related to one. But I don't see what a library has to do with it." They collect the prescription, nod good-bye to the apothecary and his well-brought-up young son, and sweep out into the street, tut-tutting a staccato of disapproval as they go.

Dubois stands dumbfounded, pondering the extraordinary story he has overheard. "Eugène," his father calls to him, "don't just stand there staring into space. If you have no schoolwork to do, then come here and help me make up prescriptions. Mijnheer Buikstra will call later for his usual, and I need to make up some more stomach tablets and some headache powders. Bring over the magnesia, that's it, the large bottle, and then I need the chamomile. . . ." The rest of the afternoon, Dubois fetches and carries and wraps for his father, all the while listening to the gossip about the lecture.

Most of the townspeople who come into the apothecary that day are shocked by the content of the lecture and think the students behaved abominably. There is talk of hooting, like ape calls, and a rhythmical stamping of feet like the beat of primitive tom-toms in the jungle. Dubois can hardly credit such behavior in a public lecture; who would dare? He, for one, would be soundly spanked and sent to bed without supper for such rudeness, he has no doubt.

"I don't know about you, Mayor," confides Mijnheer De Pauly, the owner of one of the more prosperous shops in town, "but I had to laugh when all of the noise started up and the ladies scurried out of the lecture hall like so many hens in a thunderstorm, leaving the men to argue and bluster."

"I'm afraid I was not there," replies Dubois' father politely, "though my boy here is very interested in science."

"Just as well," agrees De Pauly, nodding sagely. He is a tall man whose well-cut gray frock coat announces the success of his business endeavors. "You don't want to let him get mixed up with such things while he is still so young."

Dubois himself is agape at what he has overheard. If the ladies all ran out, was the meeting out of control, a riot? Dubois tries to imagine the scene. How could these ideas provoke such a response? And wouldn't he, too, have laughed to see all the ladies bustling out, offended, and the gentlemen pointing their fingers at one another and raising their voices, crowing for supremacy?

De Pauly goes on: "It was a shocking event, just shocking. I don't know which was worse: the vehemence of some people's views or the general unwillingness to listen and evaluate evidence calmly. I had no idea so many of the town leaders were such old ladies that they are frightened of a new theory. Well, I must thank you for these pills, Mayor," De Pauly says, taking the parcel and handing over payment. "I'm sure they'll do the trick. Good day to you! Good day to you, young Eugène."

Dubois bobs his head politely to the man as he leaves. Afterward, he thinks admiringly that a very powerful truth must lie within those deceptively simple ideas of Darwin's. Evolution is everything new, modern, and disturbing. It will change the old order, and Dubois, impatient to grow up, would like that. He can hardly wait for his chance to change things in life, to right wrongs, to improve upon the old ways. He does not yet know it, but part of the attraction he feels to evolutionary theory results from its ability to upset the old order. The other attraction is the sheer scientific power of the theory. He is drawn to it with an almost religious fervor.

Pondering the exciting events at Vogt's lecture, Dubois concludes that some people simply don't want to learn about new ideas. They are too fixed in their thinking, too old, too boring. He must never make that mistake. He must remember always to listen to new ideas.

Jean Joseph approves of Eugène's growing interest in science, for he has always planned that his elder son will become an apothecary, like him. As his twelfth year approaches, the boy asks his father to send him to the State HBS, the technical high school, at Roermond. It is an unconventional choice, and Jean Dubois is deeply conventional, but he knows the facilities for studying chemistry, zoology, mineralogy, and botany at the HBS are like those of a small university, with laboratories full of the best equipment and superb teachers. The danger is that such teachers will teach him ideas as well as facts, including that new evolutionary theory.

The elder Dubois does not consent immediately but takes the matter under advisement. What fosters young Dubois' cause is that some of the other Catholic families in Eijsden get wind of the idea almost immediately. The furor caused by Vogt's lecture is not forgotten, though it was now two years ago, and the old biddies of the village are horrified that Jean Joseph Dubois would consider sending his son to such a place. "He'll lose his religion," they predict, nodding their heads with conviction as they gossip in the shops and on street corners. "They'll teach him all those anti-Christian theories, and soon he'll believe them. He's a nice boy, a smart boy, but the mayor will be sorry if he sends his son to such a place!" In the end, they cluck and fuss so tiresomely that Jean Dubois decides to send Eugène to the HBS in part to defy them.

Eugène remains impeccably polite and studious, but attending the HBS only accelerates the process of his breaking free of convention. By the end of his first year at Roermond, when he is thirteen, he is starting to question the teachings of the Church. He does not know how to focus this uneasiness, but he begins to doubt everything, almost reflexively.

Much later he articulates these feelings. "I always knew that if I could succeed in concentrating my thoughts well on a problem, then I would live my true life. Then I am absorbed by the problem. To achieve great things, one must cast aside the unimportant and the sentimental, one must follow truth." And so I have, he thinks, and so I have.

Dubois has no intention of becoming an apothecary. He wants to study the natural sciences and evolution, which means attending medical school. Roermond is a good place to lay the foundation for such studies. It is also a crèche that nurtures evolutionary ideas and independent thinking. His new science master recognizes the keenness of Dubois' mind and suggests he read some of the great books they discuss in class: Darwin's *Origin of Species;* his new book, *Descent of Man;* Huxley's *Man's Place in Nature;* and Haeckel's masterful *History of Creation.* Darwin's work is where it all begins, where evolutionary theory is laid out for the first time, but Huxley's book is the more convincing. It is simple, short, full of facts, and impossible to refute; for anyone who reads it, there can be no further doubt that man is simply a modified ape. Huxley persuades readers that man differs less in his anatomy from the chimpanzee or the gorilla than those manlike apes differ from the lowest apes.

Haeckel's book is the second great revelation of Dubois' young life. His prose is easier for the youngster to understand than the others'. Haeckel's style is sweeping, vivid, inspiring. Early in the book, Eugène comes across a passage that leaves him gasping. It is as if Haeckel has plucked the half-formed thoughts from Eugène's own mind and crystallized them into written words.

> As a consequence of the Theory of Descent or Transmutation, we are now in a position to establish scientifically the groundwork of a non-miraculous history of the development of the human race. . . . If any person feels the necessity of conceiving the coming into existence of this matter as the work of a supernatural creative power, of the creative force of something outside of matter, we have nothing to say against it. But we must remark, that thereby not even the smallest advantage is gained for a scientific knowledge of nature. Such a conception of an immaterial force, which at first creates matter, is an article of faith which has nothing whatever to do with human science. Where faith commences, science ends.

Reading Haeckel's words is for Dubois like emerging from the dark confines of the chrysalis. He can feel his creeping wormlike self first exposed and then transformed by the light of knowledge. Religion is finished for Dubois now. Haeckel's words have burst the woolly cocoon of confused, everyday thought, revealing a theory of descent with modification that shines with truth.

Dubois makes no conscious choice; his is an irresistible metamorphosis, whose time has come. Belief without proof is blind faith, *blind* faith. It cripples the mind and the judgment, surely a sin greater than any forbidden in the Bible. From that moment on, Dubois adopts as his personal credo the need to question, to think things through for himself. He can accept nothing simply on the grounds of dutiful obedience; he reads Haeckel's words and they resonate within his mind. He becomes an evolutionist, as indelibly as one who takes holy orders becomes a priest. Though his mind is sharp, Dubois' tongue is not. He is still young, dependent, untried. Until he is grown, he will not tell anyone how he feels about religion, nor will he share his sense of revelation; it is better not to.

CHAPTER 3 THE GAME

At Roermond, Dubois invents a game in his mind, for he does not think any of his schoolfellows would play. It is a simple game. First, he asks himself, "What is the most important thing yet to be discovered? What is the best proof of evolution? Where can the biggest contribution be made?" Then he tries to formulate an answer, carefully weighing the pros and cons of different discoveries, looking for the one he shall devote his life to. He is determined to be a part of this revolution, for it is the greatest movement of his day. Science will change the world as surely as evolution changes bodies.

Some days Dubois follows Vogt in thinking that embryology will hold the most crucial answers. If he truly understood development, if he could find out the reason that a dog embryo one day stops looking like a chicken embryo and becomes a dog, surely that information would reveal a great deal about the evolutionary relationships among living organisms. That would show *how* evolution proceeded, step by step. Other days he favors Huxley's work, showing how feature after feature links man to apes and then to monkeys, demonstrating the unity of the living primates. This, too, is a strong, encyclopedic approach to the problem of evolution.

But sometimes he has a rather novel thought, one not much favored by the men of science of the day. His inspiration comes from Haeckel's *History of Creation,* which he reads over and over, it is so full of wisdom. The pages are almost falling out of their binding from use. Haeckel outlines the general course of evolution clearly, tracing man's descent by stages from the lowest life forms, through the one-celled protozoans to the worms, fishes, and lizards, and up through mammals. Like Huxley, he sees the closest group to man as the Primates: the lemurs, monkeys, and apes. And Haeckel maintains that there is a link, a link as yet unknown, between apes and man. He describes this missing link in words that Dubois nearly memorizes.

> *The Ape-like men, or Pithecanthropi, very probably existed towards the end of the Tertiary period. They originated out of the Man-like Apes, or Anthropoides, by becoming completely habituated to an upright walk.... Although these Ape-like Men must... have been much more akin to real Men than the Man-like Apes could have been, yet they did not possess the real and chief characteristic of man, namely, the articulate human language or words, the corresponding development of a higher consciousness, and the formation of ideas....*

And:

> *Those processes of development which led to the origin of the most Ape-like Men out of the most Man-like Apes must be looked for in the two adaptational changes which, above all others, contributed to the making of Man, namely* upright walk *and* articulate speech. *The two* physiological *functions necessarily originated together with two corresponding* morphological *transmutations, with which they stand in closest correlation, namely, the* differentiation of the two pairs of limbs and the differentiation of the larynx.

Haeckel even predicts that the evolution of the upright walk long preceded the evolution of speech. It is a hypothesis that could be tested, Dubois realizes, if only the fossils of that missing link, the *Pithecanthropus,* could be found. Finding the missing link would be like seeing the results of a brilliant experiment, one conducted not by himself but by the forces of natural selection and survival of the fittest, in the distant past. Such a fossil would surely prove the evolution of man, in a tangible way. Finding it would be the greatest scientific discovery ever.

He does not know where to find such a fossil, or how. These points are yet of no concern. What matters is that he becomes convinced he is the one who shall find the missing link. He is deaf to other arguments, unimpressed by other scientific questions. His eardrums vibrate to only one tune, the song of the problem he must conquer: the missing link, the missing link, the missing link. In his years at Roermond, Dubois' ambition coalesces out of the intellectual milieu like a crystal precipitating out of solution. Of his fellow students, Dubois is the most passionate, the most committed to science. He is the one who never swerves from his own road, who never deviates from his conviction. For him, learning about evolution is much more interesting than any other part of natural science. He is not drawn by physics or chemistry, important though they may be. Medicine or an apothecary career, for example—simply doling out remedies to the sick— seems mundane. It is learning that has already been discovered; it presents no challenge except that of acquiring competency. In evolution, he can make new knowledge. Evolution reveals how the universe is organized, how life itself is formed. Evolution promises someday to divulge the secrets of how man became human. Here is the strength of a unifying principle for all of the natural sciences and for Dubois' life.

As the old ladies of Eijsden predicted, Dubois' education at the HBS transforms him. He is no longer a bright boy who is interested in the sciences; he is a dedicated evolutionist. And here he parts company with his father intellectually. He cannot follow in his father's footsteps; he has not the character for it. He is incapable of following anyone else's path but his own. He wishes to please his father—earning his father's praise is one of Dubois' fondest desires—but he cannot strive for love and approval on his father's terms. He has grown up; he is now himself, inflexibly, stubbornly, and thoroughly himself. He cannot be his father's son any longer.

Before he goes home for the Christmas holidays in his last year at the HBS, Dubois worries how to explain to his father that he wants to attend medical school, not become an apothecary. He knows his father has performed invaluable service as an apothecary, and he honors that, but he seeks something more. He hopes, knowing it is probably in vain, that his father will be proud of him for choosing his own path. Anyway, Dubois has no choice. He cannot do something so small as becoming an apothecary and a mayor like his father. It is too confining. He is meant for bigger things.

The planned confrontation with his father is usurped by unexpected news that awaits him at Eijsden. Marie—his favorite sister, his special ally in the family—has decided to enter a convent. She will become a nun and devote her life to God. His parents are very pleased with Marie's decision, so proud of her that they cannot wait to spread the news to all their acquaintances.

It is a tragedy. Marie? The one who giggles and plays pranks with him? The one who can make him laugh in the midst of a solemn occasion just by the look in her eyes? A nun? Dubois feels almost physically ill at the news. Unbeknownst to him, Marie has already spent some weeks with the Ursuline Sisters, to make certain of her choice. Those who would be novices must learn difficult lessons of obedience, silence, propriety, and Marie has fared well. She is learning to walk, not run, to speak softly when spoken to and to hold her tongue otherwise, to keep her blue eyes downcast and her lively opinions to herself.

Marie is making a terrible, terrible mistake, Dubois thinks. Oh please, let her give this up. There is no truth in religion, she will waste her precious life on custom and propriety! He cannot voice these thoughts to Marie lest he wound her; he knows she is a believer. He is as gentle as he can manage to be with her.

"Marie, you cannot do this," he pleads with her earnestly once they are alone. "You can't join the convent. Please. You don't have to give up your life to be a good Catholic; there are other ways."

"*Ja*, Eugène," she replies serenely, "*ja*, I can join. The Mother Superior has already told me I will be accepted as a novice after Christmas if it is still my wish. And it will be." She smiles a little, quietly, as befits a nun.

He looks at her sadly. He does not like the new way she has restrained her hair instead of letting the curls dangle over her shoulder. "But Marie," Dubois agonizes, his face contorted, "it is the end of everything. You cannot be yourself, you cannot be clever or go to school, you cannot even marry and have children!" He tries every argument he can think of that will hold weight with her, withholding the one that truly matters: religion is false, is unproven, unprovable.

"I know," she replies gently, looking up at her dear brother in his torment. She takes his hand, as she did so often when she was a little girl and needed his protection or guidance. "I know. And I shall have to leave you and my home, too. But I will be the bride of Christ, Eugène. You cannot imagine the peace of the convent. It is so holy and so calm. I shall be very happy there."

"Calm?" Dubois repeats miserably. "You would give everything up for calm?" He turns his face away for fear she will see the thoughts in his mind.

"Listen, Eugène. Look at me. I have already chosen the name I will take upon my ordination, I think. How would you feel about Mère Marie-Angélique?"

"Are you then to be someone else?" Dubois asks softly.

"I must be. Entering the convent will make me a new person."

He hates the very idea of becoming someone else. In his life, he must struggle so hard to become who he is. "They are all so solemn and so holy, those nuns. How can you bear it? Don't you ever want to laugh and shout and disrupt the service?"

"No . . ." Marie starts to say, and then becomes more honest. "Well, maybe *ja*, a little"—she giggles—"sometimes. But nuns can still laugh. God welcomes laughter and joy. Obedience and joy balance each other out, you know."

"Obedience," repeats Dubois dully. To him the word sounds like a sentence of death. "Obedience. I do not think it is a virtue. What about thinking for yourself, finding your own way?"

"That is a man's road, Eugène," Marie points out. "You may do it, I believe you will do it. But it is not a way for me. I could never do it. My road leads to the convent, with God's will."

There is no changing her mind. Dubois' heart aches at the loss of his cherished sister. She is gone forever, his Marie Antoinette Hélène Dubois. Mère Marie-Angélique, who will take her place, can never be his friend. She will be devoted to the Church, and he is bound over to Truth and Science.

CHAPTER 4 AMBITION

In 1877, at the age of nineteen, Dubois leaves Roermond to study medicine at the University of Amsterdam. Amsterdam is a large city, a part of the world arena in science, art, and music. The buildings are old and gray, dignified, some very beautiful and ornate. He has never seen such grand architecture before. Some of the lecture halls where he listens to his professors are magnificent, oak-paneled chambers with black horsehair chairs, elaborately carved and decorated ceilings, and steeply stepped floors, so that every student can see the professor's actions and demonstrations clearly. True, he cannot study under Darwin or Huxley or Haeckel, but his professors are renowned leaders of science. There is J. D. van der Waals, who is unraveling the intricate laws of physics and motion; the botanist and evolutionist Hugo de Vries, studying inheritance of traits in plants; Thomas Place, the brilliant physiologist; and the renowned anatomist Max Fürbringer, who trained in Jena in Germany under Ernst Haeckel himself.

Under their tutelage, Dubois' mind begins to blossom. By the end of the first year—a year filled with long days and nights of reading and study—Dubois feels he has found his place in life. His brilliance is soon widely recognized, for he places top of every class. His dissections are the finest, his drawings the most detailed and meticulous, and his experiments seem always to work. His fellows sometimes tease him about his "golden hands" and his flawless memory, but it is affectionate teasing, based on admiration. They do not often tease him about being a provincial Limburger, for he is soon as cosmopolitan as any of them.

To Dubois' irritation, in the second year he is followed by his easygoing younger brother, Victor. And for Victor there will be no long nights of study, no hours bent over a microscope, no dissections that require every ounce of skill and dexterity he can muster. As ever, Victor gets by with a minimum of work and a maximum of personal charm. The family assumes they will share rooms—of course they will, they are brothers—but it is not a happy arrangement. As he matures, Dubois walks more and more along an idio-syncratic path of his own devising. He makes a point of following his own convictions; he openly flouts beliefs or behaviors he deems worthless. His intent is to achieve a certain transparency, to live his convictions through-out his entire character and soul. He has no respect for those who put on conventional beliefs like a fashionable coat, to be removed and replaced when another style comes along.

By the time his younger brother arrives in Amsterdam, Dubois has aban-doned all the trappings of Catholicism. He no longer attends Mass on Sundays or confession during the week; he does not cross himself or mutter Hail Marys at appropriate moments; he systematically excises all habits of observance from his behavior. Victor, less bold and lacking firm convictions in any case, is annoyed and embarrassed by his elder brother's open rejec-tion of religion, as perhaps Dubois intends him to be. Living in such close proximity, the contrasts between them rub the two brothers' sensibilities raw. Each becomes more fixed in his ways, less flexible, less tolerant of the other's choices.

The final conflagration starts to smolder when Victor, goaded by his brother's behavior, accuses Dubois of failing to be a good Catholic. "You bring shame on our family," Victor chastises him. "We have been good Catholics for more than three hundred years and you decide the faith is no longer enough for you! You only think of yourself, never of your effect on others."

These are bitter words to a man like Dubois. Worse yet, Victor speaks to him like a parent, when Victor is merely the younger brother. Stung by this scolding, Dubois disparages Victor's petty orthodoxy. "Better to be a thinking agnostic," he replies bitterly, using the term coined by Huxley, "better to be one who searches his conscience and admits he does not know about God and religion, than to be a submissive Catholic. You never think about anything, not God, not your family, not even yourself. You just do what is expected. I would not be like you."

"Nor I you!" comes the angry reply.

After that, they avoid each other for some weeks. In time, there comes a hesitant truce, accompanied by tentative kindnesses or small shows of consideration on both their parts. Victor obtains some especially nice tea, and shares it with Dubois. Dubois brings home a newspaper with an article in it his brother might find interesting. Slowly, they begin to take meals together again. They do not discuss their differences, but neither do they fire glances of anger and irritation at each other every time they cross paths.

At the end of a difficult exam week, they plan to go out to dinner together, a rare occurrence, and it is perhaps in both brothers' minds to seal the peace with the other. Academic pressures have shortened their tempers but have also distracted them from their differences; they are united for once in striving to do their best. They choose a favorite restaurant, a place popular with students for its filling and inexpensive meals. But the fragile reconciliation is threatened almost immediately. As soon as they order, the problem is painfully apparent. It is Friday, and Dubois requests beef while Victor, the observant Catholic, asks for haddock.

Every bite of gravy-laden beef that Dubois savors seems a personal affront to Victor, who dutifully swallows his dry and overcooked fish. Victor is uncharacteristically enraged. Dubois is himself so annoyed with Victor's mindless adherence to custom that he flaunts his enjoyment of the excellent beef. What begins as a friendly evening deteriorates into silent hostility. Surrounded by a cheerful crowd of their fellow students, exuberant at their temporary release from academic obligations, the brothers sit sullenly and eat without speaking to each other.

When they attain the privacy of their rooms, Victor explodes. "How can you embarrass me in public like that? Why is it you must insult God by eating meat on Fridays? Everyone knows we are Catholics. What do you suppose they think?" He is so angry that he does not dare to look at his infuriating brother, who busies himself hanging up their coats and hats.

"Everyone does *not* know we are Catholics," retorts Dubois hotly, turning from the hall stand. He wonders how Victor can be so stupid. "Or if they do, they are sadly mistaken, for I am no longer a Catholic."

Victor catches his breath audibly, stunned by such blasphemy. Like Victor, Dubois was born, christened, and confirmed a Catholic. Members of the Dubois family have always been Catholics, back to their oldest known relative, Bastin de Try, in 1500. Each brother pledged himself to God at confirmation, and Victor is unable to imagine breaking such a promise—not, of course, that he has any intention of abandoning the pleasures of life, not like Marie, but still . . . Victor asks himself if his brother can actually mean to refute the faith that has been part of their identities since birth. He opens his mouth to speak, but Dubois' next words come first. They are strong and clear, and they cut through Victor's confusion like a scalpel peeling back flesh in the anatomy laboratory.

"I am no longer a Catholic," Dubois repeats. "I am a scientist." He stands firm and still and upright, looking straight at his brother, daring him to object.

Even Victor cannot mistake Dubois' meaning any longer. Dubois has cast aside the religion that Victor has always accepted and followed, somewhat passively to be sure, but it is as much a part of him as breathing. Dubois has examined one of the implicit tenets of Victor's existence and found it wanting. Victor has never examined anything critically; it is too much trouble to think so hard. Victor is stunned into silence. Finally, he turns to his brother, his older brother, and begs for compromise. "Can't you just soften a little for appearances' sake?" pleads Victor, reaching out a hand but not quite daring to touch his brother's arm. "Just in public? Why must you make such an issue of your scientific beliefs? Why must you reject everything else and throw it in my face?" There is no answer from Dubois, who continues to look at his brother with a calm and open gaze. Victor drops his hand, rejected.

Then he continues, "Think of Marie, our sister." Dubois closes his eyes in pain. Victor could never comprehend that Marie's taking the veil is a tragic and misguided waste of life in Dubois' eyes. Victor thinks it a fine thing, a heroic move. "Would it be so terrible for you to eat fish on Fridays as a sign of respect for her beliefs, whatever you think?"

"Oh, no," replies Dubois, very quietly, as if explaining the obvious to a child. "I can eat fish on Fridays. I *shall* eat fish on Fridays. But I shall do it because fish is what I wish to eat—not because it is what the Pope wants me to eat. I think for myself."

And that is the end of it. After such blasphemy and disrespect from Dubois, after such unconsidered and blind obedience from Victor, there is no retreat from their entrenched positions. No compromise is possible;

indeed, no real contact between them is possible any longer. Within a few days, the brothers make other living arrangements. To be fair, few would find Dubois an easy man to live with; wisely, he takes rooms on his own.

Now that he can organize things to suit only himself, Dubois' days are much the same. He arises early, at three or four in the morning, to study. He takes only a cup of tea, which he brews up himself, and some toast browned over the gas ring before starting. At seven, he stops for an egg and more tea. Lectures and demonstrations begin at nine; he never misses one. He is always caught up on his reading and he completes his assignments on time; it is a point of pride. Afterward, he eats a simple meal prepared by his land-lady and then studies again, putting in perhaps fourteen hours of work a day. Between three A.M. and five P.M., he will not tolerate frivolity or inter-ruptions from others. His friends and colleagues soon learn not to disturb him during the day, for he does not want to come out to sit in a café, gossip-ing and drinking coffee. He never has time to discuss the day's news or the latest music, or to make plans for an outing to see the new production at the theater. He has things to learn: facts, principles, theories. He is ambitious, serious, and terribly clever.

In the evening, he reverts to being a young man like any other. At the stroke of five, he calls a halt to his work and begins to enjoy his social life. A different man emerges from Dubois' rooms than the quiet, studious one who awakens so early to get ahead in his work. This Dubois is young and handsome and fond of the company of pretty women. He has a mellow baritone voice and a light, athletic step on the dance floor. The ladies admire his strong physique and blond, blue-eyed good looks. Dubois enjoys his effect on the young ladies of Amsterdam, and he senses the approval of their parents. Being a most eligible bachelor with an oft-predicted brilliant future ahead of him is a most pleasant circumstance. He attends concerts and parties; on the weekends he indulges in elaborate Sunday afternoon teas, or picnics when the weather is fair. He rarely lacks for feminine atten-tion. His evenings are as unserious and sociable as his days are intense and solitary.

Some of his fellow students are surprised to see the charming side of his nature; others take it simply as the balance to his focused ambition and determination. Without self-discipline, Dubois knows he might be as aim-lessly happy as his brother. But Victor will never come up with a startling new idea or a clever new approach. Victor is content to drift passively, pass-ing exams but not excelling, learning facts but not too many. His laxity fills

Dubois with a kind of horror for the easy road that leads to naught of significance.

In 1881, Dubois is offered the assistantship in anatomy under Dr. Fürbringer, a coveted position. He accepts with delight. The very day after Dubois accepts Fürbringer's position, Thomas Place offers him an assistantship in physiology. No one in memory has ever been offered two such positions at the University of Amsterdam, Dubois proudly writes to his parents. This is not empty bragging; it is true.

Ironically, the second offer poses an unparalleled problem. Had the offers come simultaneously, Dubois would have preferred to take up Place's assistantship, as both the man and the field seem more naturally congenial to him than Fürbringer and anatomy. But Dubois accepted Fürbringer's position promptly; he cannot now withdraw without dishonor. Mild regret tinges Dubois' real enthusiasm for anatomy as he enters cheerfully into his new position with Fürbringer. He does not foresee the bitter struggles that will develop between them over the ownership of ideas and research. He cannot imagine that a man of Fürbringer's reputation would want to take credit for his junior's work. Why would he think such a thing? And yet, in a few years' time, Dubois will come to feel that Fürbringer is doing just that. Dubois becomes hypersensitive, like a once-scalded hand that cannot bear the slightest heat. In time, his sense of injustice forces him to leave his academic career. Later he can see the irony of his situation, for the conflict with Fürbringer will be the engine that propels Dubois into greatness. Dubois even wonders if the accident of timing that led him to work with Fürbringer was preordained. Why had Fürbringer found him first, Place second? Was he meant to work with Fürbringer, so he would be driven out of Amsterdam? Was he born to find the missing link?

These are disconcertingly mystical questions for one so thoroughly scientific, yet he cannot quiet them. They nag at him, begging for explanation, for rational cause and effect. But they have no answers. Events follow the course they take for reasons that cannot be discerned scientifically. When he thinks on it, Dubois is amused that there appears to be such a strong direction to his life: his birth at the right time; his schooling in Roermond, which exposed him to the thinking of Darwin, Huxley, Haeckel, and Vogt; his medical training at Amsterdam, where he learned so much and then was turned away from normal academic life, toward the missing link. He contents himself with the thought that trends and patterns are always more visible in retrospect than in prospect.

As he starts his new job, Dubois is too busy developing his talents as an anatomist to be introspective or to worry about the distant future. In 1884, he qualifies as a physician; his promotion to prosector in anatomy follows in the same year with breathtaking rapidity. He is in charge not only of the human anatomy course for medical students at Amsterdam but also of another course for the State School of Applied Art. He conceives a program of research on the embryological formation and development of the larynx, not forgetting Haeckel's words about the importance of the origin of articulate speech. If he becomes the man who understands the larynx, he will be the man with something to say about the evolution of human speech.

In 1883, Eugène Dubois is full of hope and ambition.

CHAPTER 5 LIGHTNING ROD

For some time, Dubois calls on Mia Cuypers. Their acquaintance begins with Dubois' friendship with her father, Petrus Cuypers, Holland's leading architect and a well-known man. His beautiful design for the new Rijksmuseum, begun in 1877 and nearing completion, is already being hailed as a brilliant example of Gothic Revival architecture. It represents a new way of thinking about buildings, about public spaces. Cuypers is the center of an informal circle of Catholic intellectuals and artists in Amsterdam, who welcome Dubois into their fellowship. The Cuyperses are also an "Indies family," one with many members who have lived and worked in the Dutch East Indies. Dubois cannot keep track of all the Cuyperses' Indies connections. There seem to be myriad photographs of brothers and uncles posed under potted palms or lounging in rattan chairs, of aunts and cousins sitting in smart carriages under huge tropical trees. Perhaps this collective experience in the colonies makes the Cuyperses a little unconventional in their thinking and behavior; perhaps the cause is simple creativity. At any rate, their home is furnished with exotic oriental carpets, and the watercolors feature sharp-edged volcanoes and placid rice paddies. Instead of a good paisley shawl, they have an intricate, floral-patterned fabric, woven with threads of gold, draped over the piano; they call it songket cloth. The Cuyperses enjoy people from an unusual range of backgrounds. In their company, the talk is of ideas and music, principles of art and design, trends and principles to guide the future. Dubois is captivated by it all.

As Dubois becomes a more regular caller, and Mia a more constant participant in his visits, the Cuyperses begin to fancy Dubois as a possible son-in-law. Subtly, they make it known to him that the thought pleases them. He is an eminently suitable young man, outspoken yet thoughtful, and they know their daughter to be neither dependably demure nor consistently sensible. She is far too opinionated and much too intelligent for most people's taste. Her parents, perhaps unwisely, have allowed her to develop some rather modern ideas and have educated her more than is common for a young woman of her class. She even looks unusual, being taller and more athletic than most young women, and caring less about fashions and hairdos than about theories and ideas. She has a disarmingly direct gaze in her large brown eyes and, upon occasion, a sharp tongue for the pompous, the self-important, or the hypocritical. She is not fashionably submissive, nor does she wish to be. In fact, she is even regarded as unwomanly by some. Others predict that a good, solid husband and a few children are just what Mia needs to settle her down and drive all those foolish ideas from her head.

Dubois fulfills the hypothetical requirements for Mia's prospective husband nicely. He comes from a respectable Catholic family, even if he does not seem terribly observant, and he is intelligent enough to suit Mia well, or so her parents think. Dubois is flattered by the Cuyperses' tacit approval and the sparkling company in their home, and he rather likes their spirited daughter. She is neither tediously unoriginal nor brainless, as are many of the young ladies of his acquaintance. Her opinions are usually backed by knowledge and thought.

"Doctor, do you think," she challenges him on one occasion, "that women ought to take up medicine?" Her dark straight brows emphasize her seriousness, not unflatteringly.

"But my dear," Dubois replies, "there are already many women who work as nurses. And some of the best midwives I have ever known are female; they have a good touch with their patients, tremendous fortitude, good skills. Of course women are highly suitable for such work! There can be no doubt about it."

"No, no, you misunderstand me, Doctor. What I mean is, should women be admitted into medical practice as physicians? Ought they attend medical school, working alongside the men?" Mia explains earnestly. "Don't you think they might have an advantage in treating female patients, a greater ease of communication, perhaps, about ladies' troubles? And examinations, you know, physical examinations of patients. Surely this is a delicate matter."

Startled, Dubois answers without thinking carefully first. "You have a point there, Miss Cuypers," he concedes. "Not that any medical man would overstep the bounds of . . . of . . . propriety with his female patients, of course not. We are trained professionals. But it might indeed be easier for a woman to divulge . . . um . . . intimate problems to another woman.

"As for becoming physicians, I am not sure. You know, the training in medical school is not for the fainthearted. Some of the things students are exposed to are not very salubrious, rather shocking even, and difficult to bear. Dissections of cadavers, some of the terrible diseases of the poor and ignorant . . . The squalor in which some patients live is dreadful, most unsuitable for the finer sex. I'm really not sure a woman should be confronted with such things."

"Oh, Doctor." Mia laughs. "Who do you think has birthed all the babies since time immemorial? Who do you think has dressed the soldiers' wounds in battle, boiled the lice off their clothes, wiped up the messes in the sickrooms, and changed the bloody sheets? Have you not heard of Florence Nightingale in England and the school she founded to train nurses? She worked on the battlefield in the Crimea, in the thick of things. We women are not such fragile creatures as you take us for! I agree, many women would prefer to be cosseted, protected, and spared the ugly realities of poverty and disease; many men, too, I suspect, if given the choice. But everyplace that men are, women are also, and we are usually doing the looking-after. And if women can be nurses, why not doctors as well?"

"You're right, Miss Cuypers," Dubois says, joining her in laughter at his own expense. "I should have thought more carefully before I spoke. Though the demands of the medical profession would be impossible to combine with the role of wife and mother, I can see no reason why an intelligent woman should not choose to become a physician. Healing, caring for the sick, helping others, is a natural talent for many women. Denying them the knowledge to do it better is foolish."

Dubois has never before met a young woman with serious views. It is a novel and slightly disquieting experience, but one that he enjoys. He begins to consider the possibility of a future with Mia at his side. It might be an asset to have a clever wife, especially one from a prominent family. When he is a full professor, she could be counted on to preside over a tea table and converse brightly at a dinner table with erudite men. Whereas some wives are always at risk of appearing uneducated fools, with Mia the problem might be to teach her not to correct their guests' errors so bluntly as she is inclined to. Mia does not conceal her wit with feminine wiles, he knows that, but neither is she unattractive. She both speaks and dresses well, if a trifle plainly. Having such a wife would show how forward-thinking and modern Dubois is.

He decides to spend an afternoon alone with Mia in some suitable setting to consider every aspect of their possible future together. He should be able to read her feelings for him from her behavior. It is a delicate agenda, not easily accomplished in others' company. Dubois does not care to ask for her hand if he must risk rejection. No, he would rather seek some subtle sign of a favorable answer before he poses the question. He thinks it is rather like conducting a few more experiments to gain a better knowledge of the situation, before publicly advancing a new theory. He certainly wants to be on solid ground before he says anything to her or her father.

He sees a notice of a concert in the park on the following Sunday afternoon. That would be the perfect occasion for his experiment in congeniality, if the weather is fair. The public gardens are at their peak and the program promises to be a pleasant entertainment. He calls in at the Cuyperses' house briefly that evening to ask whether Mia would accompany him to this concert. The Cuypereses approve of the arrangement, of course; nothing could be more proper than a daytime outing in public, even if Mia and Dubois will be unchaperoned. As for Mia, she agrees with a nod and a smile of genuine pleasure. "I should like that very much. I so enjoy good music."

As he walks briskly home, he is contented at how compatible they seem in their tastes. He believes she is more tractable than people think, less unruly. She is simply high-spirited and quick-witted, like his sister Marie when she was young, and resents being hemmed in by convention. If she could turn her energies to making a comfortable and serene home for a husband and children of her own, she might be a very pleasant wife indeed.

On Sunday, the weather dawns fine and warm. Mia looks very pretty in her pale muslin gown. The flattering little hat perched atop her piled-up hair shows off her fine eyes. Dubois is proud to have her on his arm in public. He, too, has taken extra care in his dress for the occasion, donning a top hat, pale gray gloves, and his new, very fashionable, double-breasted blue coat. They settle into good seats just before the program begins.

The first part of the concert is lighthearted and lively and the crowd is cheerful. From time to time, Dubois and Mia exchange glances of enjoyment that seem to speak of shared tastes. At the intermission, they get up to stroll around and mingle with the other concertgoers, so the ladies can display their elegant gowns and the gentlemen can tip their hats to friends and acquaintances. And then, to Dubois' astonishment, Mia smiles and nods to a Chinese man as he walks past. He is well-dressed, though his striped trousers are a little colorful, and he is behaving properly, but Dubois is taken aback nonetheless. He wonders what Mia is thinking of. Before he can speak or act, the man approaches them.

Raising his hat, he inquires, "It is Miss Cuypers? What a great pleasure it is to see you again."

She smiles graciously in return and turns to Dubois, who is struck dumb by the fact that she actually knows such a person. Before he can think how to rescue her from this unwise encounter, Mia introduces him to the other man. He is a Chinese government official, the Honorable Mr. F. G. Taen-Err-Toung, who bows in respect to Dubois at the introduction. Mia's action puts Dubois in a quandary. He has been formally introduced to Taen-Err-Toung; he cannot turn away and depart with Mia without being unutterably rude. To linger is unseemly, bound to encourage familiarity with a foreigner, but he must engage in conversation at least briefly. He takes refuge in a few comments about the concert and the musicians.

To Dubois' relief, Taen-Err-Toung is a music lover and asks if they didn't think the tempo of the second selection, the Mozart, was a little too hurried. Polite remarks flow easily and the entire situation is less awkward than

Dubois feared at first. Nonetheless, the man is a foreigner—Chinese, not even European—and clearly not someone with whom they should engage in lengthy discourse. Dubois prompts Mia to move away once or twice, with slight pressure on her elbow, but she stands her ground, prolonging the encounter. Dubois recognizes, surprised, that Mia is being stubborn, even willful. She is intentionally ignoring his wishes, even though they are in her best interest. He is startled and a little irritated, yet his curiosity is awakened. Why is she deliberately engaging in conversation with this man? What is he to her? The scientist in him emerges; he stops reacting and begins observing.

Mia is much more animated in Taen-Err-Toung's presence than she has been before. She smiles and laughs and catches the Chinese man's eye, as if they are close acquaintances of long standing. She clearly enjoys Taen-Err-Toung's company—much more, if he is honest with himself, than she seems to enjoy his own. Can Taen-Err-Toung be a friend of Mia's father? Perhaps an Indies connection? In the Indies, he supposes, Europeans might socialize with prominent Chinese officials. This is, after all, an educated man, not a coolie; perhaps that is the explanation. Dubois works a reference to the colonies into the conversation, but Taen-Err-Toung does not respond as one linked to the Cuyperses by mutual colonial acquaintances. After a few minutes, Dubois reaches the shocking but obvious conclusion: the primary connection between the Chinese man and the Cuypers family lies with Mia herself. The intermission draws to a close; they have spoken with no one else, and Dubois has much to ponder during the rest of the concert.

The music finishes about an hour later, but Dubois is inattentive. After the last note has died and the applause appreciatively rendered to the musicians, Dubois sits staring fixedly into the distance, thinking, until Mia places her hand on his arm familiarly and asks, "Shall we go now?"

"*Ja, ja,*" Dubois replies, retrieving his attention from his reveries. "Of course. Let us go."

Much of the crowd has dispersed, but Mia again walks toward Mr. Taen-Err-Toung to speak with him. Now Dubois can see the situation plainly: Mia fancies not himself, the young Dutch lecturer, but this Chinese official. Perhaps she even knew in advance that Taen-Err-Toung was likely to attend such a concert and connived to get an invitation from Dubois! There are very few places where a young lady can respectably meet with a man, much less a man of another race.

Dubois thinks indignantly for a moment that she has used him as a decoy, a dupe. Then, his mild jealousy expended, he softens and lets his good sense come to the fore. After all, he has suffered no indignity or embarrassment. She has behaved in a perfectly correct fashion, greeting an acquaintance and introducing him, as respectability demands, even if it is distinctly unusual to know such a person. Somehow she arranged to meet him here. What a thing to do! He decides that she is really a very odd girl, much more peculiar than he has taken her for.

He looks at Mia with new eyes, wondering in what other ways he has misjudged her. She appears in his eyes as she always does, attractive with a broad brow and pleasant countenance, intelligence shining from her eyes rather than flirtation, her mien serious rather than coy. Yet she is vastly different in his estimation. She is now a young woman who has sought out the companionship of a Chinese man, as a friend, perhaps even as a suitor. That, at any rate, is how she is behaving and how he must judge her.

They bid good-bye to Mr. Taen-Err-Toung and start to walk back across the park, Dubois still examining his own feelings closely. He is surprised at her actions, and his pride is wounded slightly. He was on the verge of letting himself develop deeper feelings for her. How little he understood then of her true nature! Of course, there was really nothing between them, no understanding at all, only a friendly acquaintance with the daughter of a prominent man whom he admires. No one can say otherwise. Dubois' behavior has been impeccable. What he once thought and did not express is no one's business but his own. The thought gives scant comfort, for it is not entirely honest. He was close to speaking to her father—and he might well have been refused in favor of a Chinese suitor.

His mind moves swiftly, exploring the possible scenarios and the consequences of her attachment to a Chinese man. The conclusion is always the same. However well-bred and highly placed, Taen-Err-Toung is still Chinese, as foreign and un-Dutch as anyone could be.

They walk on, Mia and Dubois, until she interrupts his thoughts with a question. As soon as she voices it, he knows his deductions are correct. He is not imagining an attraction where none exists, for the topic of her conversation is the very Chinese gentleman they have just left.

"How do you find Mr. Taen-Err-Toung?" she asks earnestly, as if she savors the mention of his very name. "Is he not unusual?"

"Yes," answers Dubois slowly. "I am surprised to find a Chinese man so cultured, so knowledgeable about our music. I have not met his like before."

"Yes"—Mia smiles—"I think so, too. He is a fine man." They stroll on, ostensibly admiring the gardens, each preoccupied with private thoughts. When she breaks the rather companionable silence again, it is to say a little shyly, "My parents would not approve of my friendship with Mr. Taen-Err-Toung."

"No," agrees Dubois, gravely and a little stiffly. "I suppose not."

"But he is a gentleman," she continues, "most interesting and educated. His views are . . . different, not Dutch. He has lived in so many places, you see, among many different peoples. He has seen parts of the world so different from our little life here in Amsterdam. Perhaps . . ." She hesitates at her own boldness, and after a moment, resumes speaking. "Perhaps we shall encounter him again another day."

Their footsteps echo on the pathway as they proceed in silence for a moment or two. Dubois nods, his eyes on her face. "Perhaps we might." She turns her gaze away from him, toward the fine display of tulips. Brilliant tulips, solid reds, demure pinks, and shining yellows—some even striped or frilled at the edges—fill the bed behind a border of glowing white tulips. Close inspection of the white blooms reveals their centers to be a deep purple-black, the darkness highlighting their paleness in contrast. In the sunlight, the mosaic of colors and shapes is so dazzling that it almost hurts the eye.

Mia speaks again, quietly. "I should dearly love to know what he will think of the public lecture this Wednesday evening at the town hall, the one about the new style of painting in France. Wouldn't you?" She holds her breath for a moment, waiting to see if Dubois has taken her meaning.

He catches her glance and returns it with a small, kindly smile. She cannot read his thoughts. Does he understand? Or is his mind opaque? Abruptly he stops walking and turns to face her, making a slight, formal bow from the waist. "Miss Cuypers," he says, on impulse, "I shall be delighted if you would care to accompany me to that lecture. Should we happen to encounter the Honorable Mr. Taen-Err-Toung on that occasion, then we may discover his opinion for ourselves."

"Oh, *ja*," she says ambiguously, "I should like that."

They walk on again in silence, past the ornamental fishponds. And then Dubois adds mischievously, "I have been told, you know, that I am quite tall and slender enough to serve as a lightning rod."

Now it is Mia who stops walking and turns to look at her companion sharply. She detects the glint of good humor in his eyes and smiles broadly in response. "Indeed," she says softly. "You will serve well to draw the fire away. Thank you."

And so the arrangement is set. In the weeks that follow, Dubois acts as the lightning rod, diverting the Cuyperses' attention while Mia and the Chinese man deepen their friendship. Before long, however, she and Dubois have been seen together so often that people begin to murmur of an announcement in the offing, an engagement. If scandal is to be avoided—if Dubois is not to appear a cad—he must either propose or cease his attentions to Mia. The latter seems the only wise choice, as he knows better than anyone else of her affection for another.

Lightning strikes quickly once the lightning rod is gone. Her parents accidentally see her in Taen-Err-Toung's company, talking so earnestly that their intimacy is obvious.

"You must break it off," her father says sternly. "It is all very well to be tolerant and to treat people of inferior races politely, they may even be very interesting acquaintances, but a permanent attachment is out of the question."

The Cuyperses' opposition is matched only by Mia's determination. "You misjudge him," Mia replies passionately. "It is not an 'attachment.' He is a fine man and I shall marry him."

Her father is startled by her bold words. He hoped things had not gone this far, that the man had not deceived his daughter so seriously. "I am not aware," replies her father, a little ponderously, "that Mr. Taen-Err-Toung has asked for your hand. I am still your father, and your future husband must receive my approval and blessing." Mia turns and leaves the room.

Within twenty-four hours, Taen-Err-Toung calls formally upon the Cuyperses. They ask the maid to admit him at once, rather than leave him standing on the doorstep in full view of the neighbors. The maid shows him into the parlor, where Mia's parents sit, stiffly, eyeing this alien person with more than a little suspicion. Within a surprisingly short time, the Chinese man openly declares his intention to make Mia his wife and asks for their approval.

"I know you have not anticipated a husband like me for your daughter," Taen-Err-Toung says directly and calmly, "but it is my wish, and hers. Please ask me what you like about my job, my family, my prospects. I beg you to give me a chance, to become acquainted with my character before making your decision about your daughter's future. You do not owe me this, I know, for I have been meeting your daughter without your knowledge. It was wrong of me, but it was not an intentional deception. We first met by accident in a public place. Once I discovered what pleasure I take in her company, I could not think how to make your acquaintance so that I might call on her properly, according to your custom."

The Cuyperses' horror at the prospect of an interracial marriage diminishes as the afternoon proceeds. The man's manner is civilized, he is obviously respectable and well-educated, and he seems genuinely fond of their difficult daughter. But their opposition to the match is too deeply rooted to be torn out so lightly. They may be tolerant of other ways and other races, but to marry a Chinese man would be exceedingly unwise. They try every stratagem at their disposal to break off the relationship: anger, extracted promises, enforced separations. There seems to be no way to prevent Mia from meeting Taen-Err-Toung, so willful is she, except by making her a prisoner in her own home. It is a distasteful tactic, soon abandoned.

After a month of conflict, Mia's parents consent to an engagement of at least two years' duration. There must be no furtive, concealed meetings. The engagement will be carried out openly and as respectably as a mixed-race engagement can be conducted. The Cuyperses are beginning to like their daughter's intended; he may well make her a good husband. His honesty and kindness are more important than his place of origin and they will say so by their behavior. If others choose to dwell on his race, then they will reap the bitter fruit of their own small-mindedness.

The couple marry in 1886, with the Cuyperses' blessing. Familiarity has done much to erase their sense of Taen-Err-Toung's foreignness and racial inferiority. He is an extraordinary Chinese man, as their daughter is an unusual young Dutch woman. Speculation about the strange married life that awaits Mia flows in Amsterdam society like water in the canals. The gossips are confounded when nothing dramatic occurs. Mia does not grow pale and complain of his "dreadful foreign ways," nor does she come running home to her parents in disgrace.

Dubois watches from a distance. He is pleased that he was not more deeply involved with Mia, but he is also a little envious. He has never experienced the sort of passionate commitment shown by this pair of determined lovers. He has yet to meet any young lady, however charming or pretty, who can inspire him to rash actions. Oh, he enjoyed the flirtations of his student days—enjoyed them very much—but his concentration and emotion are ultimately reserved for his work. It is science that stirs his passions, not romance.

CHAPTER 6 LOVE AND CONFLICT

Although at first Dubois thinks the episode with Mia Cuypers has cured him of romance, it is not long before he is attracted to one of the students who enrolls in his anatomy class at the art school in 1885. Her name is Anna Geertruida Lojenga. She is not one of the better students, for her artistic talent is limited and her memory for anatomical facts appalling. Still, Dubois has seen the danger of getting involved with a too-intelligent, opinionated young woman, so Anna's character comes as something of a relief to him. She is lively and bright-eyed, with long, dark chestnut hair wound becomingly up on her head. And while she is not clever, she is popular among the students, laughing at her own errors and making light of theirs. Anna's charm depends upon neither immodesty nor boldness; it derives from her naïve ability to see the best in whatever happens. She does possess a certain facility in drawing and watercolors, a gentle talent becoming in a young lady. She also has a sweet singing voice and a fair gift for playing the piano, as Dubois discovers when he begins to call on her. And she welcomes the attentions of her handsome young professor.

As their acquaintance deepens, Dubois finds he likes the way her frivolity lightens his own natural gravity. He knows he is sometimes accused of being too serious, too somber, and Anna teases him out of his solemnity. He enjoys her laughter, her songs, and her merriment over simple pleasures,

Dubois marries Anna Lojenga, shown here, on August 4, 1886.

even if she sometimes makes much of very little. He thinks of her as being like a butterfly or a flower: charming, decorative, but not weighty. He flatters himself that his temperament balances her natural lightness. And he is certain there is no chance he will be cast aside for a secret Chinese lover.

On the twenty-fourth of August, 1885, a sultry afternoon, he asks permission from her father to speak to Anna about marriage. Before the day is out, the couple are officially engaged. Anna is alight with joy at the proposal and Dubois himself feels an unaccustomed brightness. Anna's parents think Dubois amiable and recognize the value of a son-in-law with a promising future ahead of him. They might have preferred an upstanding young man of their own faith, but no Lutheran suitor has presented himself. Dubois is a good match.

Both sets of parents see Dubois as unalterably a Catholic, even as he is indelibly a Limburger, and they expect him to resume Catholic observance as he matures, misjudging Dubois completely. They fear that the religious differences may cause conflicts. Dubois' parents ask that Anna agree to raise the children in the Catholic faith; she dutifully pledges to do so. But no one sees that Dubois, the father of these future children, is not a Catholic himself but a scientist. Anna's parents insist upon a year-long engagement, which is announced in Amsterdam, Eijsden, and Elburg, the Lojengas' hometown. Dubois is amused at the parallel to the Cuyperses' reaction to Mr. Taen-Err-Toung. Perhaps being born a Catholic Limburger makes Dubois half as foreign as a Chinaman! But a year-long engagement will give him more time to get ahead in his career and will give Anna time to plan the wedding.

Late in 1885, Dubois is offered a lectureship, only one step below a full professorship, in anatomy at the University of Utrecht. It is a highly desirable position and so he discusses it with Fürbringer, his chairman at the University of Amsterdam, out of courtesy.

"You really ought to stay here, Dubois," Fürbringer advises him gravely. "I know Utrecht seems a step up, but if you are patient and stay here at Amsterdam, it is you who will succeed me as professor when I retire in a few years' time. They aren't going to bring in someone from outside when I go; they'd rather promote a man from within, and you're the best one. So if you want to go to the top, staying put and continuing on the course you have set is the thing." Flattered by the older man's interest, ambitious for the promised chair, Dubois turns the offer down. Turning down this opportunity awakens an ill-defined unease in Dubois' mind.

Through much of 1885, he buries himself in the study of the comparative anatomy of the organs of speech and sound production. He dissects a series of different species of animals, making detailed drawings of all their structures, comparing each species with the others, tracing the development of the larynx both in embryos as they grow and in animals according to their evolutionary relationships with one another. In the back of his mind is always the question of the evolution of speech in humans.

Anna sees Dubois' single-minded devotion to his research but never entertains any doubts about the marriage. That is the sort of thing men do, become obsessed with their professions: it is to be expected and it has nothing to do with her life, except for the promise of financial security. Dubois offers her everything a young woman could dream of. He is handsome, young, loyal, and much admired. The pragmatic realities of day-to-day living with a man in love with his work do not cross her mind.

In 1886, Dubois is promoted to lecturer at the University of Amsterdam, the very rank he turned down at Utrecht. When he tells her the news, Anna is thrilled. Two promotions—first prosector, now lecturer—in two years' time! He must be the cleverest man on the faculty of the University of Amsterdam. Maybe, Anna thinks, they will be able to afford a piano for the parlor and, in time, a horse and carriage. In every idle moment, Anna dreams of their delightful future. She will be the lovely wife of the youngest and handsomest professor in Amsterdam. They will have beautiful children and a nursemaid to attend them. She will throw splendid dinner parties for his colleagues and visiting scholars; invitations will be sought eagerly as a sign of Dr. Dubois' favor. She will have the most expensive china, the finest silver, the whitest linens, and the best flowers for her table.

Dubois is less sanguine than Anna about the significance of his promotion. He is being rewarded, ostensibly, for his excellent research on the larynx, which was published in *Anatomischer Anzeiger* early in 1886. Preparing and publishing that research was difficult.

Back in the fall of 1885, Dubois had discerned a complex pattern in the morphology of the larynx in different animals. He checked his logic, reviewed his drawings and notes, consulted all the anatomical literature on the subject. Finally he wrote a draft of his first real manuscript. His main conclusion was that the larynx in mammals is derived from the fourth and fifth branchial arches in the embryo. This sounded like an esoteric point, of interest to only a few specialists. But the derivation of the mammalian larynx from the branchial arches implied that the human voice box evolved from the gill cartilage of fishes. The structure that once filtered water to extract oxygen now "filters" air and makes sound. This was a totally new finding, and Dubois was proud of his work. It was yet another piece of evidence that showed an evolutionary link, a transition between two forms that appear to be quite distinct, such as fishes and mammals.

With some trepidation, he took the manuscript to Fürbringer, asking for his advice and comments before submitting it to a scientific journal. He hoped to have his first publication accepted before his wedding, if possible. Fürbringer understood the gravity of Dubois' maiden voyage into real scholarship and promised to give the manuscript his immediate attention. If Dubois would come to his office the next afternoon, they could discuss it in detail. Dubois awaited the interview anxiously. At the appointed hour, he presented himself in Fürbringer's august room, lined with a bookcase. A few specimens—an odd skull here, a pathological tibia, a curious tortoiseshell there—interrupted the orderly ranks of scholarly texts. The room was carpeted, comfortable, and very dignified, as befitted a full professor.

"Come in, my boy," Fürbringer said heartily, turning from his large desk and gesturing Dubois into a heavy, padded chair. He picked up the manuscript. "This is well done, Dubois, very well done. There are just a few points that need a bit of revision." He leafed through the manuscript page

by page, applauding this illustration, suggesting a minor change in that one, pointing out passages and descriptions that might benefit from greater clarity. As the minutes passed, Dubois slowly relaxed. None of Fürbringer's criticisms were serious; none detracted from or jeopardized Dubois' main conclusion. The work was good. Fürbringer's small changes would improve the manuscript and would not take much time or effort on Dubois' part. He was greatly relieved. Then, just as

Max Fürbringer, the renowned professor of anatomy at the University of Amsterdam, overshadows Dubois.

Dubois prepared to excuse himself with thanks, Fürbringer brought up a final point.

"You know, Dubois," he said, tapping the manuscript with his glasses, "this work is very important. It completely confirms what I have been saying since about 1880: that the thyroid cartilage of the human larynx is derived from the fourth branchial arch. Of course, you have heard me mention this idea in lectures. I think that you ought to add a few sentences acknowledging my work on that matter, as it is so closely related to your topic. It will strengthen your claim about the derivation of the mammalian larynx in general."

Dubois was surprised into temporary speechlessness. Credit Fürbringer for the idea? He had never heard Fürbringer say that the thyroid cartilage was derived from the fourth branchial arch. What could the man be talking about? Was he trying to claim credit for part of Dubois' discovery? Were Dubois' months of work in the dissecting room nothing but an elaboration of Fürbringer's previous idea? He was appalled. He wanted to ask Fürbringer where he had been while Dubois was working everything out, hour after hour, dissecting species after species, revealing and drawing the anatomical structures from every aspect. Why had he been all alone in the laboratory if this was really Fürbringer's work? The thoughts in his mind were so loud that he wondered why Fürbringer did not hear them.

Stammering with the sheer effort of controlling his words, he asked, "D-do you really think so, sir? That I should cite your ideas about the thyroid cartilage? I wouldn't want to slight someone else's work, especially yours, of course not, only . . ." He berated himself for sounding like an unsure schoolboy instead of a professional who had completed an important piece of research. No wonder Fürbringer did not believe he had done this on his own.

Fürbringer was patient and kindly; he had seen this reaction from young scholars before. The young always thought they had discovered the key to the universe all by themselves, when in reality they were building on the work of those who had gone before. Because Dubois controlled himself well, Fürbringer badly misjudged the depth of the younger man's emotion. "It's always a better idea, you know," Fürbringer replied genially, not meaning to patronize his junior but nonetheless enraging Dubois further. "Citing other people's previous work never detracts from your own." Dubois' face barely concealed his disappointment, so Fürbringer added, to reassure him, "You'll see. You'll see I'm right in this, in time." Then Fürbringer stood up to signal that the interview was over. He did not realize that he had sowed, watered, and fertilized the seeds of resentment in Dubois. The younger man dutifully added a mention of Fürbringer's ideas to the manuscript, but he felt as if he had been swindled.

And now the rapid promotion to lecturer seems like a payoff for agreeing to let Fürbringer claim some of his work. Dubois tries to explain to Anna why he is so ambivalent about his promotion and publication, but he barely understands his feelings himself.

*   *   *

The year's engagement affords Dubois more than a few glimpses of Anna's silliness and frivolity. Sometimes he is amazed that she thinks of his promotion and his career only in terms of furnishings for a house that they haven't even found yet and can't afford. She is utterly absorbed by choosing her trousseau and wedding gifts. For Dubois, these things rank little higher in importance than a fairy tale. He finds it unsettling that there seems not to be a serious thought in Anna's head, that she cares nothing for his work and understands even less. Why, she can't even remember if the larynx leads to the stomach or is the voice box! The only speech she thinks about is the one that will be made at her first dinner party. And then, softened by his fondness for her, he amends his criticisms. Anna is only twenty-one years old; it is surely just youth, just the tail end of her natural girlish foolishness.

As the wedding date approaches, Dubois hopes the change in his life will resolve the restlessness he has felt of late. Having a wife and a home of his own will be a sign of maturity, of having moved from the status of a student to that of a grown man with serious responsibilities. Yet sometimes the things Anna says give him pause. He begins to see how far beneath him she is mentally; intellectually, she stands not so high, like a little girl who barely comes up to one's waist. But still, she is his little Anna, his little pet, and it is enough for now that she is young and pretty. Her skin is so soft, the luster of her hair so attractive.

He ignores his occasional qualms and carries on out of a romantic chivalry. Marie Eugène François Thomas Dubois has asked for the hand of Anna Geertruida Lojenga in marriage, and they will be married. He will not prove their parents' doubts valid. They think the problem between Dubois and Anna will be religious, as if religion were important. Pah! That is nothing.

They are wed on August 4, 1886, a strikingly attractive couple who are in many ways opposites. She is as dark as he is fair, as slender as he is barrel-chested, as laughing as he is grave. In the first blush of married life, they are happy and contented in their new roles, their differences adding spice, their affection smoothing over any difficulties. Anna loves playing at being a wife, furnishing their new home as she might a doll's house, seeking just the right lace for the curtains and the right fabric for the chairs for the living room. Selecting her tea set is a task over which she spends weeks, prolonging the decision because she finds it all so enjoyable.

After the wedding, Dubois' more settled personal life gives him the courage to try to resolve the tension between Fürbringer and himself. In the fall of 1886, he sends Fürbringer a polite note, asking that he not attend Dubois' first lecture of the year. Teaching is for Dubois a nightmare; he has, in his five years as an anatomist, developed a deep aversion to this bread-and-butter task of his profession. He fears that the students respond better to other men, even those less knowledgeable than he. He does not know why they do not flock to him: he is smarter than the others; he could teach his students how to conduct an experiment or make an important observation, how to construct a good theory. But they do not love him. Perhaps he is too formal with them; he does not care for their boisterous jokes and pranks. Science is not an appropriate subject of jest. And perhaps, he thinks, he expects too much of them, because he holds them to his own standards. But he doesn't know any other way to be. He cannot compromise science, any more than he can choose not to breathe.

Dubois' request thus has a dual function. He does not want to expose his shortcomings as a teacher to Fürbringer's critical gaze, and he wants to be sure that other professors who might attend see him as a fully independent scholar. With but one publication to his name, Dubois is extremely sensitive to the possibility that he will be seen as Fürbringer's "boy," his protégé, whose every move is orchestrated by his mentor. Unfortunately, Fürbringer does not understand, or does not take seriously, the younger man's request. He replies to Dubois' request with a lighthearted note:

"So the question arises whether I am at least permitted to attend the lecture with a muzzle, or whether you are afraid that even this might damage your independent status?" Dubois' plea for understanding and protection from the older man has been completely misread. He feels mocked, and he vows never to forget these hurtful words.

His feelings of humiliation are offset to some extent by praise for his larynx work from a colleague, Max Weber. Only six years Dubois' senior, Weber belongs to the famed German school of biology, having been educated in Bonn and Berlin. He is already one of the foremost scholars in Europe. In 1879, Weber joined the University of Amsterdam as Fürbringer's prosector in anatomy, the position that Dubois recently occupied. In 1878, he was lured away to the University of Utrecht to become a lecturer, as Dubois, too, might have been. Three years later, on the verge of being promoted to full professor at Utrecht, Weber was tempted back to Amsterdam by a professorship of zoology, comparative anatomy, and comparative physiology. The deciding factor was that Amsterdam also offered access to whatever rare animals died at the Royal Zoological Society Natura Artis Magistra, one of the finest zoological gardens in Europe. It is a great research opportunity.

Weber is known in Amsterdam as a bit of a potentate. He is never seen, even in the dissecting room, in anything less formal than a frock coat and starched collar. He beard is always meticulously shaped, his luxuriant dark mustaches waxed and upturned, his speech precise and measured. He is an influential and excellent scientist, so it is an honor when he asks Dubois to contribute a section on the larynx in whales for the first volume of his enormous work on cetaceans. To everyone except Dubois, the parallel between Weber's meteoric trajectory and Dubois' is obvious. But Dubois sees only that Fürbringer blocks his path more and more. He vows that his research on the whale larynx will owe nothing to Fürbringer's influence. And he lays plans to compile all his larynx work into one comprehensive publication, as soon

In this 1886 painting by Louis Straké, Professor Dr. Max Weber (center, with beard and mustache) dissects a lion. From left to right, standing: laboratory assistant Sleking; Jacobus Janse; Johannes Oudemans; Weber; Frederich Went; the musician De Josselin de Jong; and Sleking's son. In fore-ground, "the last whaler," proba-bly a member of the crew of the Willem Barents, *in which Weber sailed in 1881.*

as the cetacean work is finished. Then no one will ever again doubt that the work is his own.

The topic Dubois has settled upon—a thorough comparative review of laryngeal anatomy and evolution—is so enormous that it is flatly unwieldy. He has too many observations and too much information to organize and present clearly. As 1886 turns into 1887, he struggles with the manuscript on his own, day after day, writing passages, striking them out, moving paragraphs here and there in an attempt to bring logic and order to the information.

Fürbringer is eager to help and asks frequently—too frequently for Dubois' comfort—about the progress of the manuscript. Dubois suspects that Fürbringer's offers of assistance are nothing but thinly disguised efforts to find out what he has discovered. Why else would he be forever urging Dubois to show him the manuscript? He resents the way Fürbringer is always passing along articles he thinks Dubois may not have read, as if he were too careless to follow the literature carefully.

The more Fürbringer tries to help, the more furtive Dubois grows. Soon he is reluctant to discuss his ideas or work with any of his colleagues. He becomes withdrawn, touchy, almost feverish, as if being literally poisoned with suspicion. Soon Dubois and Fürbringer are barely speaking—or rather, Fürbringer is speaking too much and Dubois too little. Dubois frets he will never free himself from Fürbringer's influence, yet he cannot see how to extricate himself from this sticky situation. If he stays, if Fürbringer gives his endorsement as promised, Dubois will become professor of anatomy. Will that endorsement be given if Dubois remains staunchly independent? Will people see the promotion as his due? The problem spins around and around in his brain, like a child's top, careening wildly off balance. Unlike the toy, Dubois' worries never run out of momentum and fall into blissful stillness.

CHAPTER 7 TURNING POINT

On April 4, 1887, a mere eight months after the wedding, Anna gives birth to Dubois' first child, Marie Eugénie. Dubois is proud to be the father of such a pretty child, and the husband of such a lovely wife; he is a real family man now. Still, Eugénie causes inevitable disruption in the household. They hire a nursemaid, who looks after the baby well, but they have not anticipated the social consequences of the child's early arrival. When the birth is announced, eyebrows are raised in Amsterdam society; months are counted discreetly on fingers. Anna resumes entertaining as soon after the birth as is decent, inviting friends and Amsterdam relatives for tea in her new home. She makes sure that the nursemaid is ready to bring the baby in, exquisitely dressed, to be shown off. To her surprise, there are rather many comments on Eugénie's robust size and obvious health.

"How big she is," says a stout gray-haired matron, Mevrouw Hielkema, the wife of a prominent lawyer. She tickles Eugénie under her chubby chin and strokes her sweet head. "And what lovely thick curls!" She smiles smugly.

"Oh, let me look at her, the darling," coos Mevrouw Van Boven, only a few years older than Anna. The young wife of a prosperous banker, she has undertaken to raise the children left motherless by the death of his first wife, as well as to produce more heirs at regular intervals. "Oh, *ja*"—she echoes the matron's sentiments slyly—"isn't she plump and delicious? And so big. Mine were just so small at first." And then she addresses Anna: "I don't know how you did it."

Anna's head comes up suddenly; is this a comment on the health and size of her infant, or a subtle insult based on the apparent maturity of a child born so shortly after the wedding? She blushes and looks away, unable to think of a reply that will refute the lurking implication of impropriety.

"So often," clucks another visitor, Mevrouw Hessing, a friend of Anna's mother, "early babies are thin and weak. How fortunate you are, Anna, that she is so strong." Turning to the baby and patting a plump hand, she adds, "*Ja,* and what a lovely girl you are." Eugénie gurgles appreciatively at the attention and soft voice.

Eugénie *is* a lovely baby, Anna thinks fiercely, her dark eyes flashing, and she was *not* conceived before the wedding. Anna cannot voice the thought, for it would acknowledge an accusation of wrongdoing that has not quite been made. Eugénie grows a little restive and Anna gestures for the nurse-maid to come and take the child. As they leave, Eugénie cheerful and secure in her nursemaid's arms, one more pointed remark is offered.

"What a good thing it is that you and Dr. Dubois married," clucks Mejuffrouw de Perron, a bony, large-nosed spinster, squinting after the baby. She has ruined her eyesight with too much reading and is now notoriously shortsighted. She has put her cup down in the butter dish at more than one tea party, but she refuses to wear the spectacles she so badly needs for fear they will make her look old. She still hopes vainly for a proposal of marriage. Normally, the mejuffrouw is a figure of fun to pretty young women, but Anna has been kinder to her than most. Now Anna looks at the homely woman, wondering how the mejuffrouw can so thoroughly misjudge her own attractiveness and how *she* can have so badly misjudged the woman's character. She would not have thought catty remarks to be the mejuffrouw's style.

"Oh, *ja,*" she replies sweetly, glad to have the chance to make her point. "Otherwise we should not have that darling child." She smiles blissfully at her tormentor and then, meaningfully, turns the smile on the others.

"Well, indeed . . ." sputters one of the ladies.

"I mean . . . of course not," adds the mejuffrouw in confusion. Mevrouw Van Boven, the sharpest of them all, shakes her head as if agreeing that there has been a silly misunderstanding, all the while thinking that Anna is a little more clever than she has taken her for. It is, as Shakespeare says, a palpable hit.

"Will you have more tea?" Anna asks politely. "Or perhaps"—she turns to Mevrouw Van Boven, by far the plumpest of those present—"a few more tea cakes? You seem to enjoy them so."

"Ah, thank you, delicious," the mevrouw answers swiftly, unable to fend off the unflattering implication of greediness. "Not another. I couldn't."

Soon afterward, the visiting ladies make their excuses and leave. Now Anna has time to replay in her mind the ambiguous compliments and double-edged praise she has received. No, she is not imagining insults where none exist. Apparently everyone believes that this blessedly healthy child was born nine months after her conception, which must therefore have been a full month before the wedding.

By the time a few days have passed, Anna is more than a little disturbed about what people are saying and thinking. She has rarely faced malice before; her charm and sweet nature have armored her against cattiness and jealousy. She is inexperienced in the ritual combat of society and does not know what to do. She knows, everyone knows, that early babies are usually small and sickly. No one could call Eugénie, with her fat pink cheeks and chuckling smiles, either sickly or small. Therefore she is not an early child but rather is evidence of a sin only partially concealed by a hasty wedding. In the weeks that follow the birth, Anna becomes fretful and despondent, while Eugénie thrives. Anna's characteristic sunny temperament deserts her for the first time in her life, leaving instead an anxious mother who dotes unhealthily on her daughter. She interferes too often with the nursemaid, changing her mind and countermanding her own orders hourly. The nursemaid is an experienced woman who knows her place, but she has limited patience with Anna's incessant fussing. She wishes, quite frankly, that this busybody of a mother would go away and leave her to raise Eugénie to be a placid, obedient child.

When Anna is not in the nursery interfering with the baby's routine, she is distracted and obsessed with the gossip she imagines is being traded at her darling child's expense. She works and reworks the exact interpretation of every greeting she receives, every remark made about Eugénie in her hearing, as if toiling over recalcitrant pieces of embroidery. Slowly, she begins to lose control of the household. Meals are served early, or late, or are simply ill planned; laundry is not attended to; mail sometimes gets lost or goes unanswered; and the weekly household accounts are not examined with the usual care. Anna loses weight—rapidly, unbecomingly—while Eugénie grows plump.

One morning, at a breakfast marred by burnt toast and overdone eggs, Anna fidgets ceaselessly, adjusting the silverware, rearranging the cups, placing and replacing her prized teapot at the end of the table until Dubois' nerves are almost as raw as hers. He has tried to ignore her anxiety, hoping his calm will quiet her. It is no use; her mind is so tortured that she cannot be stilled. Finally she wonders querulously, for the dozenth time, what they can say or do to dispel the rumors of Eugénie's early conception.

At this, Dubois loses his temper and puts his cup down so firmly in the saucer that it threatens to break. "I will hear no more of such nonsense," he thunders. "Why do you listen to silly chatter and idle gossip?" He thumps the top of the table with the flat of his hand for emphasis. The china and silverware rattle. "Can we conceal our child's age or birthdate? Can we lie about the date of our marriage? No, we cannot."

Anna cringes at his raised voice and does not dare to look at him. He has never been so displeased with her, so angry. Eyes downcast, she busies herself mopping up the spilled tea with her napkin rather than ringing for the girl. Dubois gets up noisily from the table and walks quickly over to the window, shaking his head impatiently. He pauses, ostensibly to look at the weather but really because he needs a moment to clear his mind.

Anna turns in her chair to watch him. She starts to speak in a small, trembling voice, knowing with one part of her mind that she is saying the wrong thing but unable to stop herself. "It is only that people say such cruel things about Eugénie . . ."

Dubois moves over to stand next to her where she is still seated at the table. The expression on his face stops her voice in mid-sentence. He is standing very still, all his attention focused on her. His pale blue eyes seem to examine her most private thoughts. Her eyes fill with tears of shame and fear and confusion. Dubois is through raising his voice to Anna. He does not wish to bully her, but she has pushed him past endurance. Now he speaks to her quietly, as if they were in church. "Do . . . you . . . suppose . . ." he asks gravely, holding her face in his hands and looking directly into her eyes, "do you suppose that anyone would think I did not behave honorably toward you during our engagement?"

The tears spill down her pale, thin cheeks, loosed by the gentleness in his voice. She cannot look away, his eyes are so compelling. "No," she answers miserably, "they could not think such a thing of you. No."

"Then," concludes Dubois, exercising his logic relentlessly, "there can be no cause for gossip, can there?"

"No," Anna replies shakily. There is nothing more she can say. She knows and fears the power of gossip, the effect that even senseless rumors can have on social position. There is no opposing Dubois, however, because Dubois is a scientist, who deals only in facts. He is older, he is bigger, he is smarter, and he is her husband. He is the unquestioned head of their young family. He is right.

Dubois pats her hair affectionately, as one might a naughty but now repentant child's, and then moves quietly across the room to the door. Before leaving, he stops and turns to speak to her once again. Standing straight and tall, the very image of "Recte et fortiter," he issues one simple command to his young wife. "Anna," he says seriously, "you must calm yourself, pull yourself together." He notes the untidy strands of hair coming loose; she has never before been slovenly.

"You are a wife, Anna, a mother; you have a daughter to look after now. If we have a well-run home, a healthy child, and a happy marriage, there will be no more talk. Making that home is your job. You must not trouble me with such foolishness. My job is to settle matters of great scientific importance. I cannot worry about you or the children or the running of the house."

He leaves the room; in a moment, she hears the front door open and close. He has left the world she inhabits with her child and the servants to go to the university, where his research awaits him.

Anna may be foolish, but she is no fool. Indeed, Dubois' outburst seems to have given her some perspective on her troubles. His certainty and strength help her regain her own; she has nothing to be ashamed of, and so she will not be ashamed. She will buy a new hat, arrange her hair in the new style, and hold her head up when she goes out. When Dubois comes home at the end of the day, she is more cheerful than she has been in weeks. Soon her appetite is better and she begins to regain the weight she has lost. She seems to have found a modicum of peace. In the evenings, she is able to sit quietly with her embroidery while Dubois reads or tries to explain the discoveries he has made that day. She does not comprehend the matters that consume him and finds them, frankly, a little disgusting. She tries to make jokes about her own ignorance of anatomy—"Perhaps I should have had a better teacher!"—but she rarely pays enough attention to understand him. From time to time, he still notices a high color in her cheeks when she is out with the baby and acquaintances stop to speak to her, but that is all. Her terrors have been vanquished by his own immunity to them.

After a month or two has passed, the household is calm and cozy once again, and Anna seems to be her old self. Dubois thinks back on the incident and is pleased at the effect of his words. Anna is learning, he thinks with satisfaction. She may worry over silly things if she likes, but she must not trouble me with them. In time, with his guidance, he believes she will develop the maturity and wisdom not to worry over such things at all. All will be well.

Sitting in the prettily furnished living room, well-fed and comfortable, Dubois reflects on the changes of the last year or two. From being an eligible bachelor, he has become a respectable family man. He is no longer an assistant with prospects, living in dingy rented rooms, but a lecturer with a lovely wife and a charming home. Anna goes upstairs and he can hear her singing softly to their beautiful daughter. He is truly content. He will soon be the professor of anatomy at the University of Amsterdam, one of the youngest ever. He will discover great scientific facts, for he has already started on his research program. It is a time of life to savor.

Dubois' sense of peace does not last long. His suspicions of Fürbringer and their awkward, even hostile, interactions continue almost daily. Then a momentous discovery sweeps aside Dubois' obsession with priority, as if it were a mere spider's web across a much-traveled path.

In July 1887, Max Lohest, a Belgian geologist, and a colleague at the University of Liège, the anatomist Julien Fraipont, publish a monograph that seizes the imagination of every person in Europe who is interested in science and natural history. The summer before, Lohest and a lawyer–cum–amateur

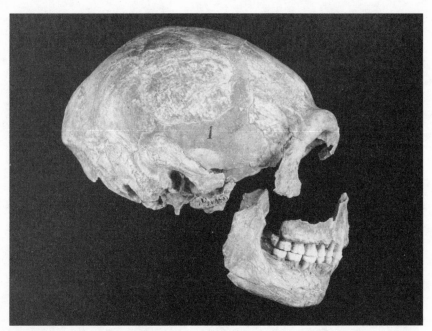

The discovery of this Neanderthal skull from Spy, Belgium, rekindles Dubois' interest in human evolution.

prehistorian, Marcel de Puydt, had found strange human fossils in a cave called Betche-aux-Rotches in Spy, Belgium. They recovered not one but two Neanderthal skeletons, as well as many finely worked stone tools and the remains of extinct animals. The newly published monograph describes and analyzes the Neanderthal fossils from Spy, which are only the third set of Neanderthal bones ever discovered.

The first heavily mineralized skeleton was found in 1856, in the Neander Valley of Germany, "born" into the world about a year and a half before Dubois himself. Perhaps the Neanderthal is truly my older brother, he muses whimsically. The significance of that first skeleton and what it said about human ancestry was much debated at the time of its discovery. Then a much more complete skull, lacking any of the rest of the skeleton, that had been found years earlier on Gibraltar, was recognized as also being Neanderthal. With the Spy finds, there are now three partial skeletons, enough evidence to resolve at least some of the debates.

Dubois takes down a favorite book, Huxley's *Man's Place in Nature,* to look again at the illustrations of the first Neanderthal to be found. He remembers the skull's jutting brow ridges, which many found shockingly apelike when the find was first announced by Johannes Fuhlrott, a school-teacher, and anatomist Hermann Schaaffhausen of the University of Bonn. After analyzing the fossils, Fuhlrott and Schaaffhausen declared the Neanderthal to be an example of an ancient, primitive race from which modern man had evolved.

Thus the pair acquired a powerful and formidable opponent, the brilliant German pathologist Rudolf Virchow. Virchow had once boasted that he *was* German science, there being no need to add that German science was the best in the world. Sometimes called the Pope or the Pasha of Science behind his back, Virchow is rigidly autocratic and formal. He is also deeply and openly antagonistic to the idea of evolution. When Fuhlrott and Schaaffhausen called the Neanderthal fossils evidence of human evolution, Virchow and his acolytes replied promptly that these were nothing other than the bones of an old, deformed human, probably a man with a bad case of rickets. Since Virchow was the world's greatest authority on rickets, few dared challenge him. As for evolutionary theory itself, Virchow pronounced it little more than unsupported speculation, entirely lacking in proof. Now, after thirty years of Virchow's unrelenting skepticism, similar fossils have been discovered at Spy and scientific opinion has reversed itself, according to the newspaper clipping Dubois pastes into a book he labels "Nota Paleontologica."

> *The supposition of Virchow . . . becomes untenable because the Spy skulls so closely resemble the Neanderthal skull and were found under such similar circumstances. We must . . . accustom ourselves to the idea that the ancestors of mankind also physically showed many resemblances with the anthropomorphic monkeys and apes.*

And since the Spy skeletons were found in the same geologic layer as the fossilized bones of extinct rhinoceros, mammoth, reindeer, horse, and cave bear—the typical animals of the ancient ice ages in Europe—any learned man would be forced to conclude that Neanderthals were contemporary with those creatures.

Reading of these finds diverts Dubois' attention forcibly from his quarrels with Fürbringer over the larynx research. He is excited, exhilarated: evolution is being proved true, in his lifetime. These fossil discoveries confirm the fact of human evolution in a new and tangible way. His boyhood game—"What would be the most important discovery?"—rings again in his mind. He once believed passionately that fossils were the key to understanding evolution, though none of the great books ever placed much emphasis on fossils. With a blinding stroke of clarity, Dubois realizes that anatomy and embryology may offer indications of evolution's working, but only fossils can prove what actually occurred, what once lived, once evolved. Perhaps he has been wasting his research efforts, chasing the wrong sort of information.

Dubois' thoughts are tinged with jealousy: others have found a compelling proof of human evolution when he himself has not. And he is the one who, from boyhood, intended to find the missing link. Why has he let himself be diverted? These new finds reawaken his love of paleontology and draw his focus almost bodily away from his resentments and obsessions. He remembers once again the larger issues of human evolution that started him on the pathway of science in the first place. One way out of his dilemma would be to find a fossil that would set him up as an expert in human evolution rather than just an anatomist.

As he ponders the new discoveries, Dubois returns to his bible of evolution, Haeckel's *History of Creation.* Here is what is central, important, worth finding. Study the larynx, unravel its mysteries, and you will understand speech. Speech is what made us human. Dubois has done that. He now knows more than any other man about the development and evolution of the larynx. He can see the continuity that stretches from the lowly fishes, with their breathing organs, to man himself and his most wondrous endowment, speech. Reading Haeckel is soothing to Dubois for a special reason, for Haeckel managed to step out of the shadow of his renowned professor, Rudolf Virchow. Once the symbol of German scientific acumen, Virchow is now being left behind because of his unwillingness to consider new ideas and new developments; ironically, Haeckel, Virchow's former student, is leading the scientific avant-garde.

The only fact that could possibly convince Virchow and his ilk would be the finding of the missing link. If Dubois could find it, the speechless *Pithecanthropus* that Haeckel wrote of, surely even Virchow could not deny it. Neanderthals are not enough, for they are only a low sort of human, a very primitive race. Only the fossil remains of the transitional form between ape and man can prove evolution irrefutably. There could be no denying evolution, with such a fossil in hand. To find the right fossil, the one with anatomy that was half-man, half-ape . . . What a grand thing it would be, to be the man who found the missing link! Dubois' boyhood ambition returns, more powerful than before.

No one knows exactly what this long-lost ancestor looked like; no one really knows who first called it the missing link. There is an old idea, the Great Chain of Being, that compares the creatures of the earth to links on a chain, each connected to the next, from lowest to highest. But the phrase itself came into popular currency more recently. In 1877, when Darwin went to Cambridge University to receive an honorary degree, the students played a joke based on the phrase. Before the ceremony began, they strung a cord from gallery to gallery across the Senate House. Then, as Darwin was called up onto the stage, they released a monkey marionette that slowly slid along the string, provoking titters and chuckles. The monkey was followed by a hoop decorated with ribbons, an object that was greeted with so much laughter as to cause Darwin to look up from his shy reverie. He seemed bewildered until he realized it was none other than the "missing link" itself. Then his solemn face broke with delight.

Dubois especially likes the boldness of Haeckel's predictions: the missing link first evolved upright posture, then the capacity for speech. Dubois is fascinated anew with this idea. Gill structures turn into larynxes, larynxes evolve into speech organs. Do words make men human? If so, then Haeckel—with his flair for writing and speaking so clearly—is perhaps more human than some of the rest of us, Dubois thinks ruefully.

Dubois knows what he will do now. The difficulties with Fürbringer may fluctuate in seriousness but they will never disappear, so Dubois must leave the university. The difference now is that he conceives of leaving his position not only to escape Fürbringer's overbearing influence but also to go in search of greater scientific glory. He will find a way to look for the missing link.

It is a breathtakingly bold and simple idea. No other man has ever set out to find the fossil record of human evolution. Antiquarians and prehistorians are always poking around in caves and rock shelters, to be sure, looking for bones and tools and signs of ancient habitation. And paleontologists and geologists are always wandering over exposures, rock hammers in hand, hoping for fossils. Still, aiming at a particular target—drawing a bead on the missing link himself—is an utterly novel idea that no one else has ever acted upon. If Dubois finds his quarry, he will never again have to worry about being recognized as an independent scientist. Prestige, positions, professorships: all will surely come his way without effort, if only he can find the missing link.

He resolves to do it.

CHAPTER 8 TO FIND THE MISSING LINK

Dubois is nothing if not logical. Now that he has decided on his goal, it remains only to think things through and decide how best to achieve success. Dubois attacks this research project methodically, even though it is still half a fantasy. First, he must synthesize into a cohesive whole every bit of information he can wring from the scientific literature; then he needs to develop a strategy for his own work. He needs a sound and workable plan of action, for the stakes are high. For weeks stretching into months, he is preoccupied with thoughts and calculations. If he goes to search for the missing link, he will be risking much. How is he to succeed where no one else has?

The first possibilities that occur to him are that the others have not looked systematically and that they have not looked in the right place. The Neanderthal remains are too advanced, too nearly human to be the missing link. Despite their hulking brow ridges and crude, heavy bones, Neanderthals are so similar to modern humans that Virchow can dismiss them as pathological and Huxley considers them merely the most primitive of human races. Dubois' opinion is that Huxley is right. What Dubois seeks is something markedly more apelike, more primitive than a Neanderthal.

Where will he find it? All the other ancient human remains that have been found (of which the Neanderthals are clearly the oldest and most primitive yet) have been discovered in Europe. Maybe Europe is the wrong place to look; maybe by the time human ancestors arrived in Europe they were already too human. So where had they come *from*?

Logically, the roots of the human family tree must be sunk into the soil of the tropics, where apes live today. Is the fundamental flaw in the work done to date that paleontologists and antiquarians have searched in Europe because that is where *they* are, where *they* live—not where apes live, and not where ape-men lived? He begins to work out his reasoning step by step; later, in 1888, he will submit for publication the arguments that he is now expanding and shaping in his mind, working like a skilled glassblower creating a fragile and complicated vessel.

Where exactly is he to look? Darwin, in *The Descent of Man*, proposes that man originated in Africa, on the grounds that the African chimpanzees and gorillas are the apes most closely related to humans. Dubois is not so sure. Male chimpanzees and gorillas are so exaggerated in their anatomy, with enormous brows, protruding faces, and large, sharp canine teeth;

sometimes they weigh twice as much as the females. No human race is like that. On the other hand, Haeckel argues that gibbons are the apes most akin to humans, because they are in many ways so generalized. They live in faithful couples and sit upright on branches with their spouses. Males and females are about equal in size, though both sexes have long sharp canines. Gibbons even walk upright on the ground, albeit with their greatly elongated arms held up on either side in a ludicrous fashion, a little like a tightrope walker. This behavior might be a precursor to true bipedality. If so, then it would be best to go to the homeland of gibbons to search for the missing link. Both the various species of gibbon and the great red ape, the orang-utan, are Asian animals.

Dubois dissects the facts and arguments in his mind like the anatomist he is; for weeks, he reads and rereads every observation that might bear on the subject. After perusing some reports from the English paleontologist Richard Lydekker about fossil mammals from the Siwalik Hills of the northern frontier of British India, he wonders if *that* would be the best hunting ground. Lydekker names one of his new fossils *Troglodytes sivalensis,* the Siwalik chimpanzee, though he later revises the name to *Anthropopithecus sivalensis* to emphasize its differences from the living chimpanzee. In any case, those fossils prove that apes lived in Asia in the distant past, as they do now. There are precious few scraps of fossil ape from anywhere, and it would not be wise to ignore such evidence. Dubois studies maps and Lydekker's drawings of the fossil. Indeed, Lydekker's finds are very suggestive.

Going to British India, however, poses a serious logistical problem. India is a long way from the Netherlands and, as a British colony, is a somewhat difficult place for Dubois to go. He doesn't suppose that finding the missing link will be an easy task, quickly achieved, so he must take Anna and Eugénie with him. How could he support them there, in a foreign land run by a rival nation? These tangled and difficult problems evaporate if he shifts his focus farther east, to the East Indies, which have been a Dutch colony for two hundred–odd years. He can take his family to the Indies; people speak Dutch there; all the educated people and rulers will be Dutch, in fact, except the native regents and princes. He knows several families like the Cuyperses, many of whose relatives have been in the Indies and prospered. All those Indies families would surely know something about the landscape and geography of the Indies; planters always do. The real question is, Is the missing link in the Indies? No point going to look for it where it isn't.

One evening, he goes to his study to set down the meager facts available to him. Of course, in one night he cannot lay out everything he knows and doesn't know. He goes back to the task, day after day, trying to pin his thoughts down in orderly rows like beetles in a natural history cabinet. He writes, crosses out, begins again, revises. Finally his ideas begin to take a definitive shape.

Point one: The missing link is a more apelike ancestor than those found in Europe. More apelike and more ancient fossils are likely to be found in the tropics, where the apes live today. Fossils of great apes have recently been found in British India.

Point two: Fossils have also been found in the Dutch East Indies on the island of Java. They were collected by the Javanese nobleman Raden Saleh, and also by the eccentric German physician-naturalist Franz Junghuhn. Junghuhn's story is especially interesting. He went to Java in 1835 as a medical officer in the Royal Dutch East Indies Army, though somehow he served for less than four years. From 1835 to 1848, Junghuhn marched from one end of Java to the other, mostly as an employee of the Commission for Natural

Wallace's line divides the Far East into two zoogeographic regions, one Asian and one Australian in nature. Since an Asian fossil chimpanzee has been found, Dubois believes the transitional form linking this ape to Homo sapiens *will be found in the western part of the East Indies.*

Sciences, learning languages, drawing and describing plants and geological formations, collecting fossils, and making the best maps available of Java. If Dubois could get a post as a naturalist in the Indies . . . He drags his mind back to the task at hand.

Point three: Logic and geological principles suggest that Sumatra, another island in the Indies, will also contain many fossils. Sumatra is said to be riddled with limestone caves, and all of the best fossil sites in Europe are in caves.

The problem is that the Indies fossils may not be of the right age; for example, Europe has many fossils, but they seem to be too recent. It is a crucial point. Here the construction of Dubois' cabinet of facts comes to an abrupt halt. He needs more raw materials, more information. It is not enough; it is not certain. He cannot risk everything for nothing.

His notebook sits closed and untouched—reproachful—on his desk for days. Then he reads some new and crucial reports about the Indies fossils found by Raden Saleh, written by the geologist Karl Martin. Martin's work is a revelation, providing just the information that Dubois needs. Martin has compared Raden Saleh's Javan fossils with those from the Siwaliks, where Lydekker's ancient chimpanzee comes from, and has concluded that the two groups of animals—the fossil faunas—are very similar and almost certainly contemporaneous. It is exactly as Dubois has hoped. He wants to find the ancestors of man who would have been of about that age, not long after man and ape had begun to evolve separately. And Martin's work suggests that those ancestors were in the Indies as well as in India.

Dubois reopens his notebook, fingers through the scribbled, crossed-over leaves, and starts a new page. Upon it he writes, happily:

Point four: The known fossils from the Dutch East Indies are both Pleistocene in age and Asian in character. It is this period and this fauna in which we expect to find the missing link.

Martin offers something more, an idea of great importance to Dubois' thinking. The similarities between the Siwalik and Javan fossils demonstrate the truth of one of Alfred Russel Wallace's great ideas, says Martin. Wallace is the natural historian who conceived of natural selection simultaneously with Darwin. Their world-shaking papers—the ones that first put forward the theory of evolution and natural selection—were read one after the other at a legendary meeting of the Linnaean Society in London, in July 1858. Little more than a year later, Darwin's *Origin of Species* appeared and rapidly became the standard explication of evolutionary theory. Wallace's greatest book came almost twenty years later. Published in 1876, it is called *The Geographical Distribution of Animals.*

On his trips through the Indies, Australia, and the Malay archipelago, collecting specimens for museums and private individuals, Wallace noticed that there is a sort of invisible line falling to the east of Java. Between the islands of Bali and Lombok is a barrier—Wallace's line, as it became known—that divides the entire region into two natural parts.

To the west of Wallace's line are the large islands of the Indies: Java, Kalimantan, and Sumatra. They are inhabited by animals and birds closely related to those in Southeast Asia, especially the Malay peninsula, India, and Ceylon. Wallace regards these island faunas as fundamentally Asian in character and asks: How did this distinction come about? The answer is a geographic one. Java, Kalimantan, and Sumatra are surrounded by relatively shallow seas; they are, in fact, nothing more than the highest points on the huge continental shelf called Sunda, which extends outward from the Southeast Asian mainland. Wallace speculates that, at one time, sea levels were low and the Sunda Shelf was solid ground. Asian animals could spread from place to place on this extended mainland. When sea levels rose later, low-lying areas were flooded and the high points became islands, where the animals were cut off from those on the new mainland. Thus the entire Sunda Shelf was once part of Asia, geographically and faunistically.

On the other side of Wallace's line, to the east of Bali, lie Lombok, Celebes, New Guinea, and the Moluccas, with very different faunas. Wallace sees the birds and animals indigenous to these eastern islands as fundamentally Australian. Here he found marsupial species that keep their young in pouches, such as bandicoots and phalangers, and Australian birds, such as cockatoos and birds of paradise. The explanation for this pattern of distribution is similar. Just as the western islands sit on the submerged Sunda Shelf, which is attached to Asia, so these eastern islands sit in shallow waters on the submerged Sahul Shelf, what might be called greater Australia. At times of low sea levels, all the eastern islands would have been connected to the Australian mainland and shared an Australian fauna.

The issue is not only what is present where, but what is missing. The western islands of the Indies have no marsupial mammals, only Asian, placental mammals. The eastern islands, for their part, lack many of the typical Asian mammals: wild cats, wild dogs, and deer.

The invisible line between Lombok and Bali coincides with a formidable geographical gap, concealed beneath the ocean, known as the Lombok

Strait. This enormous submarine trench separates the two great continental shelves, Sunda and Sahul. This deep strait is the rugged seam that stitches together the two great biogeographic regions, but the join is so treacherous and difficult to cross that it functions as a geographic barrier. Both the continental shelves and the impassable strait between them have isolated Sunda from Sahul, keeping marsupials out of the Asian–Sunda mainland while preventing the Asian mammals from crossing into the Australian–Sahul. Martin observes that the similarities between the Siwalik and Indies fossil faunas prove that this biogeographic pattern also existed in the distant past, for both the Indies and India were inhabited by mainland Asian species.

Point five: Geology, topography, climate, and geography all predict that there will be fossils of Pleistocene age in the untouched caves of Sumatra. If man's earliest apelike ancestors evolved in Asia, then their fossilized remains ought to be part of that ancient Sumatran fauna. They have not yet been found because no one has yet looked in the right place.

Dubois sits up, pleased with his work, and rubs his eyes. This is such a clear and compelling set of arguments, such a solid plan, that it must succeed. The fossils of the missing link must be in the Indies; it is overwhelmingly probable. The logic rings in Dubois' mind like the peals of an enormous church bell. He cannot ignore them, nor can he understand why no one else has come to the same conclusion.

The sheer feasibility of going to the Indies weighs heavily with Dubois. Many Dutch sons have taken up opportunities in the Indies as merchants or planters, importers or exporters, government administrators or military men. The East Indies may be a long way from Holland, but they are psychologically a great deal closer than British India. The Indies are a little bit Dutch. Not many women and children go to the colonies; Dubois has heard that there are two European men for every white woman in the Indies. But some women go, some children go, and he can't possibly leave Anna here with little Eugénie all by herself. Anna is young and strong. They will go together. Disease is the greatest danger, but he will always be there to see to his family's health. And then, after he makes his name, they can come home, like those who have made their fortunes in coffee or nutmeg and then retire to The Hague.

He feels as if he has discovered a continent, so tangible is his new conviction. He tries out his arguments the next day on Max Weber. To his surprise, Weber discourages him; indeed he treats Dubois as if he has taken leave of his senses.

"Your plan is not practical, Dubois, don't you see that? Oh, it all sounds very pretty, as you say it, very nice, but very theoretical. You can't ruin your life for an idea patched together out of this and that, one fact here and another there. This doesn't sound like you at all. This is . . . foolish."

Dubois is stunned, but the skepticism hardens his attitude. Doesn't Weber understand the brilliance of this synthesis? Doesn't he see there is no way the plan can fail? Dubois even wonders, in passing, whether Weber wants to put him off the idea so he can find the missing link himself. Later, having examined his plan once again, he goes back to his colleague. He explains it all again, as if Weber has misunderstood, as if Dubois' first explanation were not clear enough.

Weber is not swayed at all. "Don't throw away all you have for a figment of your imagination, like the foolish boy in Grimm's fairy tale," he says to Dubois, kindly. "This is real life. I know you are restless. I know you long for recognition of your hard work and intelligence. But it will come, it will come in time. Can't you be a little more patient, a little more prudent?"

He cautions Dubois against making a hasty decision. "I know you are very excited by the finds in Belgium, at Spy," Weber says, "we all are. The Neanderthals are fascinating evidence of the path of human evolution. But think how few fossils anyone has found of human ancestors. Almost nothing! Just those two skeletons from Spy, and the one from the Neander Valley, and that skull from Gibraltar. Every other fossil human ever found is anatomically modern. And mostly people find broken-up bits of fossil animals, not humans anyway. What good are animal bones for settling questions about human evolution?

"Oh, I agree with you," Weber concedes. "Martin is probably right that the Javan fossils are similar to the Siwalik ones. He is a good geologist. But how can you be sure you will find anything? And if you find fossils, why should they be anything other than more deer or pigs?"

Dubois has no answer for this, except that he knows in his heart he will find the missing link.

Weber continues: "It is obvious that human ancestors don't fossilize very often, for no one ever finds them. Setting aside your career and prospects for this . . . It is not a sound idea, Dubois. I urge you to think again."

Dubois does not know how to respond, except to repeat his step-by-step logic once again, more passionately. He knows the fossils are there; they must be there. The East Indies are the right place to find human ancestors; there are plenty of other fossils of the right age; there are lots of limestone caves to shelter and preserve fossils; they must be there. The conclusion is so obvious to Dubois, but talking to Weber is like trying to describe sunlight to a blind man. Why can't he see for himself?

Weber listens again, patiently and attentively. Finally, he sighs and says to Dubois, "For all that, you will not find the missing link, and you will give up your future."

"But if I find the missing link . . ." Dubois insists.

"Ah," says Weber, nodding and smiling. "Then you are in one strike a famous man."

It is the confirmation Dubois has been longing to hear, though Weber has not meant it as such. Dubois will go. He must.

CHAPTER 9 LOGISTICS

The very next day, Dubois makes an appointment to see the Secretary-General of the Colonial Office in The Hague, who will spare a few minutes late in the afternoon for Dr. Dubois. All day he waits impatiently for the hours to pass, rehearsing his arguments in his mind. He intends to ask the Secretary-General to subsidize a research trip to the Indies to look for the missing link, so he must present his ideas convincingly. Weber was just practice.

The Secretary-General's office is very grand, large enough to hold a desk and chairs, several armchairs, a small sofa, and a number of tables, most covered in papers and those books that are not tucked tidily into immense walnut bookshelves. The furniture is large, ornate, meant to impress; the carpet is lush and thick. The room would house two full professors, Dubois thinks, with ease. The Secretary-General is a very busy man, but he has heard of this rising young professor and is interested. Of course, he cannot imagine what the man wants with him. There are, of course, no medical schools or universities in the colonies; who would attend them? There are really no proper educational institutions at all. The colonists educate their children at home, by themselves or perhaps with a tutor or governess, or they send the children back to boarding school. So Dubois cannot be wanting to discuss education in the colonies or ask for a post. In fact, from what he has heard, Dr. Dubois is widely predicted to become one of the top professors at Amsterdam, so why would he want to leave?

The professor is shown in, his steps echoing against the marble floor of the high-ceilinged corridor as he enters. The Secretary-General is more impressed by Dubois' handsome features, proud posture, and firm stride than he expected to be. The Secretary-General stands and offers his hand, greeting Dubois cordially, and shows him to a comfortable chair away from the formality of his desk. He rings for tea to be brought in—a good sign, Dubois thinks. Their opening conversation is polite but unfocused: the weather, mutual acquaintances, matters in the news. Finally the Secretary-General asks Dubois what he has come to see him about.

"I would like," Dubois replies carefully, "a government subsidy to go to the East Indies to find the missing link. You know, the missing link is the fossil form that connects apes to man, as in Darwin's great theory. Finding it would be a major advance of scientific knowledge."

The Secretary-General's eyebrows rise and his pince-nez drop from his nose to hang tethered to his waistcoat by a chain. Is this doubt, surprise, maybe shock? Dubois cannot tell. He simply continues, as persuasively as he knows how.

"Now, this is not a mere pipe dream, sir; this is a well-supported scientific theory. There are many sound reasons why I believe this fossil will be found in the East Indies." And then he begins to build his logical edifice from the ground up.

But before Dubois has finished the preliminary sketch of his argument, the Secretary-General stops him. "I'm sorry, Dr. Dubois," he says firmly. "Surely you know the Netherlands is in a serious economic depression just now. The Colonial Office cannot possibly subsidize a scientific research expedition in the East Indies at this time, particularly one that seems so risky. Many men die of malaria in the Indies, you know, or of cholera or typhoid. Much of the Indies is simply jungle, a few villages without roads, no proper maps. It can be very difficult to travel in the countryside, much more so to keep an expedition supplied and healthy. And the expense: prohibitive. And for what?" He gestures vaguely in the air, suggesting the lack of substance in these ideas, and then clicks his tongue.

"I'm sorry to be so discouraging," the Secretary-General continues, "but I think your chances of success are slight. Logic is all very well, but . . . well, let me just say that I think you have let yourself be carried away by that crazy book of Darwin's. There is little truth in it, you know, and much speculation. Evolution is not a fact, it is a theory. I regret, Dr. Dubois, that there is nothing I can do to help you. I suggest you forget all about the missing link and concentrate on your career at the university. I have heard you are very well thought of there, with good prospects. Don't throw it all away over such foolishness: that's my recommendation."

Nothing Dubois adds will make the slightest change in the Secretary-General's opinion. In a daze of disappointment, Dubois thanks the official for seeing him and leaves before half an hour has passed. He feels . . . flattened, crushed like a top hat under the wheels of a brougham. All his hopes are soiled and lie ruined, in the gutter; all his clever synthesis is smashed and broken by the brisk tread of the Secretary-General's inflexible opinion. He had hoped for a more positive answer, had expected the logistical problems to be taken out of his hands and resolved by the bureaucracy of the Colonial Office. Instead, he has received nothing but discouragement.

Is he wrong in his reasoning? He reviews his points, one by one, again and again, searching for flaws or weaknesses. No, the flaw lies not in his theory but in the man to whom he proposed it, he decides on the way home. The Secretary-General is an administrator, a government man, not a man of science. He probably doesn't understand Darwin's theory at all. He probably hasn't even read *The Origin of Species* or Haeckel's *History of Creation.* It is easy to call these ideas nonsense if you don't know what they are. This man is judging the case simply by word of mouth. If he doesn't believe in evolution—if he doesn't know that there must be a missing link— how could he possibly support Dubois' expedition? That would be truly foolish. Ah, *ja,* Dubois realizes: "truly foolish" is how the Secretary-General thinks of him, too.

Even if he believed, as he doesn't, that there is a missing link, the Secretary-General obviously does not see what Dubois knows. The link is there, in the Indies, waiting for him; he knows it. This is not mere speculation. He is an anatomist, a learned man, an expert in evolution. He has studied the natural sciences for years. He knows what is true and how to prove it. The problem is just that no one else sees the truth as clearly as he does. By the time he arrives home, he has recovered his optimism and doubled his determination. He must go to the Indies, because the most important discovery of the century awaits him there.

Besides, his conflicts with Fürbringer are growing unbearable. Dubois has not dared to speak with Fürbringer of his intention to search for the missing link, lest that, too, will somehow be usurped by the older man. Dubois bristles like a dog defending his territory from an interloper whenever Fürbringer appears; he practically growls aloud and bares his teeth when Fürbringer offers advice or comment. The others cannot help but see Dubois' sensitivity to Fürbringer. Weber, perhaps Dubois' closest friend at the university, tries but cannot erase Dubois' suspicions. Dubois is growing desperate, frantic to escape from Fürbringer's shadow. Finally, Weber writes to Dubois, reluctantly giving his blessing to Dubois' plans to quit the university:

After thinking more about your plans it has become even clearer to me . . . that whether or not you go to the East Indies, you have lost your link with anatomy. . . . From your point of view of teaching and its value and from your pessimistic ideas of science and from your outspoken dislike of teaching I think that, for you, happiness can never lie in the theoretical explanation. Posts of this kind are so hard to come by that they ought not be occupied by people who are not completely content in this working environment. . . . You know the work demanded in anatomy and you didn't like it, unless it involved your own research; you like it so little that you want to choose another working environment. I think that you have already decided in this matter in your heart, have already made a decision, regardless of whether you go to the East Indies or establish yourself here as a doctor.

By now, the problem is not one of intent. Dubois knows what he must flee *from* and where he must go *to,* but not how to get there. His savings are woefully inadequate to pay the passage for himself, his wife, and his little daughter, much less to support them for months or even years. He does not like to ask his father for help, for Jean Joseph does not share his son's fascination with the missing link. The apothecary still wonders aloud sometimes if his son wouldn't have been better off to follow in his profession.

Dubois can think of only one way to go to the East Indies. Like Franz Junghuhn, Dubois has a skill that opens doors: he is a physician. He can sign on as a medical officer in the Royal Dutch East Indies Army. And as time passes and Dubois' suspicions of Fürbringer flourish, the idea of enlisting seems more and more attractive. He sees himself following in the footsteps of Junghuhn, a vision that takes some of the bitterness out of the Colonial Office's refusal to fund him. Besides, going to the colonies to seek greater scope and opportunity is a long and honorable tradition in the Netherlands. He, too, seeks a freer world, where discovery is not hampered by traditional thinking and narrow-minded, jealous men.

On July 29, he writes inquiring about the possibility of enlisting as a military surgeon in the colonial army. Almost immediately, he receives an answer: he will be accepted for an eight-year tour of duty if he passes a physical exam, scheduled for August 4 between nine and ten o'clock in the morning in Amsterdam. When he goes home that night, he shows Anna the letter and explains to her what he has done. She has listened to his complaints and even to his plans of going to the Indies, but she is startled to find it almost a fait accompli. Once he has passed his medical exam—and he surely will—then there will be two contracts to sign: one for him, agreeing to the terms of his eight-year enlistment; and one for her, pledging not to attempt to follow her husband on military campaigns in case of war. The sudden shift in the future is bewildering and even a little frightening to her, as Dubois can see from the expression on her face.

"Anna, you know I had no choice. You can see that. Fürbringer was taking credit for my work. You couldn't expect me to stay there in the face of that, could you?" She shakes her head no, as Dubois tries to reassure her. "Anna, you remember that I have told you of it, of my great idea. I will find the missing link in the East Indies; it will prove evolutionary theory to be true. It will make my name as a great scientist. I am certain that the missing link is in the Indies. There can be no doubt. So this is what we must do. We will go to the Indies, to make this discovery."

Uncharacteristically, Anna holds her tongue until she has sorted through the blur of thoughts in her mind. Until now, she has known where the future lies. She will be a professor's wife, living in a nice house in Amsterdam and raising children. Now she is on the verge of committing herself to go to the East Indies for eight years. Eight years seems an eternity. By the time they return to Holland, Anna will be thirty-two years old, baby Eugénie will be eight. She remembers having heard of Dutch schools set up for European children in the East Indies—perhaps in Batavia, in Java?—but her knowledge of the colonies' geography is vague. But, she thinks rather complacently, if there are Dutch children in the Indies, there will be some way to educate them. And it must not be a terrible place to live, for all those ex-colonials in The Hague speak of the Indies with such nostalgia.

Finally she speaks, an expression of almost mischievous glee on her face. "Won't they all be surprised when we tell them," she says, giggling a little at the thought. "All those silly old women who make remarks about Eugénie—and those stuffy old men at the university who expect you to let them steal your work! Oh, I wish I could see their faces when you tell them. They aren't young enough or strong enough or smart enough to do something like this. I can't wait to tell Mama, and Papa, and your family. I am so proud." Anna walks distractedly about the room as she talks, spilling over with excitement. Dubois listens and watches his wife in amazement.

"I always knew you would do something important, I knew it from the first time you walked into the classroom. I could see it in your eyes." Her mind races first in one direction and then in another. "They say it is very sunny and warm and beautiful in the Indies, not gray and cold like Amsterdam in winter. Won't it be good to get away from these cold, raw winds and the bitterness of winter? The Indies always sound like a paradise when Indies families describe it, a tropical heaven where Dutch people live like royalty, with grand houses and lots of servants. Oh, won't it be an adventure!"

Dubois had feared she would hate the idea, would dread leaving all that she knew. He has certainly underestimated her courage, her resilience. Anna turns to him abruptly, as if struck by the significance of the plan for their lives. Dubois waits for her words, waits for the fear or the protest to begin. Instead, Anna seems suddenly practical and efficient. "If this is what we must do to prove your brilliance to everyone, then I had better begin planning what we shall take with us. When do we sail?"

Dubois beams with pleasure, taking her small, feminine hands in his large, capable ones, and kissing them affectionately. She has shown so much more character than he expected. "We shall go, my little Anna, we shall go," he repeats softly. "I think we must sail before October is out. I shall check on the ships tomorrow. It will be a great adventure—just you, and me, and the baby."

Anna and Dubois pose for this portrait shortly before they leave for the East Indies.

By the next Sunday, Dubois has passed his exam; by the end of August, they have received and signed their contracts with the army. Now they go to visit Dubois' parents to tell them of their plans. Dubois' father, adamantly opposed, responds as if his son could be talked out of this decision, as if an irrevocable step had not already been taken. As for Dubois' unhappiness at the university, that is the result of unworthy thoughts and ridiculous imaginings. Jean Joseph dismisses his son's problems with Fürbringer completely.

"You cannot believe," the father blusters, his normally florid face turning choleric, "that Fürbringer—that a man of his reputation—has any interest in stealing your ideas. What folly!" He shakes his finger at his son, as if at a naughty child. "You think you are smarter than everyone else, but this is only sinful pride. Fürbringer is a great man, a brilliant man. Why would he need to take your ideas? If only you will put aside your absurd concerns, in a few years' time you can be the professor of anatomy, with a fine house overlooking the canal in Amsterdam. What could be better than that?"

Anna squirms in her chair like a schoolgirl while her father-in-law scolds her husband. She hates it when they disagree. She hoped her in-laws might be pleased and excited for them, happy that Dubois was going to fulfill his dream at last. She does not know what to do except keep silent and pray for this to be over. Fortunately, Dubois makes no reply. He simply stares at his father and waits. His very silence contrasts the virtues of a "fine house overlooking the canal" with those of making a discovery of lasting importance to science. What a narrow vision his father has. If Jean Joseph Balthasar Dubois has ever known what it is to discover something new, he has forgotten how it feels.

The room is filled with disharmony and anxiety as the two men look at each other and weigh each other's character and determination.

"Will you have more tea, Father? Eugène? Anna?" Trinette asks nervously, to fill the void. The men do not show they have heard her question; Anna gratefully accepts more tea. Dubois' mother waves a plate vaguely, trying to smooth over the situation. "Another cake? Just one more, Anna; they are your favorites. You must keep your strength up. You have a baby girl to look after." The clatter of cups, the exercise of passing plates, the ordinary politeness, seem to ease the tension a little.

In a moment, Jean Joseph advances another argument, seeing that his first has not changed his son's mind. "Of course the professor puts his name on your papers, sometimes," he says, thinking he is making a large concession to his touchy son by even admitting the practice. "That is because

Fürbringer is the professor, the supervisor of all those who work under him. He is like the father, who is the head of the family and responsible for all its members. Your name is Dubois, isn't it? Like my own? It shows the relationship between us. So of course your work should bear Fürbringer's name, too, from time to time. It is only fitting. If Fürbringer puts his name on your papers, it is like offering his imprimatur. It is the way things are done. There is nothing wrong with it, except to you because you are so proud and stubborn."

The elder Dubois shifts in his chair, recalling with irritation almost thirty years of his son's stubbornness and idiosyncrasy. The memories prod him like a knitting needle buried among the sofa cushions. "You can never go someone else's way, Eugène, can you? You have no grace, no gratitude, no ability to compromise. It must always be your own way or not at all. Well, this attitude will not do, not in this world."

"You raised me to be honest, Father," Dubois replies stiffly. "To claim another man's work as your own is surely a dishonesty. It is theft. Property of the mind is no different from material property. If a man takes your hat, and says it is his own, he is both stealing and lying: that is plain. If a man takes my ideas and passes them off as his own, that, too, is stealing and lying. Would you have me compromise my honesty?"

His father makes a noise of disgust but says nothing in reply. It is no use talking to Eugène; he is so unyielding. Jean Joseph will not accept his son's point of view, nor will the son heed his father's pragmatic advice. Fürbringer's actions aside, there is the missing link to be found. Dubois is not only leaving the university, he is also going to the East Indies. There is push and pull.

Dubois is filled with sadness as he and Anna leave his parents' home to journey back to their own. Before coming, he hoped to make them understand what a fine thing he was attempting; he hoped for his father's praise and encouragement in this bold undertaking. Now he knows they understand nothing, do not share his vision at all. One day, he will return to Eijsden with the missing link, in triumph. Then Jean Joseph will see what it means to the world of science. The name of Marie Eugène François Thomas Dubois will be on everyone's lips; his missing link will be in all the newspapers and journals and books. Then his father will see that he was right to pursue his vision. Then his father will be proud. Maybe Jean Joseph is just too old and too tired to understand, to think the goal worth the risk.

As for Anna's parents, they are distraught when they learn of the plan to go to the East Indies. Her mother breaks down and sobs at the thought of her daughter and baby granddaughter being taken so far away for so long. "The East Indies?" she cries in horror, pressing her crumpled handkerchief to her eyes. "The Indies are full of tropical diseases and black men. Who knows what might befall you? There are snakes and tigers and poisonous plants, a thousand things to hurt you. Even good Dutchmen lose their morals there and go native—I have heard the stories many times. And what of Eugénie? She is just a baby. You can't take her to such a place. It is too dangerous. Oh no, no, you can't go."

While Anna pats her mother's trembling hand and dries her tears—the daughter and mother reversing roles—Dubois tries to calm his mother-in-law's fears with logic. Thousands of Hollanders have lived in the Indies over the last two centuries. Truly, two out of three used to die of tropical diseases but not anymore, not now. The new treatments are wonderfully effective.

"Besides," Dubois says reassuringly, "I am a doctor. Anna and Eugénie will be with me, and I will look after them. You know I will. As for what happens to Europeans in the colonies, going native, that is all nonsense based on a few people who had no morals when they lived in Europe either. You have seen pictures in the homes of Indies families. They don't live in grass huts, running around naked except for a few beads and trinkets, like savages. We shall live in a proper house, with servants and a garden. We shall wear proper clothes, and Anna will see that the house is kept clean and sanitary. I am a professional man; this is my family. I will see that they are safe. After all, I am not some young rake shipped off to the colonies to escape his creditors or a woman he has wronged. No harm will come to us, I assure you."

Anna's mother takes little consolation in these words; she can think only of the separation, eight long years. Her father says little but sits close to his wife, as if his physical nearness might give comfort. That he, too, has doubts is obvious from the expression on his face, though he does not voice them as a woman would. He likes Dubois—thinks him a good man, a sound man—and Anna is his wife. She must go with him, if he is going.

Dubois tries to smile and restore a pleasant atmosphere of optimism. In response, Anna's mother makes a visible effort to stop her tears and asks after Eugénie. With relief, they speak about the baby for some minutes. Eugénie is their only granddaughter, so every event in her young life seems important. Soon after, Dubois and Anna excuse themselves, saying they mustn't be late returning home.

With the parental trials over, there is only one more group to be told the news. In some ways, this is the interview that Dubois dreads most. He chooses a moment when most of his colleagues are together to announce his intentions, so he will not have to face Fürbringer in a private interview. There is much consternation at his words. Clearly Fürbringer does not understand what Dubois has done and why, and he only reluctantly accepts Dubois' formal resignation. "I advise against this, you know," the professor says heavily, putting a hand on Dubois' shoulder. "You have a great future ahead of you here. I have always planned that you will succeed me when I retire; surely you know that. You are already so well thought of. You must not throw the chance away."

Dubois is implacable. "I shall go," he replies. "I shall go and I shall find the missing link. Staying here, carrying on my dissections, publishing articles in scientific journals, moving up the academic ladder cannot compare with finding the missing link. I must go."

Fürbringer sighs and offers his hand; Dubois hesitates for an awkward moment, and then takes it firmly. The man was good to him, but he left Dubois no room to breathe. Opportunities will be less limited in the Indies. Weber and De Vries are there too, wishing him luck, expressing their sorrow at losing a good friend and colleague. Perhaps they think Dubois crazy, risking everything for one of Haeckel's ideas. No matter; they are his friends, they have worked with him for years, they will miss him.

"I was afraid you'd do this," Weber admits, "you've been talking about your ideas so long and so ardently." Fürbringer, surprised to hear that Weber has known of Dubois' plan in advance, shoots a sharp look first at Weber, then at Dubois. "Well, I wish you the best of luck," Weber continues innocently, not realizing that he has said something startling. "May you find your missing link! If anyone can do it, you can, that's for sure. If you need something sent to you—books, maybe, or instruments—let me know. But you'd better take most of what you need, you know. The mail ships may come twice a week, but they don't travel very fast."

On one of his last days at the university, Dubois gives a final lecture, explaining his resignation and his research plans in the Indies. It is one of his best lectures, full of passionate conviction and brilliance. The students seem awed, some envious, others shocked. Dubois is very pleased with himself when it is over.

A few days later, he receives a letter from Fürbringer that leaves him incredulous. Fürbringer writes as if he is trying to salvage Dubois' career, which is about to be ruined. He offers to edit the huge larynx manuscript and see to its publication; Dubois is dumbfounded at the man's audacity. Does Fürbringer think this, too, is his work, not Dubois'? Or does he think Dubois is a dead man, about to succumb to typhoid or malaria? Dubois has no intention of dying. He is going to live his true life. He will take that manuscript with him and work on it in the evenings. He will never turn it over to Fübringer; better to leave the work unpublished than to publish it with Fürbringer's name on it. Dubois does not answer the letter, not even to tell Fürbringer when he and his family will sail. He does not want Fürbringer's face to be one of the last sights he sees as he leaves Holland. He must look forward, not back.

But as the steamship S.S. *Prinses Amalia* pulls out of Amsterdam harbor on October 29, 1887, Dubois regrets his pettiness. Hastily he writes a farewell letter to his former mentor and dispatches it from the first port of call. Better to leave on a friendly note, he thinks, than to take small-minded revenge.

The ship carries mail, parcels, other officers headed for Sumatra and Java, a few colonials returning from home leave, and a large group of nuns

Anna (seated, bottom right), Dubois (with beard, standing, second row),
and baby Eugénie (not shown) leave for the Indies on October 29, 1887,
on the S.S. Prinses Amalia.

going to work in a mission. For weeks, these few people are the entire world. Before leaving Amsterdam, Dubois decided to grow a beard to symbolize the change in his identity. He is no longer an academic professor but a medical man and fossil hunter. Now that the beard is past the short and prickly stage, Anna thinks her husband looks very distinguished. She also admires his second lieutenant's uniform, with its smart tunic with brass buttons up the front and a high collar. How handsome he is! He is fit and strong and young, on the verge of a great adventure. She is glad to be married to him and not to one of the others.

Baby Eugénie is a pretty thing, almost seven months old, and she is a great favorite on the ship. The nuns cluck and fuss over her endlessly. In this company, no one compares Eugénie's birthdate with her parents' wedding date. All they see is her thick curly hair, fair Dutch complexion, and sunny disposition.

The voyage is long and sometimes rough. Anna is with child again, a few months gone, and she suffers from nausea whenever the sea is less than glassy calm. She is naturally cheerful, but she cannot remember ever before feeling so poorly. She spends many days as close as possible to a convenient basin or bucket. Since Eugénie will not sit quietly for long, even to listen to her mother singing, it is a blessing to have the nuns as substitute mothers. They are always happy to walk the child around the deck or to play simple games with her. When Anna's nausea persists, Dubois insists on giving her a full medical exam, for he fears something may be going awry with the pregnancy. To his relief, he can find nothing but the wretched combination of seasickness and morning sickness. He advises her to rest, to sip water and nibble dry biscuits when she can. She suffers miserably for the first few weeks of the voyage, but sometimes in the afternoons she seems her old gay self.

Fortunately, Dubois is robust as always, with a sound stomach and a good appetite. Only the roughest of seas upsets his constitution. Since Anna often needs to lie quietly in the darkened cabin, her husband spends a good deal of time with his fellow officers, learning Malay, which they tell him will be invaluable in dealing with servants and other natives. It is a challenge Dubois enjoys, for he is gifted at languages. He already has a command of English and French, considerable knowledge of Latin and Greek, and a good deal of German. Basic Malay, simplified into a trading language by the seafaring natives and the colonists, is not very difficult for him.

The old Indies hands teach the first-timers with good-natured humor. In Malay, how something is said may be as important as what words are said. The subtlety of a yes that means agreement, a yes that acknowledges the other's point without agreeing, and a yes that means "I seriously doubt what you say" can be crucial. Still, Malay is far simpler than High Javanese, the language used in the courts of the regents, sultans, and native princes: it has intricate rules of address and manner, based on the relative status of the speaker and the listener and the circumstances under which they converse. Fortunately, Dubois will have little need of High Javanese.

When the men's Malay-language classes progress far enough, they start making jokes with one another. They practice simple conversations, giving mock orders and competing with one another to think up silly things to ask for: "Boy, bring me a baked coconut with elephant gravy, on the best china." "Have you planted the piano in the garden, as I told you to? Did you water it well?" The sense of boyish fun makes the learning pleasant and polishes the participants' fluency. Dubois tries to help Anna learn Malay, too, for their servants will speak little Dutch and she will be in charge of the household while he is at work or away searching for fossils. Perhaps he hopes, too, that the mental exercise will distract her from her nausea. He gives her lessons in the afternoon, when she is stronger and feels better, but Anna finds it hard to concentrate. Too, Eugénie often demands her mother's attention in the middle of a lesson.

Taught in isolation, with only her terribly clever husband present, Anna is very shy about speaking Malay. She fears she will mispronounce the words or say something indelicate. The old-timers tell cruel stories of the misguided and garbled commands of newcomers to the Indies and some stories get back to Anna. She imagines gruff voices making fun of her as she struggles to learn Malay. "And she said to the houseboy," one story (too vulgar to be told to a lady, but Anna learns of it nonetheless) goes, "taking him from door to door that night, 'Make water here! Make water there!' instead of telling him to lock up!" Another tale features the young mother who told the babu "The baby is rotting," like bad meat, instead of telling her the child was spoiled. "Can you imagine?" the old Indies hands cackle in glee. "No wonder the babu ran away; she probably thought the child had leprosy." Anna fears becoming another humorous tale. Instead of babbling glibly as thoughts occur to her, she develops the habit of thinking a sentence out carefully before she speaks it. The jolly camaraderie of the lounge is very different: Dubois and the other men boldly plunge into sentences, making all sorts of grammatical and pro-nunciation errors with unshakable confidence. They can laugh at them-selves; the natives will figure out what they mean; they have no fears. They are Dutch, they are male, they are superior.

As the ship travels south and east, the healthy, brisk sea air of the European coast gives way to the much sultrier climate near the equator. Anna's fashionable long-sleeved, high-necked dresses are first warm and then downright stifling. She relies on a parasol or seeks the shade when she appears on deck, which is more frequently as the trip progresses. Likewise, Dubois' woolen military uniform is exceedingly uncomfortable. The high collar becomes soaked with sweat and rubs against the skin of his neck, already tender with sunburn. When the ladies are not around, the military men unbutton their tunics or even remove them to get more air, but they cannot, of course, appear in company in such a state of undress. If some of the soldiers lack natural delicacy, the scandalized expressions of the nuns are enough to refine the coarsest temperament. They all look forward to passing through the Suez Canal and arriving at Port Said, where traditionally the tropics begin. There, Europeans on their way to the colonies usually flock ashore to the Simons Artz Emporium to have lightweight tropical garments made up.

Dubois spends as much time as possible on deck, in the open air and breeze, despite the burning sun. He marvels that Europeans can live and work in a tropical climate for long, and envies the young unmarried men, who move their mattresses up onto the deck at night to sleep. The unabated heat saps his vigor and blurs his concentration. He mops the sweat from his face constantly, turning clean, crisp handkerchiefs into sodden messes at an astonishing rate. He begins to understand all the stories about colonials who give in to the seductive languor of the tropics—who "go native" and lose their Dutchness. In this heat, moving or even thinking energetically requires a tremendous effort. The effect is immensely irritating to Dubois, who prides himself on his character, sharpness of mind, and physical vigor.

He will not go native, he pledges. He will adapt and accustom himself to the heat, dressing in lighter clothing, working earlier in the mornings. He will do whatever he has to in order to be able to work hard. No one will work harder than he. He will not lower his standards, nor will he sink to behaving like a native.

He wonders whether, before leaving Amsterdam, he should have asked more questions about life in the Indies, but what difference would it have made? Knowing more would not change the weather. At the time he was too busy packing up books and taking meticulous, detailed notes on every scientific article he thought would be of use during his sojourn. The only library he will have in the Indies is what he brings with him, and books are heavy and costly to transport. If he is lucky, he will find a few colleagues who have brought out their own scientific libraries, and he may be able to borrow books from them now and again. But on the whole, he will be truly independent, at last. He uses much of the time on the ship to fill the gaps in his knowledge about life in the Indies. He listens endlessly to those who have been out before about the best way to manage in the colonies, and asks every question his fertile brain gives birth to.

Anna simply lies in her cabin or in a deck chair, reclining in whatever shade or breeze she can find. She is miserably hot and uncomfortable, no longer so sick to her stomach but far from her usual self. She feels like some spineless sea creature exposed by the low tide: soft, shapeless, hardly able to support the weight of her own body. The nuns bring cool cloths to place on her forehead and sit with her through the sweltering afternoons, sometimes reading aloud to her from soothing books. They urge her to swallow cooling drinks and to try to eat a little bland food, especially in the evenings when it is less hot. She must keep up her strength; as the wife and mother, she is the mainstay of the family and it is up to her to keep them safe and virtuous.

She thinks to herself gratefully that their sympathetic attentions are probably as much help as any of their remedies. What, after all, would a group of aging nuns know about the trials of pregnancy? She cannot mention it, of course, but she feels as if the baby inside her is generating as much heat as the relentless sun overhead. It is as if there is some mystical connection between the sun and the child. This will be a tropical baby, conceived in the cold of Europe but carried and born and raised in the sun.

CHAPTER 10 PADANG

At long last, on December 11, 1887, the ship arrives in Padang, Sumatra. The Dubois family has been on board for forty-four days and half a world.

Padang—for this day, it is all of Sumatra to them—is the most astonishingly different place they have ever seen. Even the places they passed coming through the Suez Canal, with natives in long, loose shirts strolling among palm trees and camels, were not as foreign as this. Despite the Dutch buildings—or perhaps because of them, they look so out of place— no one could mistake this port for anyplace in the Netherlands, or for anywhere in Europe, not for a moment. The air is hot and damp and full of strange scents. Leaving the ship, walking down the docks, Dubois is suddenly enveloped in an invisible cloud of cinnamon, and then it is gone, replaced by the rich smell of some flowering plant he cannot identify.

Gazing down the unpaved streets that surround the harbor, Dubois sees they are broad, tree-lined and pleasant. But the trees themselves are wrong, bizarre umbrella-shaped creations with leaves blue-green or silvery, not the deep green of Dutch trees. The people, the smells, the sounds, the vegetation, even the air itself are foreign. Nothing is tidy, constrained, neat; nothing is *Dutch.* Here in Padang there are comparatively many Europeans and quite a number of proper European buildings, but it all looks wrong. The lush, overblown tropical plants and trees look somehow indecent next to the European buildings. The native dwellings look more suitable, but utterly strange: long houses of bamboo with matting walls and upward-curving thatched roofs mimicking the line of a pair of bull's horns. And the heat and the closeness of all of those natives are daunting.

He and Anna are surrounded by brown-skinned natives: uniformed but barefoot servants in crisp white, awaiting their masters; minor officials carrying out their Byzantine duties; sailors, laborers, porters, and who-knows-

The Duboises live in Padang and Pajakombo on Sumatra before Dubois moves his operations to Java, where they live at Toeloeng Agoeng. Other major towns (Batavia, Solo, Malang) and fossil sites (Wadjak, Trinil) are also shown.

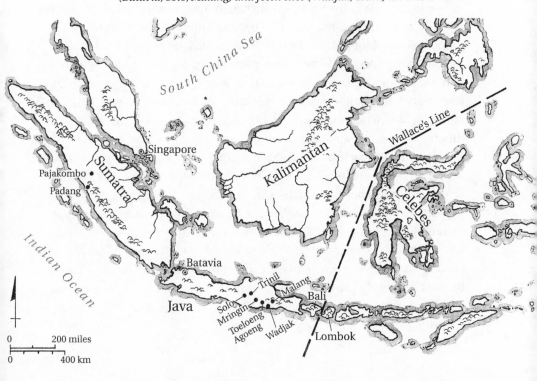

what. Traders offer their wares, whatever those might be, calling out from the small, thatch-roofed stalls that line the roadside. Dubois calls to mind the Malay word for those small shops, warungs, with satisfaction. All around them are sweating coolies, dark-skinned, dark-eyed, with exotic headwraps, naked chests, and plaid sarongs tied around their waists. These men carry all the crates and packages off the ship and load smaller piles into two-wheeled carts, dokars. Bigger items go into the larger, four-wheeled vehicles, a new design called a delman. Even the small, wiry ponies that pull the carriages look exotic. They seem to come in every color, from dun to bay to a glossy chestnut. Most are decorated with horse brasses, with colorful tassels of red and green or yellow tied to their forelocks. The largest, heaviest goods are transported in enormous crude wooden oxcarts, pulled slowly by huge, pale-skinned, wet-nosed cattle with massive humps on their necks. They must weigh well over a thousand pounds, Dubois judges.

The natives, men and women both, are small and graceful, and nearly all wear ankle-length sarongs. Somehow, the garment imparts to the most muscular of men a strangely flowing movement, almost feminine. The natives seem less solid than the Europeans; they are certainly slighter and shorter and, of course, much darker. Amid the crush of workers, colonials, and arrivals, Dubois stands out as a striking figure, a big man, strong and self-confident. He is five feet nine inches in his stocking feet, a full head taller than most of the natives. In his hat and uniform—for, of course, he is properly dressed for arrival—he is over six feet tall, taller even than most of the Europeans. And he is muscular, perhaps 175 pounds in weight with a barrel chest and big arms. His impressive solidity is set off by his coloring. Both his beard and his thick wavy hair are a golden reddish-blond; his eyes are palest blue, a rare shade even in Holland. He is about as unnative and as thoroughly Dutch in appearance as any person that can be imagined.

Dubois is intrigued by the natives' exaggerated, high cheekbones and strong jaws. These features are especially pronounced on the elderly, whose skin seems translucent, as if he can see straight through to the bones beneath. He admires their skulls greatly, for he has never seen such a shape before. He thinks he should obtain some for study. He must make a few notes on them, in any case. He is less attracted by the betel-stained, blackened teeth and red gums that so many natives have and he is sickened by their filthy habit of spitting a mixture of saliva and betel juice on the ground. He wonders why anyone would chew betel when the results are so disgusting.

He slowly becomes aware that he is the object of much attention. Some stare overtly, others more discreetly, for the natives are naturally polite. To them, Dubois exudes power and wealth and authority. These attributes, as much as his unusual physical characteristics, mean that dark eyes will follow his actions wherever he goes in the Indies, for the entire eight years. The surveillance is never-ending. He is examined as if he is a fierce exotic bird, one that has sharp talons and a powerful beak, one of which it is wise to be careful.

The heavy air is filled with a cacophony of languages: High and Low Javanese, Madurese, Malay, Acehnese, Minangkabau, Buginese, and endless dialects. But their voices are generally soft and their faces friendly. Dubois can spot few signs of argument or conflict as brown hands gesture gracefully, postures change, heads tilt at peculiar angles as if the necks that support them are made of rubber. Dubois is riveted by the scene in front of him, hardly knowing where to look first. There is more color and activity here than in the busiest market day at home.

He becomes aware of small islands of Dutch and English language amidst the noise. The longtime European residents of the Indies, the Indische, have a strong accent and an unusual pattern of speech; they drop the "I"s out of their sentences and elide some of the consonants, shortening and softening the words. This seems to be part of the tropical languor, as if it is just too much trouble—or maybe too hot—to speak correctly.

Anna is at first a little giddy and then a touch frightened by the strangeness of the activity surrounding them. A wave of light-headedness sweeps over her and she clings to Dubois' arm for support. "Eugène," she whispers urgently to him, "are these strange brown people to be our servants, in our home? However will I speak to them?" She clutches a hot and squirming Eugénie closer to her side. "How can I turn Eugénie over to a native nursemaid? Shall those brown hands touch her soft pink skin, wash her hair? Eugène, where have you brought me? What shall we do?" She is close to fainting.

"It is all right, Anna," Dubois replies quietly. He holds her arm firmly, encouraging her to stand still. His calm, solid presence reassures Anna and quiets Eugénie, too. "We shall find an experienced babu at the military base."

"Babu? A babu?" replies Anna, her eyes rolling a little hysterically.

"Yes, Anna, a babu, a nursemaid." He keeps his voice low and calm. "You remember, Anna. We learned those words together. We shall hire a babu that someone Dutch has employed before, a woman with references. She will know just how to care for Eugénie. It will be perfectly safe," he whispers comfortingly to her. "You remember our plan. We shall need a babu for Eugénie, a kokkie to prepare the meals, a djongas or houseboy, a ladies' maid for you, and a garden boy—they call them tukang kebun, remember?"

These few words of Malay somehow reassure Anna. She remembers them. She could not have summoned them up out of the confusion of her brain by herself, but she remembers them and they constitute a sort of plan. The words themselves repeat in her brain like a spell of protection: a babu, a kokkie, a djongas, another babu for me, and a tukang kebun; a babu, a kokkie, a djongas . . .

Before long, their piles of luggage are unloaded from the ship and a stocky uniformed man appears at their side. "Lieutenant Dr. Dubois?" he inquires, saluting.

"Yes," replies Dubois. "I am Dr. Dubois. This is my wife, Anna, and my daughter."

"Isn't she a beauty?" the man responds admiringly, patting Eugénie's disarranged curls. Then he remembers himself and his duty. "Captain Hendrik Krull," he says, bowing his head. "Doctor, Mevrouw. Sent to collect you. This your luggage here?" He gestures toward the enormous pile of their belongings.

"Yes, that is it," Dubois says, a little embarrassed.

"Certain you've got ever'thin'? No point leavin' barang-barang here, you know. Natives'll only steal it."

"Yes, yes, that's all of it."

"Ajo!" This is the first time they have heard this common Indies exclamation, an expression roughly equivalent to "Let's get going!" The captain turns away from the Duboises and calls to his men. "Coolie! Adik! Here. That's their barang-barang. Careful now! Hati-hati! Don't break ever'thin'! Keep that trunk the right way up. Don't drop it."

Krull is a wonder of efficiency. He somehow gathers up the Duboises and all their belongings and shepherds them effortlessly through the chaos. He hands them into a smart delman in a matter of moments. Anna sinks gratefully into the feeling of being taken care of, while the coolies pack the luggage into a large oxcart that follows them down the broad, dusty street.

"Don't want to hang around here all day," Krull confides. "Need a rest, no doubt, and a meal before startin' to unpack. Quarters are ready for you. Sent my people over this mornin' to give it a bit of a clear-out."

"Thank you," says Anna vaguely; she hardly takes in what he says, she is so distracted with looking around at the people and buildings and vegetation.

At the base, they are a little disappointed by the quarters assigned to them; the house is drab and small and poorly furnished. But it is clean, the walls are thick, and the roof looks sound. Still, Anna expected something better for a second lieutenant's family. She has not yet seen the cramped, tin-roofed barracks where the enlisted men live, or she would know how good these accommodations are.

The officers' quarters are arranged in a long row, one after another, separated from the avenue by a low whitewashed cement wall. Each building is almost indistinguishable from the next but for differences in the plantings in the forecourts and gardens, if there are any. Where there are plantings, there is a European or Eurasian wife who tends them lovingly—or rather, who sees to it that the tukang kebun, the garden boy, tends them. No njonja, no married European lady, would stoop to actual digging, planting, or weeding; that is natives' work. The njonja gives the orders, plans the garden, and oversees the work, but she does not do it herself. But there are very few European or even Indo wives at the military base in Padang at all, although all the officers are Pures, people of European blood with no native admixture. The military discourages marriage except among the senior men who have served honorably for years.

It is the rainy season when they arrive and every afternoon it rains for a few hours with a fury that the Duboises have never experienced. How can *rain* be foreign? wonders Anna. But it is. Water pours out of the sky. It is like standing in a waterfall. And everything is swimming in mud before the day is out. Mud gets on everyone's boots, clothes, hands, and face; there is no way to avoid it. Mud spreads itself insidiously throughout the house. A full staff of servants is an absolute necessity. Besides, as soon as they arrive at the base, there are dozens of would-be servants crouched in the yard in front of their house. Even the rain does not drive them away. They squat on their heels and wait, silently, for the hiring to begin. Each hopeful employee bears his or her precious references, creased and crumpled pieces of paper that have been folded and refolded many times. As the natives are nearly all illiterate, and even the literate ones read no Dutch, they have no idea what the mysterious runes on these aged pieces of paper say. But they know they must have them if they are to gain employment from a totok, a newly arrived European.

Waking the next morning and seeing that the silent, dark-skinned, expectant crowd is still there, Anna knows she must do something. She is nervous about dealing with natives, but they have been there all day and all night, and they show no signs of leaving until she hires her staff. They simply wait and watch. She is unable to bear their collective inspection for another moment, although her dark hair spares her some of the incessant scrutiny meted out to Dubois.

Anna takes a deep breath and begins. She needs servants to unpack, to clean, to cook, to look after Eugénie. The very idea of having half a dozen people to look after the three of them delights her. What a fine lady she has become, all of a sudden! Trying to project an aura of firm dignity, and failing utterly, Anna steps out onto the front veranda. She is only twenty-five years old, not long a wife, and completely inexperienced in the way of things in the Indies. She has almost no Malay at her command, so she gestures to one respectable-looking middle-aged man. She calls him forward in Dutch and, to her pleasure, he seems to understand. Slowly he stands up and walks forward, stopping respectfully a few paces away from her. Anna smiles triumphantly. He understands her! She is elated; maybe she won't have to learn Malay after all. In this, she is completely mistaken.

The man holds out his paper to her. "Referentie," he says, using almost his only word of Dutch.

Anna takes the paper solemnly and starts to read. "This is to introduce Ahmad," the writing says. "He is an old rogue and will steal from you regularly." Anna is astonished. She reads the sentence again; it has not changed. She composes her face as well as she can and continues to read. "When he is not stealing, he is as lazy as the day is long. But he keeps the other servants from stealing quite so much and he has a good appearance. He makes an acceptable djongas if you don't expect him to do much." Anna cannot prevent a smile forming on her lips. Here, apparently, reference-writing is a comic art. She struggles to maintain the dignity befitting a njonja. Finally she just nods and smiles tightly at Ahmad. He steps confidently to one side, turning proudly to face the crowd of applicants. A number of the other waiting men sigh, stand up, and walk away. They can see that Ahmad has been hired, even if the njonja doesn't know it yet.

Slowly, uncertainly, Anna proceeds to hire the others: first the babu for Eugénie, then her ladies' maid, then the kokkie, and finally the tukang kebun, the garden boy. She still cannot read much on their faces, nor is she particularly concerned with their feelings. Even the ladies' maid, who will be in intimate contact with her, is unimaginably distant. The invisible barriers of race and station mean that no European ever thinks of his or her servants as fully human, though the aristocracy of the princes and other noble Javanese families is recognized. Like everyone else in her position, Anna never asks whether servants find it insulting to be called by terms used for youngsters, like "adik"—"boy"—or to have their names replaced by their occupations. Does a laborer dislike being called "Coolie"? Who knows? Who would think of such a question? European superiority is unquestioned. To be sure, there are many of mixed race in the Indies, but the "whiter" someone appears, the higher his or her status.

By the end of the afternoon, Anna has hired her household staff and the crowd of natives has dispersed. She feels relieved, lightened. How oppressive their dark, watchful presence was to her! Of the new servants, the babu gives Anna the greatest sense of security. As promised by her excellent references, Sanikem is a quiet and graceful young woman, with a soft voice and a smile. Very quickly, Eugénie gives Sanikem her complete trust and devoted love; she is as happy and safe in her babu's arms as in her mother's or father's. The ladies' maid, Lilik, also has a gentle way about her that makes her easy company for Anna. She never intrudes, never seems in the way, but never strays very far from her mistress, lest she need something. The djongas, Ahmad, holds himself with an air of superiority and, in his crisp white uniform and spotless headwrap, adds a style to the new household that Anna and—even more—Dubois approve of. They do not forget his reputation as a rogue, but they enjoy his elegance and agree that he is one of nature's gentlemen.

The servants must maintain a constant vigilance to keep the floors clean and polished in the rainy season, attending to them each time anyone comes in or out. The rain seems to bring every sort of creeping creature into the house, into the cupboards and beds, under the cushions on the chairs. Insects crawl behind and beneath the books, silverfish making themselves at home between the pages of Dubois' precious library. Cockroaches of nightmarish size hide in the crevices in the furniture and behind the sideboard, beneath every box or receptacle placed on the floor. Anna cannot help crying out every time she lifts an innocent object only to find a miniature terror lurking beneath it; Dubois fears for the safety of his books.

Small, spotted, splay-toed geckos climb up and down the walls, hiding behind pictures and mirrors. They unnerve Anna at first, but she learns to appreciate their insect-eating ways. The servants refuse to chase the geckos away, for they are considered lucky. Damp and mildew are more potent enemies, relentless, ubiquitous. They spoil everything: books, photographs, linens, papers, shoes, clothes that touch each other as they hang in the wardrobe, even the folds in the curtains. The only way to dry anything is to hang it next to a fire or to spread it out over the bushes on a rare sunny morning, in the breeze.

Each day, when the rain starts, Anna rushes outside to revel in it, like a child. Babu and Eugénie usually follow her and they dance for joy as the blissfully cooling drops strike their skin. After a few minutes, perhaps even half an hour on a lucky day, the initial sprinkle turns into a blatant downpour. And then for two or three hours the rain pounds angrily, constantly on the roof, falling down in torrents that beat holes and tiny canyons into the ground. The incessant noise seems to invade the brain. Rain assaults the senses, bringing temporary deafness. Inside the house, Anna's ears are pounded by wave after wave of noise until she can hear nothing at all; the same happens to Dubois at the hospital, where he must shout at patients to make himself understood. The doctors and aides fall into a sort of dumb show, waving their hands and miming actions rather than trying to speak. Rain makes people blind, too, for it is impossible to see across the avenue or to the end of the garden. All that is visible, all that is audible, is rain. The smell of rain, the constant wetness, gets everywhere, creeping up the steps onto the veranda, forcing its way through the bamboo blinds and finally into the house itself, invading every drawer and box and piece of furniture. Sometimes Anna wonders if Indies people are so fond of hot spices and curries because they want something to break the sensory monotony of the monsoon rains.

And yet, and yet . . . the rain is not all bad. The air is much cooler and fresher afterward, and the rain always stops. In the late afternoons, there is often a period of almost pleasant weather, when the world seems new and lovely. Dubois walks slowly home from the hospital through the puddles and mud and the fresh, sweet smell of wet flowers and plants. This is the time for the evening bath, taken in the tiled mandi room out behind the main building. There is something wonderfully sensuous about ladling the cool water out of the enormous jar and pouring it slowly over oneself, washing away the sweat and the grit and the cares of the day.

The second bath is an important event in the set pattern of the day, which begins with early coffee and the first bath. A solid breakfast on the veranda readies Dubois to go off to the hospital. There is a full lunch at midday, rice and meat and vegetables, with hot spicy condiments on the side, followed by a welcome nap; "writing a letter" is the Indies euphemism for a brief sleep in the afternoon. Later, he returns to the hospital, coming home as the sun sinks rapidly, when it is time for the second bath and a change of clothes. Uniforms and formal wear are put away; looser, cooler clothing is adopted for evening. Dubois puts on lightweight batik or silk pajama pants with a white tutup, the high-necked, button-fronted jacket popular in the tropics; Anna wears a loose kimono or a long, lace-edged jacket called a kabayah over a blouse, with a sarong wrapped about her waist. They put soft slippers on their feet, and Anna loosens her hair. Drinks and dinner are taken in comfort on the back veranda, in privacy. After dinner, it is time to sit in the soft glow of the paraffin lamps, listening to the lonely cries of insects and the shriek of soft-winged bats. Intermittently, there is the cheerful "chi-chak, chi-chak" of a gecko in search of prey or fiercely defending its territory, some nondescript section of wall. Smoke drifts in from the cooking fires of the kampong, mixing with the fragrance of night-blooming melati bushes and the scent of the peppers and spices used in the dinner. Somewhere soft Indonesian voices murmur indistinctly; perhaps a sleepy child laughs, something rustles in the trees. Some nights, the melodious gonging of the gamelan, the traditional orchestra of the Indies, rings softly across the darkened landscape for hours.

The Indies seem a gentle and welcoming world at times like this. Dubois and Anna often pass the evening companionably side by side, she in her rocker, he in a long rattan chair, the panjang kursi. They listen to the night noises and talk of the day's events. Sometimes she embroiders while he reads, until the insects get so bad it is impossible to sit near the lamps. From time to time, a strange barking cry rings out—some nocturnal animal, in one of the large trees that line the avenue? Some nights Anna hears an eerie keening sound, full of sorrow and mystery, that makes her shiver. Dubois asks around, but he never learns what creatures make these sounds. He and Anna come to accept them as part of the Indies. Sometimes the night is still, with only the rustling of leaves or the rhythmical patter of the last raindrops to disturb the velvet silence.

During the day, Anna is hot and uncomfortable in her European clothes, and the mildew and mold soon damage them. She cannot, at first, bring herself to abandon the clothes from her trousseau and her first year of married life in Amsterdam, but then she sees how most of the other Dutch women in the Indies dress. Even Pure women—not just half-caste Indos—

often wear a sarong and kabayah during the day as well as in the evening. The clothes look a little strange to Anna at first, so loose and shapeless on the corsetless women, but they are admirably practical, so cool and light. She has been in the colonies only a few weeks, and already she isn't even dressing properly. Her mother would be shocked. The elaborate European fashions are cut to emphasize a buxom hourglass figure, which Anna hasn't got, and are much too hot. Though she tries to preserve at least some of her favorite dresses for more formal occasions, Indies clothes give her a sense of bodily freedom that is rather exhilarating.

Anna cannot help but wonder why there are plenty of children on the base although wives are in short supply. Often, the children's hair is even darker than Eugénie's, and their complexions are tellingly light, sometimes almost white. Piecing together what she sees with bits of gossip, Anna learns that many of the officers have native concubines, nyais, who run their houses and bear their children. There is a well-understood code of conduct. A man doesn't parade his nyai; she is not introduced to the European wives, nor will she ever be seen at dinner with Europeans. The nyai's world is inside the house; she is a silent, graceful shadow who does not appear in the public sphere. A nyai's presence is obvious nonetheless, once Anna knows what to look for. There is an unmistakable woman's touch about the house: better, well-served food; beautiful furnishings; flowers; and a man who is contented and meticulously turned out. A nyai may not be openly acknowledged, but neither is she concealed.

Anna is bewildered by the matter-of-fact acceptance of nyais and mixed-blood Indo children. Is *this* what people mean when they speak of immorality in the colonies? Anna finally works up the courage to ask Clara, one of the few other wives on the base, about it. Clara's first response is to reveal that her own family is an old Indies one with many "darker" branches. Anna blushes, feeling somehow she has accused her new friend of something.

"Oh, Anna," Clara teases her, "you are such a totok! To be shocked at the presence of nyais . . ." She clucks and shakes her head in wonderment. "Very common, you know, for a young man to take up with a 'walking dictionary.' There have never been many women out here among the Pures, not even many Indo women. So whom is a man to marry? Who is to look after him, to make his home? The best thing a young man can do for his career in the Indies is take a nyai, a pretty, clever girl to teach him the language and the local customs. She runs his household, makes him comfortable, gives him children. In return, she lives very well: a nice house, servants, maybe jewelry and beautiful clothes. . . . Of course, there are always a lot of Indo babies. Sometimes the fathers acknowledge them, even pay for a European education, if the relationship is a long-standin' one."

"They take their father's name?" Anna is incredulous. In Holland, she heard whispers of bastard children, but such offspring were hidden away. There might be furtive payments to the mother, who was of course ruined for life. This open recognition is shocking.

"*Ja*," replies Clara, as simply as if describing the flowers in her garden. "Happens fairly often. Sometimes a nyai is with a young man for only a brief spell, a few months or a year. Y'know how young men are, fall in love with one woman in June and another in September. Kassian!" she sighs, using the common Indies expression of sympathy and drawing out each syllable. "Men can be so fickle. In a situation like that he won't acknowledge the children, but if he is with her for years and years . . . Even then, takin' them away from their mother so they can be raised properly and educated usually causes a real susa, a big fuss. Much more expensive than lettin' them run free in the kampongs! An Indo boy with a good education can hold a responsible job, do well. And a girl, if she's a beauty and not too dark, she can make a good marriage—maybe even marry a Pure."

Unable to help herself, Anna blurts in wonderment, "Things are very different here!"

Clara looks at the expression on Anna's face and collapses into laughter. "Oh, Anna!" she says through her giggles. "You are such a child."

Anna laughs, too, but she has an unsettled feeling, as if she has looked at a familiar room through a kaleidoscope. Nothing holds its shape here; everything shifts and transmogrifies into something else. She does not know where she is in this alien world. "But sometimes, surely," Anna says, growing serious, "a man later marries—I mean, really marries—a woman he has met at home on leave, a totok like me. What happens when she comes out to the Indies and finds her husband's brown-skinned children?" Her face grows somber as she imagines the sense of betrayal.

"You have to understand life in the Indies," Clara answers. "That is the way things are here. Young officers have nyais, and most of the administrators and the planters, too. Anyone who marries a man in the Indies expects there was a nyai before her. And the children, they are just children. All of them are precious, whatever color they are; so many children die here. Oh, everyone wants a fair child, like your Eugénie, but it doesn't matter much once they're here. Some people even say that the Indo babies are healthier than the Pures. Sometimes a woman whose baby has died adopts a native baby and raises it as her own: an anak mas, a golden child, we call that one."

Anna has a great deal to think about and adjust to in her new life. Having a nursemaid to look after Eugénie in Amsterdam was one thing—and, she remembers suddenly, that nursemaid was an opinionated thing, always on the verge of getting above herself. Babu is another matter entirely. Eugénie plays happily with children from the kampong, sitting in the garden with only flowers, pebbles, and twigs for toys. She is admired for her gay charm—her mother's legacy—and for the fair complexion she inherited from her father. The striking Dubois family is more often talked of in the kampongs and military bases of west Sumatra than they realize.

After a month or two, Anna begins to feel at home in Padang. She is no longer frightened to step outside the house or even, for that matter, to walk around inside her own home and lift objects beneath which insects might lurk. Her Malay is much improved, though she does not speak like a native—never that, of course. The kokkie remains a bit of a problem to Anna, for she cooks whatever she wants. Strange dishes appear on the table, full of rice and unfamiliar fruits and hot spices. Some of them are quite alarming and leave Dubois and Anna coughing and spluttering from hot peppers. Others are tasty and appealing, but Anna does not know what they are called and so she cannot ask for them again until she learns their names from a friend. Sometimes she wonders who is the employer and who the servant.

While Anna settles slowly into her new world, Dubois is plunged into his. He is inundated with hospital work almost immediately, for there are a great many patients. Malaria, typhoid, and typhus are daily fare, along with sup-purating tropical ulcers that will not heal, tuberculosis, cholera, and myriad nameless but deadly fevers. Dubois is perfectly efficient with the usual ailments: the broken bones and gunshot wounds, the septic blisters from bad boots, the rotting teeth, the stomach cramps, bilious livers, and heart conditions. At first he must often ask other physicians for advice, because there are medical problems here that he has never seen before. He learns quickly how to manage them; he must. Infections and fevers appear, ravage the patient, and kill within days. This is life-and-death medicine with an urgency that Dubois has never witnessed before.

Almost daily, he comes home to warn Anna about something: the need for vigilant hygiene; the poisonous snakes and insects, to be avoided and removed from the garden by the garden boy; the horrifying diseases that come like a wind out of the west in the morning and can kill a child before nightfall. Not all is nightmare, though. Almost daily, he comes home to tell her about something wonderful he has discovered, too: a new flower, like the melati, a sweet-smelling jasmine that perfumes the evenings; a fabulous bird he has seen, with violet wings and shiny green tail feathers; a strange new animal that one of the boys has caught up a tree. It is a new world, this colony, both frightening and inspiring.

The press of hospital work does not diminish with the passage of the wet season; the dry season is as bad or worse. The stream of patients is so continuous that Dubois soon realizes his ambitions for fossil-hunting will be pushed aside for years, if not forever, if he is not persistent. He has not come this far to practice medicine, he reminds himself. He could have set up a practice in Amsterdam if that were his goal. He has more important things to do than treat fevers and sew up wounds. No, he has come to the East Indies to find the missing link, and he means to do it.

Dubois offers to lecture to his fellow officers about the missing link. Much as he detests teaching, talking about his missing link will be a pleasure. Nearly everyone attends, to take the measure of this new doctor from Amsterdam. And there is a dearth of entertainment at the base; any novelty is treasured. Dubois hopes his lecture will be the first step in persuading the commanding officers that he should be released from some of his hospital duties to search for fossils. He wants to find the best way to frame his arguments, the most persuasive path through his logic. Then he will compose an article on the subject for publication and stake his claim to the search for the missing link. Everyone will know what he is trying to do and why.

The reception of his lecture is mixed. Some of the officers are keenly interested and come up afterward to ply him with questions. Dubois counts these as successes, for he can see he has awakened their curiosity and stirred their scientific instincts. Some leave wordlessly after the lecture, with an odd expression on their faces. Perhaps they think he is just another eccentric who has sought refuge in the colonies, like some ne'er-do-well son of a prominent family. A few, exhausted by their duties, fall asleep and snore gently through the lecture, never hearing Dubois' inspired words.

*   *   *

When Dubois squeezes out a day here or there to explore the area immediately surrounding the base, he finds it unpromising. Padang sits on a flat coastal plain with few hills and no caves. It takes hot hours of travel toward the interior to find caves; even there he finds no fossils.

He tries to get the Minangkabau, the local tribe, to lead him to better places. "I am looking for caves with bones," he says, "old, old bones that are like rocks. I want the bones of old animals that do not live here anymore. Can you take me to such a place?" The brown faces look back at him with guarded eyes. Have they failed to understand? He tries again, with no better success. Dubois thinks sometimes the natives are being downright evasive. They are very polite, they call him "Tuan Dokter" respectfully, but they don't offer any help. Maybe they have never seen any fossils. Or maybe they think he is asking about the graves of their ancestors. They must have feeling for the dead, they are so superstitious about ghosts and spirits. They believe so strongly in the old stories that they even make their houses with curving roofs, upturned at the ends, to symbolize the horns of the bull buffalo that long ago saved them from Javan domination. He wonders if they think he seeks the bones of that animal. Do they believe that if he found such bones, disaster would befall the Minangkabau?

Frustration lends a bitter flavor to Dubois' days, and the old feeling of desperation returns. He must find a way out. He cannot waste his life here. He does not know what to do. Anna can offer no help. By now she is so heavily pregnant that she cannot concentrate at all. She seems to him like some dull, placid broodmare, content enough but unable to think or speak cogently. It is up to him to solve this problem. He brought his family to Padang, so surely he can get them out again. But where shall they go, and how? He turns to his colleagues for advice. It is not, he explains, that he wants to go back to the Netherlands. No, what he wants to find is an Army post where the countryside is more promising, with more caves. He asks his colleagues what they know about the various regions of Sumatra: Where are the hospitals? What is the landscape like? How densely settled is this area, or that? He requests a transfer to an upcountry convalescent hospital near Pajakombo. The area is more remote, less settled, and there is no huge Army barracks full of potential patients nearby.

In April, his frustration is broken by some good news. The article based on the lecture he gave to his colleagues is accepted and published in the *Natuurkundig Tijdschrift voor Nederlandsch-Indië* (Journal of the Natural History of the Netherlands Indies). Now the colonial government cannot afford to ignore his ideas. The body of the text presents all his arguments in logical order, cementing them into a veritable wall of reasons why the East Indies will yield up the fossils of the missing link. When he finally reads his words in print, he thinks he has made a masterpiece. It is the best thing he has written yet. He is especially pleased with his closing words:

It is obvious that scholars from other countries will soon realize the promise of the East Indies. They will come and search for important fossils here and will find them, unless the Dutch authorities do something more to support such scientific work. And will the Netherlands, which has done so much for the natural sciences of the East Indian Colonies, remain indifferent when such important questions are concerned, while the road to their solution has been signposted?

With these words, he issues his challenge. If the authorities ignore his pleas for help, men from other countries will move in to find the missing link on Dutch territory; the implicit threat plays nicely on the long-standing rivalry between the Dutch and the English. The Dutch have never forgiven the English for taking control of Java between 1811 and 1816, and they remain jealous of the English settlements throughout Southeast Asia. The Dutch would hate to be beaten to an important discovery by the English, who are very interested in fossils and evolution. Those words of his will put a weasel in the henhouse, he thinks eagerly. Only fools, having been warned, could fail to respond now to his requests for a proper Dutch expedition to find the missing link.

CHAPTER 11 PAJAKOMBO

Dubois sends a copy of the article to the Governor of the west coast of Sumatra, R. C. Kroesen, with a letter. Kroesen is a thoughtful man, who reads the article carefully and understands Dubois' arguments. He can see the wisdom of helping this earnest young man in his endeavors. He can see, too, the danger of ignoring him and being blamed if someone else finds the missing link first.

Kroesen takes two actions. First, he lets Dubois know that once he finds a good site, he will, as Governor, make forced laborers available for the work. This amounts to substantial assistance. By law, natives are required to pay taxes to the colonial government. A farmer with good land can pay with specified export crops (indigo, sugar, tea, or coffee) in an amount equal to one-fifth of his yield. This is not a popular alternative and the natives prefer to grow rice to feed their families. The other alternative is to spend a month or two a year laboring on public works, such as maintaining roads, digging wells, or clearing land. Some officials abuse the system, using forced laborers to tend their extensive personal gardens, build new houses, or work fields for the official's profit. Dutch administrators argue that their salaries are insufficient to live in the grand style that befits a tuan besar, a big man. Since natives expect their rulers to be both rich and generous, living modestly would cause a loss of face. Whether Dubois' scientific expeditions are a proper use of forced laborers is never asked. "Proper use" is what the Governor says it is.

But Kroesen does not stop there. His also lets it be known to Dubois' military superiors that he, the Governor of West Sumatra, is taking a personal interest in Dubois' search for the missing link. Kroesen's support smooths the ruts from Dubois' road. Perhaps as a direct result—who can ever tell how these things come about? —Dubois' requested transfer to the hospital in the highlands at Pajakombo comes through. He will be in charge of the entire small hospital, with no one to answer to except his own conscience, and there are not many patients.

The Dubois family leaves for Pajakombo in May. Moving is an elaborate process, complete with numerous bearers, packhorses, oxcarts piled high with possessions, and sedan chairs, one for Eugénie and another for Anna, who is now almost completely incapacitated by her pregnancy. Most of the servants, who have worked for the family for five months, decide to move to Pajakombo also, for Anna and Eugénie are much admired and Dubois is considered a good tuan, a dignified, learned man of impressive stature.

Anna passes the long, arduous trip pleasantly, lying in the sedan chair and playing with Eugénie from time to time when Babu brings her over for a brief visit, napping, looking at the new scenery, and enjoying the steady change from the steamy coast to the brisker highlands. Every step along the way seems to bring cooler air and higher elevation, a true blessing in her condition. The servants do most of the unpacking and settling into the new house; Anna reclines in a long chair, her feet up, and feebly directs the placement of objects.

Anna gives birth to their first son, Jean Marie François Dubois, at home in Pajakombo, on June 15, 1888. She is attended in labor by her husband and her maid. There are experienced native orderlies in the hospital, but they are all male and it does not seem proper to have them present at the birth of a white child. Besides, Dubois is certain that he needs no other help. The baby boy is healthy and robust, dark-haired like Anna but fair-skinned and strong, like his father. That very evening, while his wife and new son sleep after their exhausting day, Dubois pens a letter to his parents, to tell them of the safe arrival of their first grandson, named Jean after his grandfather and Marie after his grandmother.

Pajakombo is far from the sweltering coast and the busy port of Padang. The Duboises are more isolated in the highlands than before. The Army outpost in Pajakombo is so small that the only real reason to keep a European physician there is to run the convalescent hospital. Dubois is the sole qualified physician on staff. Men recovering from serious illnesses or wounds are sent to this hill station, away from the coast's heat and its heightened danger of fever or festering infection. Despite the pleasantly cool climate and the light workload, the place is so lonely that most medical officers request a transfer out almost as soon as they arrive. Most of the other Europeans at Pajakombo are the patients themselves, some of whom stay for weeks or months. Sometimes Anna yearns for her women friends in Padang, but Babu and Lilik, her maid, are a comfort. Babu is pleased to have another beautiful white baby to look after. She calls Jean her little sinjo, or master, providing him with loving care and attention without upsetting Eugénie.

Dubois is proud to have a son; he has always hoped for a boy like he was, intelligent, active, and strong. He holds Jean in his arms, looking with amazement at his miniature eyelashes and translucent eyelids, admiring his curly dark hair, the beauty of his tiny hands. He gazes proudly at his son, blessed with the peaceful sleep of the innocent, and imagines his future. If Dubois' search for the missing link is successful, then Jean will grow up the son of a famous man, a scientist respected worldwide for his fossil discoveries. As a boy, he will have a position even greater than Dubois had as the son of Eijsden's mayor and pharmacist. Doors will open for Jean; opportunities will present themselves for education, work, learning, maybe even adventure . . . Dubois feels wonderfully optimistic. He has a wife, two fine children, and the governor's support. He is poised to make his great find, to reach his goal. His ambitions seem markedly closer to becoming reality. He has much more time to explore the countryside for caves, and the highlands are full of caves. Some of them are bound to contain fossils.

In September, another physician arrives, Dr. Pollak. He is not an addition to the staff; he is recuperating from a bad case of chronic malaria that has left him thin and weak. Repeated doses of quinine have finally defeated the fevers, but Pollak needs time to rebuild his strength and stamina. He is luckier than most: the graveyards are full of young men who have succumbed to malaria in their first year in the Indies. Still, he will be in the hospital at

Pajakombo for too long, feeling too well to do nothing but not yet strong enough to leave.

Dubois often stops by to sit with Pollak for a while, telling him of his ambitions and his conviction that the missing link lies hidden in a cave nearby. As another scientifically trained professional, Pollak soon grasps the essence of Dubois' arguments. He is intrigued; he knew very little of the missing link before now. Dubois leaves him a copy of his article. He is gratified, the next day, to see his new friend poring over the pages and making notes in the margin. "This is a fine piece of work, Doctor," Pollak says to him, looking up. "You have opened my mind. I do hope you'll stop by later when you have time and answer some of my questions." The missing link becomes a regular topic of conversation between the two, who like nothing better than to speculate where Dubois will find it, how he should search for it, and what anatomical features it will turn out to possess. In this isolated place, Dubois has found a friend who is willing to help him plan and evaluate and has not the least interest in usurping his ideas. Pollak is no Weber or De Vries, for he is not so knowledgeable about evolution as they, but neither is there any risk of his turning into an idea thief.

"You know," Pollak offers thoughtfully one day, "you could make small expeditions of several days at a time, if you turned some of your responsibilities over to me. I am quite well enough for light medical duties. There are hardly any serious cases in the hospital just now; nothing I couldn't handle with the native orderlies. Why don't you take a few days and go searching some of the caves nearby? I can always send a runner for you if something arises that is too much for me. But things are quiet here, very quiet, and the patients don't need much attention."

Dubois is genuinely surprised. Ambitious as he is, he has not considered asking Pollak to assume his duties. "That's a very handsome offer," he replies to his friend, "very kind of you. As I am in charge, there is no one to approve or disapprove. Are you sure you feel strong enough?"

Pollak nods, eager to do something, anything useful, and happy to be of help; he has never met another man with such a fertile mind and such determination.

"Then," continues Dubois, "perhaps we could try it for a few days at the end of this week. I could get on with my bone-hunting, maybe locate a site worth excavating. Once I find the right place, the Governor will let me have those laborers, and then it will be much easier to keep everything going even if I can't be there in person. If all goes well while I am away from the hospital—don't work yourself into a relapse, now—then we might make it a regular thing."

The plan is put into effect. Dubois begins to search the Pajakombo region systematically, checking every cave and rock shelter, making small test excavations, searching for any small fragment of tooth or bone. Most of the caves are sadly empty of fossils, as is only to be expected. But some contain fossils, and Dubois collects them all, carefully marking and annotating his map. For weeks he finds nothing but miserable, broken scraps. As Pollak grows stronger, Dubois' forays grow longer. Being back at work improves Pollak's spirits, and it pleases him to see Dubois come striding back from his trips, spilling bits and pieces of fossils out of bags and boxes and showing him the markings on his maps.

But Dubois' is difficult work, and lonely: plodding through the countryside, climbing up absurdly steep hills, cutting paths through the prickly, recalcitrant underbrush, searching for caves. He brings along the teenaged son of the tukang kebun to carry equipment, gather firewood, and provide general assistance, but the boy's presence cannot alleviate his profound isolation of the mind and spirit, painful but not enough to blunt the razor's edge of his perseverance.

CHAPTER 12 FOSSILS

Within a few weeks, Dubois has accumulated a sizable pile of fossils. His map is full of marks recording what he has found where; his notebook is cluttered with notes, calculations, and drawings. Soon he begins to draw new maps of his own, taking great pleasure in the exercise of making each one beautiful, lettering it carefully, coloring in the different areas. His maps are both science and art, a tangible symbol of his growing knowledge.

The fossils he gathers are not complete skeletons or even whole bones; they are simply teeth and splinters of bone. A good museum would reject all of them. However, they represent the first collection of fossils from Sumatra, an island whose fossil history was completely unknown before Dubois began his work. And, incomplete though the specimens are, they convey much to Dubois' learned eye.

Nearly all of his fossils are teeth, long since fallen out of the jaws that contained them. But even these isolated teeth are very informative. As an anatomist, Dubois can usually identify the type of animal and sometimes even its species from a single tooth, but some of these are in worse shape than most. They have been gnawed—determinedly, repeatedly—by porcupines, leaving ridged and gouged surfaces. The fragments of limb bone, when Dubois finds any, are similarly scored with broad, squared-off grooves left by chisel-shaped teeth. There is not much of the original anatomy left for him to study.

Still, Dubois can already say something useful about the fauna of ancient Sumatra. He fingers the better specimens, cataloguing them in his mind. There were elephants here once, that is certain: these great, brick-sized lumps of dentin and enamel could belong to nothing else. And rhinos, too, with that F-shaped pattern on the crowns of their teeth, maybe not so very different from the living Sumatran rhinoceros. Here are a few tiger's teeth, sharp and long and shiny, and a pile of molars from pigs and deer and wild cattle. And here, here is the culprit himself, Dubois thinks, picking up a few teeth of a porcupine, the animal that chewed and damaged nearly every other fossil he has collected. The right animals lived here in Sumatra in ancient times, there can be no doubt.

Dubois starts exploring the highlands the month after Jean's birth. In August 1888, he embarks on prolonged field explorations and finally meets with success: the cave called Lida Adjer has a few fossilized bones—intact, complete bones of a good variety of Pleistocene fauna, not just teeth and scraps. For once, Dubois has gotten there first, before those wretched porcupines. Is this *his* cave, the cave of the missing link? Maybe. The thought lends urgency to his preliminary visit to the cave.

He collects an impressive pile of fossil riches and heads back to Pajakombo for more supplies. When he gets home, he runs up the stairs of the front veranda to share the news with Anna while the boy follows with the crates and bags.

"Anna? Anna, come quickly!" he calls.

She is there in a moment, having heard the sound of his horse's hoofbeats. "Are you back then, Eugène?" she greets him, smiling. "Ahmad, get the Tuan some cool lime juice."

"I have found my cave, Anna," he blurts in a rush, "I think I have found it. This is a really good site. You should see it: there are rhinos and pigs and deer bones everywhere. I shall write to Governor Kroesen immediately to tell him of this important find. I am so close now, so close. This might be the right place."

Still recovering from the trial of childbirth, Anna must rally her strength to respond to his excitement. She duly admires his fossils—dusty, oddly shaped things, but she knows they are all the world to him—and asks a few questions, but mostly she just smiles as he tells her of his hopes. She thinks perhaps it is all coming true, just as Dubois predicted. She sits and adjusts the loose pillow on the rattan rocking chair for comfort. Even dirty and tired, Dubois is a handsome, clever man, and she is glad to be his wife. Babu hears the commotion and shyly brings the children out on the veranda to greet their father. Dubois is so exuberant that he lifts Eugénie up and whirls her around until she giggles and giggles. Then he goes over to little Jean, in Babu's arms, to softly stroke his new son's porcelain cheek.

"Your papa has found a great thing today," he tells them. They do not understand, of course, they are far too young, but they know their father is joyful and their mother is sweetly contented. Babu takes them to bathe and change so they will not be underfoot. The Dubois family settles in for a happy evening, there in remote Pajakombo, up in the Padang Highlands of Sumatra.

Dubois writes to Kroesen, telling him of the wonderful fossils at Lida Adjer and asking for the promised laborers. The Governor replies promptly.

> *September 8, 1888*
>
> *We are really going to look forward to a hopeful future. . . . To start, I will place officially at your disposal six forced laborers. . . .*

Reading Kroesen's encouraging words fills Dubois with energy. Now it begins, he thinks. Now I will find the missing link.

It takes him a few weeks to get everything organized, and he must wait for the coolies to appear, but Dubois is on fire and cannot slow down. His moment has come; he can feel it. When everything is ready, the procession sets off for Lida Adjer. He rides on horseback at the head of the small column, the coolies in their conical hats following. Sometimes he trots back to urge on the sluggards at the back.

They set up camp near the entrance to Lida Adjer. They have brought along a woman to cook; Dubois appoints one of the more intelligent men to be foreman, or mandur, while others with more muscle and less brain will do the digging, gather firewood, and haul water. Lida Adjer is the first cave where systematic work has been justified, so Dubois is forced to improvise his procedures and train his workers in them at the same time. The men are peasant farmers; they know how to dig the soil to plant crops and they know how to dig irrigation ditches and wells. This is neither of those tasks, and they are confused. They could understand if the Tuan Dokter ordered, "Dig here, plant this crop." They could understand if he said, "I need a well here"—or a drainage canal, or a ditch for irrigating the field. But he wants something different from them. He wants digging done with care, like digging up sweet potatoes when there is no remnant of the stem or leaves but the ground is like rock and needs a pickax. How can they work gently with pickaxes? Normally they work for a while and then take a break to smoke a cigarette, chew betel, drink water, relieve themselves, exchange a little gossip, or rest awhile. Digging slowly is perfectly sensible, but they do not want to dig ceaselessly.

Inside the cave it is so cold that the coolies complain. The Tuan Dokter shows them some old bones, fossils he calls them, that are heavy like stones. He tells them that is what he is looking for, but they know he is lying. What good would old bones be to anyone? Why would a Hollander go to all this trouble for something worthless? The Tuan Dokter's strange ideas do not end there. He wants them to keep the excavated area level at the bottom and he wants them to maintain a regular shape, with corners. He draws the shape on the ground with a stick. If they do not follow it, the Tuan Dokter shouts. They would rather follow a pleasing curve or round an edge than make a straight-sided hole. What is the use of that? It is not even beautiful to dig a hole that way.

Dubois explains it all again to the mandur, and the mandur instructs the men. This is the way the Tuan Dokter wants you to dig; like this, not like that; keep this corner square, keep the wall of the excavation vertical. This instruction is repeated many, many times over the subsequent days. Limited progress is made in turning the ragtag assembly of peasants into a competent, if not efficient, excavation team.

What helps is that they realize the Tuan Dokter is completely convinced of the way he wants them to work and absolutely immovable in his convictions. He wants the hole dug the way he tells them, and he will make them do it again if it is not right. They do not understand his drive or his determination, but they begin to respect this unusual orang belanda. Hollanders are not usually so hardworking, and he is physically impressive. They have never seen such a man, so tall, so straight, so strong, so fair-haired. Most strange of all are his pale, clear, blue eyes. Sumatrans, of course, all have brown eyes; so do nearly all Indos and most of the Europeans they see. But not the Tuan Dokter. It is the subject of much discussion. Eventually they decide among themselves that his eyes are like those of Garuda, the eagle-god. He can probably see right into their souls with those eyes, they are so piercing, so sharp. Perhaps he can see even the future and the past. He always knows whether they are trying to shirk a difficult task or pretending to be sick. Anyway, in the presence of the Tuan Dokter, no one dares to steal food or to sneak off to the nearest village to drink rice wine and find a girl.

Unlike other Europeans they have worked for, this one does not wander off to take a nap or sit in the shade smoking while they work. The coolies find his motivations incomprehensible. Some of the men suppose the work involves one of the secrets of the orang belanda, the Hollanders, that they will never understand. Others think it is some special Garuda secret. Whatever its source, they recognize and respect his determination. He is very strong—in his body, obviously, but also in his mind and in his will.

Before long the men do start to find the strange, rocklike bones that the Tuan Dokter wants. How did he know they were there? Every find is given to the mandur, who in turn takes it to the Tuan Dokter. He examines each specimen by paraffin lamp or more often takes it to the cave entrance, where there is more light. All work stops as Dubois inspects each find. The men wait for the pronouncement: Is it good? Is this what they have been seeking? Can they stop digging in this cold cave? The Tuan Dokter always tells them what they have found: the tooth of an elephant, he says, or a frag-ment of an ancient deer antler; this one is a tooth of a tiger, that one a buffalo. There are parts of stranger animals too, ones they hardly ever see, like the rhino and the tapir. Sometimes he draws a small picture to show them what they have found and they stare at it, amazed. Dubois is pleased with each find. Soon he has a good and diverse fauna; he can list the ani-mals that lived together in the Pleistocene of Sumatra.

Normally dignified and calm, Dubois cannot contain his exultation when the men find some fossil apes, rare orang-utans and gibbons. These are new species. Nothing is known of the ancestry of modern Asian apes, nothing! Lydekker can have his Siwalik Hills chimpanzee, for Dubois has the orang-utan and the gibbon of the Padang Highlands. This proves that they are finding fossils from a tropical forest, a wet jungle like the ones that cover much of Sumatra. And where there are apes, there will be ape-men. He is buoyant, floating in the pure joy of vindication. He promises the men that he will go hunting that afternoon, to bring meat for a feast, for he is an excellent shot and an avid hunter.

That night after dinner, he sits in his rattan chair on the small oriental car-pet in front of his tent, making notes on his accomplishments. He is proud.

He has proven that fossils of the right age can be found in Sumatra. He congratulates himself, picking up each of the specimens from the small table at his elbow. Here are the ancient apes that lived with the ape-man; here are the rhino, the tapir, the buffalo. No one has ever discovered so much about the prehistory of Sumatra before. He has done this completely on his own initiative, with no resources except the laborers the Governor has kindly sent him. All this is his own work, his inspiration. Perhaps he should name one of the new species after the Governor; maybe this fine antelope. Kroesen would like that, having a new animal named after him. It would be a good way to acknowledge his help. He will call the species *Tetraceros kroesenii*.

In a few days, the men have exhausted the fossil deposits in Lida Adjer. There is no missing link here. He has found all he is going to find, despite his intuition to the contrary. Back home in Pajakombo, he wastes no time in writing up a provisional report of his finds and sending it to the Dutch East Indies Government. A special copy goes to Governor Kroesen, with a letter pointing out the importance of the fossils and the immense assistance provided by the laborers he supplied. He also sends a copy to another influential supporter, Willem Groeneveldt, the Director of the Department of Education, Religion, and Industry, who oversees scientific research in the Indies.

Dubois is hungry for success. Lida Adjer was surely the first bite of the meal; he is sure to find his transitional ape-man in the next cave, or the next after that. For a long time he has felt like Sisyphus, doomed to push a rock eternally uphill, and now his luck has changed. He corrects his thoughts. No, it has not simply changed: *he* has changed it. His work is gaining speed and mass as it rushes downhill to become a landslide, surging toward his destiny, the missing link.

*   *   *

At home in the Netherlands, the political climate for scientific research is changing rapidly. Max Weber, recently returned from an extended voyage to Sumatra, Java, Celebes, and Flores to collect freshwater fishes, now understands for himself the enormous unexploited potential for research represented by the Indies. His new wife, the botanist Anna Antoinette de Bosse, has discovered many new species of marine algae on the voyage. Weber sees that Dubois' arguments about the urgency of scientific research in the Indies can be extended to include all kinds of natural-historical research. To support and encourage such studies, Weber and some colleagues persuade the government to form a Committee for the Promotion of Research in the Natural Sciences in the Dutch Colonies.

One of the first formal proposals brought before the committee is Dubois', concerning his search for the missing link. Though Weber once listened to Dubois' plans with a certain skepticism, now he is a wholehearted believer. The geologist Karl Martin is in favor of Dubois' research too. Having studied Raden Saleh's fossils, Martin is sure that many more treasures lie undiscovered in the Indies. With Dubois' report on Lida Adjer and his published article as fuel, Weber and Martin make a persuasive case. The fossils Dubois has already found show that his theory about the home of man's origin is probably correct. He has already produced Sumatran fossils about which there was no information before—has revealed new, tangible, scientific knowledge—though the man has had not one guilder of government support. Martin emphasizes how difficult this must have been, what a triumph it represents. Now the government must provide material assistance to Dubois before it is too late. Dubois can no longer be dismissed as a lunatic follower of Darwin's "crazy book." He is a visionary natural historian carrying out significant and successful inquiries into the origin of man, in one of the least studied regions of the world. The committee agrees.

This additional support is especially sweet to Dubois. He has been so long an outsider, so long a seer in the land of the blind speaking of things invisible to others. He receives a still larger crew, sometimes laborers and sometimes convicts, plus Corporals Franke and Van den Nesse from the engineering corps, who act as supervisors. Dubois ponders strategies for searching for new caves. How much should he rely on existing maps in deciding which areas to visit? There are no really thorough, accurate maps of the Padang Highlands of Sumatra. He must rely on word of mouth, instinct, and his knowledge of geology. There is one clue, one precious clue that he clings to: the mountain range in the Padang Highlands is known as Boekit Ngalau Sariboe, the "mountains of a thousand caves."

In March, Dubois receives the best reward of all for his dedication and perseverance.

REGISTER OF RESOLUTIONS OF THE GOVERNORS-GENERAL
OF THE DUTCH-INDIES   *Buitenzorg (March 6, 1889)*

Firstly.
The military surgeon of the 2nd class M. E. F. T. Dubois, provisionally for the period of one year, to receive a grant of f250 — (two hundred and fifty

guilders) a month above his income and to be placed at the disposal of the Director of Education, Religion, and Industry, in order to be charged with paleontological research in caves in the government of Sumatra's West Coast and contingently in Java; with the order to report on the results of his research in a timely fashion to the above-mentioned Director and to place the obtained fossils at the disposition of the government.

and with the regulation:

a. that for traveling related to carrying out this research, he will be allotted a free use of means of transport or compensation for the costs of transport according to the existing regulations:

b. that as long as he stays at the disposal of the Director of Education, Religion, and Industry, he will be relieved of his medical duties.

Secondly.

To authorize the Commandant of the Army and Chief of the Department of War in the Dutch-Indies for the benefit of the above indicated research to retain two workmen from the Engineers; with regulation that they will receive a grant of f25— (twenty-five guilders) every month.

Thirdly.

To invite the Governor of Sumatra's West Coast, by this article, with reference to the indicated researches as far as they take place within the region of his administrative control, to dispose of as many forced laborers as will be desired by the Master M. E. F. T. Dubois, provided the total is not more than fifty at the same time; with regulation that the control and supervision of these forced laborers, conforming to the prescriptions of the Regulation of order and discipline among the prisoners in the Dutch-Indies, will be practiced as much as possible by the leaders of the local authorities at the places where they are put to work.

Fourthly.

An abstract of this document will be given to the military surgeon Dubois for his information.

Now he is officially searching for the missing link. The drudgery of hospital work is behind him. All he can see ahead of him, as he pierces the veil of the future with his keen Garuda eyes, is research, fame, and glory. Now he will explore not just the immediate vicinity of Pajakombo but every one of the thousand caves of the Padang Highlands.

CHAPTER 13 GARUDA

Dubois is sure that better, richer caves with more fossils and better specimens lie ahead. When the rains stop, in April 1889, he starts his explorations
in a much grander and more comfortable style. But most of the caves and
rock shelters prove empty. It is deeply frustrating to clamber up hillsides
and mountains in the heat, fighting his way through the vegetation, badger-
ing the coolies forward, to find only bat droppings and dirt. Dubois is
muscular and fit, yet he finds the unrewarded climbs very taxing. The predictable emptiness of the caves is discouraging. Even worse, some are
empty of fossils but inhabited by a tiger or a bear. More than once, delega-
tions of locals beg Dubois to shoot marauding tigers that are killing valuable
water buffaloes and beloved small children.

The tiger-hunting episodes make the coolies very reluctant to venture
into new caves. Dubois sees their hesitation as cowardice, for he has both a
rifle and a fearless nature. He will show them his courage, as an example,
when the next occasion presents itself. He does not have to wait long. They
come to a large cave on a steep hill, with a low and narrow entrance. It is
impossible to see into the interior by the light of a candle or paraffin lamp;
this suggests the passageway is narrow for some distance. Dubois gestures
and then orders them to enter the cave. The men stand chattering quietly
and incomprehensibly among themselves with worried faces. No one will
step forward to be the first inside.

Now is the time to show them how a European behaves, Dubois thinks.
"Very well," he says. "I shall go."

He kneels down and crawls into the cave, leaving his rifle with the men.
He carries a gas lamp in one hand, leaving the other free. He squirms his
way forward on his belly and elbows until the passageway enlarges. He is
relieved when both the sides and the roof slope away from him into a much
larger and more comfortable space. He crawls the last few feet and then
stands and looks around, holding the lamp aloft. Before his eyes can focus
in the dim light, he recognizes the fetid smell that assails his nostrils: the
stench of cat urine combined with the indescribable odor of rotting meat.
He is in a tiger's lair. It is littered with partially eaten bones and deer legs in
varying stages of decay. Over there, in the shadows, Dubois can just make
out a large dark shape. To his immense relief, when he lifts the lantern high,
the shape proves to be a pile of large bones, perhaps those of a buffalo. It
occurs to him forcibly that the tiger may return at any moment.

There is no bravery in sitting in a tiger's lair. Dubois flings himself to the floor and decides to crawl backward out of the chamber. If he meets the tiger coming in, the tiger will be faced with his stout boots, not his naked face. The difficulty is that he cannot see where he is going. Unable to see anything except where he has been, he tries to feel the way with his feet. In his anxiety, he fails to follow a turn of the passageway and becomes firmly wedged. He cannot free himself. It is a frightening realization. Worse yet, he feels faintly ridiculous. He thinks he is close to the entrance, so he calls to his men, but his voice is muffled by the bulk of his own body. There is no answer.

"Boy!" he cries again. Then louder, *"Boy! Adik!"* Soon he is bellowing like a wounded buffalo. Surely the laborers must hear him. Why are they not waiting obediently by the entrance to the cave, as he imagined they would be? Have they all wandered off, leaving him there in a tiger's lair?

"Mandur!" he shouts again, using all the power of his lungs to call the foreman. The dust and dried debris on the cave floor, stirred up by his exertions, cling to the sweat on his face, making him more uncomfortable than ever. He is embarrassed, enraged, frightened.

"Mandur, toeloeng! Toeloeng degan cepat!" "Foreman, help me! Help me quickly," he yells. Is that a noise behind him? Yes, the mandur is answering.

"Tuan Dokter?" the mandur asks politely, addressing his master's booted feet. "I am here. What assistance can I give?"

"Help me," Dubois replies crossly. "Cepat! Hurry up. I am stuck in this passageway. Have the men pull on my legs. *Carefully!* Hati-hati!"

The mandur gestures to the men to do as the Tuan Dokter asks. Two lie down and reach into the passageway to grab his feet; two more seize the feet of the first. Together they heave on his legs, freeing him suddenly, like a cork from a bottle. The men conceal their smiles at the sight of their great, blond Tuan, so disheveled, stuck headfirst in a tiger's den. They knew it was a tiger's den, but the Tuan Dokter insisted on going in. Who were they to tell him otherwise?

When Dubois returns home, none the worse for wear, he writes to his family in the Netherlands, telling them of his reassignment. He had hoped to have news of a new fossil site to tell them; instead there is only that embarrassing story of the tiger's den. Well, the caves full of fossils will come in time.

Communicating from the Indies with people at home has an odd rhythm. Letters are dutifully sent to Holland, letters from home regularly arrive in the Indies, but the arrivals reply to missives dispatched months before. Writing from the colonies is a little like putting a letter into a bottle and flinging it out to sea. By the time an answer comes, if it does, the moment has passed. An entire life can be expended in the gap between letters. More than one Indies family has received letters of congratulation on the birth of a new child, letters that arrive on the very ship that will carry back news of that child's death.

Dubois is impatient to hear from his parents. When the looked-for letter comes, it is bitterly disappointing. They do not understand how significant his achievements are; they do not see that he has already succeeded where everyone predicted failure. All they see is that he has committed himself even more deeply to a foolish and useless quest. As he reads their letter, devoid of the approval and praise he had hoped for, Dubois' despair rises uncontrollably, flooding over the banks of his reason. The endorsements of the Governor of West Sumatra and the Director of the Department of Education, Religion, and Industry seem to mean nothing to his parents.

It is in this state that Anna finds him, sitting on the back veranda staring into the garden at nothing, clutching his parents' letter in his hand. She can see from his demeanor that something is terribly wrong. She sits down on the arm of his chair and gently takes the crumpled pages from his hand.

"Oh," she says softly, after skimming them. "It doesn't matter, Eugène. I know what you have done. Max Weber and Karl Martin and all the others

Dubois photographs the village of Giring-Giring, near Padang,
after collecting fossils from nearby caves.

know what you have done. Your parents just can't see it yet. Your mother is not very educated and your father, though he is a big man in Eijsden, is only a country apothecary. He doesn't understand about the missing link." She strokes his beautiful blond hair and lays her cheek against the top of his head. "When we go home, and you are famous among all the great men of science, then your father will see."

Anna is no better educated than her mother-in-law. But she knows a good man with an aching heart when she sees one, and she knows her husband has given up much to come here. So has she, for that matter. But it is now all beginning to turn out right. He is finding fossils, and their children are healthy, and the highlands are so much cooler than that awful swamp of a coastal army base. . . . A nap and a bath and a good dinner will surely put him right.

CHAPTER 14 FEVERS AND SPELLS

The mood of despair marks the onset of Dubois' first battle with malaria. He falls prey to enervating, fluctuating fevers that recur and recur like an evil force. A few hours after reading his father's letter, he is taken over, body and soul. Sweat puddles beneath his body, soaking the sheets and the mattress and any number of towels. The servants lift him, change the mattress, remake the bed, repeat it all again a few hours later. Anna applies cool, wet cloths to his forehead and wipes his burning body. Servants bring cool drinks and bowls of broth, and stand fanning him for hours. None of these treatments brings any relief; malaria has its own schedule. The burning heat is followed by teeth-chattering chills, a sensation of deep and final cold, as if death has invaded Dubois' bones. He cannot quite believe that his heart is still pumping, his blood still warm. Surely his blood has turned to slush, like the gray half-melted ice in the canals of Amsterdam in winter. The servants bring blankets and quilts to warm him, they move his bed near the fire, but it is not enough. He shakes uncontrollably, like a pitiful leaf in a thunderstorm, pelted and pounded, wrung this way and that. This, too, subsides with time, leaving Dubois weak and pale.

The relentless rise and fall of body temperature over the hours and days that follow is more exhausting than any physical task Dubois has ever known—not that he is fully aware of his own state, for he cannot think clearly. Merely opening his eyes or permitting Anna to spoon liquid down his throat is almost more exertion than he can muster the strength for. Breathing seems so difficult, sitting up a trial, eating beyond conception. In his more lucid moments, he knows he must take sustenance, especially salty soups and broths, for he is losing quarts, maybe gallons, of salt water during the sweating episodes. But giving in to the illness, letting the malaria sweep him away, seems so easy, so inviting. After two days, Anna has him carried to the hospital, where she can be relieved by Pollak and the orderlies.

"You must," she whispers to him fiercely, spooning broth into his slack mouth. "You must. You have things to do. You cannot give in now. You have a wife and two children. You have a destiny. You will not give in."

Close as he is to unconsciousness, her urgent whispers penetrate the disordered gray fog of illness and reawaken his character. He begins to fight, to prod his own brain into working, to will his body into resisting. He takes a turn for the better.

"I think his case will prove to be a mild one after all," Pollak tells Anna, relieved when he checks on the patient later. "I thought for a while there . . . but no matter, I may have misjudged him. The next few hours should tell us. You go get some rest. The orderlies can look after him now."

Too tired to show her gratitude properly, Anna simply rises and heads back to their house where she collapses in her room. This is the first time she has wrestled hand-to-hand with death. Without Pollak's help, Dubois would probably be dead already; she could not have carried on much longer. The orderlies are experienced at nursing but will not take responsibility or make decisions; she can do that, but she lacks knowledge. She does not know what to do for these tropical ailments, so wickedly virulent. Dubois was no help, for he has been delirious half the time and too weak to think the rest.

If they were alone, on some coffee plantation on some remote mountain, and her husband contracted malaria, what would Anna do? Despite the heat, she shivers a little from fear as she sees for the first time the reality of life for most Indies families. She'd know what was wrong, but she wouldn't know what to do about it. There would be no doctor nearby to turn to, no hospital, maybe no supplies of quinine. She couldn't have had him transported any distance, even by sedan chair, in that weakened state; it would probably have killed him outright. So it would be just nursing and comforting, that would be all she could do, while her husband or maybe her child weakened and shook and sweated and died.

People die here every day, she realizes suddenly, horribly. Death is everywhere, malaria just one of his servants. And malaria is a genial illness, almost companionable, reluctant to leave prematurely. Malaria never hesitates to pay a second visit, or a third. Sometimes it kills rapidly, impatiently; other times it lingers, goes, then returns again and again to wreak slow destruction upon its chosen. This is a hard, hard place, the Indies. Now Anna understands what coming here might have meant, might yet mean. Dubois has known and faced this horror silently from their first few days here, but Anna has not. She is aghast.

However, Dubois' health continues to improve. There is no question he will recover fairly rapidly, as malaria goes, Doubtless his robust constitution is an asset. Before long, he is sitting up in bed, cheerfully organizing another expedition, consulting his maps, ordering supplies and equipment, asking the Governor for more laborers. As he gains strength, he makes hospital rounds and checks on the patients, discussing treatments and plans with Pollak, who will once again take over for him when he leaves to fossil-hunt.

*　　*　　*

Dubois searches an ever-widening area of the Padang Highlands. Weeks accumulate into months without success. The new areas are thick with forest and very precipitous in slope, sometimes nearly vertical. There are few established tracks and no roads. The thorns of acacias and the barbs of alang-alang grass clutch relentlessly at clothes and skin. The only recourse is to send a few men in advance, using machetes to clear a pathway and to cut steps. No one wants to do this for long. Tigers lurk and wild pigs attack viciously, charging out of the forest as if their young were threatened. Dubois shoots them, which improves his meals but not those of his workers, who are mostly Muslim. One coolie is badly wounded by a boar and Dubois sends him back to the hospital with gangrene. Pollak cannot save him.

The highland people are suspicious of his expedition, as if he were searching for valuables to steal, and he has to keep a tight rein on the men to ward off misbehavior. Some of the coolies are considered ruffians even in their own villages, where the power of public opinion and rules of expected behavior are strongest. Villagers rarely offer any useful information and sometimes refuse to sell them supplies.

Even the caves seem to reject them, being more and more difficult to find, more and more consistently devoid of specimens. Only a few yield fossils. Dubois' sense of futility and frustration becomes entrenched. He is alert for attacks from animals or snakes but bored with finding nothing, seeing nothing, learning nothing. His health begins to suffer, or perhaps it is a return of malaria. He is short-tempered with the men, demanding of the engineers, merciless toward himself.

They will continue. They will work harder. They will explore more caves. They will find something. They must be more diligent, that is all. The relentlessness would bring some of the men to the edge of rebellion, if they had enough energy left to make trouble. About a quarter of them run away, sneaking off in the night and melting into the shadows. They return to their villages, Dubois supposes, or take refuge with relatives. He doesn't know; he doesn't care. They are gone. Those that remain are more intelligent, more loyal, more dutiful, or simply more exhausted. As the days pass, Dubois drives his coolies harder, certain he can break the spell of bad luck. The men know things are going badly. Sores refuse to heal, every wound festers, food spoils and makes them sick, water sources are found inexplicably dry in this, one of the wettest of countries. Among themselves, the men say that the Tuan Dokter is losing his power, his vision. He is no longer seeing what will happen now. His Garuda eyes grow cloudy, or maybe he is bewitched by some powerful enemy. It is a dangerous time and they must be careful, very careful. Hati-hati!

Then one day they find a deep and promising cave and the Tuan Dokter seems to be his old self, full of enthusiasm and energy. "Come," he says to several of the men. "It is a big cave. We will all go in with lamps and look for the bones like rocks." They follow him willingly enough, for he sounds again like a man with power. It is an exercise in self-discipline for Dubois. He has summoned his enthusiasm and his faith in himself, putting on a good face for the men. They will just go a little farther in, bring a few more lights so they can see properly. Perhaps this cave merits a test excavation, he thinks. Then he begins to have an odd feeling about it. A slight noise, a small sound, penetrates his concentration because it is wrong.

"Quiet! Diam!" he calls to the men imperiously. For once, they obey instantly; all talking ceases. There is no sound but breathing, not even the shuffling of feet. The men can hear nothing, so in a moment they start to move again, exploring, looking.

"No, wait, tunggu," orders Dubois, and they fall silent again, out of respect. The noise—a tiny weak thing, almost a stillborn infant of a noise—comes again. It is a little rattle and then a dull thud, then silence again. Dubois knows what it signifies, this unimportant little noise. "Turn and walk out!" he commands the men. "Get out *now*! But walk. *Do not run.* Do not touch anything."

They respond to the tone of his voice, to the sound of authority. They do not know all of the Dutch words he uses, but they understand the urgency. They turn in fear and walk out in terror, borrowing what courage and dignity they can from Dubois' stony face and rigid bearing. It is like walking away from a tiger, moving slowly for fear of provoking an attack and yet longing to run. Dubois leaves last of all.

Almost as soon as the cave is cleared, there is a terrible roar and a resonating impact. Dust and small pebbles explode out of the cave mouth, coating them all from head to toe. The ceiling of the cave has collapsed. The laborers begin to shriek and gabble in terror as they realize what has happened. The cave's interior, the place in which they stood moments before, is now a solid mass of rubble. Every one of them would lie buried beneath that pile of rock, but for one thing: the vision of their Tuan. Their Tuan Dokter has foreseen the danger. He heard the noise—the first faint noise, which no one else could hear, the noise that maybe was not for the ears but for the spirit—and it told him that the stones would drop. The evil spell upon him is broken, or perhaps he has defeated the enemy who works against him. The Tuan has saved them all. They look at him with new admiration, and with fear, thinking of the great power at his disposal. He can hear the future.

Dubois is more than a little shaken, though he tries to disguise the trembling of his knees. "It is early, but we will return to camp now," he says calmly, then turns to lead the way back. He has to concentrate to walk without stumbling, without collapsing in terror at what might have been. No one would have ever found their bodies. Anna would never have known what became of him; the children would never have known. He would be . . . gone, that is all. He would be the mad Dutchman who went off looking for the missing link in the Padang Highlands and was never heard of again. He would have been nothing, just dust, more crushed bones, a meal for the insects. He says nothing more to the men.

The incident becomes an integral part of the elaborate mythology the coolies are building about the Tuan Dokter, how he is stronger than other men, more powerful, and fiercer. He works the men very hard. But he sees things even other totoks cannot. Maybe he has an amulet that protects him from harm. How else does he escape the tiger's lair and the cave-in? Who else even heard the noise he noticed?

Though the men are awed by his prescience, tentacles of fear and despair curl and wind their way through Dubois' thoughts. Unwelcome images of failure, humiliation, and death haunt him. It is a long and lonely night for him, a tiny speck of Dutch genius lost in the middle of Sumatra. By the next day, he is laid flat with fever.

There is something malevolent about it all. The thought spins around and around his disordered mind; he cannot concentrate on anything. It is fever, malaria, that is all, he tells himself. It is not a spirit or a curse. He has a fever. He must treat himself, as he treats the men. Call Pollak. No, he is not at the hospital, not at home, no Pollak. He must get to his medical supplies, must take quinine. But he cannot stand, cannot even sit up. When his boy comes in to see why he is late for breakfast, he gestures to the boy to bring the medicine chest over. "Obat," he croaks faintly. "Medicine, I need obat." Leaning over the side of the bed, Dubois unlocks the case slowly, fumbling with the key. He opens the chest and extracts the precious bottle of quinine. "Bring coffee and bread," he tells the boy. Exhausted, he lies back down and waits. The boy returns and holds him up, so that he can drink the coffee, chew the bread, and swallow the quinine. Without food in his stomach the bitterness of the quinine will make him vomit. "There," he whispers. "Bring soup later, biscuits. Get me a fresh towel." He gestures feebly at his sweat-soaked face and chest.

While the fever does its dance, Dubois lies in his tent and waits. Waiting is not difficult, for his mind drifts off into confused, cloudy fantasies. The boy is loyal to the Tuan Dokter, for he is a special man. The Tuan is exhausted from his struggle against evil forces yesterday, so now he is vulnerable. The boy will guard him. He sits quietly outside Dubois' tent, listening and waiting to be of use. He takes the blankets off when the Tuan is restless and sweating; he puts them on again when the chills come, and even lies on top of the Tuan, outside the blankets, to add his body's heat and weight. He brings coffee and soup and tea and helps the Tuan eat. Dubois' days collapse into nights; his nights warp into formless nightmares. Even a strong, young man is helpless in the grip of malaria. He cannot shake it off by an effort of will. Fortunately, the boy grasps the idea of keeping his master quiet, clean, and comfortable and he sees that the water is thoroughly boiled to make tea or soup. He holds the basin for the Tuan to urinate into and empties it far from the tent. He learns how much quinine Dubois is to take, and when, and makes him swallow it.

When the fever abates, the expedition continues. The coolies have been glad of the rest. It is almost mid-October; the rains are threatening. Dubois is still weaker than he is willing to show the men, and he is discouraged. He has accomplished very little this season. Perhaps he was unusually lucky to find Lida Adjer early on, a rich site and relatively easy to get to. He must report to those who have supported him, educate them as this season has educated him, about the realities of fossil-hunting in Sumatra. He writes to F. A. Jentink, the director of the National Museum of Natural History in Leiden and a member of the research committee:

October 17, 1889

Everything here has gone against me, and even with the utmost effort on my part, I have not achieved a hundredth part of what I had visualized. Where the cave explorations are concerned, the reverses began right at the start, with my coming here in the poeaza (the period of fasting) when the Malays are as indolent as frogs in winter. . . . A survey of the caves I was provided with seemed fitted only to put me off the scent because there were very few real caves among them. Yet these did exist, as I saw later, and people had simply concealed their existence from me . . . because they thought that the "Company" would appropriate the gold and saltpeter the inlanders get out of these caves. . . . After these experiences I went searching without guides and in this way I have found a few very useful caves, but still never the best one could wish for. What's more, it was necessary to live out in the forest for weeks on end, usually under an overhanging rock or in an improvised hut, and it turns out that in the long run I can't stand up to that, however well I was able to bear the fatigue at first. Having now come back, with my third bout of high fever, which nearly finished me for all searching for "diluvialia," I have had to give it up for good. . . .

The trouble with the personnel was even worse. To begin with, one of the two engineers assigned to me to supervise the forced laborers was totally useless, and after repeated warnings and exhortations to carry out his duties properly he was transferred at my request. . . . Meanwhile the other engineer died of fever. . . . The number of forced laborers I have placed at my disposal is 50. Of these, some (7) have run away or have been turned out because of misbehavior; at present (in the rains) 50% are sick. There are also foremen and cooks among them so that at the moment—it is sad, but true—only 12–15 laborers are working. A third drawback lies in the site itself. That it is overgrown with forest would not be so bad in itself, if there were only roads or at least paths, and if the steep limestone mountains did not make communication with most points as good as impossible, and if there were not a total lack of water in many of the mountain places.

As he composes the letter, he muses about the difference between Sumatra as it is and Sumatra as he envisioned it, sitting in Amsterdam. He

In 1887, two corporals from the Engineering Corps, Gerardus Kriele (left) and Anthonie de Winter (right), join Dubois to supervise the day-to-day work of the forced laborers.

thought finding the missing link would be a relatively simple task: all he had to do was look in the right place. He sees now that looking and finding are more grueling than he anticipated. He does not want to die of fever before he finds the missing link. He jestingly calls the fossils "diluvialia" in his letter, as if they were remnants of the biblical Flood. The only flood they are associated with is a flood of sweat: sweat from digging, sweat from fever, sweat from climbing up and down these wretched mountains.

He hopes Jentink and the others will understand. Dubois will not give up the search, no indeed, but his sponsors need to understand how very taxing it is. Fossil-hunting is not a stroll across a meadow to inspect the exposures, or even a vigorous day's hike, as in Holland. No, fossil-hunting here is tigers and fevers and closemouthed villagers who do not know their own area or will not tell a tuan what they know. It is having to teach the men the most elementary work habits. And the first pair of engineers was worse than useless. Maybe the two who have come to replace them, Anthonie de Winter and Gerardus Kriele, will be better. He hopes so. Still, all of it is up to him: all the decisions, all the evaluations, all the procedures and strategies must be created by Dubois alone. If the prize were not so important, it would be less urgent to seek it.

He needs a better plan.

CHAPTER 15 TO JAVA

The expedition is halted by the rainy season, but Dubois is still preoccupied with his fossils. He writes for advice to the geologist Rogier D. M. Verbeek, who knows the Indies well. Verbeek replies promptly and kindly, suggesting Dubois may have more luck in Java, where Raden Saleh's fossils came from. Verbeek has seen a lot of limestone there, which is good for preserving fossils, and Java is not so wild as Sumatra. There are more roads, more people, and more cultivated fields and villages. While villagers might be a nuisance, they can also be very helpful. And the existence of roads and well-traveled paths would lessen Dubois' transportation problems. The expedition could search more ground if the men were not always having to make roads and paths; the supplies could be hauled by oxcart rather than carried by the coolies.

Dubois weighs the advice. Is Java the answer? Ease of access would be a real advantage. He could start by trying to relocate Raden Saleh's find-spots. Too, Junghuhn's maps of Java are much better than the vague, inaccurate ones of Sumatra he has access to. Verbeek's suggestion pricks Dubois' memory, reminding him of a piece of valuable information that he has completely neglected. Almost a year ago exactly, on October 24, 1888, a mining engineer named B. D. van Rietschoten was searching for marble outcrops near a Javan hamlet called Wadjak. He found instead an old, petrified skull and picked it up as a curiosity. Eventually, he dispatched the skull to C. P. Sluiter, curator of the Royal East Indies Society of Natural Science in Batavia. After a brief correspondence, Sluiter generously forwarded the skull to Dubois in Sumatra with a letter, which arrived like an unexpected gift.

December 21, 1888

I received your letter of 15 November already some days ago, but I waited to answer until I received the box with the skull from Mr. Van Rietschoten from Kediri. I have received it and looked at the contents. This is, however, in a very sad state. The skull is broken in a number of pieces, and everything is embedded with a tremendous incrustation of limestone. Because an examination of this species is completely outside my usual work, I think it would be best if I send the complete collection, as it is, to you. I have tried to remove the limestone from a small piece with hydrochloric acid and this works excellently. Another question is, however, whether the skull is a fossil or not. I myself doubt it is, judging from the

condition of the bone, but I am not knowledgeable about such matters.
Also the situation of the cave makes the age perhaps somewhat doubtful.
So you will be able to make your own judgment about these things, I
enclose the letter of Mr. Van Rietschoten, although it is rather insufficient.

Dubois was astonished that a man who wasn't even looking for fossils could stumble across a fossil skull, when he himself had spent months scrambling up mountains deliberately searching for fossils with modest results. What exactly had this mining engineer found? He set the box down on a table in his study, extracting each fragment carefully from its wrappings and laying them out upon a pad of folded cloth. What a thing this specimen was! Though it was in pieces, he could see easily that it *was* a skull, a type of human skull. And it was certainly fossilized; it weighed several pounds or more. It was worth cleaning off and fitting back together, but he already knew it wasn't his missing link. The braincase was much too large and the face too small. It reminded him of some of those old natives, with not an ounce of fat left on their faces. But it was nothing like an ape, still not strong enough in the face, no big canine teeth, the brain much too big. There were brow ridges, but not massive ones like a Neanderthal's or a gorilla's. No, this Wadjak man was primitive—very interesting—but human.

Because of that judgment about the skull's humanity, Dubois cleaned, glued, and examined the fossil for only a few days. He knew the typical shapes of the skulls of various races of mankind and this skull was surely neither European nor mainland Chinese. There was something a little . . . a little crude about it. He sent Sluiter a prompt evaluation of the specimen.

In every respect . . . the skull differs so greatly from the type of the
present-day inhabitants of the West-Malay islands that there can be
absolutely no doubt that it has nothing to do with that present-day
race. . . . While I don't dare to make a definite statement before I have had
the opportunity of classifying it precisely through close study and compar-
ison, the skull seems to me to show the greatest similarities with the
Papuan type, and I am virtually certain that the first representative of the
primordial people of Java has now been discovered.

With the dispatch of this letter, Dubois pushed the Wadjak skull to the back of his mind. He kept it, of course, but he placed the box on a shelf that was rather hard to reach, for his attention was then focused on the fine fossils from Lida Adjer. Now, in October 1889, Verbeek's letter reminds Dubois that

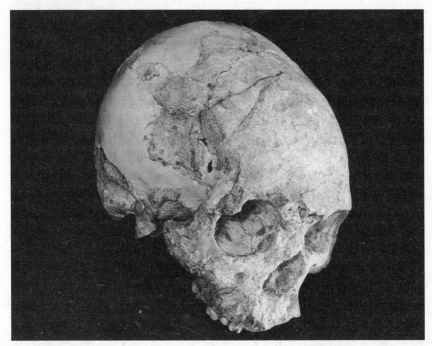

This skull from Wadjak, Java, is found in 1888 by B. D. van Rietschoten, a mining engineer.

he has paid far too little attention to the only human fossil ever to come from Java, which sits on a shelf in his own study. What a fool he is!

During the rainy season of 1889–1890, the Wadjak skull draws Dubois back again and again. It is a fine specimen. Of all the caves on Sumatra that he has explored, only Lida Adjer yielded fossils as good as this. This Wadjak skull proves that the preservation of fossils is very good in at least one find-spot in Java. He looks up Wadjak in Junghuhn's Map of the Island of Java. The town is in East Java . . . ah! near the volcano Mount Willis, very near the town of Toeloeng Agoeng, Kediri Residency. People say that East Java enjoys a much healthier climate, with much less malaria than Sumatra, so there would not be such a cruel risk to his health. There are Anna and the children to think of, too. Maybe he should start again in Java.

He broaches the subject with Anna one quiet evening, after the children have been put to bed. The servants pad silently about, lighting the lamps, cleaning up after dinner, and the two of them sit side by side in long chairs. Feet up, comfortable, contented, Dubois looks over at his comely wife and thinks that life is good.

"Anna," he says softly, reaching out a hand to touch hers, "do you like it here?"

She is surprised by the question; she has not considered whether she liked it here or not. They are here, she and her husband; her children are here; the servants are here; it is the Indies, not Amsterdam, and that is all. She is a good-tempered woman, happy in most circumstances. "Why yes, I suppose so," she replies, smiling. "Why do you ask?"

"I thought that a move might be in order. I have not found a good cave since Lida Adjer, that July just after Jean was born. That was more than a year ago! I think that the caves of Sumatra are less promising than I once believed. Besides, there is so much fever here, so much; I didn't know how dangerous it would be. The children are still so small and helpless. . . ."

"Have you noticed something?" Anna interrupts, alarmed. "I thought Jean was a little fractious today, but just tired, in need of a nap. Do you think he was feverish? Or that redness on Eugénie's cheek. That is just an ordinary rash, isn't it? Do you think they . . ." She could not bring herself to finish the sentence, so awful was the prospect of her darling children coming down with some wicked tropical disease.

"No, no, not that," he reassures her. "They seem fine and healthy. I am sure they are well. But they might be safer somewhere else, somewhere where there would be better fossils, too, and where maybe I wouldn't have to be gone so much of the time. I shall propose to Groeneveldt that I move my expedition to East Java."

"Java?" says Anna, vaguely. "Will it take us long to get there? By boat, I suppose, from Padang. Show me on the map. Do we know anyone there? Would we be near Batavia? I hear there is quite a social life in Batavia."

In the early spring of 1890, Dubois petitions Governor Kroesen and Director Groeneveldt for permission to move his operations to East Java. Characteristically, he includes a detailed, logical list of scientific reasons why the new plan is likely to meet with success. Permission is duly granted by government resolution on April 14, 1890. He will take Kriele and De Winter with him.

Once again, the Dubois family pack their belongings, which seem to have grown hugely, fertilized perhaps by the rich tropical climate. They sail from Padang for a new island, Java, and a new fossil-hunting ground. Before reaching Tanjung Priok, the harbor at Batavia, they pass the old harbor, Sunda Kelapa. Dubois spies a load of tropical lumber from Kalimantan being unloaded from a wooden Makassar schooner and points it out to his children. The swarthy Buginese sailors look like pirates, and perhaps they are. Small but muscular, they heave massive loads of fresh red wood up onto their naked shoulders, two men to a load, using only a folded rag or two to protect their skin from the rough-cut planks. Then the pair walk slowly down the gangplank, a massive tree trunk rudely shaped with an adze to give a grip to horny bare feet. Thus the entire ship is unloaded, hundreds upon hundreds of pounds of lumber, until the hold is empty. The old harbor is packed with perhaps thirty of these wooden schooners, sleek and fast, painted in vibrant stripes of red and green, white, sunny yellow, and brilliant blue. Their dull, rust-colored sails look heavy, furled with stout ropes greasy with use. It is like encountering something from the eighteenth century, Dubois thinks, when the Dutch ruled all the Spice Islands.

Before long, their steamer—it is also the mail boat—docks a little farther up the shore at Tanjung Priok. The Duboises are received by the European population of Batavia as eagerly as a package from home. Welcoming a new, handsome couple is an exciting prospect. And such an exotic thing Dr. Dubois is here for: to find fossils! The more scientifically inclined among the men have some appreciation of what he is looking for, as do a few of the more educated women. No one has ever come to Java before to look for fossils except that eccentric German Junghuhn, who bothered himself with all kinds of curiosities, and that was years ago.

The Duboises are fêted and dined and introduced to all of the important people in the capital. They are caught up in a social whirl the like of which they have not seen since their courting days in Amsterdam. Dubois takes the opportunity to explain his scientific aims to the men, seeking out Sluiter at the Natural History Museum in particular. The ladies show Anna where to have new stylish dresses made up and order clothes for the family, too. She revels in the opportunity to make women friends, to shop and talk of fashions at home. She had not realized how sorely she missed female company. There are so many goods available in the capital that she has not seen for a long time: china and fabric and furniture, paintings and books and music, imported foods from home that her mouth longs for. There are even potatoes, true Dutch potatoes, though they are in cans. Her head spins with the opulence of Batavia compared to their quiet, simple life in Pajakombo. For the first time in their short lives, the children play with other Dutch children, not natives.

The Dubois family enjoys the sojourn in Batavia. Anna and Dubois promenade in smart delmans along the Konigsplein, the broad avenue in the fashionable district, and go to receptions at the Governor's house. They attend musical evenings, reviving some of the songs they used to sing together when they were courting, to general acclaim. Anna's clear soprano and even her out-of-practice skill at playing the piano are much praised. Of course, the pianos here are generally out of tune anyway; the impossible heat and constant humidity ruin musical instruments in a few months. But to play even an out-of-tune piano is a treat for Anna. Dubois' handsome features and proud stature draw the eyes of more than one colonial lady bored with her husband and the predictability of Batavia society. Dubois will do no more than flirt, and very mildly at that. He is not immune to the sidelong glances of the prettiest young matrons; he knows his novelty and athleticism make him a popular partner on the dance floor. But he is still more interested in science than in romance, as the ladies discover to their regret.

Anna is a great success, a lovely totok wife not yet been ruined in face or figure by the tropics and childbearing. She is gay, amusing, and new; no one has heard her stories before or grown weary of her personality. One eligible bachelor about ten years Anna's senior, Willem, pays particular attention to her, bringing her cups of punch or tea, pulling the chair out for her to sit, fetching her fan when it is too warm, fluffing the pillows to ensure her comfort. It is Willem, not Dubois, who stands by her side through social occasions, asking with courtly attentiveness whether the lamp is too near or the evening too chilly. He seemingly asks no more than to be near Anna and to serve her. Dubois sees nothing wrong in this; it is a usual thing for a lady in the colonies to have a chaste admirer. He appreciates the compliment to his wife. Willem is a good chap, but why he has nothing better to do than listen to Anna's chatter and dance attendance on her, Dubois cannot fathom. Anna glows with the attention. After a month or two, Dubois declares it is time for them to go. He must start up his expeditions again before the dry season is over. Or perhaps, just perhaps, he feels Willem's intentions are growing less innocent.

Dubois decides to settle in Toeloeng Agoeng, a town in East Java whose name means "Noble Help": a good omen, he thinks. Located at the foot of majestic Mount Willis, Toeloeng Agoeng is close to Wadjak. But the real reason for moving there is that the house formerly occupied by the Dutch Assistant Resident, the second administrative official in Kediri province, can be leased at a very good price. There are few houses so fine in the eastern part of the island.

The family spend two long days on the train from Batavia to Toeloeng

Agoeng, traveling with all of their belongings and no servants to help them. The Resident of Kediri province kindly sends a carriage and an oxcart to meet them at the station and take them to the house, which is elegant and spacious. It is a large one-story building with a red tile roof and generous verandas at both the front and rear. Substantial white pillars support the roof over the veranda, and a decorative wooden fretwork joins each pillar to its neighbor, except in the center where the broad stairs rise from the yard. It is a fine Indies house, both in the arrangement of its rooms and in its style of architecture.

The two deep, shady verandas, front and back, form the foot and cap of a letter I, connected to each other by a long central hallway that is broad enough to be considered a room in its own right. Anna will arrange rows and rows of flowers, palms, and other plants in Chinese blue-and-white flowerpots right up the front steps and onto the veranda, as if to extend the garden into the house. Scrolled rattan furniture, rocking chairs and long chairs, are interspersed with conveniently placed small tables and small, ornate rugs. Anna takes a critical look at the bamboo blinds of the front gallery, which have been spoiled by mildew and will need replacing.

From the front veranda, a central pair of large carved doors lead to the inner gallery, a long narrow room that parallels the verandas. The inner gallery is a more formal reception or entertaining area for guests, forming the transition between the public front gallery and the inner, more intimate rooms. Anna will hang large mirrors along the walls to reflect the light without lending heat; among the mirrors hang lithographs, and her prized collection of good Delft plates. Maybe she will be able to get one of those colorful paintings by Dezentjé, the native artist who is all the rage in Batavia; they are so dramatic, full of pinky-orange sunsets and purple volcanoes. Potted palms, hibiscus, and the like will soften the stiff arrangement of chairs and tables set against the walls.

If the verandas are the cap and foot of the letter I, then the central hallway is the vertical stem. It is reached by passing through a second wonderfully carved set of enormous doors. With the doors open, Anna can see through the entire house from back to front. It is beautiful: cool and softly lit by the diffuse sunlight that reflects from the mirrors, the plates, and the gleaming marble floors. The interior is a vast, elegant expanse broken only by clusters of furniture and flowers. The smooth stone floors feel deliciously cool to bare feet. The children and servants sometimes lie flat upon the floors, to soak up the cool. Along each wall of the central hallway are doors that open into smaller, rectangular rooms, four to a side. On one side is Dubois' study, a gentlemen's sitting room, a parlor, and a ladies' withdrawing room; on the other are the bedrooms for Anna and Dubois, Eugénie, and Jean, as well as the nursery proper. At the foot of the hallway, another set of doors opens onto the back veranda. It is a mirror image of the front veranda but it is a far more intimate and private space.

When they first arrive, they find a lovely mahogany sideboard with glass-fronted shelves in the central hallway; it is by far the nicest piece of furniture in the house. Anna thinks she will display her fine tea set in it, and the best china, perhaps some crystal. But upon unpacking, she finds her teapot—a prized wedding present—is broken beyond repair. It is to be expected, she thinks a little sadly, when we move about so much. She soon learns that the shops in Toeloeng Agoeng do not carry fine European china. Perhaps the Njonja may be able to find a comparable teapot in Batavia, she is told.

In the meantime, whimsically, she decides to buy a plain, ivory-colored stoneware teapot from a Chinese merchant. The teapot has a humorous shape, like a plump Indies njonja, that appeals to Anna. It is well-rounded and broad at the bottom, narrowing to a sort of waist just above the lower attachment of the handle, and swelling generously again at the top to receive the lid. The only decoration is a pair of incised parallel lines that demarcate a stripe around the widest part—that, and the clever angling of the handle, just where a thumb-rest is needed to balance the weight of a full pot. The spout is stout but rather elegantly curved, with angles that repeat the one on the handle. The overall effect is of a European teapot crossed with Javanese chinoiserie. Anna thinks of it as "Njonja Dubois' teapot." She is becoming a little bit Indische.

The Toeloeng Agoeng house sits on a property that, like the house, is much deeper than wide. As she explores it, Anna is pleased by the layout of the house and the compound. Behind the main house lies the large and private garden, surrounded by a low wall of whitewashed stone. The garden was once lovely, she can see, and will be an excellent place for the children to play. It is planted with flowering bushes—frangipani, melati, and tjempaka, a variety of magnolia—and fruit trees. There is a coconut tree, a few feathery casuarinas, some papayas, and a small group of mango trees, though everything is in need of a good deal of attention. She shall set the new tukang kebun to pruning and fertilizing and ridding the garden of snakes and scorpions at once. Set off by itself is one very large waringin tree, the habitual roost of a small flock of cooing, brightly colored doves. Anna thinks she may have a circular bench built around it, to remind Dubois of the one in his parents' garden in Eijsden.

Behind the garden is a separate building with the mandi room for bathing, the toilets, the laundry room, the kitchen, and storage rooms. Opposite are the stables, with stalls for four horses. The servants' quarters make up a small kampong or village farther back on the property, where the noise of their children and chickens and the smoke from their cooking fires will not bother the Tuan Dokter or the Njonja. Like every Indies household, this one is a miniature of the colony itself, a hybrid born of the mingling of Javanese and Dutch cultures under tropical skies.

Anna must hire new servants, but she feels much more confident after living in the Indies for three years. She does rather well this time, she believes, especially with the four most important servants: the djongas, the kokkie, the children's babu, and her own babu. Dubois' pay has increased with his secondment to the Department of Education, Religion, and Industry, so Anna hires additional servants: a syce, who will look after Dubois' horse and will also help the head gardener, and a laundry woman and seamstress. Anna feels very grand, having such a large staff and a fine, high-ceilinged house with large rooms and a marble floor. Soon after arriving in Toeloeng Agoeng, she realizes she is with child again and her happiness is complete.

Toeloeng Agoeng is ideally placed for Dubois' explorations. He relishes being free of interference. The closest military outpost is Fort Van Den Bosch in Ngawi, almost 150 miles away. At last Dubois is accountable to no one, no one but himself and his director, Willem Groeneveldt. What makes him feel most free is the certainty that there will be no summons to come help out at the hospital or to turn up for some military event or other. He can devote himself entirely to science.

Dubois returns home from his first brief expedition bearing not boxes and crates of fossils, as Anna expects, but a huge, grotesque bird in an enormous bamboo cage shaped like a minaret. It is a marabou stork, to be a pet. Anna cannot imagine why he has brought this creature home. Like all its kind, the stork is ugly, almost five feet tall with long, knobbly legs. Its repulsive head is nearly bald, mottled pink and gray in color. From its neck hangs an obscenely naked, wrinkled, pink pouch like an old man's scrotum, not that Anna allows herself to think of such things. For all its homeliness, the marabou is a formidable creature, with a wingspan of nearly nine feet and a long, heavy beak. Its beady eyes assess constantly whether anything within its view is edible. Indeed, the loathsome bird soon exhibits a fondness for carrion and rotting meat, as well as a deadly tendency to attack anything smaller than itself. Lizards, geckos, snakes, mice, worms, insects, fruits, and even unwary chickens are fodder for this nightmarish apparition.

Still something about its stately pacing around the yard amuses Dubois; Anna can see the humor it in, when he points it out. The bird walks slowly, peering down its long beak as if looking down its nose at everything. Its angular, stiff-jointed walk is a ghastly parody of a self-important military man's. "Do you see?" says Dubois to Anna, chuckling. "It is the very image of that pompous colonel in Batavia. Let us call him the Adjutant. That is the nickname for these birds in Africa, I believe." It pleases Dubois to look out from his study window and see the Adjutant lurching around the garden, unaware of its own ungainliness. Secretly, Anna hates it and the children are terrified. The kokkie complains that the Adjutant snatches food from her as she carries it from the kitchen to the house, and Babu is certain it will eat the children's fingers or toes. One day Dubois sees her beating the stork with a fallen branch, defending the children from its vicious beak. After that he has a pen constructed to confine the stork to a corner of the garden.

In the back garden at Toeloeng Agoeng, one of Dubois' servants will build a pen for the marabou stork named the Adjutant.

CHAPTER 16 JAVA FOSSILS

Dubois plans first to reconnoiter the caves and rock shelters near Wadjak, where van Rietschoten found the skull. As before, the greatest problem he faces is in training the laborers, but his assistants Corporals Kriele and De Winter from the engineering corps far surpass their predecessors in their ability to understand the work, supervise the coolies, and map the excavations.

They meet with almost immediate success at Wadjak. There are more extinct mammals there: fossils of antelope, rhino, pig, two sorts of monkey, and even Dubois' nemesis, a porcupine. After a few days' work, they discover a second human skull, more fragmentary than van Rietschoten's, but very similar in shape and size: confirmation that they have relocated the find-spot. Once the second skull is prepared and cleaned, Dubois is unsurprised to find it is of the same race as the first. Later work in the same area yields a few fragments of a skeleton belonging to one of the skulls. As he explains to Groeneveldt in his first report, these remains are human: no missing link but scientifically important examples of an extinct Australian race at a higher level of evolutionary development than the Neanderthal fossils of Europe and thus undoubtedly fairly recent.

He decides to explore farther inland, along the entire range of the Kendeng Hills where Raden Saleh's fossils came from. If the Kendeng Hills fauna is really like that of the Siwaliks of British India, Dubois may find something as good as Lydekker's *Anthropopithecus*. From June until the beginning of October 1890, Dubois' men survey the entire length of the range, tramping more than a hundred miles from near the town of Semarang on the central north coast almost to Soerabaja, the thriving port at the eastern end of Java. It is a demanding but profitable endeavor; they find numerous sites, like Kedoeng Loeboe and Kedoeng Broebus, with mammalian fossils more abundant and more complete than those of Sumatra. Yet Dubois is still looking in the wrong places. Looking in caves and rock shelters is a poor strategy in Java. During the dry season the hills and slopes are carpeted in thick, leathery djati or teak leaves, each big as a dinner plate and tough as a poorly cured hide. The djati leaves make walking awkward and almost completely obscure the ground and anything lying on its surface.

Frustrated, Dubois leads the men away from the mountains and along the riverbanks, where movement is easier. Unexpectedly, they spy fossils in the exposed riverbeds. This is a completely new type of find-spot, out in the open air away from caves or natural overhangs. With a flash of insight, Dubois realizes that he has been hampered by a European notion of where fossils will be found, a notion that does not apply here because the topography is different. Java has few lowland caves; most of the land is dead flat and covered with luxuriant vegetation or cultivated fields. The mountains—volcanoes really—constitute the only relief and they rise abruptly out of the plains, without foothills, like cups that a child has upturned on a tabletop in play. Everything in Java is horizontal or vertical.

To find fossils, Dubois needs a glimpse of what lies entombed in the rock beneath the surface. Rivers are his best excavators, more valuable than a thousand coolies, for rivers have been cutting their way through the rocks and soil of Java for eons. In the dry season, when the water is low, the underlying sediments are exposed to anyone who has eyes to see what is there. Rivers are Dubois' window into the past. He will search where no one else has looked before and he will find what no one has found before. He explains his new strategy to Kriele and De Winter, who understand almost immediately and are grateful to be spared the endless climbing up and down mountains. From now on, they will search the rivers in the dry season, walking along the bluffs and looking down at the exposed sediments,

Dubois sorts his fossils on the veranda at Toeloeng Agoeng,
leaving a narrow walkway to the front door.

being especially careful in searching point bars where sand and debris accumulate at the bends in the river.

Together, Dubois and the engineers lay out a plan of exploration focusing on one of East Java's major rivers, the Bengawan Solo. They will work west from the fort at Ngawi, where they can obtain provisions and supplies. Dubois will select likely spots for excavation, leaving the day-to-day work in the hands of Kriele and De Winter. They will make sketch maps, take precise notes on anything they find, and send frequent reports and crates of fossils back to him at Toeloeng Agoeng. They are to label fossils according to where they are found. If they find anything exceptional, he will join them as soon as possible.

After starting the men off, Dubois settles into family life at Toeloeng Agoeng. Whenever a crate of fossils arrives, he investigates it with enormous anticipation, unwrapping specimens one by one from the djati leaves that cushion them: who knows what lies within? He lays out the fossils to make preliminary identifications. The larger ones—a nine-foot-long elephant tusk, the two-foot-long skull of an extinct buffalo with a six-foot spread of horns—simply will not fit into his study. Before long he takes over the front veranda as a sorting space. There, he can work protected against rain and sun, with a pleasant breeze to keep his head clear as he arranges, cleans, compares, and labels specimens. Soon the fossils usurp nearly the whole veranda, like a tiresome guest who calls at teatime, settles into a comfortable chair, and will not leave. There is only an ever-narrowing bone-free path from the front steps to the inside of the house. It is awkward, Dubois can see that, and unconventional, but he is a man of science and these specimens are his life.

Despite his intentions to stay at home more and guard his health, Dubois is drawn to the sites like an ant to sugar. When exciting fossils begin to appear, he must go see for himself the place where they were found. And when Kriele or De Winter writes that they are about to move camp, Dubois feels compelled to join them and pick the next spot himself. But as her third pregnancy advances, Anna asks him to stay at home more. He agrees. He needs more time to prepare, identify, and study the fossils, for no one else in all of the Indies can be trusted to do this exacting work.

At the end of 1890, Dubois writes, a trifle smugly, to tell Groeneveldt that he and his men have collected hundreds of fossils from several different sites in East Java. In his quarterly report, he describes the fossils from the Kendeng Hills and those found along the banks of the Bengawan Solo. He can already prove that there are two different types of elephants (*Stegodon*, an extinct type, and a relative of the modern *Elephas*) in the ancient fauna, as well as extinct hippos, rhinos, hyenas, felines, and pigs. Some of the species or genera still exist, but not in the Indies. Clearly the Javan fossil fauna is derived from that of mainland Asia, in accordance with Wallace's brilliant notion. Establishing the precise age of paleontological specimens is a tricky matter, but faunas can be correlated from one region to another and Dubois' conclusion is that the Javan fauna he is collecting now is a little older than Sumatra's and a little younger than Siwalik's. The Javan fossils are from the end of the Pliocene or even from the Pleistocene, the period when the missing link must have lived. Thus, these sites are the right age and have the right kind of Asian animals. He will surely find the missing link soon.

In his final report of the year, Dubois is able to boast of an exceptional find made on November 24, 1890.

> *Amidst the remains of typical representations of the fauna con-*
> *cerned, and in the same layer of sandstone-like andesitic [volcanic]*
> *tufa, a human fossil was found, the right side of the chin of a lower*
> *jaw with the sockets of the canine tooth and of the first and second*
> *premolar. . . . When it came to me it still had on it a small piece of*
> *sandstone-like andesitic tufa, which is the chief sediment among*
> *the bones found.*

At last he has found an ancient human fossil, older than the Wadjak remains. Admittedly, it is a small and rather unimpressive fragment, only a bit of jaw with two teeth in it. Its incompleteness makes it difficult to classify, leaving Dubois elated at its discovery, frustrated at its ambiguity. About the most that he can say is that its chin is minimal—even less marked than the Neanderthal's—

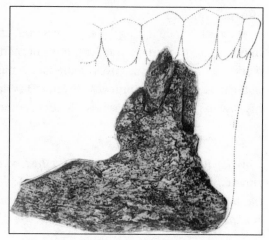

This mandible or lower jaw from Kedoeng Broebus is the first good find in Java. The dotted line indicates the placement of this fragment on a human jaw.

and that it is distinctly human. He lists the specimen as *Homo spec. indet.*, paleontologist's jargon for an indeterminate or unknown species of the genus *Homo.* If only there were more of it!

Until now, Dubois' secondment to the Department of Education, Religion, and Industry has needed annual renewal. On the basis of the many fine fossils he has collected in Java in 1890, Dubois asks Groeneveldt for a three-year commitment. During 1891, 1892, and 1893, he proposes to focus on two areas, caves in the Kendeng Hills in the wet season and the exposed riverbanks of the Bengawan Solo in the dry. To Dubois' great pleasure, Groeneveldt heartily approves the plans.

It has taken Dubois more than three long years of persuasion and exploration and fever, but the effort has been worth it. He now knows how to search effectively for fossils, and where. Groeneveldt is impressed, offering his support and congratulations. From afar, by letter, Dubois' father expresses only pessimism and doubt.

CHAPTER 17 COOLIES

Dubois' decision to stay close to home during the end of 1890 and the beginning of 1891 has almost immediate repercussions. While the coolies, whether laborers or convicts, have great respect for the Tuan Dokter (the legend of his magical powers has followed him here), they do not honor Kriele and De Winter. With Dubois too long absent from the site, the coolies become surly, lazy, and insolent. The letters from the engineers ring peal after peal of complaints.

Dubois tries separating the engineers from each other, thinking they may find it easier to manage smaller groups of men. He sends De Winter off to work in rock shelters and caves on the south coast of Java, while Kriele works in the north. Nothing improves. By June, De Winter is so desperate to make his case to Dubois that he sends a letter of complaint addressed to Anna.

> *Bessole   June 28, 1890*
> *Will you also be so kind as to say to the Doctor that I have sent away three forced laborers because of illness, but I suppose that it is laziness, because they have asked me three times to go back to Toeloeng Agoeng because the work here is too heavy.*

The next month, De Winter writes again: "All the forced laborers ask me for clothes because they come back from the cave and they have been drilling and their clothes are full of drill-spatter."

Kriele fares no better:

Kedoeng Broebus   July 11, 1890
Now I have here 25 forced laborers but they are such a strange people: 4 have already run away, 6 are ill, 1 is dead. It would not be a bad thing if you would send some belly-drink because they are all troubled with their bellies.

Soon the two engineers fall out with each other; Dubois does not know why. De Winter writes, "So it is also very hard for me without a mandur, therefore I ask you politely if I may have the mandur that works with Kriele because Kriele has only 8 men and I have 15."

Dubois cannot spare the time to go to the sites, for on January 16, 1891, Victor Marie Dubois greets the world with a bellow, another healthy, vigorous baby boy safely delivered by his father. He is perfect, tiny, and Dubois is very proud. And yet, before long, Anna's complete absorption in the three children leaves Dubois feeling a little neglected. He is no longer the center of the family. He is peripheral to that indissoluble unit of mother and children. Still, he cannot leave and, despite the engineers' complaints, the work seems to be going fairly well.

Dubois longs for someone to discuss the new fossils with. Maybe Sluiter, that fellow in Batavia who sent him the Wadjak skull, would be interested. Dubois writes, telling Sluiter of the progress of the digs and the finding of the fragmentary jaw at Kedoeng Broebus. He also asks Sluiter to send various books and articles he needs for analysis of the jaw. Sluiter replies cheerfully enough but, to Dubois' astonishment, demurs over sending the books, citing the "perils" of trusting such items to the unreliable postal service.

Dubois is deeply offended; after all, some of the books he has requested are his very own, lent to Sluiter when he was in Batavia. He cannot help but imagine some ulterior motive on Sluiter's part. The edifice of their friendship, walls barely in place, is already developing serious cracks. Sluiter writes again,

January 6, 1891
I really would like to send you back the boxes by return of post, to prove to you that the black suspicions that you felt about me are completely

groundless. . . . Even if you need the books <u>slightly</u> for Heaven's sake write me then, then I would send back immediately what you want. It would be like a nail in my coffin if I suspected in the slightest that I had wheedled the books out of you through my clumsy writing.

A fine letter, but he does not send the books and Dubois is not pacified. How is he to work as a scientist in such a society? Sluiter, a man he liked and trusted, now will not return Dubois' own books!

Sluiter writes again at the end of January.

January 23, 1891

Why are you so disconsolate in your letters? I believe that your isolation there in Toeloeng Agoeng makes you melancholy. It is certainly a pity that you cannot dash over to Batavia more often to convince yourself that the sympathy we developed for each other during your brief visit has certainly not lessened. . . . Write me one day and tell me what is really the matter. I don't mean to be a busybody, but maybe I can contribute something from the sidelines that will improve the relationships among the small number of scientific men that the Indies can boast of.

Indeed, Dubois has sunk into a monsoon-season gloom. He thinks nostalgically about his colleagues at Padang and Pajakombo, especially Pollak—a good man, Pollak! But at Toeloeng Agoeng there is no one, and Sluiter is a bitter disappointment. The government officials are preoccupied with miles of road mended, bridges built, crops taxed; the planters, when they come into town, are generally drinking too heavily to hold any sort of complex thought in their mind; the merchants are small-minded and pedestrian. Toeloeng Agoeng may be a major railroad stop and, compared with Pajakombo, a thriving metropolis, but Dubois feels utterly isolated.

Late in March 1891, the problems with the coolies take an alarming turn. De Winter, working in the north in Kantjilan, catches one of his laborers throwing fossils into the alang-alang grass, trying to conceal them and bring the burdensome work of excavation to a faster close. De Winter cannot tell how long the sabotage has been going on. He knows only that the man has committed a grave offense, a scientific mutiny of the vilest sort. On March 29, he writes to Dubois telling him that he has taken the miscreant directly to the head-jaksa, the native magistrate, who had the man flogged.

Dubois is appalled at the news, both of the man's iniquity and of his harsh punishment. What is he to do? The laws governing the treatment of forced laborers are clear. The sentence was ordered by a native magistrate and is therefore legal, though De Winter is strictly forbidden to beat his coolies on his own authority. Dubois considers flogging an inhuman treatment, suitable only for stubborn mules or lazy oxen. There are other ways of maintaining firm discipline.

The coolies dig for fossils at the same unhurried pace that they plant crops, Dubois thinks, working with pickaxes or short-handled hoes called patjols. It is completely meaningless labor in the hot sun, as far as the coolies are concerned. Now that he thinks of it, Dubois is a little surprised that the men remain as good-natured as they do. But he must find a way to get them to work better, more reliably. He does not understand the coolies at all. Before he came to the Indies, Taen-Err-Toung was the only Asian whom Dubois had ever met, and he was no peasant working off his tax burden or serving out his sentence.

Even Dubois' own workers, men he thinks he knows, are capable of shocking barbarities. He learns this one day when a party of his coolies happens upon a village where only the women are at home. As if it were a perfectly normal response to the situation, his coolies begin looting the village and assaulting the unprotected women. The women scream and hammer on the village gong; Dubois hears it from a considerable distance and so do the men of the village, who rush back to protect their homes and families. A huge mêlée ensues, locals against forced laborers. A number of men on both sides are seriously hurt.

The incident comes to official attention and the Resident orders a formal hearing to determine guilt. Of course Dubois attends, ashamed of his coolies' part in the affair. One of his crew is sentenced to fifty strokes with a cane.

"You shall have to keep more of an eye on the coolies," he reprimands Kriele and De Winter back in camp. "We cannot permit such behavior."

The engineers protest loudly: The raid was not carried out with their approval or their knowledge. They would never allow the men to do such a thing if they knew of it.

"Of course not, I know," Dubois concedes. "But it cannot happen again. We are in charge of these men. *You* are in charge of them. They will not break the law when they are under my authority, especially not in such a barbaric way. And why is that man, whatever his name is, not unhappy at his sentence? Fifty lashes is a severe beating."

"Tuan Dokter," De Winter explains, trying to be tactful, "he does not expect to receive the sentence. The natives bribe the guards to let them off easily, with light strokes and fewer strokes. He has been convicted before and he knows he will not suffer much. That is the way things are done here. He was convicted only because he was so badly wounded that no one would believe he was somewhere else at the time, even if he bribed witnesses to say so."

"Is this true?" demands Dubois of the other engineer, De Winter.

"Yes, sir," De Winter confirms reluctantly, fingering his dark mustache uneasily. He can sense Dubois' disapproval. "Most of those convicted never receive their full sentence, and some of them are not punished at all. It is the way in the Indies."

"Well, that is not the way it is going to be in my camp," Dubois insists. "This was an unprovoked, vicious crime. Men for whom I am responsible cannot go around thieving and assaulting unprotected women."

Dubois insists on supervising the caning personally. It is a horrible ordeal. Long before the fifty lashes are completed, the man's back is welted and bleeding, muscle and bone exposed. Nonetheless, Dubois sees to it that the man receives his full sentence. Afterward, Dubois treats the man with salves and releases him from further work. He is repulsed by both crime and punishment.

A few days later, Dubois spies another worker, a venal, greasy man, sporting some new rings on his fat fingers: booty from the village, Dubois suspects. Something must be done; he must find out the truth. Apparently innocently, Dubois selects the man to go with him to explore a deep sinkhole in the limestone, where the villagers say fossils can sometimes be found. He has the men lower the two of them down on ropes to the floor of the pit, well below ground level. Of course, the pit is barren of fossils. Dubois is not disappointed, for he wanted only a good place for a private confrontation. He turns to the man and begins to question him about the rings, using his powerful physique and formidable presence to intimidate the man into telling the truth.

"Where did you get those rings?" he asks. There is no answer; the coolie squats at his feet, eyes averted. "If you do not tell me where you have obtained them," Dubois threatens, looking down at the man, "you will not leave this pit alive. There will be a terrible accident, most unfortunate."

The coolie pleads his innocence, his fat cheeks trembling, his eyes bulging in fear. "No, Tuan Dokter, bok'n, bok'n, not stolen," he stammers. But he does not meet Dubois' pale eyes with his dark ones.

"You did not have those rings before," continues Dubois sternly, ignoring the protestations of innocence. "You took them from that village, didn't you?"

"Bok'n, Tuan Dokter, bok'n . . ." the man repeats earnestly, sweat dribbling down his forehead and falling like raindrops onto the dusty ground.

"You are not listening." Dubois takes hold of the man's shoulder and shakes him a little. The man's flesh seems spongy, like the meat of an over-ripe papaya, and Dubois is disgusted at the feel of him. "Look at me!" Reluctantly, the coolie lifts his eyes from the ground. "Now, tell me where you got the rings. I want the truth. You know I can see the truth when others cannot. Where did those rings come from?"

The man confesses: he stole them during the raid on the village. He has not the moral fiber to resist Dubois' assertions or the piercing look in his Garuda eyes. "But do not turn me over to the head-jaksa, please, Tuan Dokter," the man begs, knowing now that any sentence of flogging will be carried out in full. "I work for you; I am part of your gang of men. You are my tuan. You punish me, as you see fit. That is right."

Dubois has seen enough of the harsh justice of the native magistrate to agree. He will not allow flogging, but the man must be punished. Dubois returns the rings to the village headman. The next morning, he carries out the punishment he has conceived. He straps an enormously heavy fossil, an elephant bone, onto the miscreant's back. It is a four-foot-long cylinder of solid stone. Dubois announces to all the workers that this man is being punished for theft. He will carry this burden all day long while he performs his normal daily work. He is not to be helped by others. Anyone who helps him will receive the elephant punishment himself. The natives are wide-eyed and serious-faced. They nod their understanding of the punishment and the rules of behavior. By the end of the day, the thief is bowed and utterly exhausted, but he makes no protest and shows no resentment.

The incident increases Dubois' prestige further. He is surely supernatural. No one ever sees Dubois drinking liquor, like an ordinary man. If they clamber up a mountain, he clambers up faster than any of them. If he orders them to carry firewood, it is less than he can carry himself. As a matter of principle, he never resorts to coarse language and he is never seen unkempt or dirty. He can see the truth when it is hidden and he knows what lies underneath the ground. He moves through life like a man on a holy mission, and the coolies revere him.

In contrast, Kriele and De Winter are physically less impressive, being thin-chested and long-limbed. They are naturally more disheveled, their uniforms cheap and often rumpled, their mustaches poorly trimmed, their hair a little long. And their concentration wanders noticeably during long and unprofitable excavations. Sometimes, too, Dubois finds altogether too many empty beer bottles when he visits the camp. The engineers are not bad men or stupid ones. They simply lack Dubois' moral fiber and intensity of purpose. The natives think the engineers perfectly ordinary orang belandas. Like all Dutch, they are sometimes arrogant, usually unfair, frequently insulting, and often incomprehensible. They drink, they smoke, they shout, and they are sometimes seen in the company of the prettiest of Javanese girls. The engineers they understand; they recognize the type.

But Dubois is something else, something extraordinary, the coolies conclude. He is a chosen man. Clearly he has been given a special role or dharma through a divine revelation: that is the only possible explanation. They accord him a place in their pantheon and agree that he operates above the strictures of law or customary behavior. Dubois is destined for greatness.

CHAPTER 18   DISCOVERIES AT TRINIL

Whatever the workers whisper among themselves, Dubois never mistakes himself for a god. He is a dedicated man of science, smarter and harder-working than the others, that is all. Yet sometimes he has an intuition that amounts to a stroke of genius, like coming to the Indies in the first place, like searching the riverbanks instead of looking for caves. He has never

Dubois selects a site along the Bengawan Solo near the
village of Trinil for a major excavation.

believed in destiny, at least not before coming to Java. But things are different here, inexplicable; even a strong-minded Dutchman sometimes feels the workings of the supernatural.

Dubois' uncanny ability to reach beyond the readily observable manifests itself again in the choice of a new site for excavation. Working down the Bengawan Solo from the fort, looking for a glint of bone and a likely place to dig, they pass many bends in the large, slow, brown river. The point bar near the tiny village of Trinil is undistinguished, except for a thumb-shaped piece of the bank that projects above the ground. Dubois looks at that shape and feels a strange enthusiasm creep over him. Why? He does not know. It is not rational; it is intuition, or . . . something else. He has a strong feeling at this place; there is something mystical about it, though he cannot say what. A religious man might even call it God's providence, he supposes, but he is not a religious man.

"Let us dig here," he says, and they do.

Almost at once, the men begin to find fossils. Dubois leaves the crew there, working hard at the excavation, and returns to Toeloeng Agoeng. By the end of August 1891, the team has found many species in the sandstone at Trinil: the remains of the axis deer, an extinct buffalo, the elephant *Stegodon,* and others. Dubois starts calling the site his Charnel House, believing that the skeletons are those of animals killed and preserved by the volcanic eruptions in this region. Not for nothing is the region of Ngawi known as the hellhole of Java. It is full of ancient lavas and lapilli, and some days it seems as hot as the inside of an active volcano. The fossils are a kind of salvation.

In September 1891, the men make their next really important find on Java. It is the right third molar, the last cheek tooth, from the upper jaw of an apelike primate. When the fossil arrives at Toeloeng Agoeng, wrapped in the usual djati leaf, Dubois is jubilant. The tooth is something very like Lydekker's chimpanzee from India, *Anthropopithecus sivalensis.* Dubois has only a single tooth, not a jaw, and he believes his specimen is not quite as ancient as Lydekker's. He knows now that he will have to travel to India to make detailed comparisons between the two.

Dubois' third quarterly report to Groeneveldt in 1891 celebrates the Trinil site and its fossil fauna. Specimens are abundant, largely complete, and often beautiful in their stony way. He is sure he is approaching the missing link. There is no doubt in his mind that his quarry is fossilized and waiting, here, on Java.

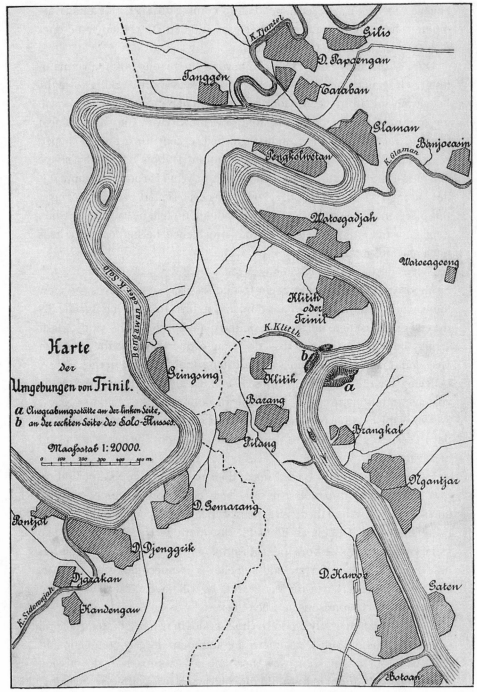

Dubois' map of the area shows the exact location of the site at Trinil.

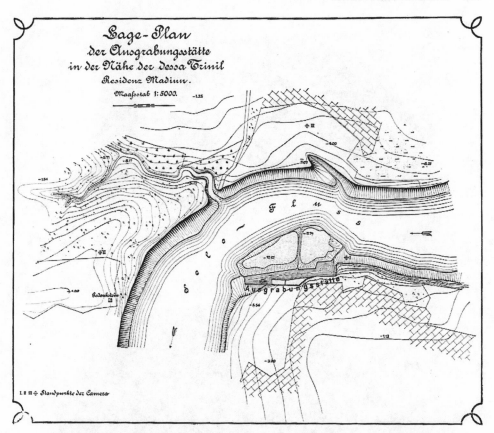

Lage-Plan
der Ausgrabungsstätte
in der Nähe der dessa Trinil
Residenz Madiun.
Maaßstab 1:5000.

Fluss

Colon

Ausgrabungsstätte

I. II. III. ◊ Standpunkte der Camera

A second, more detailed map records additional information about the excavation and the location of standpoints (Standpunkt I, II, III) from which Dubois photographs the dig.

> *The most important find was a molar (the third molar of the upper right side) of a chimpanzee* (Anthropopithecus). *The genus of anthropoid apes, occurring only in <u>West-</u> and <u>Central-equatorial Africa</u> today, lived in British India in the Pliocene and, as we can see from this discovery, during the Pleistocene in Java.*

This remarkable find is soon eclipsed. In October, the engineers turn up a strange bone; it is about the size of a large coconut—not the green outer part, but the dense, hairy seed itself—and it is similar in shape to half of a coconut that has been split longitudinally, except that the fossil is more pear-shaped than ovoid. At first, the engineers think it may be the carapace of a turtle. It is a dark rich chocolate brown in color and thoroughly fossilized, heavy with stony matrix that encrusts many surfaces. The corporals

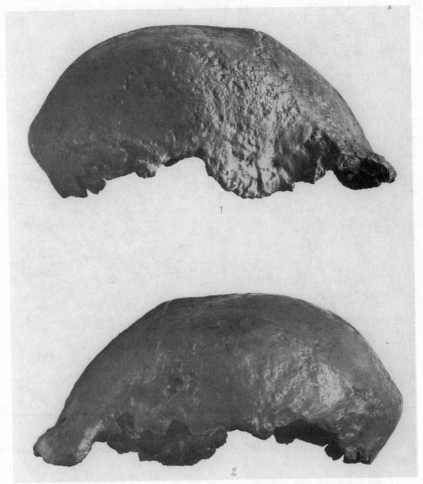

In October 1891, the skullcap of Pithecanthropus *is found at Trinil. (Top, right side: bottom, left side.)*

do not dare try to clean off the matrix for fear of damaging the specimen. They pack the fossil as it is in djati leaves, with the others from the last few weeks' work, and send it off to Dubois.

The specimen is not a turtle carapace but the skullcap of a higher primate, something big-brained. The face and the bottom of the skull are broken away, but there can be no doubt of what it is. The engineers' notes and letters indicate it comes from the same level as the *Anthropopithecus* tooth. There is only the one other primate fossil at Trinil, so Dubois thinks the skullcap is probably from the same individual as the tooth. The skullcap has a chimpanzee look to it, too, with a big brow ridge and, behind it, a pinched-in place, a postorbital constriction before the swelling of the braincase. It is a stunning fossil, showing the approximate size of the braincase and much of the shape of the skull. Dubois has nothing with which to compare the skullcap directly, for there is no such thing as a chimpanzee skull in all of the Indies. His last quarterly report of 1891 to Groeneveldt has a triumphal tone.

> Near the place on the left bank of the river where the molar was found, a beautiful skull vault has been excavated that, undoubtedly (like the molar), has to be ascribed to the genus Anthropopithecus (Troglodytes)....
>
> As far as the species is concerned, the skull can be distinguished from the living chimpanzees: first because it is larger, second because of its higher vault.... The height of the frontal part is not lower than the human skull from Neanderthal or the first skull from Spy; but the fossil chimpanzee deviates from these because its parietals are flatter and the occipital is less developed. With the Pliocene Anthropopithecus sivalensis, of which there is only an incomplete mandible known, only the molar can be compared directly. Probably there is a close affinity between them.

Following protocol, Dubois compares his find to the other known genera of fossil apes. His fossil is neither the "Oak Ape," *Dryopithecus fontani*, from France, known mostly from jaws and teeth, nor the gibbonlike *Pliopithecus antiquus* from Europe. Those fossil apes are much smaller; this skull is too big. He can also see plainly that this is not a modern chimpanzee.

> Besides, the most important fact is that the living chimpanzee, in his teeth, approaches humans more closely than does the gorilla or the orang-utan, which is found in the same region of Java, while this Pleistocene chimpanzee approaches the human more closely because of its skull.

He has found an exceptional fossil, more primitive than a Neanderthal, apelike in its teeth and yet most humanlike in its skull. He closes the report by observing proudly that, before this find, Lydekker's jaw of a Siwalik chimpanzee was "all that we possessed in the way of fossil higher anthropoids, so it is clear from this what a gain for science the Javanese fossil is."

Dubois is enormously tempted to assign a scientific name to the new find. Naming a new species of this importance would establish his place in the annals of science forever, much more so than the other new fossil mammals he has already found. The obvious name, *Anthropopithecus javanensis*, rolls nicely off his tongue. This name would demonstrate the close evolutionary relationship between his fossil and Lydekker's Siwalik *Anthropopithecus*, while also indicating the place of origin as Java. But it is premature, he decides; more study is needed before he formally names the fossil. On December 30, 1891, he sends a letter to Governor Kroesen in West Sumatra, informing him of the find. "The creature to which this skullcap had once belonged," he writes, "was truly a new and closer link in the largely buried chain connecting us to the 'lower' mammals."

The skullcap was found, unfortunately, late in the dry season, in October when the heavy afternoon rains were already beginning. Excavation cannot continue much longer, so Dubois orders the men to work more intensively for a few weeks, to no avail. Whatever else lies beneath the waters of the Bengawan Solo, within those buried sediments, must wait until April or May when the next dry season commences.

Dubois uses the rainy season to study the new fossils carefully and to begin the nerve-racking task of picking away the stony matrix from inside the braincase. Using pins, sharpened bamboo sticks, pointed medical instruments, flattened nails, fine chisels, Anna's large embroidery needles—anything, in fact, that seems vaguely suitable—Dubois works a little every day to clean the skullcap, cradling it on a sand-filled muslin bag on his desk. He stops from time to time, rubbing his hands to relieve the tension and peering at his progress through his large magnifying glass. Sometimes he simply stands, swinging his arms to relax his muscles, and looking out the window at the pouring rain. It is close, concentrated work. One slip, one push with too much pressure, one tap too hard on a fine chisel, and the fossilized vault bones may shatter. It is a job requiring tremendous manual dexterity and infinite patience, the latter not being one of Dubois' usual virtues. The work also requires a detailed knowledge of anatomy; he must anticipate the curvature and shape of as-yet-invisible structures so as not to damage them. Every day he can see a little bit more; every day, he inspects the braincase anew and marvels at its shape.

He is deeply frustrated by his lack of comparative material and literature, feels himself awash with specific anatomical questions that cannot be answered with his resources. Dubois stares at the few images of chimpanzee skulls in his books, hoping to see more than is shown, until he thinks he is going blind. He acquires three gibbon skulls and a few human ones, but they are not enough. He already knows his skull is neither human nor gibbon. He feels a terrible need to discuss his find, but there is no one.

CHAPTER 19 GATHERING RESOURCES

Early in 1892, Dubois writes to Max Weber in Holland, pleading with him to send a chimpanzee skull. Surely with Weber's connections at the zoological gardens he can lay his hands on a chimpanzee skull! Weber understands Dubois' pressing need, but a chimpanzee skull is not easy to come by in Europe. He searches the catalogues of those who supply curiosities to the public and specimens to the museums, writes in vain to other anatomists and natural historians. All he has to report to Dubois is that neither Amsterdam or Leiden possesses a chimpanzee skull that can be sent to Java. Months pass while Dubois waits anxiously in Toeloeng Agoeng for word from Weber: impatient at the delay, grateful for Weber's efforts, despairing of their success.

In July, Dubois and Anna are invited to a reception at the Resident's house in Kediri. Anna dearly wants to go, and Dubois agrees. He is doing no one any good waiting at home for a skull that never arrives. They pack up their best clothes and depart for Kediri. It is one of the few trips out of Toeloeng Agoeng that they have made since settling there; Anna is as excited as a girl. Perhaps the reception, the society, will take Dubois' mind off his troubles. Besides, it is a good opportunity to speak to the Resident, the Assistant Resident, the Regent, and other important officials.

Dubois expects nothing of the reception until he spies two men he does not know: a pair of strangers in the stiflingly limited society of Kediri province. One is about his own age, one considerably older. His spirits lift suddenly and inexplicably, like those of a schoolboy who sees at close quarters some girl he has worshiped ardently from afar. The young man has an odd look on his intelligent face; perhaps he is just as bored as Dubois by idle gossip and polite chitchat. He stands next to the dignified older man. Dubois finds someone to introduce him to them.

The older man is Robert Boyd, a planter of Scots extraction born in Batavia, the son of another Robert Boyd. The elder Boyd came to Java from Aberdeenshire during Stamford Raffles' brief rule and stayed on after the colony reverted to the Dutch in 1816. That Robert Boyd established a successful trading business in the port towns of Semarang and Soerabaja,

The "Old Warrior," Robert Boyd, becomes one of Dubois' closest friends.

marrying a Javanese woman. The second of his seven children, the son born February 20, 1828, is the Robert Boyd who stands in front of Dubois at the reception. Now sixty-four years old, Boyd is a dignified patriarch, with an aquiline face accentuated by a pair of fine white drooping mustaches and a shock of snow-white hair. He is well-read, despite or perhaps because of his isolated life on the plantation, and well-traveled within Java. His intelligence is uncompromising, but there is great kindness in his manner. Dubois likes him immediately.

Boyd owns a large upland plantation not far from Toeloeng Agoeng, where the fertile soils of Mount Willis are excellent for coffee-growing. The plantation is named Mringin, after the Javanese name for the tall banyan trees that grow there in abundance. The main house is called Ngrodjo, Javanese for "central headquarters." In 1848, at twenty, Boyd married an Indo woman, Embok Maas Warsina. When she died after almost twenty years of marriage, he was still in his prime. He remarried at forty, this time choosing as his bride a real princess from one of the old, aristocratic Javanese families in Solo, Raden Roro Samira. She was a beautiful woman, naturally elegant and refined as only a Solo princess can be. At their marriage, she took the Christian name Grace, to please Boyd. She was ten years younger than he and yet she, too, died well before him, only four years after the wedding. That was in 1872, twenty years ago, and Ngrodjo has been largely a bachelor establishment ever since. Grace's daughter, Anna Grace Penelope, survives, and the household is enlivened by Boyd's nyai and their children, but he has not married again.

The younger man is Adam Prentice, a Scotsman of thirty-two years, Dubois' own age. He is a handsome man, fair and blue-eyed like Dubois, just an inch or so taller and leaner. Prentice's broad-shouldered build and physical self-confidence betray his years of living a demanding outdoor life on plantations. It is Prentice who walks or rides the entire plantation from end to end, following the rough roads once a month. He checks the trees and inspects them for coffee leaf disease and other pests; he listens to the mandurs' problems and suggests solutions; he settles minor and major disputes among workers; he plans the week-to-week work on different parts of the plantation. He is up every day before dawn, tending to a thousand diverse duties that keep the plantation running and the crop good. He has an uncanny skill at handling the men and, as time goes on, Boyd has come to trust him with more and more responsibility.

By nature a confident man, Prentice has been shaken by a recent bereavement. Two years before meeting Dubois, Prentice was working on another plantation; there he received a promotion, which finally gave him a salary sufficient to marry. He promptly proposed to his sweetheart, Jane de Clonie MacLennan, and was accepted. The daughter of another planter in East Java, Jane was a pretty woman with an enviably fair Scottish complexion and auburn-tinged, luxuriant hair. Her good looks made her seem fragile, but she was a strong and practical young woman. She could shoot a gun as well as any of her brothers, break a horse, and plant a flower garden; she could teach a child to read, write, calculate, and dance a credible waltz; she could dress a wound or make a dress. More than once she helped with picking and sorting the beans on her father's plantation, and with keeping the books. She and Prentice loved each other dearly.

Their wedding was a gala weekend party on her father's plantation in Pasoeroean on January 16, 1890. All their friends and relations attended, including Jane's eldest and favorite brother, Theo, and his wife, Isabella. Isabella's father, Robert Boyd, was invited to the wedding too, in the open-handed colonial way. It was their first meeting, but Boyd and Prentice immediately took to each other. Before the weekend was over, Boyd had impulsively offered Prentice the job of managing Mringin, as a sort of wedding present. There would be another raise in pay, and a brand-new house for the newlyweds at Tempoersarie, at the far end of the plantation.

Prentice took to the job at Mringin as if born to it. He liked Boyd and the way he treated his workers. He was good to them, honest and firm, and they were fiercely loyal to the Tuan Boyd. The small house at Tempoersarie seemed perfect to the young couple. Jane was so proud to furnish and decorate her first home as a njonja and soon had a garden started. In August 1891, she was delighted when a visit to the doctor in Kediri confirmed that she was pregnant. She and Prentice were filled with hopes and plans for their family. They both expected an easy pregnancy and birth, for Jane was wide-hipped and healthy. She had been born at home with only a midwife in attendance, and so had all her siblings. "Easy as popping a banana out of its peel for women built like us," her mother, Doortje, always said.

Jane's labor started on the morning of May 20 and nothing went as expected. The contractions were severe, she bled and bled, and still the baby did not come out. After twenty-four hours, the midwife could do nothing more to help. The baby was still alive, she thought, but it would not come out, and Jane was weak and pale. Prentice sent a man with a swift horse for the doctor. Still the contractions continued; still the blood flowed, until Jane was barely conscious. By the time the doctor arrived, Jane was in a bad way.

"Mr. Prentice," the doctor said gravely, after examining her, "I do not think I can save your wife. She has lost too much blood. If you agree, I can try to save the child, but it will surely kill your wife."

And so it was. By the evening of the twenty-first, Prentice's world was shattered. His beloved wife, his tender, funny, wonderful wife, was dead in childbirth. The baby, a boy, survived, weakened and feeble, but breathing and crying. Prentice was so overwhelmed by the unfairness of it all, by the tragedy that had so suddenly befallen him, he did not know what to do. From somewhere, someone obtained a wet nurse, a Javanese woman whose full breasts offered plenty of milk for little Gerard. In his grief, Prentice never really knew where she came from or who found her; someone did, and she saved the child. But nothing would bring back his Jane or his hopes for the future. Later, he remembered little of the first week after Jane's death. Did he eat? Did he sleep? Did he bathe? He has no idea. All he can remember is the descent of a blackness so profound there seemed no possibility of light.

What pulls him out of his despair is the child. With good milk flowing into his rosebud mouth, Gerard strengthens and grows impossibly quickly. His laugh one day is Prentice's first new memory: a little gurgling chuckle, a sound of pure joy, that penetrates his father's grief. Here is a reason to live: an enchanting, lovely child with Jane's eyes and a hint of reddish curls. Prentice's love for Gerard is not without pain, however. The baby is a living reminder of Jane, and of her death; he is the star that shows up the empty night of Prentice's existence. Worse yet, in only weeks Prentice realizes it is impossible for him to raise the child on his own. The wet nurse is good with Gerard, very good, and after nursing is finished she will stay on as babu. But Prentice's job demands that he travel throughout the plantation for weeks on end. He cannot leave the baby alone with a wet nurse in remote Tempoersarie, with no Europeans for miles. Anything might happen. And as the child grows, who will be there to teach him and look after him? He will not have his only son raised solely by a babu, however loving and devoted. That would not be right.

It takes great courage, for the act is like cutting off his own hand, but Prentice arranges to send the child away, to Jane's father and stepmother in the East Java hill station of Malang. They will raise him until Prentice's situation changes. The wet nurse will go with him. The journey down the mountain to Malang is one of terrible duty. Prentice takes his son, the only person he has left to love, to the child's grandparents. His other task is to see to the placing of the stone on Jane's grave in Malang. There is no pleasure in either.

> *Jane de Clonie MacLennan*
> *died 21 May 1892*
> *beloved wife of Adam Prentice*
> *Blessed are the pure in heart for*
> *they shall see God.*

Malang is only half a day's travel away from Tempoersarie and is one of the prettiest towns in Java, every planter's retirement dream. There are white-pillared, gracious houses, with cool verandas and large gardens set off by neatly whitewashed walls. The avenues are long and broad, lined with trees and flowering bushes; the climate is pleasantly cool for Java. It will be a good place for the boy to grow up.

Boyd suggests that Prentice stay in Malang for a while: "Help the baby settle in, y'know." It is a kind suggestion; Boyd can see that Prentice is overwhelmed by his loss, and the plantation can survive without him for a while. Prentice cannot focus his thoughts, cannot muster enthusiasm for anything. Sometimes he feels indignant—angry—as if he has been robbed, but he cannot identify a thief to loose his anger upon. At other times he is simply empty, lifeless and colorless as a dead rice stalk. He thinks bitterly from time to time that the trouble is that he has mislaid his hopes, his dreams. Jane always used to tease him about his carelessness; he would lose his boots, his watch, his record book repeatedly. When he asked her where the missing item had gone to, she would reply with a smile, "Now think, where can you have left it?" Then they would both laugh and Jane would tell him where he had left whatever he had lost. This time he knows where he has left his dreams: in the graveyard, with her. Mostly it is brute determination and the routine of life that hold him together. And there are reasons to try, to go on: the boy, the job, the gratitude he owes Boyd, his good friend and employer.

A month or so later, Boyd comes down the mountain to attend the Resident's reception in Kediri, hoping that Prentice is sufficiently recovered to come back to Mringin with him. Though most Indies planters are hard-drinking, rough, uncivilized men, unused to citified ways, Boyd and Prentice are cast from a different mold. They are country gentlemen with a love of the land and a passion for natural history. They are both kind and clever, not a common combination, and these qualities make them welcome in any home in the colonies. The Resident's reception is a trying occasion for Prentice, one of his first ventures into society since his wife's death. He cannot make trivial conversation any longer. He grows furious that others can, that people can waste their lives in such idle chatter. It takes a profound effort of will to remain polite.

His new acquaintance, Dr. Dubois, seems different. He, too, has little use for empty pleasantries; he prefers to talk of something real, something important. He asks real questions, listens thoughtfully to their replies. Prentice does not say much at first, but there is a look in his eye that Dubois responds to. Gossip in the colonies spreads news efficiently, so Boyd and Prentice know that Dubois is a physician seconded to the Department of Education, Religion, and Industry from the Army. What they have not heard before this evening is the true reason why Dubois is in Java. Dubois tells them the story of his scientific quest; unable to disguise his passion, he will judge their worth by their reactions.

Boyd is full of suggestions of places he might search: an intelligent, knowledgeable man, Dubois thinks. He understands what Dubois means, and he immediately addresses himself to the problem.

Prentice's reaction is extraordinary for one in his circumstances. The idea of searching for an ancient, missing link so captures his imagination that he forgets his own grief and sorrow. He has so many questions to ask of the good doctor. He is stunned by the notion that there is this one thing, this one creature, that will weave together all of life into a unified whole. He is full of admiration for Dubois' vision and passion. Something in Dubois' life calls to the part of Prentice that has been quiescent, shrouded in pain and mourning. Prentice loses track of what Dubois is saying for a moment, observing with surprise his own interest. He thought that part of him had died with Jane, but he sees it is not true. There are people in the world worth knowing, worth being with, and he has just met one.

Prentice has no reason to trust Dubois with information so personal, but he does, instinctively. Almost telegraphically, Prentice shares the facts of his situation with his new friend, as if in apology.

"You will have heard, perhaps," he says in a low, intimate voice, "of my circumstances." He looks quizzically at Dubois, who shakes his head, no, and waits.

"I have recently lost my wife, Jane. She died giving birth to our only child, a son. It has left me . . ." He turns his face away, unable to find the words to describe his condition. He pauses for a painful moment before continuing. "My son will live in Malang, with my in-laws, until I can care for him myself. I . . . well, let me say I am not myself. I once had a deep interest in natural history, though, and your undertaking intrigues me. I should be pleased to learn more of it, to get to know you better. But you must forgive me if I seem from time to time a little . . . rude, distracted. I am still in mourning."

After a long moment of silence, Dubois speaks, his eyes full of sympathy. "Of course; I see. It is a hard thing," he says, lightly touching Prentice's elbow, "a hard, hard thing. I cannot imagine . . ." He falls silent and then continues, "I suppose that you must persevere. It will take much courage, but you must do it, for the boy's sake. For your own."

Prentice nods tightly, not trusting himself to speak. Dubois has spoken straight to his heart. He and Dubois turn and walk out of the inner gallery onto the front veranda. They stand there in silence, looking out at the darkness. In age, temperament, and looks, they might be brothers. Dubois has intuited more about how Prentice feels than anyone, despite the briefness of their acquaintance. His kindness and sympathy are so painful and so welcome to Prentice that he fears he will weep. As they stand there companionably, quietly, Prentice becomes aware of an easing in his chest, as if a wire snare that has bound him to sorrow has been released.

A few minutes later, Boyd joins them on the veranda. He lights a cigarette and sits with Prentice and Dubois without speaking. Outside are the night noises, the dark calls and hoots and the eerie creaking of bamboo that sounds like a soul in torment. The sweet smells of woodsmoke and spices drift through the night air. Inside are the bright, tinkling chatter of an Indies reception, the clanking of glasses, a few snippets of music here and there, falsetto laughter. Out on the veranda, they inhabit a crepuscular zone, neither light nor dark. They are suspended between worlds in a region where the mysticism of Java meets the practicality of Holland, weaving the disparate into a new whole.

Prentice takes a deep breath. "Tell us," he says, "how you came to be here, Doctor. How is it you made such a bold decision?"

Dubois answers with an account of his own past, giving more than the facts of his life this time: his promising start at Amsterdam, his need to escape Fürbringer's influence, his hope of earning his father's approval by making a great discovery.

"I left all that—my position at the university, my parents, my country—to come here, because it is here that I will find the missing link. I know it must exist, for I have read the writings of Darwin and Haeckel and Huxley, all the true men of science. I know it must be here. So I have dedicated myself to finding the form that links apes and man, that proves evolution once and for all," Dubois explains simply. "I only hope I shall not lose my life in the endeavor. The malaria has nearly killed me a few times, but not yet, not yet."

"Dreadful disease," agrees Boyd solemnly. "Even with quinine, many a man dies of it in a few days. I suppose you carry a full medical kit with you on your expeditions?" Dubois nods. "Yes, you'd need to. No telling what troubles you might encounter. We have to do a lot of doctoring on the plantation. Everything from machete wounds to fevers and childbirth: the lot."

These are dangerous topics for Prentice; they threaten to mire him once again in thoughts of death and dying. He deliberately turns away, allowing his intellect to draw him in another direction. "But what about this missing link, Doctor? What exactly do you think it will be like? I am not so learned as you in these matters, but I have read a little about evolutionary theory. When I was a young man and we went back to England on leave, people were still arguing about Darwin's book. My father used to love to tell the story of Bishop Samuel Wilberforce being bested in debate by the anatomist, Thomas Huxley. 'Tell me, sir,'" Prentice mimics, putting on resonant, pompous tones and clutching his lapels as if they were a clergyman's vestments, "'is it from your grandmother or your grandfather that you are descended from the apes?' That's the Bishop," he clarifies. "And then Huxley says righteously, 'I would rather have an ape for a grandfather, sir, than a man who uses his position to conceal the truth!'" Boyd and Dubois share a hearty laugh at his reenactment of the debate; even Prentice smiles wanly.

"Huxley said something like that, in any case," Prentice continues. "But even though that was thirty years ago, not everyone accepts the theory of evolution at home. I think it is the same in Holland? Then the man who finds the missing link will make a great contribution to science, possibly the greatest ever. He would prove Darwin's ideas true."

Dubois' handsome face flushes with pleasure at Prentice's remarks. This is an educated man, a man like him. *"Ja,"* he says warmly. *"Ja.* I am here because there is nothing—nothing—more important in science to do." There is a touch of embarrassment in his voice at admitting to the audacity of his goal.

"Then we are doubly honored to make your acquaintance, Doctor," says Boyd, with a slight bow of his head. He gestures in the air with the hand that holds his cigarette, as if sketching a bright future with its glowing end. "To meet a man of such vision, a dedicated man of science is a rare thing in the Indies."

Boyd does not know how deeply true his words are. Dubois does. He has never been like anyone else, never met anyone who truly understood him. He has always been one of a kind, an outsider, different. It is not always a comfortable role to play, but he has never played another. He does not know how.

CHAPTER 20 FRIENDSHIP

Prentice and Boyd come to see Dubois' specimens the next morning at Dubois' invitation; he can explain better what he has been finding if he can show them the fossils. Perhaps they would care to take the midday meal with his family, as well. They are delighted to accept.

They arrive promptly for mid-morning coffee, planning to stay into the afternoon. The three fall quickly into happy companionship: inspecting the fossils; listening while Dubois points out this feature and that; comparing specimens to drawings in books; looking over the maps and diagrams from the excavations. The planters are not ignorant of anatomy, for they have butchered many an animal and are keenly interested in natural history. They have never before thought about anatomy analytically, as Dubois does. He shows them how an animal's anatomy reveals its purpose and function in the world:

"Now, this antelope tooth here, you see how those teeth are ridged? Those enamel ridges cut up the grass it eats. And these, now, these are a pig: quite different, *ja*? Little bumps on the teeth, like little volcanoes. Different food: not grass. The same sort of thing works for the limbs. The antelope has long, slender limbs, for fast running; that big bump of bone there is for the attachment of a powerful muscle, to move the leg. Not so the pig. You see how different it is?"

Dubois can read an animal's bones as if they were a written record of its ancestry and habits, whether he has ever seen the beast alive or not. Even long-extinct creatures tell him their secrets. Prentice and Boyd are stunned, fascinated. Before the morning is out, they begin to acquire a little of the knowledge that lets them do as Dubois does. It is like learning to read. They understand the principle, but their vocabulary of shapes and bony bumps and depressions is still very limited. They are like children compared to Dubois. The hours pass so pleasurably that Anna has to call them three times to come to the meal before they hear her.

It is the beginning of a remarkable friendship. The relationships are not symmetrical, for Boyd is Prentice's superior and much older than the others. He is less interested in physical challenges, less likely to propose a hike to the top of the mountain than a calm cigarette in a rocking chair on the shady veranda. His curiosity is tempered by his greater maturity; he lends a helpful balance and stability to their interactions. He offers fatherly advice to the younger men when he can do so without insult, for both Prentice and Dubois are proud and he would not break their spirits. He enjoys listening to Dubois' ideas, according him an unquestioning respect for his learning that the younger man treasures. Boyd's mind has stayed flexible and he never lacks for passionate convictions. In private, the younger men call him the Old Warrior.

Prentice and Dubois are like a pair of boys playing a game, pushing each other to greater heights, faster accomplishments, bigger dreams. The immediate congeniality they felt upon meeting deepens into real intimacy within weeks. In the tropics, acquaintance progresses to friendship with a speed that would be remarkable in Europe. But here, distances are great and the infrequency of visits lends them greater import. Relationships sprout, blossom, fruit, and ripen—or spoil—in a few weeks, like exotic tropical plants.

Of the two, Prentice is the practical one, the man who can organize a workforce or build a road, who knows where the best provisions can be obtained and the fastest horses bought. Dubois is strong and bold, Prentice's match in most physical endeavors, his superior in book knowledge, his inferior in knowledge of Javanese ways. They share the same sort of independent mind and painful honesty. Dubois trusts Prentice completely, more than any man he has ever known. It is an odd friendship, for they were raised in different cultures with vastly different educations, and yet they seem to have everything that matters in common. Prentice is soon closer to Dubois than his brother Victor has ever been. Dubois knows without asking that, with Prentice, there will never be any shirking or cheating, no drifting along thoughtlessly in that infuriating way of Victor's. Dubois sees that Prentice is honest all the way through: clear as a piece of glass, deep as the ocean. His character shines out of his eyes. It is what Prentice is, not so much what he says or does, that Dubois values. He thinks it is his great good fortune to meet such a man, here of all places.

When they are together, he and Prentice, Dubois notices that something almost magical happens, something unprecedented. Dubois cannot explain it. It is as if he can think faster and more clearly when Prentice is there. As the friendship deepens, they develop a style of communicating that seems to operate as much through intuition as by words. They are united in a bond of understanding that Dubois has never before experienced.

The gift of Dubois' close friendship enables Prentice to reduce his grief to manageable proportions; he learns to live again. In a few more months, he is yearning once more to achieve and accomplish, to make a difference to the world. His passion matches Dubois' own. It does not matter that one pursues a sort of holy grail of coffee—better strains, better treatment of diseased plants, better techniques for planting and tending the trees—and the other chases the missing link. What matters is that they share the intensity of purpose, the purity of desire.

Dubois soon becomes a regular visitor at Mringin, a haven of calm and civilization in the unsettled wilderness of Java. It is for him a center of intellectual stimulation, for the genuine give-and-take that has been so lacking in his life. Half a day's ride on horseback up the steep roads and trails takes him to Ngrodjo. Boyd's house stands alone on the mountain in a clearing fringed by waringin trees and surrounded by hundreds of acres of coffee trees. The trees are equally spaced, each with its own irrigation ditch, each pruned short to facilitate picking. To Boyd and Prentice's already warm friendship, Dubois adds the spice of his genius, his own way of seeing and questioning. It is a blessing. They have not been bored—there is always something to worry about or plan for on a plantation—but they have not been challenged like this before. Dubois leads them into a world of the mind to which they have been strangers. As the weeks pass, he rides up the mountain to Ngrodjo more and more often. Anna actually comes to expect him to pack up a small bag and go whenever the fossils from Kriele and De Winter are especially interesting.

Ngrodjo was built to be the home that Boyd would share with Grace, his second wife, the Solo princess. Few of Grace's feminine touches have survived the decades since her death. Ngrodjo is a masculine place—functional, comfortable, expansive—with a natural, not a planned, beauty. It is still a family home, too, but the family itself is ever-changing. The first crop of Boyd children is gone—the three sons, William, Alexander, and Robert, working on other plantations; the first daughter, Janet, killed in a horse-back-riding accident; and the last girl, Isabella, married to Jane's brother Theo. There were only three children of Boyd's marriage to Grace: Erroll, who died before he was three months old, and the two girls, Anna Grace and another Janet. Their half-sister, Elena, the child of Boyd's nyai, is also at Ngrodjo. There always seem to be more than three girl children around, however. Big-eyed, sweet-faced, dark or fair or in-between, they are Boyd's grandchildren or the offspring of servants or workers. There are even a few children of no particular parentage who have adopted Mringin as their home. No matter; they are welcome, whoever their parents might be. Ngrodjo is the heart of Mringin, with a life and joy all its own, and the heart of Ngrodjo is her people.

As the dry season grows hotter, Dubois occasionally takes his entire family up to Ngrodjo. It is an enormous trouble, a real susa, to take all of them and the appropriate servants up the mountain. Dubois rides ahead on his horse, his luggage strapped to the saddle, while Anna, Babu, and the children follow in sedan chairs carried by coolies with a rhythmic bounce. The syce comes behind with pack animals and coolies carrying still more luggage. The steep sinuous road from Toeloeng Agoeng to Ngrodjo twists and wriggles back upon itself again and again like an eel, gaining only a few feet with each pass across the mountain. It is slow going, but the oppressive heat lifts as they gain altitude step by step. Halfway up, a delicious coolness envelops them, like a silken cloak. There is a freshness in the air that they have felt nowhere else in the Indies. The secret must reside in the trees, in the mountain itself. Perhaps the sensation is a part of the spirit of the place; they do not know or care.

At night, they revel in the luxury of needing light blankets on their beds. After dark, the leaves on the trees rustle in the breeze, the night birds call, the fruit bats flap silently on golden wings. The locusts in the forest begin to drone so loudly that the children complain they will never get to sleep, but they do, of course, almost as soon as they put their heads down. The children know they are as safe here as at home, for Mama is here, and Papa, and their faithful Babu lies on a mat across the doorway.

In the morning, the valleys are obscured by a thick, opaque cloud, as if the rest of the world, outside of Ngrodjo, has disappeared while they slept. Only the tops of a few tall trees are left behind, poking through the mist, to show where there was a forest. The gardens of Ngrodjo are indistinct and the mist turns familiar objects into fantasies—is that tree a minaret, that bush a dragon? After breakfast, the sun slowly burns off the cloud and children's laughter echoes as they explore the garden and the edges of the forest. The little ones play themselves into happy exhaustion, running with the dogs, riding ponies, chasing birds, catching geckos, watched over by their babus. On one visit, they romp with a marvelous tiger cub that Boyd has acquired as a pet, but it soon grows too large and fierce to be their playmate. Boyd keeps it, though, training the animal to pull the cart that carries the monthly payroll up the winding road through the plantation to Mringin. The tiger does not particularly enjoy the task, but its slow, soft-pawed menace puts a stop to all attempts to intercept the cart and steal the payroll. One day the tiger is found dead in its large enclosure in the garden, having eaten poisoned meat. The miscreant is never found.

Anna has no other women to talk to at Ngrodjo, but she doesn't mind. She supposes there must be a nyai somewhere about, mother to some of

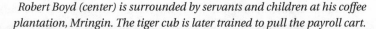

Robert Boyd (center) is surrounded by servants and children at his coffee plantation, Mringin. The tiger cub is later trained to pull the payroll cart.

those café-au-lait children, but she never catches sight of her and she never asks. What would she talk to a nyai about, anyway, if they did meet? Fashions at home? The latest magazines in the "traveling library," the box from Europe that slowly makes its way from home to home in the Indies? It seems unlikely. Anna is able to enjoy being a guest in someone else's comfortable and well-run house, with no responsibilities.

Prentice loves to see Dubois' family. Anna and the children remind him of what might have been, of the life he and Jane might have had, but it is not any longer a painful vision. He finds Anna charming, graceful, and soothing, as Jane used to be. Eugénie and Jean are nice children, but Victor is Prentice's especial favorite. He is only six months older than Gerard, whom he misses badly. Being with Victor is a sort of substitute for being with his own child. Sometimes Prentice closes his eyes and thinks: This is what my boy is like. Gerard will be as tall as this now, maybe able to throw a ball like Victor, running and laughing in the garden in Malang. . . . It eases the pain of separation.

But Dubois enjoys his stays at Ngrodjo better if he goes alone. His family distracts him from what he comes for—the companionship of Prentice and Boyd, the discussions, the exercise of brainpower. Soon Dubois comes to realize that Ngrodjo is more his home than the house in Toeloeng Agoeng; Boyd and Prentice are closer to him than his own wife and children. It is a sobering thought, but true. Anna and the children do not need him and they never challenge him. He must always move at their snaillike speed, with endless distractions; there is no getting anywhere if the children are involved. And Anna does not even really listen to him anymore when he talks about ideas or theories. Maybe she never did; he cannot be sure. But at Ngrodjo, he feels a sense of complete belonging, of fulfillment and even love. Sitting on the veranda with these two men, talking about ideas, looking out over the spectacular view and admiring the mottled bluey-green of acres and acres of coffee trees: it gives him such peace. He has never before felt so at home, so accepted for who and what he is.

The three lend one another books or articles cut from newspapers and discuss them through letters or at later meetings. How unlike Dubois' stilted friendship with Sluiter! Though his fossils are in Toeloeng Agoeng, cluttering up the front veranda, he finds an emotional and intellectual satisfaction at Ngrodjo that is unmatched. No tentative theory he formulates about the fossils, no observation, has any reality until he has discussed it thoroughly with Boyd and Prentice. These two men melt his isolation with the warmth of their friendship. In their company, he suffers no doubts about his eventual success, for *they* have no doubts. He will find the missing link— probably at Trinil, the most promising site he has found yet. If the missing link is not there, it will be somewhere, soon. Their faith in Dubois renews his courage, replenishes his determination.

CHAPTER 21 TRINIL

In mid-May 1892, it is dry enough to resume work at Trinil. The coolies remove the soft silt deposited by the river during the rainy season and then wait for the sun to dry everything out so that it is possible to start digging. Kriele and De Winter have resolved their problems and use the drying time to make a new plot of the excavation, measuring the position of the squared-off hole and its depth. Dubois reminds them to keep the best records they can of the excavation; it is crucial to recognize the different geologic layers and to separate their contents.

Word soon reaches Dubois in Toeloeng Agoeng that the coolies are again uncovering many excellent fossils. He decides to visit the site. He will stay for a while, to remind the laborers of the way they are to dig and to help the engineers establish a regular daily schedule. After two weeks, Dubois' enthusiasm for the site has vanished, and he writes in his diary:

> *July 28, 1892. A few cool windy days seduced me into setting up my tent nearby the excavation. After a stay of 14 days I discover that there is no more unsuitable place available in Java, because of health and malaria, for the study of fossils than this hell (which I, being born Roman Catholic, had mistaken for purgatory).*

It is relentlessly hot and airless at Trinil. Although the big, brown Bengawan Solo laps lazily at the base of the point bar, only a few feet below the edge of the excavation, there is no breeze. Any movement of air is blocked by the high bluffs or is dissipated by the sinuous bends of the riverbed.

Dubois is again stricken with malaria. For some days, he believes he will die there at Trinil, adding his bones to the growing pile of fossils, his Charnel House. They could just lay his body in the excavation and cover him up, he thinks miserably. He wonders idly how long it would take his skeleton to fossilize. Years? Millennia? And the cool, the lovely cool of the river . . . He wouldn't mind dying so much if he could only lie in that cool, cool river forever with all his fossils around him. He does not die, nor does he linger at Trinil long after his recovery. He packs up the fossils excavated to date and heads back to the shady house at Toeloeng Agoeng to take more quinine and await further shipments of specimens.

In August, the engineers make a discovery of tremendous significance, though it takes some weeks to reach Toeloeng Agoeng. The new specimen is a fine and nearly complete femur of a large and long-legged animal. It is found by a coolie, hoeing with unusual vigor after being told off by Kriele for laziness. The *ping* of the metal blade of his patjol as it strikes the fossilized bone is unmistakable, so the coolie stops digging immediately and notifies Kriele. Thinking it might be something important, Kriele finishes excavating the specimen himself. His energy is rewarded; when the femur is completely uncovered, he can see it is an unusual specimen. He can also see it is no longer complete—the initial patjol blow broke some fragments off the end of the bone—so he orders all the coolies to cease excavating immediately. They are instead to search every inch of ground surrounding the find-spot. With their fingers, they are to sift every handful, every crumb of sediment until they find the missing bits. It is dark when they complete the task.

After dinner, Kriele shows De Winter the femur. He agrees with Kriele that they have not had anything like this before. Maybe Dubois will be pleased with it; maybe it is his missing link. Neither knows much anatomy, but they know this is not the bone of a pig, an antelope, or a tiger. It must be something new. Together they plan how to glue the missing pieces back into place. This is a delicate task, better done in daylight than in the wavering half-light of a paraffin lamp. De Winter agrees to see to the repair job in the morning while Kriele starts the coolies digging. Unfortunately, the luck that led the coolie to the femur does not hold, as Kriele explains to Dubois in a letter.

September 7, 1892

De Winter told me that those small pieces missing from the thigh bone were blown away by a strong wind while being glued together on a djati leaf, and could never be found again.

Dubois reprimands the engineers for their carelessness. Still, he cannot be too severe with them for it is such a good find. The femur surely comes from a large, apelike primate and it was excavated from the same geologic layer as the tooth and the skullcap. The bone is too large and strongly built to come from either a fossil orang-utan or a gibbon, like those he found on Sumatra, so it must be part of the Javan chimpanzee, *Anthropopithecus,* just like the skullcap and tooth. Otherwise, Dubois thinks, laughing to himself at the folly of such a proposition, we have to believe that two such rare apes died and were preserved within yards of each other, one leaving only its head and tooth, the other only its leg.

He is anxious to go up to Mringin to show the new fossil to Prentice and Boyd. Before then, though, he must understand the bone better. Its shape is fascinating. It is amazingly similar to a human femur, a bone Dubois knows as well as he knows his own thigh. He has seen dozens, nay hundreds, of femurs in the anatomy lab. He would never have predicted that the animal of the tooth and pear-shaped, ape-browed skullcap would have such a femur. Of course, he could be mistaken about the skullcap. There is still matrix to be removed. When that is done and the chimpanzee skull arrives—if it ever arrives—perhaps he will find the skullcap less apelike and more human than he thinks. If not, the creature must have been a chimera indeed.

The most obvious difference between the fossil and any other femur Dubois has seen is a large growth of pathological bone tissue on the inside of the thigh, up toward the hip joint. The growth is the aftermath of an old, healed injury, obviously, but what injury? He cannot remember anything in the anatomical collections at Amsterdam with such a growth. The growth is very large, the size of a woman's fist, which suggests that walking was very painful for a long time after this injury. How did the creature survive long enough for the new bone to form? He cannot say. This bone is a curious record of the past.

He sketches the bone, using the exercise of drawing to make himself see in greater detail. Eventually, he will produce an exact drawing using the specialized instrument called a camera lucida, which projects an image of the specimen onto the page so that it can be traced with great precision. He has improved on the design of the camera lucida a little himself; he can turn out exquisitely accurate drawings. For now, he examines the specimen by drawing it roughly and analyzes it by taking detailed measurements and notes. He finds only a few features that strike him as more apelike than human; it is a very humanlike femur, certainly. However *Anthropopithecus* lived, it was adapted to walking upright like a man. That fact raises disturbing possibilities.

After a few days, he packs up his precious fossils in padded wooden boxes he has had made for them, straps the boxes behind his saddle, and rides up to Mringin to share his good fortune with his friends.

"Halloooo!" he calls as soon as Ngrodjo comes into view. "I've got something extraordinary."

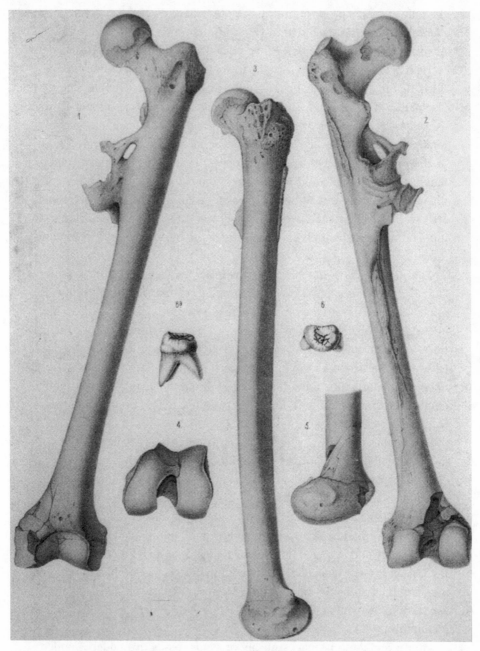

Dubois draws the upper third molar together with the left femur or thigh bone of Pithecanthropus erectus *for his monograph. The femur is shown in (1) anterior, (2) lateral, (3) posterior,
and (4) distal views; also shown is (5) the medial side of the distal or knee end. The molar is
shown in distal (6a) and occlusal (6b) views.*

His timing is good; Prentice has finished a tour of the plantation and will be staying at Ngrodjo for some few days, consulting with Boyd over the plans for picking the beans. They set aside their work immediately: the Doctor is here, with a new discovery.

"Let me show it to you," Dubois says proudly, unstrapping his boxes. He hands his wiry bay gelding to the waiting syce, who will give the horse a good brushing and see to its needs. Boyd and Prentice hastily clear the table of cups and record books, making space for Dubois to open his boxes.

"Djongas," Boyd calls, waving a cigarette-laden hand above his head, "bring cool lime juice for the Tuan Dokter, and more hot water for the tea." But Dubois cannot wait to quench his thirst before showing his treasures to his friends.

He begins immediately. "You know about the tooth," Dubois says, opening a box and taking out a small tin soap dish. He opens the tin, within which the molar rests on a bed of cotton wool like a baby in a cradle. "That was the first find, September of last year. It definitely belongs to an ape or something nearly human, I am certain of that. And then"—he moves to the next box, bigger, deeper, with more padding covered in velvet—"then there is the skullcap. We found that last October." He extracts the fragile specimen and hands it to Boyd, who holds it gently and carefully.

"Isn't it a beauty?" Boyd says admiringly, while Prentice nods in agreement. "I'd forgotten how fine it is. And you've cleaned out much more of the matrix. Do these indentations on the inside tell you anything about its brain?"

"I think so," replies Dubois, a little surprised at the Old Warrior's acumen. "They would if it were a man. But I haven't got my chimpanzee skull yet, for comparison."

"Ah, kassian," says Prentice sympathetically, clucking his tongue. "Not yet? Weber will find you one though, before long."

"And now," announces Dubois dramatically, opening the largest, longest box, "now there is *this*. They found it a few weeks ago, in August, though it didn't reach me until early September." He withdraws the long brown femur and displays it across both of his palms with a flourish. Prentice puts down the soap dish that holds the tooth and stretches out a hand to take the femur. Then he stops his movement and looks at Dubois, questioningly. "No need for great care, Prentice, no need," Dubois reassures him. "The thing is as solid as an oak tree."

"So it is," Prentices agrees, examining the curious object. Then he points at the bone's upper end. "This is the ball joint for the hip, isn't it? But Doctor, what is this here, this growth on the bone? Surely that is not normal."

"You will not guess again!" cheers Dubois, using a Dutch expression. "You are too clever. *Ja*, the creature had some sort of injury, a massive wound I think, that bled and clotted and eventually turned into this pathological bone. Very puzzling thing, that. What I can't get over is how like a man's femur it is. The ball of the hip joint is really spherical, with this pit for the attachment of ligamentum teres." He remembers himself and adds a hasty aside, gesturing. "The ligamentum teres carries the blood supply to the head of the femur." Boyd and Prentice nod their understanding, so he continues.

"The shape of that ball is just like yours or mine. The strong, straight shaft is very human too, and the knee joint. Why, the condyles, these things like the rockers on a chair at the distal end, are practically the twin of every set I've ever seen in the anatomy room!"

His friends share his evident delight in the specimen. Then Boyd asks the important question. "Doctor, tell us, is this the same thing as your skullcap? The bones look alike to me, both that same shiny brown color and very hard."

"I think so," says Dubois carefully, "I think so. The tooth is that of an ape, definitely. The skullcap has brow ridges like a large ape's but too big a braincase for an ape, and the femur is undoubtedly from a large apelike animal that walked on two legs like a man. The specimens are just too complete for me to be wrong about this. The most telling thing is that the engineers say they all came from the same geologic stratum, within maybe twelve or fifteen meters of one another. I believe they must come from one individual. It is terribly important, that."

They fall silent, awed at the wonderful thing that has befallen one of them. Three bones—the head, the tooth, and the leg—are from one animal, a completely unknown apelike beast with puzzlingly manlike features. What a triumph! Over the next few days, they examine the fossils again and again, Dubois pointing out anatomical details, Boyd and Prentice acting as sounding boards for his ideas and raising new possibilities. They cannot offer Dubois scientific expertise in anatomy, for he far surpasses them in that regard. What they can and do offer him is their devoted attention and interest, their validation of his discovery.

"Doctor," asks Prentice quietly the next afternoon, when they have run out of questions and observations to share, "is this *it?* Is this your missing link? Have you found it at last?"

Dubois doesn't answer right away. He doesn't know what to say to this crucial question, asked by his dearest friend. "I don't know, Prentice," Dubois replies hesitantly. "I just don't know yet. I must be sure before I say a thing like that. I must be *certain*. But this femur is a big surprise to me, and the size of that braincase . . ." His voice trails off. He starts again in a moment, trying to express his thoughts, "And then the brow ridges . . . You see, I can't help thinking that there is a lot of ape in that skull, and a lot of man in that femur. Doesn't that sound like a transitional species, a link between apes and man?"

He looks at Prentice with his pale blue eyes and Prentice looks back at him, believing in him utterly. Dubois doesn't yet dare to declare that this is the missing link, that his long search is over. But they both can see the idea dancing at the edge of their consciousness, like a shy woodland creature that hesitates at the edge of a clearing.

CHAPTER 22 THE BIRTH OF *PITHECANTHROPUS*

The rains start again before the workers can find any more primate fossils at Trinil, and Dubois closes down the excavation for the season. What a pity, but there it is: the river is rising. Whatever fossils hide in the sediments have waited a long time and will have to wait for next year.

In November, he is elated to hear that Weber has procured a chimpanzee skull for him in Berlin, only the month before. The skull should be arriving in Java soon. Dubois takes to meeting the mail train from Batavia almost every day. The rest of the time he works on framing the sentences of his third quarterly report of 1892, which will be sent to his supporters: Groeneveldt at the Department of Education, Religion, and Industry, Jentink in Holland, and Kroesen in Sumatra.

> *The most important find of the month of August is the left femur of* Anthropopithecus, *of which the existence was shown a year ago by a molar and a skullcap. This femur lay in the same geological level as the other two fossils, although it lay upstream from the others by about 15 meters along the course of the ancient river that had deposited the volcanic material. From the conditions under which they were found and from comparative research it appears that the three skeletal elements belonged to the same individual, probably a female, and were very ancient.*

The next part of his report gives the interpretation and significance of the find. He writes and rewrites this section many times before submitting it. He would not deny the implications of the femur's shape, nor does he wish to overstate them.

> *This being was in no way equipped to climb trees in the manner of the chimpanzee, the gorilla, and the orang-utan. On the contrary, it is obvious from the entire construction of the femur that this bone fulfilled the same mechanical role as in the human body. Taking this view of the thigh bone, one can say with absolute certainty that* Anthropopithecus *of Java stood upright and moved like a human.*

Then he estimates the cranial capacity of the skullcap, following a key publication by Theodor Bischoff that he has recently obtained. Bischoff documents the mathematical relationship between various linear measurements of a chimpanzee skull and the volume of its braincase. If the relationship holds true for this creature, then—while Bischoff's ape has a brain volume of 410 cc—the Trinil skullcap has an impressive 700 cc of brain in its skull. This volume is still much smaller than a human's, about 1,250 cc in Europeans, less in other races usually thought to be inferior.

Dubois' concluding paragraph is bold. The femur proves he has a new species of ancient ape, not Lydekker's *Anthropopithecus sivalensis* but something of his own, an upright ape.

> *Because of this find, a surprising and important fact has been brought to light. The Javanese* Anthropopithecus, *which in its skull is more human than any other known anthropoid ape, already had an upright, erect posture, which has always been considered to be the exclusive privilege of humans. Thus this ancient Pleistocene ape from our island is <u>the first known transitional form</u> linking Man more closely with his next of kin among the mammals.*
> Anthropopithecus erectus *Eug. Dubois, through each of its known skeletal elements, more closely approaches the human condition than any other anthropoid ape, especially in the femur—a fact that is totally in accord with what Lamarck proclaimed and which was explained later by Darwin and others: that the first step on the road to becoming human taken by our ancestors was acquiring upright posture. Consequently, the factual evidence is now in hand that, as some have already suspected, the East Indies was the cradle of mankind.*

Dubois copies the report in a good hand and submits it to Groeneveldt on November 27.

Within a week, he realizes he has made a terrible mistake: he has misunderstood how Bischoff measured his chimpanzee skull. If Dubois' measurements are taken differently, then his estimate of brain size is bound to be wrong. The very idea makes his heart pound and his head ache. How could he have been so careless? He was in too much of a hurry, too impatient. It is his worst fault. Hands shaking, he measures and remeasures the fossil skullcap several times, confirming his initial error. Each time he calculates the size of the braincase anew, using the corrected measurements—and its volume is even greater than he initially estimated. In fact, it is enormous. After the fourth repetition, Dubois accepts that the skullcap has a cranial capacity of nearly 1,000 cc.

This is an amazing value, more than twice the size of Bischoff's chimpanzee. No ape's brain is that large, he thinks; it is not possible. This Javan skull is comparable in brain size to some human races, like the Andaman Islanders or the Australian Aborigines. It is an astounding discovery. Can it be true? Dubois measures yet again, repeats his calculations, arrives at the same figure. It is true.

He writes to Groenveldt immediately, asking that the passage be corrected when the report is published. There is no way to disguise his stupid mistake. The key thing is to find out the truth and make it known.

December 4, 1892

In calculating the relative volume of the brain of Anthropopithecus erectus, *the results being mentioned in the report for the third quarter recently submitted, I made a regrettable mistake. The brain of this transitional form was considerably larger than one would gather from the report . . . nearly 1,000 cc.*

This fact tips the balance of the creature from ape to . . . almost human. This must be something very like the missing link: an upright-walking ape with a brain as big as some humans'. The thought that his search is over makes Dubois' chest feel too small to contain his heart. His thoughts whirl in his skull like dust devils, fast, faster, so much to do . . . Both elated and apprehensive, he works feverishly on the formal description of these new fossils. He scribbles a quick note and sends it up to Ngrodjo with a boy, to let Prentice and Boyd know what is happening: "I have made a terrible mistake. Skullcap 1,000 cc, not 700. You see what this means. Please excuse my temporary absence. I will visit and explain fully as soon as possible."

And then, two weeks later—on December 18—the chimpanzee skull finally arrives from Weber. "Anna, it has come, it has come!" Dubois calls joyfully as he bounds up the steps of the house. "Here it is at last!" Anna and the children and most of the servants come running. It has been more than a year since Dubois first wrote to Weber with his request. The entire household has been tensely awaiting the arrival of the precious skull, which for some strange reason the Tuan Dokter needs so badly. He opens the parcel carefully and unwraps the skull. "Look, Eugénie," he says, bending down to show the prize to his daughter, "look. This is the skull of a chimpanzee all the way from Africa!"

"Is it very 'portant, Father?" asks Eugénie dubiously, poking at the skull with a small, chubby finger. At five, she understands little of what this means; Jean, at four, still less; and Victor, cuddled in Babu's arms, senses only the happy tone of his father's voice.

"*Ja,*" says her father contentedly, turning the skull over in his hands. He pulls his magnifying glass from his pocket to stare at a minute anatomical detail. "*Ja,* Eugénie. With this, I can understand my fossil skullcap from Trinil. With this, I can learn how close to the apes and how close to man he was. This is very important."

Eugénie nods and smiles at her tall, handsome father. She adores him and often wins his praise with clever questions.

"That's good, Father," offers little Jean seriously, pretending that he understands. Anna hugs the children and Babu hustles them into the garden while Dubois heads for his study. They are not to disturb their father; he has scientific work to do now. But even the little ones know that the household will be a happy place for days to come. Dubois settles down to write a note of thanks to Weber.

December 19, 1892
I received your letter of October 21st, saying that you had finally found me a chimpanzee skull, on the 28th of November. Yesterday the postal parcel, sent at the same time as your letter in October, finally arrived.
Especially the <u>chimpanzee skull</u> I have awaited with desire and on the tip-toe of expectation.

With the chimp skull and the gibbon and human skulls he collected previously, Dubois can make his own observations and be sure of them. He compares the gibbon to the human, the gibbon to the chimp, the chimp to the human, and all three species to the fossil. The anatomy of the Trinil skullcap is strikingly similar to that of the chimpanzee skull except that the fossil is so much larger-brained. The sheer brain size makes the fossil resemble the human skull in shape, too, despite those apelike brow ridges. And then there is the femur from Trinil to be considered, so like a human thigh bone, so unlike a chimpanzee's.

Dubois works in a frenzy over the next few days. He takes his meals on a tray in the study and barely sees his wife, his children, or even the light of day. All he can see is his skulls. He starts work at first light and stops at night only when Anna comes in and insists that he needs his rest. There is so much to do, to learn; it is so imperative to get it right.

Christmas morning, Dubois arises very early. He carries his cup of coffee into the study in the pearly light of dawn; it is a special time of day, when he feels like the only person awake in the world. There the fossils sit, solid and brown and hard as rocks, on his desk, waiting for him.

"Good morning," he says to them fancifully, bowing at the waist, "and Merry Christmas to you." The modern skulls, yellowish white and a little greasy to the touch, sit patiently on another table. He nods politely in their direction and sits down at his desk. For the next few hours, he checks his measurements again and reads through his pages of notes and comments. He reviews his sketches of the fossil skullcap, comparing them with sketches of the modern skulls, blocking with his hand those parts that are broken away on the fossil. Right lateral view; posterior view; left lateral view; anterior view; superior view; inferior view: yes, the observations are all correct. Then he proceeds to the tooth and the femur. Point by point, he compares the anatomy of the fossils with modern teeth and femurs once again. His mind is absolutely calm. The conclusion is clear. When he retired last night, he was convinced. This morning, he has corroborated everything once again. Now he is completely certain.

He packs his specimens away carefully, each on its own cushioned velvet bed within its own wooden box, fastening each lid securely with a small brass clasp. He places each box carefully on the shelf where it belongs and puts away the modern skulls, too. Then he tidies his desk, putting away books and organizing his papers and notebooks until all is straight.

He walks out of the study, closing the door quietly behind him, and finds Anna eating breakfast on the back veranda. "It is done, Anna," he says, contented but as weary as if he has been up all night. He seats himself at the table opposite her. "I have found it. Those bones are the missing link. It is true, at last." He smiles a slow, wise smile, not a boy's enthusiastic grin. He is in this moment a man who knows exactly where he stands.

Anna is uncertain; his mood is an enigma to her. "You do not sound very pleased," she offers, hesitantly. "Are you sure?"

"Oh, *ja, ja,*" Dubois says with a nod, helping himself to tea and toast. The djongas has seen him come in and will bring his boiled egg and fruit shortly. "It is certain. I shall spend today with you and the children. Tomorrow I will finalize my report. But I have it, Anna, I have found the missing link. Everyone will see now; everyone will understand I am not just a crazy man who ran off to the Indies in search of an idea."

Anna stands up and comes around the table to embrace him. "I am very proud of you, Eugène," she says fondly. "I knew you would do this. I never doubted it. What a Christmas gift to us all!" It is a lovely morning, she thinks, her eyes shining with pride.

Dubois has barely finished his breakfast when they hear the children awakening. "Mama! Papa!" little voices call excitedly. "Are you up? It is Christmas! Come on, Babu, it is Christmas! Don't be so slow!" The holiday is upon them.

The next day, Dubois returns to concentrated work on the fossils. That day, and all the next, working late into the night, he organizes and polishes his facts, getting the report ready to send to Groeneveldt. By the morning of the twenty-eighth, he is well satisfied with what he has done. He does not expect to make further revisions now. In the very act of writing the covering letter to Groeneveldt, he has an epiphany. He intended to keep the name *Anthropopithecus erectus* with which he endowed the fossils in his third quarterly report, written but a month earlier. But as he pens the letter "A," for *"Anthropopithecus,"* the truth of the matter overwhelms him. It is wrong. He lifts his pen from the paper. No, this is an injustice, he thinks. These fossils are not Lydekker's chimpanzee from the Siwaliks, transported to Java; they are not a chimpanzee at all. They are something entirely new. They are the transitional form, the missing link that joins apes to man, and he knows it. He cannot call the fossils *Anthropopithecus* any longer.

There is only one name that can be given to this creature. He superimposes a "P" over the "A" and writes *"Pithecanthropus,"* Ernst Haeckel's name for the hypothetical missing link. That name will make his position clear to any man of science. Now, what species name should he attach to the genus *Pithecanthropus*? He cannot use Haeckel's proposed name, *alalus*—"speechless"—without some evidence that it is correct. Who can tell if this creature could talk, from what is known of it? There is no face, no jaw, and even the inside of the braincase is not yet entirely cleaned of matrix. It would be wrong to name this fossil based on information he does not have.

He will call it *erectus, Pithecanthropus erectus,* a name that will honor both the marvelously manlike femur and the capacious but strangely apelike skull. His missing link will be known as the ape-man who walks erect, *Pithecanthropus erectus.* The name has a good ring to it. And to himself, and a few friends, the creature shall be known by its initials, *P.e.*

He writes to Groeneveldt:

> *December 28, 1892*
>
> *I have, Your Excellency, the honor of offering the first installment of the description of some of the fossils I have collected. This installment deals with only one species,* Pithecanthropus erectus.

It is five years, two weeks, and three days after Dubois' arrival in the Indies. He has abandoned his job in Amsterdam and sailed halfway around the world; he has kept his wife and daughter safe, and fathered two sons and brought them into the world; he has searched two islands for fossils; he has survived three terrible bouts of malaria, tigers' lairs, and cave-ins; he has begged and argued and persuaded until the government itself has provided help; and he has collected and cleaned and identified thousands of mammalian fossils. And he, Marie Eugène François Thomas Dubois, has found the missing link.

CHAPTER 23 1893

Dubois longs to visit Boyd and Prentice at Ngrodjo in the new year as soon as possible, to go over every scrap of evidence with them. Before he can, Prentice stops in Toeloeng Agoeng unexpectedly on his way back from Malang. It has been a painful Christmas for Prentice, the first since Jane's death. His boy is well, thriving, but Prentice is able to spend little time with him. The boy instinctively turns to his grandfather, not to his father. That fact burns in Prentice's memory like a flame.

And so Prentice comes to Toeloeng Agoeng seeking comfort and companionship in the home of his best friend. How he envies Dubois the comfortable, familial air of his household! He treasures the echo of Eugénie's and Jean's voices as they play boisterous games in the garden, the sound of Anna's clear soprano singing lullabies to little Victor. And he sees the warmth of a woman's touch about the house: the flowers, the pretty tablecloth, the homey meals, the silver-framed photographs of Holland—how he misses all that! Anna's beauty and femininity round the sharp edges of reality, soften the harshness of life in the Indies. Prentice badly misses having a wife.

Talking with Dubois about the fossils and their new name takes Prentice's mind off his losses. How fine *P.e.* is! Dubois' open enthusiasm is contagious. Prentice is proud to be among the first to know what this extraordinary man has done. Only much later does Prentice suspect that his friend has deliberately spent extra time explaining and displaying the fossils in order to lighten Prentice's melancholy mood. He is grateful that the doctor understands him so well and so kindly. And the doctor's medicine works; his shared joy brings Prentice back to the here and now of life in Java, drawing him away from the sad might-have-beens. He is able to return to Mringin with renewed hope for the future. Dubois has succeeded against tremendous odds; so will Prentice. He may be able to start up a plantation of his own before too long. Then he can make a real home and a life for his son. Hard work and perseverance and strength are what it takes.

Before Prentice leaves, Dubois asks him for advice about a problem he has just come to apprehend. Dubois has found the missing link. What is he now to do? It is Dubois' responsibility to write a monograph, to make the specimen known to the scientific community in Europe. Yet he is hampered by the paucity of comparative material. He has no osteological collection, only three gibbon skulls, a few humans, and a chimpanzee skull at his disposal—and that last required heroic efforts on Weber's part. Compounding the difficulty is the lack of a good, up-to-date scientific library; not even in Batavia is there such a thing.

Can Dubois write a detailed scientific analysis of the fossils that will stand for all time under such conditions? Should he try? To turn out a work that is not thorough or up-to-date would be tragic, and his circumstances are certainly difficult. On the other hand, if he waits until his return to Holland to write, then publication will be sorely delayed. Dubois is obligated by the terms of his enlistment to stay in Java until late in 1895, almost three years hence.

Dubois and Prentice play devil's advocate with each other, marshaling the arguments for and against each course of action, switching roles, exploring all alternatives. Nothing is decided. Later, when Dubois travels up to Ngrodjo, the two of them review it all for Boyd.

"Doctor," Boyd asks carefully, feeling his way toward the crux of the problem, "what is it you have to do in this monograph? What is *in* books like this one?"

Dubois is taken aback; he has not considered this practical but central issue. "Well, Boyd," he begins, "you know, no one has really ever written one before, not like this. Fraipont and Lohest wrote a monograph on the Spy Neanderthals, that's the closest thing. Let's see. They talk about the site a bit; they describe the fossils, and then they compare the specimens to the various races of man to see which are the descendants of Neanderthals. But that's not what I'd have to do. *Ja,* I must describe the fossils, give all the measurements and figures and so on, and I have to explain what they mean. But the proper comparisons here are with apes, not with living humans. The question is not 'Which race is closest to *P.e.*?' but 'Where in the evolutionary transition from ape to man does *P.e.* lie?'"

"Would you have to discuss many other fossil specimens to answer that question?" queries Boyd. "Do you have to go study fossils collected from other parts of the world to write your monograph?"

"No," says Dubois, breaking into a broad grin as he realizes the truth, "I don't, because there isn't anything else like this. Even Lydekker's *Anthropopithecus* in India: I certainly ought to go see it, but it is only a piece of upper jaw. How can I compare a fragment of upper jaw with my skullcap or my femur?"

"You ought to stand on your own two feet, then," says Boyd forcefully, swallowing the last of his drink and thumping the empty glass on the table. "Use your own experience and good judgment. You've got some skulls to compare it with; you know what femurs are like and could lay your hands on a few without much trouble. You've got a pretty good little library there in Toeloeng Agoeng in your own study. I don't see any real reason to wait. And there's every reason not to delay," Boyd concludes sensibly.

"Good point," agrees Prentice, nodding. "The doctor has no road to follow. He is cutting his own path into new territory. He has something entirely new. *P.e.* lies between apes and man; it's not just some primitive sort of human. If I can see that with what the doctor has, then so can the men of science in Europe. Doctor, all you need to do is lay out the evidence of what you've found, honest and straight, without any fancy comparisons."

Boyd raises another point in favor of writing now. "It's not very likely," he admits, his voice deepening as the subject grows more serious, "but delay might be a fatal mistake. What if some other fellow has read your article on the promise of the Indies and comes and finds a fossil for himself? Now that there's a committee in Holland to promote scientific research in the Indies, someone else could come looking—and finding."

Dubois pales at his friend's words. He has not considered such a possibility, at least not since he found his own fossils, but of course the danger is there. Martin the geologist, for one, is terribly interested in the Indies; so is Verbeek. They would probably have an easy time gaining support from the committee for an expedition now that Dubois has pointed the way. What if they found something like *P.e.* and published first?

"No, Dubois," says Prentice firmly, "you must do it now. It is your duty and your right. You have done all the work and you have succeeded with so little help. Now you deserve the glory."

"Yes," concurs Boyd. "You owe it to science and to yourself to publish as soon as you can. Don't hurry now, you don't want to 'marry in haste and repent at leisure,' as the proverb says. But you must start now and do all that you can from here."

"You both advise that course of action?" asks Dubois, looking at each man in turn. Prentice nods; Boyd does too. "Then I shall do it. I think you are right. And since this monograph is a new thing, I shall do it as I like. After all, I have carried out the entire project to my own design from the beginning."

The next morning, Dubois rides back down the mountain to Toeloeng Agoeng and his fossils. It is a new year and a new phase in his life. He is now the man who found the missing link. He has only to tell the world.

The task of writing the monograph proves far from simple. Replicating someone else's pattern of organization would be far easier than inventing one. He struggles to order his observations and to describe the finds clearly and simply. He must decide not only what needs to be illustrated but also how to convey the fossils' features accurately. Meanwhile, Kriele and De Winter and the laborers are still at work in the mountains. The endless litany of complaints about the coolies continues. The engineers have offered a small monetary reward for good specimens as a means of curbing the coolies' tendency to throw away or hide fossils. Boxes of fossils continue to arrive at Toeloeng Agoeng, but the offered reward goes unclaimed.

In February, something extraordinary occurs. The first Dubois knows of it is when the Assistant Resident drives up to his house in Toeloeng Agoeng unexpectedly in a delman. "Have you seen this yet, Dubois?" the man asks, waving a copy of the *Bataviaasch Nieuwsblad* as he picks his way up the bone-laden steps to the veranda.

The *Bataviaasch Nieuwsblad* is a new type of newspaper, an invention that can be laid squarely at the feet of its editor and publisher, P. A. Daum. It is an opposition press, a paper that is regularly and outspokenly critical of the government's colonial policies. Daum's sarcastic "tropical style" is well-known and has been the cause of his being fired from every paper he has ever worked for, save those he owns himself. Still, Daum's great ability as a writer makes him a force to be reckoned with in the Indies, and he says what no one else dares to. To boost sales and fill pages, Daum has begun to serialize scandalous novels about the Indies. They are enormously popular, for the characters are often based on recognizable colonial figures and revolve around situations closely resembling life. The novels are innovative, too, for they are written in a simple, conversational style—parlando, it is called—very different from the formal language of the European novel.

The little-known truth is that the novels, like almost everything else in the paper, from the editorials to the news reports and the advertisements, are by Daum's own hand. He disguises his role behind a constantly changing array of pseudonyms, but the truth is that few other journalists will write for him. The issue of the paper that the Assistant Resident brings to Dubois' door is dated Monday, February 6, 1893.

"Come sit in a comfortable chair, in the shade," says Dubois genially to the Assistant Resident, indicating the only chair on the veranda that can be reached without stepping over fossils. "Can the boy get you a cup of tea, or a

Dubois' fossils cover the shaded veranda and soon march down the steps to the garden.

cold juice perhaps?" Dubois beckons for the djongas. Once the official is settled comfortably, Dubois returns to the business at hand. "Now what is it you think I should read?"

"It is this, on the third page," the Assistant Resident replies, handing over the newspaper.

IN PURSUANCE OF PALAEONTOLOGICAL
INVESTIGATIONS OF JAVA
(Report to the Mining Works, third quarter 1892)

It is well known that etymologists sometimes explain the origin of words through most surprising derivations and on such occasions produce results that are astounding to the uninitiated, but if we compare these attempts with what some paleontologists dare to do with reference to fossils, then their endeavor appears to be child's play.

This flashed across my mind as I read the article mentioned at the head of this piece.

For some time, the military surgeon Eug. Dubois busies himself with paleontological investigations and, because of his predilection for such studies, works with exceptional diligence and assiduity. He is totally absorbed by his work, so to speak. As a firm Darwinist, he dreams of making a discovery which the great master of evolution will greet with joy. Namely, the discovery of the until recently missing link between the animal world and man.

Should this be taken amiss? I do not believe so. At present Darwinism is the backbone of the education of most high school graduates. The heavy facts that are brought up against Darwin's theory by the most competent authorities—these leave them cold. Examine their libraries and, ten to one, you will not find a single paper in which Darwin's theory is opposed. It is old-fashioned to think differently, their teachers have told them.

I fear, however, that this time the Darwinian outlook of the esteemed Mr. Dubois has played a trick on him, a danger that an impartial observer would have escaped.

What then is the case?

In the Kediri region, in volcanic beds, he found some time ago a skullcap, not an undamaged skull, but a fragment, along with a loose molar.

In August 1892 he found, 15 meters from the place where the
skullcap and tooth had been lying, a thighbone with a striking
resemblance in measurements and shape to the bone of the leg that
supports humans.

Instead of thinking of the remains of a human skeleton (Dubois
had already demonstrated the presence of a primitive race here in
Java, more like the Papuan race than the present population, in a
previous report) he makes comparisons (the details of which are not
known yet) with a more surprising result than the most abstruse
etymologist has ever put forward!

Given: a molar and one skull fragment found together as well as
a left femur, that has been found 15 meters further away.

Until now, Dubois has been skimming the article but he stops when he
reaches these words, and reads them aloud. "A 'skull fragment'?" he repeats,
outraged. "A *fragment*? I have the better part of a skull, lacking only the face.
I have as much of the skull as the original Neanderthal fossil, more than
Lydekker's *Anthropopithecus*. This is no fragment! I have shown it to you,
Assistant Resident, haven't I?"

"*Ja, ja,* two times," the official replies a little anxiously. He fears Dubois
will insist on showing him all of the thousands of fossils crowded onto the
veranda.

"This author has set out to make me sound like a fool! What else does he
say? Here, *ja,* here is the place." Dubois reads on aloud.

Question: What does this mean?

A non-Darwinist would scratch himself through his fur before he
would propose a genetic link between the monkey skull and the
monkey molar and the femur, which has a close speaking acquain-
tance with a human femur.

Not so the esteemed Mr. Dubois.

"His sarcasm is really too much," protests Dubois, and the Assistant
Resident nods in agreement. "'The *esteemed* Mr. Dubois'! What a back-
handed compliment. Now, what else does this scoundrel say?"

With the data mentioned above, he thinks that the theories of
Darwin and Lamarck have been confirmed, that the first step on the
road of the incarnation of our ancestors has been acquired with
erect posture (page 11);

That molar, skull fragment, and left thigh bone were once part of an upright female apeman, named Anthropopithecus erectus *Eug. Dubois (page 11).*

That this creature fed differently from the present-day anthropomorphic apes, despising tree-climbing and even carrying artificial weapons (page 14).

Finally, according to the esteemed Mr. Dubois, this is the actual evidence that the Indies were the cradle of the human genus (page 14).

Whew, that's it!

Why is this animal given such a beautiful name? Why wasn't he named Hanoman (*) Communis *of the family Hanomanaceae . . . ? The historian would have gained something from the find, but* Anthropopithecus erectus!

No, I am afraid that the esteemed Mr. Dubois, prejudiced because he has completely swallowed Darwinism, has gone too far, and has constructed a connection between the human femur and the monkey skull and molar where none has ever existed. If humans also lived, at the time when the cataclysmic eruption buried so many dead animals in volcanic tuff, no one would be surprised that humans also would be killed and their skeletons or remains of severed body parts would be preserved in the tuff.

In the meantime, this publication of Dr. Dubois will create a furor, especially in the "Land of Intellectuals," and it appears to me that the facts must be reviewed by an impartial committee of experts, before the government should endorse such a report.

"Oh, the man is sly, very sly indeed," Dubois asserts. "And who is he, anyway?" Dubois glances down at the bottom of the page, to the signature, and reads it, too, aloud.

Signed,
Homo Erectus
Batavia 3 February 1893

(*) Hanoman is a white monkey, according to the Javanese mythology of Dhewi Handjani, that is also called Dhoyopati or Boyosoeto.

Note of the editorial staff.

"Ha!" exclaims Dubois, stabbing at the page with his index finger. "Ha! Homo Erectus! That's a good joke. So this Homo Erectus thinks my fossil species is nothing but a monkey, does he? Nearly one thousand cc in its brain and it is a monkey, like Hanoman? Oh, *ja, ja,* of course. Why should the Amsterdam anatomist know anything about anatomy?" Dubois laughs ironically and then continues. "And of course I—who studied anatomy for years, who picked the spot for the excavation, who has examined every find from Sumatra and Java—I am too foolish to understand that these three pieces are from different animals. Somehow, this Homo Erectus thinks that three different primates—all the same size, mind you—have died and been buried within a few meters of each other at Trinil. When there is only a handful of fossil primates in all of Asia, I have the good fortune to find a site with three species, one of which has left only its head, one its leg, and one its tooth." He places his hand on his chest, dramatically to announce, "I am a lucky man!" Then he turns to his visitor. "Really, Assistant Resident, I don't know whether to laugh at this fellow or be outraged. What does he know about my fossils? He hasn't looked at them, not for five minutes, that much is sure. Is *he* an anatomist? I wasn't aware of there being another in Java. He can't be a trained man of science or he wouldn't be so skeptical about Darwinism." Unconsciously, Dubois rests his hand on one of his specimens and strokes it gently, comfortingly, as if it were a beloved pet.

"And this person, whoever he is, is already out of date: he's read my third quarterly report, but not the fourth. I wonder what he'd make of my calling the beast *Pithecanthropus erectus* to show that it is Haeckel's missing link. I daresay he won't like that one bit."

"Funny thing, *ja?*" asks the Assistant Resident, a little relieved by Dubois' relatively moderate reaction. He had feared a tidal wave of outrage, a demand for prosecution for libel or some such. "He's awfully scathing for not knowing anything about the subject. I expect he's one of these opinionated colonial types, thinks he knows everything because he's got a big plantation or a prosperous business. Can't say I think much of a man who hides behind an assumed name, though, d'you? It's not the behavior of a gentleman."

"Exactly!" agrees Dubois. "If he's got an opinion about my bones, why doesn't he say so and come here and look at them? You know me, Assistant Resident. Have I ever denied anyone a look at my fossils who wanted to see them?"

"Certainly not," replies the Assistant Resident. He is not the only person in East Java who has been shown the fossils at length when they hadn't the least interest in them. "It wouldn't be fitting behavior for a scientist to hide the fossils away, and we all know, Doctor, that you're a scientist all the way through." He pauses and then adds another point. "You know, this *Bataviaasch Nieuwsblad* is a bit of a rag, not a respectable paper. Everyone reads it, but that editor, Daum, he's always going off on some tirade or other. Why, he's already been thrown in jail once for publicly criticizing the government, and he started this paper up as soon as he got out! He's just the sort of fellow who'd publish a scurrilous, anonymous letter like this." His face shows a good deal of righteous indignation.

Dubois appreciates the display of loyalty. Suddenly his face changes and he breaks into a mischievous grin as a new thought crosses his mind. "Ah, Assistant Resident, this is good news after all. He has drawn attention to my find, made it important. One thing is for sure: my *Pithecanthropus erectus* won't be born into the world unnoticed after this!" The two men burst into good-natured laughter.

"Will you answer him in the next issue, Doctor, show him up?" asks the Assistant Resident, curious.

Dubois considers. "No, I don't think so. I've got a lot of scientific work to do, writing my monograph. There is no point in conversing with idiots who have no interest in the facts." He decides to ignore Homo Erectus' absurd essay and get on with his writing.

Dubois works his way methodically through the material as best he can, relying on his personal library, his notes of the published literature, and his own thorough knowledge of anatomy. How hard is this, compared with finding the fossils in the first place? Dubois puts in an official request to be sent to British India to examine Lydekker's *Anthropopithecus* specimen, but it is not granted. Apparently the Army feels it has done enough for him already.

Before excavation can start up again, *P.e.* is once again thrust into the spotlight. The *Tijdschrift van het Koninklijk Nederlandsch Aardrijkskundig Genootschap* (Journal of the Royal Dutch Geographical Society) reprints Dubois' third quarterly report in its entirety, including the name *Anthropopithecus erectus*. That is perhaps agreeable, but the action of the Society's secretary, J. A. C. A. Timmerman, is not, for Timmerman adds a note that Dubois' conclusion that the Indies were the cradle of the human race has been reached "rather hastily."

Dubois is confounded and insulted. What does Timmerman know of Dubois' labors and expertise? Does he know that Dubois sacrificed a fat professorship in Amsterdam to come to the Indies to find the missing link? Does he know how Dubois has labored, fighting against uninformed opinion, skeptical colleagues and family, reluctant coolies, difficult geography, and disease itself to find the missing link? Dubois has done all that; Timmerman has done nothing. Does he know that Dubois has been looking for the missing link since 1887, nearly six years? "Hasty"? How can Timmerman call him hasty? Timmerman sits in his comfortable chair and reads one tiny report about the fossils, and he has the audacity to call Dubois hasty.

The bitter taste of these attacks lingers, making Dubois uneasy and unsettled. His mood fluctuates from optimism to despair and back again. Sometimes he has high hopes for 1893: Anna is with child again; his monograph is proceeding well; there may still be more bones from Trinil. Other times the entire endeavor seems useless, if ignorant men can discount his conclusions before they have even seen his evidence. Prentice and Boyd are of enormous help to him. Their view is that these criticisms will evaporate as soon as Dubois presents the facts, so he must ignore them and get on with the business of writing up his fossils. Dubois leans heavily on their friendship and good sense.

The illustrations are of paramount importance, for they show the anatomy of the fossils. Dubois works carefully with the camera lucida and experiments endlessly with photographing the fossils, too. When his monograph is published, he hopes, his fossils will be so exquisitely documented that no one will be able to doubt his conclusions. The words must be correct too: systematic, well-organized, clear. Under the strain of all of this work, Dubois sometimes loses heart and becomes short-tempered, especially when the children are noisy and boisterous. And there is soon to be a fourth! He cannot imagine how the chaos of four children will ever be subdued. Babu simply must learn to keep them quiet and happy and out from underfoot.

Dubois becomes obsessed with the idea that the monograph itself is as important as the fossils. This will be the official announcement of his *Pithecanthropus erectus;* this will be the publication that justifies all the risks and hardships he has endured; this will establish his reputation as a top man of science. The monograph has to be of the highest scientific quality, or Dubois will fail. To have found the missing link and still to be disbelieved—ah, that would be a painful punishment indeed!

CHAPTER 24 DISASTER

One of the thoughts that sustains Dubois through the first half of 1893 is the hope that Trinil will yield even more fossils of *P.e.* Kriele and De Winter start the coolies excavating there again in April. There is a renewed urgency about the work, now that success is so close, but no more fossils of *P.e.* are found.

At the beginning of May, Dubois' world is shattered like a fossil struck by a patjol. He receives a letter from his mother telling him that his father died, on April 11, at the age of fifty-nine. There was no warning, no long illness, only unexpected finality. Nothing in Dubois' life has ever been so terrible as this blow—not the disappointments, the rejections, the fevers, the critics, the empty caves, the coolies' sabotage, the loneliness. Dubois never imagined that his father might die before he could return home in triumph. It is impossible, unfair, untrue . . . no, it is true. Jean Joseph Balthasar Dubois will never see his son's great gamble pay off. He will never see the name of Marie Eugène François Thomas Dubois become famous in scientific circles throughout Europe. And he will never see Jean or Victor or this new baby, Dubois' fourth, the one that grows and waits to be born.

It is too cruel a fate. Dubois' devotion to science separated him from his father. Now, as the moment of reconciliation has come within reach, his father has died. The rift between them will never be healed.

Dubois cannot work, nor can he see anyone. He sends a brief note to Prentice but leaves it to Anna to explain to everyone else. He does not care what she says, to whom. His thoughts are only of his terrible, terrible loss. It is a black time; he can hardly bear the presence of anyone, even his family. Of all his friends and acquaintances, Prentice is the one who knows best the ways of grieving and the road to healing. And he silently returns the kindness and sympathy that Dubois once showed to him.

Ngrodjo, Saturday morning

Dear Doctor,

We have been examining the forest these few days. I have not called to see you as we have a visitor and I understand from your letter that you are seeking privacy. I fear you have very bad news from Europe, though I would fain hope it were otherwise.

As to Lalie Djuvo, the cottage is at your disposal whenever you wish to go there. You take the train to Sorrong.

Dog cart to Kasrie   *1/50*

Dog cart Kasrie to Pringen   *2/50*

Horse Pringen to Trettes   *–/50*

Horse next morning Trettes to L.D. *5/00*
*   (sometimes only f4/00 for horse)*

You better stay the night at Trettes, either in the big house (if it not be let) or in the school room (if the big house be let). The mandu Sarieman at Trettes has the keys of L.D. Tell him you come from Toeloeng Agoeng, are a friend of mine & wish to stay at L.D. Ask him to get a horse, coolies & rice & etc. for you to go there & to send Pak Termoh or some other trustworthy coolie (wages 75c or f1/00 per day but better pay only 75c which is quite enough to cook & watch the house, keys, & etc.) There is also an orang djager at L.D. & if you take a boy with you then you don't require to keep Pak Termoh or another coolie, but just as you like.

Sincerely yours,
Adam Prentice

If you take a boy with you give him money & tell him how much to pay. The coolies always ask more, but do not expect it!

The next day, May 5, Dubois journeys up to the Lalie Djuvo plateau where the bungalow sits. It is small but comfortable, beautifully sited with a breathtaking view: remote, peaceful, silent. He can be alone there with his thoughts and his grief in this lovely place. There is no bustle, no children, no callers, no noise. The boy sees to Dubois' meals, when the Tuan Dokter wishes to eat, and stays out of the way the rest of the time. Dubois walks, rides, sometimes reads, but usually he cannot concentrate enough to know what words pass before his eyes. Mostly he endures. He remembers his father, the things not said and the things he wishes unsaid. It cannot be done. There is no hope.

For days, he observes the sky, the birds, the trees; he watches the sun rise and set. The sun and moon seem to move at a pace not unlike his own; he slowly begins to heal. In time, the hills and the sky remind Dubois there is something bigger than his loss. He remembers the endless, magnificent unrolling of evolution, a twisting path that leads from time immemorial to time unimaginable. And on it, he is like a small rock or tiny frond of fern, just one insignificant being in the long course of human evolution.

After five days, Prentice comes up to Lalie Djuvo, ostensibly to bring

Dubois more supplies. He comes prepared to leave again promptly if Dubois wishes to be alone, but he can see that his mute companionship and devoted friendship are welcome. He does not jar Dubois' sensibilities, does not intrude in any way. He offers only himself, his kindness, his warmth. Prentice and Dubois spend many hours together, speaking little, doing much. Physical activity is healing somehow.

On the eleventh, they decide to trek up to the peak of Gunung Ardjoena, the mountain that looms over the plateau. They pack water, a little bread and cheese, and some fruit, and begin to climb just after dawn. It is an intimidating slope, but they are young and strong and used to exertion. It is cool when they set out. They climb higher, paralleling the course of the sun itself. There is no way to escape the baking rays; they must simply be accepted. By late morning they have reached the summit and the world lies at their feet. They find a shady spot to rest in, sitting back to back and looking down on the vista of green and trees and jungle and more green, as far as the eye can see. Here and there, dotted against the landscape are little villages, thatched roofs visible like tufts of dried grass against the green, a curl of smoke from a fire, a few square green rice paddies outlined against the wilderness.

"How trivial humans are," says Dubois quietly.

"Yes," says Prentice. "Our lives are very small compared to all this." He gestures with a sweep of his arm. "And yet, we are part of it, part of it all."

"*Ja*," answers Dubois after a moment. "*Ja*. We are a little bit, not much, but there."

They sit a while longer, drinking their water, eating their food, each enjoying the physical closeness of the other and the harmony of their thoughts.

"It is a hard thing," offers Prentice in a soft voice. "I remember what you told me, when we met: it is a hard, hard thing." Dubois does not answer. He does not need to. "I think," Prentice says slowly, "I think I will mark our presence on this occasion." He gets up and goes over to a large boulder. He selects a round fist-sized stone to use as a hammer against his pocketknife, the improvised chisel. After half an hour or so of determined hammering, the letters of his name are visible: Adam Prentice. His presence is inscribed in the rock.

Prentice walks back over to Dubois, who sits silently on his own, not having moved during Prentice's labors. Prentice stops in front of his friend, mutely offering the tools with an outstretched hand. Dubois looks up, his eyes full of his sorrows but not so blind that he fails to see the answering warmth in his friend's eyes. Dubois takes the stone in his right hand, reaching up his left to grasp Prentice's own strong, brown arm for assistance in rising. The two friends walk together the few paces back to the boulder. Prentice points to a place above his own name, looking at Dubois with questioning eyes.

"No," says Dubois, almost inaudibly. "Next to yours, *with* yours, my friend, not above." And there he carves his name upon the boulder, a monument to stand for all time. Completing his task, he stands back to look at their work, satisfied. The two men brush their fingers over the carvings, freeing small pieces of dirt and debris, and smile. It is done.

Later they make their way back down the mountain, jolting their knees with every step, watching for loose rocks and soil that will send them careening down the slope. They are tired but enjoying the stretch and bend of muscles and joints, the pleasures of youth and strength. It is a slow return from the heights to mere reality, but a pleasant one. They stop to wash the sweat from their hands and faces in a clear stream halfway down. The water is so inviting that they throw off their clothes and swim in the shallow pool. It makes them feel wonderful, alive and free. They reach the bungalow at Lalie Djuvo in the afternoon, in time for a nap and a bath and the evening meal on the veranda. The stars seem especially brilliant, the sky unusually clear that night.

Prentice stays about a week and then goes back to Mringin, while Dubois travels home to Toeloeng Agoeng and his family. He is ready to return, ready for the warm chaos of family life, ready to engage with his fossils and monograph once again. He throws himself back into his work with renewed energy. None of the crates that have come from Kriele and De Winter hold any new treasures, but Dubois is not so disappointed. He does not need anything more. He is writing this monograph for his father, for himself, for the new child that grows in Anna's womb, and for P.e. He makes swift progress now, as if the interval away from the work has enabled his ideas to ripen without his being aware of the process.

He sees a great deal of Prentice and Boyd during that year, more than ever. And sometimes as he works alone in his study he muses about the time at Lalie Djuvo. Prentice saved my life, he thinks, smiling at the memory. He saved my life.

CHAPTER 25 LETTERS FROM A FRIEND

Mringin   July 9, 1893

Dear Doctor,

Herewith I have much pleasure in presenting you a volume entitled "A Popular History of Science." I also send you an ink bottle I obtained for you at Soerabaja. I do not flatter myself that it is exactly what you wanted, but I could not get a more suitable one here.

I further send you a photograph of my child taken last month when he had completed his first year.

By this opportunity I return with thanks Mr. Opzoomer's pamphlet "De Vrucht der Godsdienst" ["The Fruit of Religion"]. I have read it carefully through and to a great extent would agree with his views, but from a Darwinian standpoint of evolution I should, I think, look at the matter in a different light. Therefore much of what Opzoomer writes seems to me to be overdrawn, still to a great extent I would be happy to endorse his views.

I still have yours of Tyndall's work entitled "New Fragments." I have not yet finished its perusal.

With my kind regards to all your household, hoping one and all are in good health and that your work is progressing to your satisfaction. Believe me to be

Very sincerely yours,
Adam Prentice

How very thoughtful of Prentice, to send the book and to buy him the ink bottle he has needed, Dubois thinks. And the photo of his child! Dubois looks at it closely. He can see Prentice in the boy's face, there is no doubt of it: something about the eyes and the set of the nose. Dubois is pleased with the photo, knowing how Prentice treasures Gerard. In his response, he asks Prentice to send one of himself, as well. He is a rare friend indeed. He and Dubois correspond every few days now, unless they see each other, so close have they become.

The monograph is coming along nicely. Dubois feels now that he can see the shape of the whole, which makes the information fall neatly into place. His experiments with the camera lucida are producing good results—high-quality illustrations of the third molar and the thigh bone—that, accompanied by some full-sized photographs, should make for a handsome volume. Dubois has decided to write it in German, the language of science in

Europe. To publish in Dutch would limit his audience too much, and the English morphologists are not yet as highly regarded as the Germans. The latter's supremacy, at least, owes much to Virchow and Haeckel and their influence. Writing in German is a struggle for Dubois because every word, every sentence, must be absolutely correct. It all takes a great deal of time. Perhaps, though, he also will publish a translation of the monograph, to make it more available to the English, who are often so poorly versed in other languages.

He outlines his plans in a letter to Prentice, to see what his reaction is. His first task in the monograph is to describe the defining characteristics of his new species, *Pithecanthropus erectus* Eug. Dubois. This description must meet with strict rules of nomenclature and must be sufficiently exacting that a trained observer, faced with another fossil, could readily decide whether it is or is not the same thing as *Pithecanthropus erectus*. He must do nothing less than characterize, justify, and make recognizable the essence of *P.e.*, as it is known from his specimens. Then he plans to recount the history of discovery—briefly, for this is not a geological treatise but a morphological and anatomical one. The meat of the monograph will be the description and interpretation of the three fossils themselves, in minute detail, with comparisons with bones of closely related species: the human, gibbon, and chimpanzee material. Since *P.e.* lacks the huge bony crests and muscle markings of gorillas or orang-utans, these more massive ape species need not be discussed at all. And since *P.e.* is certainly not human, there is no need to examine the variations within the races of mankind. He can close with a summary of his general conclusions about the evolutionary place and importance of *P.e.*

He expects the manuscript to be completed this year, in 1893, or early 1894 at the latest. He will take it to Batavia to be printed—that will take a few weeks or even a month, he supposes—and then he will mail it to everyone of importance in Europe. That way the monograph will reach the scientific community six months or so before his return to Europe at the end of his tour of duty. He may even, he thinks, undertake that trip to British India to see Lydekker's fossils before sailing for home. After all, he is already halfway around the world from Europe. He is not likely ever to be closer to India than here.

Tempoersarie   13 July 1893

My dear Doctor,
It was with great pleasure I received this morning your very kind letter

of yesterday with the excellent work by Newcomb which I shall highly prize not merely for the interesting subject matter of the book, but yet more for the sake of the giver.

I am glad you are so pleased with the photograph of my child. His name is Gerard Alexander, *and he has to bear the surname of* Prentice-MacLennan. *The portrait was taken when he was fully 12 months old. His birthday was 21 May 1892.*

So you are to return to Europe in a year's time! I earnestly trust the conditions under which you return will be such as you could hope from Government. Your visit to Calcutta will interest you, I am sure, and now you are in the East it is an opportunity that won't again occur.

I am happy the publication of your book in English will soon be accomplished. I shall be very happy indeed to be favored with a copy and shall regard it with peculiar affection on account of the associations connected with it. That affection will be still increased if, as I hope, the work will make a name in Europe for its author.

It will perhaps surprise you to learn that I am leaving Mringin. I shall be in Toeloeng Agoeng on the fourth of August and on the morning of the fifth. I proceed with the fast train to Semarang to administer the coffee estate of "Geboegan" at Oenarang for one year as the owner is going to Europe to see his children. His name is P. J. van der Leeuw. His wife died two months ago of measles, in Europe. We all have our troubles! When the year is finished I expect another administration. Thus you see there is no chance of my going to Europe soon, but the time will yet come, I hope. As I before informed you I am interested in making an application for undeveloped land on the Yang (Besoekie) & hoped ere now to hear something about it but, although fully two years ago, no reply to the land application has yet been received. Government are however now commencing the new railway in that direction & I daresay will give out grounds there in order to make traffic for the railway. If that application is given out, and I live, I shall expect to be in an independent position after 6 or 7 years as my friends have money to plant coffee & will give me the administration and a third share—very generous, indeed! The question however is, "will Gov't give out the grounds or refuse them?"

I have much pleasure in sending you my portrait taken just after our marriage—in happier moments, gone for evermore! You will favor me with one of your own when you have the opportunity?

Re Semarang—At first I disliked the idea of going out there and did not decide until the 7th of this month. I am now glad however that I am going, as Mr. V/d Leeuw is a man of energy & method, and I shall have the opportunity of learning much while in charge of his estate. In answer to my telegram accepting the situation, he wrote that I had taken a load off his heart, and he could now go to Europe & remain there with an easy mind.

At present, I have stopped all reading for a time (though not without a struggle), and am working hard at Javanese as it is imperative I speak that language fluently. Afterwards, however, I shall return with ardor to those fascinating subjects that engross my thoughts.

Hoping to have the pleasure of seeing you soon.

Believe me with kind regards to Mrs. Dubois and yourself,

Yours very sincerely,

Adam Prentice

Prentice is to leave Mringin! Dubois is deeply saddened. He had all but forgotten how lonely he was before he met Prentice and Boyd. Mringin without Prentice will not be the same. Prentice has meant so much to him, especially during those days at Lalie Djuvo. He had expected to continue seeing Prentice often until he returned to Holland, maybe even to persuade Prentice to return to Europe at the same time. Prentice is a rare man and now he, too, will be taken from Dubois, or at least placed well out of easy reach. Still, he must think of his friend and not of his own selfish desires. He can only wish Prentice well. It is a good opportunity, taking over Van der Leeuw's plantation. Prentice has suffered such a terrible loss, with such good grace and courage; perhaps it is his turn now for success.

Dubois conceives of the idea of taking Prentice and Boyd to Trinil for a few days. It is not long before Prentice will leave for Oenerang, so they must make plans quickly. Dubois wants very much to share the place of his discovery with these, his closest companions, his most ardent supporters. To be sure, Trinil is no highland bungalow or pretty hill station; it is hot and dreary and conditions are primitive. But a few days there should do them no harm, and they have often expressed an interest in seeing his excavations for themselves.

Who knows? Perhaps the engineers will have a surprise for him when they get to Trinil. He has one for them, in any case. Unbeknownst to Kriele and De Winter, Dubois has recommended them for promotion to sergeant for their stalwart work. He has just heard from the Governor that their promotions have been granted, so he can take their new stripes with him. That should please them. Besides, it is time to see how the coolies are behaving—or rather, to see if they are behaving any worse than usual.

Tempoersarie   July 16, 1893

My dear Doctor,

My best thanks for your warm, hearty letter. I hope to merit by the fulfillment of my duties all the good wishes you entertain regarding me and I know and feel you sincerely mean all you write. Yes, I wish a happier, a much happier future for both of us than we have experienced during the past year, and if fate has not brighter days in store for us, we have at least sounded its depths of misery, and much worse than we have had in the past year—personal bereavements, anxiety of mind, uncertainty for the future—cannot be meted out to us. I think I shall now go on progressing (in a worldly sense) until I have obtained independency, but if reverses overtake me I shall endeavor to bear them philosophically. If you have been enabled in any way to profit by me during the days we spent here together, the debt is much much larger on my side, for I have borrowed from you besides a mass of actual facts, new ideas and a broader insight into the things of the universe—acquisitions above all mere worldly wealth; and that being so, you have no thanks to give me, but rather thanks to receive <u>from</u> me.

Now those days of fellowship are drawing to a close their value is enhanced. Like so many other advantages we esteem them more fully when they have passed away & are lost to us.

I shall be happy to accompany you to Trinil on Tuesday together with Mr. Boyd.* I shall set out for Toeloeng Agoeng tomorrow afternoon & reach there tomorrow evening.

I am in haste to close as I am pressed for time.
With kind regards to Mrs. Dubois and yourself
Believe me to be
Sincerely yours,
Adam Prentice.

*It is perhaps the last opportunity we shall have—the three of us—of going together in company. A.P.

The excursion to Trinil is nearly perfect. Regrettably, there are no new remains of *P.e.* from the diggings, but there are plenty of good mammal fossils and the weather is relatively cool and cooperative. The engineers, Kriele and De Winter, are most pleased with their unexpected promotions and take their stripes immediately to their tents, to sew them onto their uniforms. An hour later, Kriele, the bolder of the two, approaches Dubois.

"Sorry to bother you, Doctor, when you have your guests here," he says a little hesitantly, "but could I ask you something?"

"*Ja*, Sergeant, it is no trouble. What do you wish?" Dubois answers jovially.

"De Winter and I, we wondered if you might take our photographs. We would like to have a picture to send to our families. And it is such a rare occasion to have distinguished visitors at the site, we thought you might want to celebrate a little. I have taken the liberty of asking the kokkie for an especially fine dinner tonight." Kriele smiles, a little embarrassed at his forwardness.

"An excellent idea!" Dubois responds. "What do you think, Boyd, Prentice?" he asks jokingly. "Shall we take a photograph of these fine fellows?"

"Yes indeed," Prentice answers. "They have surely been instrumental in your success here. A portrait is certainly called for!"

"You look pretty handsome in those uniforms, boys," teases Boyd gruffly. "Will you send the photos to your sweethearts?"

Kriele grins, for that is exactly what he intends to do with the photograph. De Winter, who has no sweetheart, will send it only to his mother.

Gerardus Kriele and Anthonie de Winter pose for these photographs on July 19, 1893, in their new sergeant's stripes.

Dubois and his friends inspect the excavation and the new fossils until it is time for the evening bath and dinner. To everyone's surprise, Dubois produces a bottle of wine from his luggage. He has never before been observed to drink any sort of alcohol in the field, so determined is he to set an example for the coolies. But this is a special occasion, a celebration, as Kriele has indicated, and possibly the only time he will be in Trinil with the people he cares for most.

After the Old Warrior retires to bed, Prentice and Dubois sit up late into the night talking. No other person has ever been such easy company for Dubois. He fears he has no gift for friendship, but Prentice is as true a friend as a man could wish for. In return, Prentice sees no fault in Dubois, no shortcomings, only the good—perhaps merely the reflection of his own goodness, mirrored in Dubois' eyes. Around Prentice, Dubois feels free to be himself, to be exactly what he is: brilliant, quick, impatient, with a mind so finely focused that he sometimes appears deeply selfish. Yet Dubois is a good and courageous man and Prentice loves him for it.

The trip to Trinil stays in Dubois' memory as a brief moment of peace and happiness in a dreadful year. Soon after their return, Prentice leaves for Oenerang, some hours away. Dubois is rather lonely and Anna seems unwell, not herself. Dubois passes it off as an adjustment to the lack of Prentice's welcome company, but soon he realizes something more serious is wrong. Anna rushes back from her bath in the mandi room one day, badly frightened. She is bleeding from her womb; the water is stained pink. It is not yet time for the baby to be born; she has had no contractions, and her waters have not broken. What is going on? Concerned, Dubois examines her thoroughly. He can detect nothing wrong, except the insidious bleeding which stops, starts again, stops over the next few hours. They both hope for the best, but it is the worst that comes.

Early on the morning of August 30, 1893, Anna goes into labor and is delivered of a stillborn fetus. Dubois receives the tiny, lifeless corpse into his hands. It was a girl, a miniature, perfect child with blond hair . . . a child who does not breathe or cry or open her eyes. She is dead, and there is no reason why.

Anna is hysterical with grief and pain. "How could you bring us to this cursed place?" she accuses him. "What are we doing here in this foreign land where children die as easily as leaves drop from the trees? And you, you run away, up the mountain to your friends at Mringin, abandoning me and the children. You know how badly I feel the heat when I am with child! How could you leave me?"

They are bitter words and Dubois has no answer for them. He does not know what went wrong or why the little girl died, but he is filled with guilt. It is true, he has devoted more time and attention to Prentice than to Anna in recent months. But he did not think she needed him any longer, she is so consumed by the children and their needs. There seems to be no room for him in the family at Toeloeng Agoeng, but he is welcome and needed at Ngrodjo.

Is it his fault she lost the child? He does not know. He searches his memory for something, some small ailment of Anna's that he might have treated, some symptom. . . . He finds nothing, only that the child is dead. They would have named her Anna Jeannette, in honor of Anna and Dubois' late father, had she lived. Perhaps it is no one's fault. Some things are incomprehensible in the Indies, and death is chief among them. Death strikes rich and poor, loved and unloved, European and Indo alike, without mercy. What might that child have been? It is a cruel question, for now she is nothing but a searing wound to the heart. Anna feels she has been torn to pieces. Dubois is overwhelmed with grief and guilt. And sometimes he asks himself, in the deepest, most private portion of his mind: Was this one *my* child? Was this the one who would have been like me?

CHAPTER 26 AFTERMATH

The Indies is a cruel country. One day little Anna Jeannette dies, the next Dubois must see to her burial, before ghastly contagion follows, before rot sets in. Funerals cannot be postponed in this climate. Anna is still confined to bed, weak and weeping inconsolably, exhausted from venting her anger at God and Dubois and anyone else who might have decreed this bitter fate.

Babu approaches Dubois, eyes downcast and very serious. "Tuan Dokter," she says quietly, "I am sorry but you must come. The Njonja is crying. She will not let me take the dead baby from the room. She says the little one will awaken and call for her."

"I will come, Babu," replies Dubois, stifling his own sorrow. When he enters his wife's bedroom, he sees the pitcher lying shattered on the floor and water everywhere. "What is this?" he asks.

"The Njonja threw it at me," Babu whispers. "She shouted at me for stealing the baby."

"Oh," says Dubois, shaken. This is not like Anna at all. She is even worse than he thought. "The Njonja is ill from the baby dying. She does not know what she is doing," he explains. It is as close to an apology as a tuan can offer a babu. Babu understands. "You may leave us now, Babu," Dubois says. "Clean up the mess later."

He goes over to the bedside and sits down, taking his wife's hand in his. "Anna," he says softly, "Anna." She wakens from her exhausted sleep, opening her red and swollen eyes, her lovely hair now greasy with sweat and snot and tears.

"Does the baby need me?" she asks, confused, trying to sit up.

"No Anna, no," he replies. "I am sorry, Anna. The baby is dead. She could not live, I don't know why. I must take her to the cemetery and bury her today."

"No," Anna whimpers. "No, please." She clutches at his hand restlessly.

"*Ja*, Anna, it must be. You know it. You know how things are here. I am so sorry, Anna, so sorry." He wipes her face with a damp cloth. He does not like to look at her directly, for fear she will burst into bitter accusations once again.

"No, Eugène," Anna insists, pounding the bed with a feeble fist. "You cannot take her! You cannot leave her in that horrible old cemetery all alone. I will not let you."

"I must, Anna. The weather is very hot. We cannot leave her here. These things must be . . . taken care of. We shall put up a lovely stone for her, our little girl. We can go visit it every day if you like."

"*No!*" Anna screams wildly, trying to get out of bed and go to her child. "You cannot take her away! She is a baby. You cannot leave her alone. She must stay with me!"

"Anna, Anna." Dubois holds her, comfortingly. "You are very tired. We are all upset." He settles her gently back in bed, then gets up to mix a powder into a glass of water. "Anna, I will give you a little sedative. You need to sleep. Now drink this, there's a good girl." He helps her sit up and swallow the potion.

"Yes, Eugène, I am tired, so tired," she replies slowly. She looks over at the still little form, lying in a cradle on the whitest linens. "Anna Jeannette is dead, isn't she?"

Dubois nods slowly, carefully. "I am so sorry, Anna. There was nothing I could do. I am very, very sorry. She must be buried today."

"If I let you take Anna Jeannette," Anna offers, "you must promise me something."

"What is it?"

Anna looks at him steadily and says, "You will bury her across the street, in that little Javanese cemetery behind the wall. She won't be alone there; there are other children, and she can hear her brothers and sister playing."

"Anna—" Dubois starts to protest. And then he thinks, Why not? Who will care but Anna and himself? "Very well, Anna, I promise. I will keep her there, near us, where she won't be alone."

And it is done, that very day. The Javanese are confused and frightened. They do not understand why the Tuan Dokter and the Njonja do not bury their baby in their own place, with all the other European babies. Why is she here, among their people? In the end, they decide it does not matter. A dead child is a dead child, loved and mourned whoever its mother. The grave is so tiny, not much more than a foot long, surrounded with thin Javanese bricks. Dubois orders a piece of marble to be carved as a gravestone.

Here rests Anna Jeannette Dubois
Daughter of Marie Eugène François Thomas Dubois
and Anna Geertruida Lojenga Dubois
August 30, 1893
* Toeloeng Agoeng
+ Toeloeng Agoeng

In keeping with local custom, the place of birth is preceded by a * and the place of death by a +.

It is many weeks before Anna recovers in mind or body. She is, for a time, perhaps a little mad. The other children are bewildered. Babu still takes care of them, loves them; Kokkie still cooks for them; but Mama cries and looks sad. She will not play with them or sing to them, like before, and she will not let them out of her sight. They do not know what to do to make things better.

Dubois too is bewildered, haunted by his own grief, confused by hers. He cannot forget Anna's reproaches. In his mind, he knows that babies die inexplicably in Amsterdam as well as in Toeloeng Agoeng. In his mind, he is not responsible for the child's death. In his heart, he feels that perhaps he is. He placed his science and his intellect above his family. That was the path of betrayal, the one that led him away from them. Somehow that choice caused this tragedy. He cannot bear to look at Anna's tear-stained eyes and ravaged face. They do not speak very often anymore.

A week later, he decides he must take decisive action. The grave is not enough, it is too impersonal. So that he will never forget the day on which

Anna Jeannette died, Dubois writes in his pocket calendar: "Anna abortus." Every time he opens it, he sees the words, like a cold, factual public decree: "Anna abortus." There are no words written on any other day of that week. Later, he crosses the words out, using deep strokes of his pencil as if he could alter the reality by obscuring them. It changes nothing, neither the grief nor the guilt nor the anger. The next day, to force himself to face the truth, he writes it again: "Anna abortus."

He labors on, working on his monograph when he can concentrate. His father and child both lie dead in the ground, but he has resurrected *P.e.,* dead these many, many years. It seems a hollow exchange. Sometimes he thinks his work is trivial, not worth the cost. But they might both have died anyway, and his soul too, had he stayed in Amsterdam. All he can do is turn to science, increase his devotion to the truth. He redoubles his efforts, like a nun saying extra novenas for the dead. He must do even better now. He will continue to place science above everything, for the truths he discovers will last forever, and people only die and leave him alone.

He practices science as a priest practices religion. If he were religious, he would say his life was dedicated to the glory of God, as his sister Marie said when she joined the convent so many years ago. He searches for truth, for the essence of things. However burdened his heart is, it is his duty to carry on with his scientific work. In truth, there will be peace, there will be rest. There is none anywhere else.

CHAPTER 27 PERSEVERANCE

Dubois is deeply alone.

His father is dead. There is an awful finality about it. His father never gave his approval of Dubois' aims and never will. His mother, thinking to comfort her son, writes that she is saying extra prayers and lighting candles for Dubois' soul. He presumes she prays to lessen his punishment for the sins of pride and of failing to honor his father. Dubois cannot decide which gives him greater pain: his mother's well-meaning violation of his deepest convictions, or his feeling of abandonment at his father's death. The child he would have named in his father's memory is dead too, a chapter closed before its first words were written. Anna is dead to him now, too, of course. The youthful affection and goodwill that fueled the early days of this marriage of opposites have been expended. Whatever deeper ties grew in their place have withered under the parching burden of grief. The loss of this child has left between Anna and Dubois only a barren wasteland where nothing can grow.

Even Prentice is lost to Dubois. In this time of despair, letters are not enough. He needs his friend's presence, his solidity and sympathy. But Dubois cannot travel as far as Oenarang now. It is his duty, and he always honors his duty, to stay at home with the grieving stranger to whom he is married. He cannot even go up to Ngrodjo, to seek solace from Boyd. Every time he conceives of going, Anna's angry words echo in his mind: "You run away, up the mountain . . ." and he knows he cannot go.

Anna does not think of Dubois' state; she only wonders if she will survive this trial. Little Anna's death is the embodiment of a mother's greatest fear in the Indies. Anna supposes she must survive, for who will look after the other children if she does not? But it is a long time before she laughs or sings again. She gets out of bed in the morning as an act of sheer courage. She bathes and dresses properly when she is able to, which is not always. She walks across the road to sit by the pathetic little grave in the Javanese cemetery for hours, one hand resting on the gravestone, turning into a stone herself. She does not know whether she weeps or not, the difference between the two states being so trivial. Sometimes Dubois thinks he hears her talking to the dead child, or even crooning a soft lullaby, but he does not dare go over to see what she is doing. What can he do for her? What comfort can he offer, what apology? There is none.

The Javanese who come to leave offerings at their own families' graves are a little frightened of Anna at first. This strange, pale njonja has no business being in their cemetery. And they can see that her soul is in danger of quitting her body. She does not speak to them; she may not even see them. They take care not to come to her notice.

Dubois has a bed made up in his study and starts to sleep there every night. He tells the servants it is so he will not disturb Anna, who needs rest to recover from the aftereffects of the birth. No one is fooled, not even the children. It is just one more way in which he withdraws from Anna and retreats into his work. He restricts his world for a time to the two rooms, his study and what was formerly the gentlemen's sitting room, coming out mostly to take meals. Yet the work seems sterile. He waits for more crates of fossils to come from Kriele and De Winter at Trinil and receives only complaints of cold, of sickness, of bellyaches. Before the baby's death, he had notes from Kriele and De Winter, both complaining. Kriele's was written first.

Trinil   August 8, 1893

A very friendly request to separate me as soon as possible from De Winter; since we have had yesterday such a dreadful conflict.

De Winter's is next:

August 25, 1893

The Resident from Solo has not been able to give me prisoners, as he is afraid they will run away here, and besides there are so few prisoners in Solo. Now they provide me with 10 free-men, without payment, who will be picked up with the supplies today. In my view this work can never go right with those free people without payment, for every morning those people must come from different desas or villages, and every day others, and besides to order those people or to instruct or to keep them working the whole day just like prisoners will be difficult. I will try to work with the free-people and will send you later a message about that.

On September 3, De Winter writes again. Things are no better.

Regarding the free people, their work leaves much to be desired.

You can imagine yourself that those people come to work from villages hours away in the neighborhood, and thus they find it impossible to be with me in the morning. When I want to start, around 7 o'clock in the morning, there is as usual no one, for example; someone comes at 7 o'clock, two at half past 7, three at 9 o'clock, etc: then at 11 o'clock I let them go to their homes to eat, and I tell them that they must be back at 1 o'clock, but then at 3 o'clock there is still no one. You will see how difficult this is.

The engineers will have to work out their difficulties without intervention from Dubois. He carries on with the tedious work of cleaning and examining specimens, on good days finding something he believes is a new species. Then he often makes notes for an article about the species. He is proud that his is an outstanding collection of Javan fossil animals, even if they are so far mostly stored on his veranda. They are mute, beautiful in their dusty way. The fossils do not accuse him of wrongdoing, of neglect. They only speak to him passionately of the past, of their adaptations and ways of living and ways of dying. The fossils are very forgiving. They are his most constant companions now.

He knows he should be working on his monograph, but it is hard to keep focused on the work when there are so many tangled emotions to obscure the facts. Sometimes he simply sets that project aside, telling himself he will complete it once the dry season's excavations at Trinil are finished. More fossils of *P.e.* might turn up at any time, after all, and they will need to be incorporated into the work.

Other, less sensitive projects help fill his time and distract him from his problems. Last year, in 1892, he published an article about ancient climates and environments in the *Natuurkundig Tijdschrift voor Nederlandsch-Indië* (Journal of the Natural Sciences of the Netherlands Indies). Now he expands the work, elaborating his ideas, and submits it as a small book to a publisher in Nijmegen. Checking and correcting the proofs in the last few weeks before publication is relatively mindless work, which he finishes during the worst of his despair. There is something numbingly abstract, soothingly unemotional about looking over the pages, altering a word here or correcting a misspelling there. He also works halfheartedly on the English translation, for Swan Sonnenschine & Co. have expressed interest in *The Climates of the Geological Past and Their Relation to the Evolution of the Sun.*

At the end of September, there is a letter from Prentice.

Geboegan, Oenarang   September 27, 1893

My dear Doctor,

How has time been dealing with you lately? With me it has been far from agreeable. You know how comfortable I was in a sense at Mringin and on what a friendly footing I stood with my employer Mr. Boyd, our mutual friend? Well, here everything turned out exactly the reverse. During the past 7 weeks I have had nothing but very hard work and little or no friendship. My master has been in very bad health and worse temper. My working hours have been from 4 a.m. till 6 p.m. I have never had a pleasant meal here til today. We did not eat our food, we merely gulped it down, & never, or hardly ever, was a word spoken at table. Mr. Boyd can tell you how he found it when he brought me here. Very few people could have stood it out, and many and many a time did I regret much ever having left the friendly shelter of Mringin for the inhospitable surroundings of Geboegan. Yesterday however I did get some relief as my chief left for Semarang and does not return here. On 3rd October I go down to Semarang to see him away. He sails for Europe per S.S. Gedeh on 5 October.

So far for myself and how is it with you and yours at Toeloeng Agoeng, and how is your work progressing at Trinil? I have often looked back with pleasure to the excursion we made three together, & to our bathing in the river & etc. & etc. Have you found anything further of the missing link, or other interesting fossils? Are you going to Calcutta at the end of the year, or what plans have you got?

Needless to say I have read <u>nothing</u> since my arrival here—reading has been altogether out of the question, the matter of the precise color and form of coffee beans, and the cost of cleaning 100 coffee trees, are of more importance to those here than the highest scientific investigation or discoveries. My time is all taken up in work, work, work.

It is warmer here than at Mringin. The parcel indeed runs up to 4,000 feet, but the house is situated at 1,200 feet elevation and has a large zinc roof.

How has your book on climates been received in Europe? (I have drank [sic] *nothing since coming here.)*

With best wishes to yourself and family, hoping you are well, & trusting to hear from you soon. Believe me to be,

Sincerely yours,
Adam Prentice

Ah, Prentice too is unhappy, isolated. Dubois would not wish these feelings on his dearest friend. If only Van de Leeuw were a genial man, a good man, like Boyd: it is too much to expect, Dubois supposes. How he wishes he could join Prentice for a few days or a week, as Prentice joined him at the house at Lalie Djuvo, when they read and walked and rode and climbed Gunung Ardjoena together. They will not have another such time for healing and companionship. It was a luxury not to be repeated. Still, he will write and try to tell Prentice of little Anna's death and his wife's suffering, and his own.

With the dry season coming to an end, Dubois makes a very short trip to Trinil to inspect the site one last time. No more fossils of *P.e.* have been forthcoming, though the rhinos and elephants and tigers and buffalo continue selflessly to donate their bones to his Indies museum. He does not believe that there is more of *P.e.* to be found at Trinil; the thing is done, over. He decides to erect a monument at the site, something permanent. He designs a cement marker and tells Kriele and De Winter to place it on the high bluff opposite the site itself in 1894. Attached to it will be a bronze plaque. In Dutch it reads:

P.e.
◄━*175m ONO*━◄
1891/93

There it stands, for all time: *P.e.* was found 175 meters east-northeast of *here,* with an arrow pointing to the bend in the river where the fossils were found, between 1891 and 1893. Neither Dubois' name nor his initials appear anywhere on the marker. This fact puzzles some later scientists: why would he not lay open claim to his wonderful find? But to think that is to misunderstand the purpose of the plinth. It is not a monument to vanity. It is not

Dubois designs a monument, which is erected at Trinil on September 5, 1894, to mark the location of the site for all time.

there to honor Marie Eugène François Thomas Dubois or to speak of the prospects he left in Amsterdam, the families grieved and separated, the bouts of malaria, and the deaths of his father and daughter. This is an eternal witness to the missing link, missing no longer: to *P.e.*

CHAPTER 28 THE MONOGRAPH

The monograph is a different sort of monument to the work. The pages begin to pile up now, taking their final form and needing no further revisions. The illustrations are done, and they are handsome. Dubois is proud of them. There will be photographs and camera lucida drawings of the bones; tables of measurements and comparisons with the ape skulls; and a few diagrams.

The text begins with a description of the geographic location of the find-spot along the Bengawan Solo and a brief mention of the important Pleistocene mammals excavated there. But the monograph is not the place for detailed descriptions or analyses of the fauna; those will come in later, more focused publications. He describes the general geological setting and explains that the tooth and skullcap were found within about one meter of each other, the femur being some fifteen meters farther upstream from those two but in the same layer. He has a great deal more information about the site itself and the plan of digging, but those details do not belong here. This monograph concerns the morphology of *P.e.*, not the geology of Java.

Dubois struggles to make clear the relationship among the three fossils of *P.e.*, to forestall specious arguments like those raised by that fool who called himself Homo Erectus. Dubois writes simply, "It would be foolish to doubt the three relics belong together on the basis of this slight distance between their locations." He uses the German word *thöricht* for "foolish." The choice will come back to haunt him.

He offers the following evidence in support of his conclusion. First, the incomplete condition of the skeleton of *P.e.* is unremarkable, for no complete skeleton of any animal has been found throughout the entire excavation, although thousands of fossils have been recovered. Another proof is the condition of other species, in which skeletal elements from a single individual have been found separated by as much as twenty or thirty meters. Finally, since no other specimen has been found in all of Java that could possibly represent *P.e.*—with the possible exception of the very fragmentary jaw from Kedoeng Broebus—it seems obvious that these three bones, buried so close to one another in the same sedimentary layer on the same point bar, are from one individual.

This introduction sets the stage for the anatomical and morphological description of the fossils. Dubois begins with the skullcap, undoubtedly the most important and impressive specimen. The outside of the skullcap is now completely clean of matrix, but some of the interior is still obscured with sediment. It is too dangerous to proceed further in cleaning the specimen without better conditions and tools. Nor can he measure the volume of the skullcap directly by pouring in seeds or water, as is commonly done. But the volume is obviously large and obviously important, so Dubois uses the method he invented in his third quarterly report of 1892 to estimate the volume of the skullcap from its linear measurements. It does not seem so difficult to him, with his talents in mathematics.

Clearly the volume enclosed by a shape is determined by the dimensions of that shape, whether it is a skullcap, a sphere, or a box. Because the skullcap of *P.e.* most closely resembles a chimpanzee's, Dubois first uses the chimpanzee as a standard of comparison. The total external length of the skullcap of *P.e.* is 185 mm, 1.33 times the recorded length of Bischoff's chimpanzee skull. Similarly, the total breadth of the skullcap of *P.e.* is about 1.33 times greater than that of Bischoff's chimpanzee. All other things being equal, this implies that the volume of the skullcap is $1.33^3 \times 410$, the volume of Bischoff's chimpanzee skull, or 2.35×410, which equals 963.5 cc.

The problem with this estimate is that "all other things" are not equal, for the skullcap of *P.e.* has a much higher vault than that of a chimpanzee. How much difference does vault height make to skull volume? Dubois investigates the question by comparing the measurements of male and female chimpanzee skulls, and finds that a higher vault increases the capacity of the braincase by at least one-seventh of the total. Adjusting his calculations for *P.e.* accordingly, Dubois arrives at a minimum volume estimate of 984 cc. When he repeats these calculations using a gibbon rather than a chimpanzee as the standard, the result is 991 cc. The two estimates are gratifyingly similar. There can be no doubt that the braincase of *P.e.* encloses a volume close to 1,000 cc, as big as the brains of some human races. No chimpanzee or gibbon has a cranial capacity approaching this value; the chimpanzee Dubois has in his possession, thanks to Weber, contains 365 cc; two gibbon skulls in his collection, both male, contain 135 and 140 cc, while a third, of a different species of gibbon, is 133 cc. Braincase volumes reported in the literature for other apes fall far short, never ranging higher than 465 cc, measured by Bischoff for a male gorilla. Thus for Dubois the truly impressive size of the braincase of *P.e.* is a strong link to humans, an indication that *P.e.* is well along the journey toward humanity, while not yet having arrived at its evolutionary destination.

Because its braincase is so large, Dubois suggests that the missing face of *P.e.* must also have been more human than apelike. And other aspects of the skullcap are reminiscent of human crania. The brow ridges over the bony orbits resemble those on the fossil Neanderthal skulls from Europe. Dubois, like Huxley, considers the Neanderthals a low and primitive race of human, so he does not compare *P.e.* with them in any detail. His interest is in placing *P.e.* along the continuum that links apes and humans, not in ordering it within the range of human races. Besides, he does not have casts of the Neanderthal skulls to work with and he decides not even to attempt to procure them. It might take much too long and the comparison is irrelevant.

There is Virchow's vehement and influential opposition to Neanderthals to be considered, too. In his monograph, Dubois remarks that the Neanderthal skulls are perhaps deformed; their femurs certainly are, for the femur of *P.e.* is much more humanlike than the thigh bones of Neander-

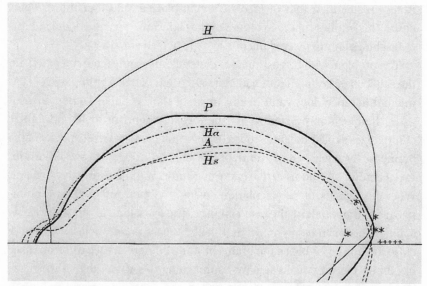

Dubois compares the skull profiles of (P) Pithecanthropus, (H) a modern human, (Ha) an agile gibbon, (A) a chimpanzee, and (Hs) a siamang, another type of gibbon. The human skull is the largest, but Pithecanthropus' *skull vault is bigger than the apes'.*

thals. There is no point in comparing a normal specimen with a diseased one, and *P.e.*'s femur is normal except for that lump of pathological bone where an injury healed.

While the skullcap has humanlike features, notably its size and volume and the remarkable height of its vault, it still also resembles a chimpanzee skull. Part of the resemblance can be attributed to the shape of the back of the skullcap. Partway down the posterior profile of the skullcap is a distinct protuberance called by anatomists the occipital torus. Below the torus, the skull slants steeply inward, a combination of features common in gibbons and young chimpanzees. The torus and change of profile may be exaggerated in *P.e.*, Dubois suggests, because the species had an upright posture (attested to by the shape of the femur). This humanlike posture places the skull of *P.e.* atop its spine like an apple balancing on a pencil: a most unapelike position for a rather apelike skull to be in.

Dubois also notes that the sutures between the bones of the skullcap are so completely fused that they are no longer readily visible. This is a sign of physical maturity, even perhaps of old age. Too, the general gracility of the skullcap and the striking lack of bony crests suggest that this individual *P.e.* was probably female. It is strangely satisfying to be able to deduce such fine details of his missing link's life!

When he arrives at this point in the writing, Dubois feels that he has accomplished a great deal. To describe the skull and present his subtle views of its mixed ape and human features is difficult. He strives always for clarity, for precision, for irrefutable logic. It is concentrated work and complex, sapping his fragile strength while it grounds him anew in his work.

One of Dubois' last cogent thoughts before he sinks yet again into a malarial fever is that Anna seems somewhat better and the children less fractious. When he emerges once again to consciousness, he finds that Prentice has called at Toeloeng Agoeng. Prentice! And he left again, without seeing Dubois, who lay sweating and delirious in his bed, not a hundred feet away. What a precious chance has been lost.

"He is a very nice man, your friend Prentice," Anna says, smiling at the memory of the visit. "He stayed all afternoon. He was so sympathetic and understanding about the loss of our daughter, he quite eased my pain with his kind way and gentle conversation. Too bad that you were too ill to see him. I know he was sorry to miss you. If only he were still at Mringin, as in the old days, we might see him more often."

Dubois feels a stab of jealousy. It seems grossly unfair that Anna saw

Prentice and he did not, that Anna had the benefit of Prentice's kindness and he did not. He must have some relief, some comfort.

Boyd's daughter, Anna Grace, is about to be married, to M. G. de Witte, a man in the sugar trade. They will settle in Blitar afterward, near the sugar mill. The wedding itself is to be in October, at Toeloeng Agoeng, with an enormous colonial party up at Mringin to follow. They had, of course, planned to attend the wedding, but now Dubois proposes to Anna that they travel up to Ngrodjo early and stay on for the celebration afterward. Normally this is the sort of event Dubois detests, with far too much drink and far too many people behaving idiotically, but on this occasion it sounds more like a welcome escape. Most of the guests will drink and dance and flirt all night, then sleep late into the day before starting in again on picnics and excursions and games. As long as the servants continue to clean up the messes and produce more food and drink at frequent intervals, the party will bounce raucously along with little help from the host. Dubois is counting on that to allow him to steal some quiet time with Boyd. They could go off on horseback to inspect the coffee or something, just to get away from the party.

Anna hesitates at first, for attending will mean abandoning her daily vigil at little Anna's grave. But she agrees nonetheless. She has always enjoyed parties, and the prospect of music and conversation and maybe a few handsome men to flirt with seems to lift her black mood.

Before they set off, Dubois writes to Prentice. When they return from Mringin, the reply from his faithful friend is waiting for him.

Oenarang   October 31, 1893

Dear Doctor,

I was sorry not to have met you during my brief visit to Ngrodjo, but owing to your fever and my work outside examining ground we missed each other.

I hope however your general health is good and trust you will yet pay me a visit here ere going to Calcutta. It is rather a long way, to be sure, yet I hope time and circumstances will permit you to come to Geboegan. My own visit to Ngrodjo was wholly unexpected, or I should have advised you beforehand. I was suddenly called to Pasoeroean on business of my father-in-law's, &, being in East Java, took advantage of the opportunity to call both at Malang and at Toeloeng Agoeng. My child enjoys perfect health and seems to me to be growing wonderfully. He can walk about the room

already and understands whatever is said to him but can't speak yet, or
won't speak. He is now 17 (seventeen) months of age.

What news have you now about your climate book? The critics have
had time to peruse its pages. What opinions do they express respecting it?
I trust it has been well received and has enhanced your reputation. Have
you had it printed in English or not?

With my kindest regards to Mrs. Dubois and the family, trusting you are
yet at Mringin enjoying a cool climate and open air bathing.

> *Believe me to be*
> *Sincerely yours,*
> *Adam Prentice*

CHAPTER 29 WRITING UP

The days at Ngrodjo give Dubois a little peace to add to his small store. He is able to resume work on his monograph in November. The introduction is completed, and the section on the skullcap. The next section must describe the molar tooth.

There is not much to say. In some aspects the tooth seems humanlike; in others—especially in the wrinkling and folding of the shiny enamel—apelike. There is nothing remarkable about it and it shows little wear. Dubois must admit that the tooth by itself cannot be firmly identified. It is certainly from a higher primate, but nothing rules out its belonging to a chimpanzee or an orang-utan or a human. Still, the tooth is clearly from an adult, a fact consistent with its belonging to the same individual as the skullcap and femur found so close by.

The femur is another story entirely. In all its principal features, the thigh bone is human. Its size and shape would not seem out of place in the anatomy laboratory, though it is twice as heavy as a modern femur of similar size on account of being fossilized. And the shaft itself, so straight and strong, is appropriately angled to pass the body weight from the joint at the hip in toward midline to reach the knee, as in man. The ball of the femur, where it meets the hip bone, and the condyles, where it meets the upper end of the tibia at the knee, are thoroughly human. In all its mechanical details, this femur is functionally identical to that of any modern, upright-walking human. With a bone like this, so thoroughly different from the thigh bone of apes, *P.e.* must have walked like a man.

Only in small details, like the somewhat poorly developed ridges where various muscles attached to the bone, is the femur somewhat more apelike than humanlike. Dubois does not want these differences to be overlooked:

> The points suffice, however—as I would emphatically point out—to separate the species in question from Man, who always shows differences in this respect. On these points, minor ones from a mechanical point of view, the femur resembles that of the anthropoid apes.

Then there is the pathology on the femur to deal with. As a physician, Dubois has seen such growths—exostoses—before. It must be the result of a ghastly wound to the thigh. How exactly was *P.e.* injured? It is not an easy question to answer. The most likely alternatives, to Dubois' fertile mind, involve weapons or tools, perhaps a wooden arrow or the tip of a lance driven deep, penetrating the muscle and the bone of the thigh. This scenario implies that *P.e.* might have used and made tools, which Dubois believes likely but, on the available evidence, unprovable. In any case, the wound occurred and bled copiously. The huge, irregular clot that resulted eventually ossified, turning into the strangely sculptured excrescence visible now on the femur. Ossification, though, is not a rapid process. How did an individual so badly wounded survive long enough for healing to occur? Did someone take care of her? Who?

These questions are too theoretical and emotional for Dubois' taste. He moves on, instead, to deal with more concrete matters. He calls the femur the "pillar, girder, and siphon" of the body; it is the main support of the body weight, the main lever that thrusts the body forward into movement. Thus the substantial size of the bone is very revealing. His first impulse, upon receiving the femur from Kriele and De Winter, was to hold it in front of his own, for comparison. *P.e.*'s thigh bone is very nearly as long as Dubois': a remarkable fact. A femur this size could support—was built to support—a large man. No one has ever before tried to determine the stature of an individual from the length of his femur, but it does not take Dubois long to work out a rough estimate. With a femur as long as this one, *P.e.* stood perhaps 1.7 meters (5'7") high, only a few inches shorter than Dubois himself.

For a creature of this size to move by swinging from its arms through the trees, like a gibbon, seems ridiculous, especially with a femur shaped for upright walking as this one is. Even the angle of the shaft from condyle to hip joint is much closer to a human's than an ape's. Dubois measures the angle of the femur of *P.e.* as 125 degrees from the horizontal plane of the knee; he cites published reports on human femurs ranging from 112 to 135 degrees, while ape femurs are nearly vertical (180 degrees) in orientation. Too, the torsion on the shaft of the femur indicates that the ape-man walked flat-footed, not on the outer edge of the foot like an orang-utan or a chimpanzee.

How long ago did this surprising creature attain such height? There is no precise means of dating the fossils or the beds from which they were extracted. The Siwalik fauna from India is considered to be Miocene, for it contains very archaic animals and none that survive today. Such faunas must be older than the Trinil fauna, for the mammals at Trinil are not so archaic and even include a few species, like the gaur (a type of wild cattle), that still exist. Thus both the higher primates—*Anthropopithecus* from the Siwaliks and *Pithecanthropus* from Trinil—and the other mammals suggest the Javan fossils are more recent than the Siwalik ones. How recent? If the Siwalik fauna is late Miocene, Dubois guesses that the Trinil fossils come from the subsequent Pliocene period or perhaps even later, from the early Pleistocene.

As for the place of *Pithecanthropus* along the evolutionary road from ape to human, Dubois has no doubt—it is an intermediate: "An Anthropo-pithekos has become a Pithekanthropos. . . . *Pithecanthropus erectus* is the transitional form which, according to the theory of evolution, must have existed between Man and the anthropoid apes: he is Man's ancestor."

By its anatomy, *P.e.* demonstrates the proof of the proposition put forward by Darwin and the French scholar Lamarck. Clearly the femur, with its almost completely humanlike form, shows that upright walking evolved first, while the skullcap attests to the later development of the large brain that characterizes humans.

If *Anthropopithecus* evolved into *Pithecanthropus*, did *P.e.* in turn evolve into man? It is a weighty question. Dubois believes she might have, although the evolution must have happened rapidly. By the time of the great ice ages, in the early Pleistocene, true but primitive humans like Neanderthals already inhabited Europe and survived by hunting new species like the reindeer, horse, woolly rhino, and so on. Thus the preceding era, during which these new forms evolved, must have been a time of rapidly changing conditions. Many archaic mammals became extinct and a few new ones arose and evolved rapidly, like the clever and upright ape-man that was ancestral to man. As its brain and body improved, its selective advantages became ever greater. Man's ancestors experienced a period of accelerating change, evolving *per saltum,* by leaps. Evolution, Dubois becomes convinced, is not a matter of *gradual* changes.

And what about the other apes? Where did they evolve from? Dubois sighs and thinks again. He must work out the answer to that problem separately. He tries to construct from the scant fossil record of ape evolution a series of evolutionary steps that seem logical. Gibbons are the only living apes to regularly walk upright, although of course this is neither their habitual means of locomotion nor the gait to which they are most strongly adapted. Still—Dubois nods to himself—it must be taken into account; this fact is a clue to the past history of apes. If a gibbonlike form is ancestral to all of the apes, it could have imparted to its descendants some of the features of tooth and skull that are seen in *Anthropopithecus* and *Pithecanthropus,* as well as a limited ability to walk upright. But *Anthropopithecus* is a chimpanzee, an ape, and *Pithecanthropus* is just as clearly neither ape nor man. Perhaps *Anthropopithecus* was the progenitor of both lineages: the dividing point, from which some descendants evolved into apes and others into *P.e.* and humans. That would account for the gibbonlike features found in all three. Nothing is known of the ancestor of the Siwalik *Anthropopithecus* so, following Haeckel's lead, Dubois proposes it was a gibbonlike form and names it *Prothylobates,* meaning "forerunner of the gibbon *Hylobates.*"

The final question is to what zoological family *P.e.* belongs. Dubois cannot justify putting the ape-man into the human family, the Hominidae, not with that apelike molar and ape-shaped skull. Nor can he accept that *P.e.* is simply an ape, an ancient member of the Simiidae. How could an ape have a femur that so closely resembles a human's? What ape has a skull that houses nearly a thousand cubic centimeters of brain? No, *P.e.* is not an ape, either. There is only one logical place for *P.e.*

> *Although already quite advanced in the development of the human type . . . , this Pleistocene form had not yet reached it. —It occupied a position between this type and the type of the great apes. . . . Considering all these circumstances I feel compelled to place the species in a new genus—*Pithecanthropus*—but also in a new family—Pithecanthropidae—between the Hominidae and the Simiidae. . . .*

Dubois knows this bold move may cause a furor—not many men have the opportunity to name a new species, much less to create a new genus and recognize a whole new family of mammals—but he believes the decision to be honest and correct. *P.e.* is neither ape nor human. It is so distinctive as to warrant a unique taxonomic position. He has found something totally new, and his classification must reflect that. If the world is surprised . . . well, let it be surprised. He has the fossils to prove his contention.

It is late in 1893 when he finishes the manuscript. He has been as handicapped by Prentice's absence as by the lack of comparative osteological samples, but he has finished his work on his own and he is proud. Completing it gives him a feeling of peace and satisfaction that nothing else has ever offered. He has found it—found his missing link—and proven Darwin's theory correct when everyone around him said the project was impossible and mad.

CHAPTER 30 SEPARATION AND LOSS

Dubois is pleased with life again, happy to be able to tell Prentice in his Christmas letter that he will take the manuscript to Batavia to be printed in the new year. He will lengthen his trip, making a detour through Semarang on the way back if Prentice can come there to see him. It would let him share with his friend the satisfaction of having finished his job, even if they could only meet for a few hours.

As this awful year, punctuated by tragedies, comes to a close, Anna seems to improve, become more like her old self. Yet a shadow flickers in her eyes from time to time, attesting to a wound that will never heal completely. Once she sat for hours on the back veranda after dinner, embroidering and listening to the night sounds while Dubois read. Now a new, disquieting sound invades the nighttime peace of their garden. It is a dreadful call, an unearthly groaning sound full of suffering. They have never heard its like before. As its first anguished notes split the night, Anna shivers despite the heat and gathers up her things and goes to her room. She cannot bear to be outside in the night with whatever it is that cries so pitifully.

Her babu, who has been squatting on the veranda a discreet distance away, arises and follows close behind her. "Ah, Njonja," says Babu approvingly, "better to go in. Do not listen to pontianaks."

"What do you mean, Babu?" Anna asks sharply. "What is that noise?"

"Njonja, it is pontianaks, female demons, because of the death of the baby. Her soul is crying, up in the trees. You hear it; you know. But do not listen. Pontianaks only bring disaster and doom."

The wailing starts up again, penetrating even to the interior of the house.

Anna begins to cry and shake, whether from fear or sorrow she does not know. The call pierces her body, bringing forth an echoing cry of loss from her very marrow, a cry that no one else can hear. It is as if her dead baby is being wrenched from her body over and over again.

"Will they go away, Babu?" she pleads helplessly. "Will they always be there now?"

"Ah," replies Babu evasively, "no one can know. Mungkin, mungkin tidak." She makes a Javanese gesture of ambiguity with her eyebrows. The Njonja does not seem to understand, so she repeats herself, using English this time. "Perhaps yes; perhaps no."

"Is there nothing we can do?" begs Anna. She knows she will die if this torture continues every night. Life in such pain cannot be possible.

"If the Tuan Dokter gives a sedaka, an offering, to the pontianaks," Babu suggests hesitantly. "Perhaps then they go."

In her state of agitation, Anna thinks Dubois on the veranda is oblivious. She is mistaken. The terrible, anguished cry seems to claw at his soul. As a matter of principle, he tries to ignore it, to explain it away. With the onset of each wail, his jaw clenches and his spine stiffens, as if he is physically resisting its power. He exerts a tremendous effort of will to concentrate on his book. He is a man of great fortitude and he will not be driven from his own veranda. *Recte et fortiter, Recte et fortiter*: he repeats the family motto to himself like a prayer. But no European prayer can banish this Javanese thing that screams in the night.

Anna decides she cannot go back out on the veranda to speak to Eugène. What is there to say, anyway? "Do you hear that sound, the moaning of our dead daughter's soul in torment?" It is not the sort of question one asks Dr. Dubois, even if one is his wife. Anna goes to her room and tries to sleep with the doors and windows closed and a pillow over her head. It is stifling, airless, and still the cry slices through her flimsy defenses like a tiger's claws ripping through a mosquito net. She dozes briefly, to awake with a start, covered in sweat, her heart pounding in fear, her body curled in a self-protective fetal position. She does not know what to do or to believe. All that she knows is that her baby lies across the road in a shallow grave and some . . . *being* . . . in the garden is expressing her anguish.

In the morning, Anna tells Dubois what Babu said about the noise. He does not want to believe it. "This is superstitious nonsense, Anna, just the silly tales of an uneducated babu," he blusters to cover his own unease. It took a lot of willpower, a lot of his sheer Dutchness, to stay out on the veranda the night before after Anna went to bed, pretending not to notice the awful cries.

She turns her face away, so tired and anxious that she looks as if she has been physically pummeled. She does not argue with him; she has never argued with him. She does not expect sympathy but hopes for it. "Babu says," she remarks quietly, "that perhaps if we got a dukun in, a native priest, to hold a special ceremony, the pontianaks would go away." She cannot bring herself to beg him openly but the tone of her voice is enough.

"Anna," he says more tenderly, "Anna. It will probably be gone tonight. It was just some animal, maybe a civet cat in heat. You will never hear it again."

But it is not gone, whatever it may be. That night it repeats its heartrending moans and wails, and the night after, and the night after that, until Dubois stops counting. The question is not "Will it cry tonight?" but only "When will it begin?" The expectation, the waiting, is as demoralizing as the sound itself. Anna no longer sits with him in the evenings at all, though she cannot escape the cries and she cannot sleep once they have started. The only time she can rest is in the afternoons, when *it* is silent, and even then she has terrible dreams.

The djongas, Nassi, arranges for a dukun to come drive away the demons in the garden.

Dubois determines to hold his ground on the veranda. No one can tell him what makes the noise, nor can he imagine a creature that could produce a sound of such longing and misery. But he realizes that he must do something, soon. Anna is deteriorating again, losing contact with reality, growing pale and fretful from lack of sleep. She never asks him again about getting in a dukun to perform a ceremony, but he can see the question in her frightened eyes, day after day.

Finally he gives in and asks the djongas, Nassi, to arrange it. There is a palpable air of relaxation in the household, as if all the inhabitants have been holding their breaths, hoping and praying for the Tuan Dokter to see sense. The servants know well that the Dutch do not always—do not often— understand the importance of such things. It is a peculiarity of the Dutch— perhaps a failure in their upbringing?—that they are so often blind to the spirit world. This trait does not mean they are exempt from its punishments: see how many of them die, babies and strong young men alike! See how often their plantations and businesses collapse in financial ruin! And no wonder. No Javanese would go through life without taking steps to protect against disaster. Still, it is difficult always to foresee what will be the right amulets, offerings, or ceremonies. Even a careful Javanese makes mistakes.

The next day, an old, old man comes to the garden. He is nearly toothless, his back bent into a C, his hair thin. There seems to be no flesh on him at all, just a nearly transparent covering of tissue-thin, brown, wrinkled skin. He walks painfully, with a stilted, shuffling gait, as if his knees and ankles cannot be trusted to change their position without breaking. Dubois thinks he has never seen such an old man before in his life. What can his age be? Ninety? One hundred? More? Could this ancient wreck have seen the decline and fall of the V.O.C., the Vereenigde Oost-Indische Compagnie or Dutch East India Company, at the close of the eighteenth century? However old the dukun may be, the servants treat him with the greatest respect—as if he were the Pope himself, Dubois thinks ironically. Yet, for all his decrepitude, the old man has a certain presence that even Dubois can feel; there is a power about him, a strength that has nothing to do with his pitiful physical condition.

Dubois watches out the window but cannot see most of what the man does. For a long time, he seems to wander aimlessly through the garden, mumbling and muttering in falsetto, scattering something—herbs? dried remains of something?—here and there. Sometimes he lights small fires and adds to them something that he has brought with him, carefully folded in a fragment of beautifully batiked cloth. Whatever it is makes the fires smoke, and then he puffs the smoke here and there with the pitiful remnants of his lung power. He extinguishes each fire by pouring a thin liquid onto it from a little flask that he also extracts from his cloth sack, waiting until each spot is cold before addressing the next. For a long time afterward, he sits rocking under the trees with his eyes closed, humming or perhaps singing to himself in a quavering voice. When the song is finished, he sits unmoving for such a long time that Dubois begins to worry that he has gone to sleep, or died. What a susa *that* would be: the dukun dying in the middle of the ceremony, in their very own garden! The Dubois family would probably have to leave this house, maybe leave Java, if such a thing happened.

Eventually the old man slowly gets up and shuffles to the foot of the stairs, up to the back veranda. In a moment, the djongas appears, carrying his payment, as prescribed, on the palm of his hand, wrapped in a clean white linen cloth covered in red blossoms. As he comes forward, the djongas bows his head to avoid the old man's eyes. He takes care not to touch the dukun as he hands over the parcel of linen. The flowers tumble to the ground as the transfer is effected, and they are left where they fall. The old man speaks to the djongas in a whisper, warning him that no one is to disturb the remains of the fires for three days, no matter what. The djongas nods his understanding. No one sweeps the garden for a full week. The servants, when they cannot avoid passing through it, scurry, eyes down, placing their feet very carefully. Babu will not allow the children to play there either.

The ceremony looks to Dubois like the rankest mumbo-jumbo, a combination of overacted mysteriousness and arbitrary improvisation that can have no possible effect on anything living in the garden. He does not interfere, for Anna's sake, but he is very skeptical. Still, he has been here long enough to know that more than distance and culture separate Java from the Netherlands. A strange, indefinable quality flavors the very air in Java; forces are at work here that cannot be grasped or understood. In this place, even this tough-minded scientist, who denies that which cannot be measured, must acknowledge the possibility of spirits and magic. He does not know how this can be so. Something about Java defies logic. Maybe dead babies have a voice in Java. Maybe it *is* Anna Jeannette, calling to them from the little graveyard across the road. He does not know. His scientific techniques for deducing the truth do not apply to such things. What he knows is that the night cries stop after the old dukun's visit, and he is grateful.

In January 1894 he packs the long, handwritten pages, the precious photographs and drawings carefully for the train trip to the capital. His first duty upon arriving in Batavia is to go over each page in excruciating detail with the printers. They must understand everything: the unfamiliar scientific terms, where this illustration goes and that one, how the captions are to be printed. There is only one firm in Batavia whose work is of high enough quality, Dubois thinks, and so he entrusts the work to them. As he walks away from their offices, he feels a terrible pang of regret. This must be the way Prentice felt, leaving Gerard with his in-laws for the first time, Dubois thinks.

While in Batavia, he takes time to report to his superiors in the Army and to the Governor on his progress. They are pleased that the monograph will soon be forthcoming although, he warns them, it deals only with *P.e.* and not with the abundant fossil fauna he has collected. That will have to wait for later publications. His superiors suggest that he consider reenlisting— his scientific work has gone so well, produced such impressive results—but he cannot spend another tour of duty in Java. Another eight years might kill them all. He fears for Anna's sanity and, always, for the children's health. His own health is surely compromised, for he is plagued by intermittent fevers that make him as useless and limp as an empty snakeskin. Besides, he has already found what he came for. No, his tour of duty will end in 1895 and he will go home. Once again, he asks for permission to travel to British India, to inspect the Siwalik fossils for himself and perhaps visit some of the sites where they come from. This time, he is successful. He may go at the end of 1894.

Dubois tends somewhat absentmindedly to some shopping Anna has asked him to do, an endless list. And then, laden with more luggage than he came with, he reboards the train, this time taking the line that travels the northern route, through Cirebon to Semarang. He has only a few hours with Prentice in Semarang before he must catch the late train that runs south across the island to Solo. There, he will rejoin the line that goes through Magelang and Kediri before turning south to Toeloeng Agoeng.

Dubois finds himself foolishly excited; it has been so long since he and his friend saw each other often, but the goodwill is as strong as ever. Disembarking at Semarang, Dubois pauses a moment on the step to look around at the confusion and chaos of the station. There are Dutch travelers in their dusty suits, weary and travel-stained as he himself must be; military men in sweat-soaked uniforms; a few pretty njonjas returning upcountry from a holiday in Batavia in their new outfits; hard-bitten, red-faced planters coming back to their plantations. Swarms of porters appear with a wave of the stationmaster's arm, seizing everyone's luggage and barang-barang with enthusiasm. Dozens of vendors come forward as the passengers disembark, calling out what drinks and snacks or newspapers they have to offer. Entire Javanese families squat with their lumpy bundles of possessions on the platform, waiting for a train or perhaps simply settling in to beg for a living. Small brown-skinned boys dodge through the crowds, celebrating the sheer excitement of the hordes of people and the magnificent puff-puff-puff of the gleaming steam engines. The sheer activity and randomness of it all are pure Indies, Dubois thinks.

Through the crowds and the colors and the noise comes Prentice, that good honest face, those clear blue eyes, shining with sincerity. He strides along like the embodiment of some god, a blond Prince Ardjoena out of the Javanese sagas perhaps: strong and young and handsome. Dubois drops his luggage and surges forward to greet his friend. Their handshake turns into a clasping bear-hug, so delighted is each with the other's presence.

"Let's get a boy to watch your luggage—just leave it there—and we'll get out of this and go have a decent meal," Prentice proposes. They hire a trust-worthy-looking boy and leave the station, striding side by side out of the clamor to find a quiet restaurant where they can talk. They speak from their hearts, of their worries and struggles and triumphs, with no hint of jealousy or misunderstanding to mar their accord. They are as close as ever, as close as they were on Gunung Ardjoena, and they reminisce over that wonderful time, and the trip to Trinil, too. It is the blessing of their friendship that they can be silent together in perfect harmony, too. Soon it is time to return to the station and find the boy, who is dutifully waiting. Dubois hates to say good-bye to Prentice yet again. He boards the train with a feeling of fullness tinged with regret; what wonderful days those were, when he and Prentice could see each other nearly every day. To be separated once again from Prentice is like leaving a part of himself behind.

No, Dubois corrects himself, Prentice is not a part of him. He is outside, a different man altogether, a better one. But he is what enables Dubois to be himself, his true self. Somehow he frees Dubois, shows him the right way to go, to behave, to think. Everything is so clear in his presence. It is a rare gift to know such a man.

CHAPTER 31 INTERMISSION

As the dry season begins, Dubois decides not to send Kriele and De Winter and their motley crew of laborers back to Trinil for a full season. He does not expect more specimens of *P.e.* to turn up; indeed, it would be inconvenient to find more with the monograph already at the printers. But they will place the monument at Trinil before returning to various collection sites in order to take more detailed notes on the geology and improve their sketch maps. The engineers complain ceaselessly about the coolies, the coolies complain ceaselessly about the work, they all bemoan the difficult living conditions. The troubles now seem as timeless and cyclical as the rise and fall of the Bengawan Solo itself. Dubois is more concerned with making sure the mammalian fauna already in his possession is properly cleaned, packed, and crated for shipment, and that his notebooks are up-to-date. With a great sense of anticipation, he waits for the proofs from the printers in Batavia, for his first glimpse of his own, glorious, triumphal monograph.

With Dubois' scientific quest over, the tensions in the family ease. Anna grows better and better daily; she and Dubois resume cordial and pleasant relations, if not the affectionate intimacy they once enjoyed. The children grow calmer and more placid as their mother's hours at the cemetery

By September 1894, the excavation at Trinil is huge but no more Pithecanthropus *remains are uncovered.*

diminish. Time is repairing the wounds left by the deaths, though the scars are permanent.

Dubois is focused on his monograph, compiling lists of those to whom he will send it for maximum effect, and planning his trip to India and the family's long-awaited return to Holland. After seven years in the Indies, the last year seems to fly by unnaturally swiftly. He and Prentice resume their intellectual dialogue via letters, as of old. Dubois is not so burdened with troubles as he once was, and Prentice has begun to succeed handily at running the plantation now that the owner, Van der Leeuw, is no longer disagreeably underfoot. They correspond about many topics, including John Tyndall's disproof of the idea that life can generate spontaneously from inanimate ingredients, a crucial advance in biology. It is the sort of issue they would once have argued on the veranda at Mringin.

Geboegan, Oenarang   June 30, 1894

My dear Doctor,

I was very agreeably surprised to receive the package you sent by post today. Please accept my heartiest thanks for the books which I doubly prize for the importance of their contents and for the goodwill of the giver.

You need not fear I shall peruse them with any hostile feeling. Having renounced my early beliefs—not without a hard struggle, and reluctantly, yet decisively—and having come to regard in a more scientific and truer light the mysterious universe in which I live and the problems bound up in it—this latter owing to my acquaintance with you, to what I have learned from you in conversation, and what I have gained by the perusal of the books you have already favored me with—my mind is now open, I trust, to receive any truth without bias or prejudice.

Rest assured I shall carefully and attentively and profitably peruse the books now so kindly given me. I too was pained by reading a cold, unsympathetic criticism of Tyndall and his work. He was indeed the apostle of science, and the quickening of your mind received from a perusal of what he wrote was no doubt felt by many, many more searchers after truth & knowledge. That a great mind like [the physiologist] Helmholtz should admire & esteem the eloquent & devoted priest of science is alone a sufficient reward for Tyndall's labors, and a mighty rebuke to the carpings of lesser critics.

I shall now draw to a close as I am wearied & my hand tremulous. I have been 7 hours on horseback in the sun today. We are getting more

coffee than our taxation, and my work is appreciated both by the bank at
Semarang and by the owner of this estate in Holland.

> *With best thanks & best wishes,*
> *Believe me,*
> *Sincerely yours,*
> *Adam Prentice*
> *Compliments to Mrs. Dubois*

At the beginning of August, the copies of the monograph are ready. The printers pack them up and send them by fast train to Toeloeng Agoeng. Dubois cannot keep his hands from shaking as he unwraps them, even though he has corrected the proofs with the greatest of care and knows the result must be fine. The book looks to his eyes like a miracle. It is a handsome volume, oversized, and bound in a dignified buff-colored board. As he pages through it, he pronounces the illustrations to be excellent, the paper good, the typeface strong. It is a work anyone might be proud of, especially one who has nearly died to accomplish it.

He shows the work off to his uncomprehending family—the children seem more genuinely interested than Anna is—and to anyone else who has expressed the least interest in his work. He carries a copy up to Ngrodjo, to share it with Boyd. Then he carefully wraps dozens of other copies to send back to Europe. These are his envoys, who will carry his ideas across the oceans back to the scholars of Europe. Soon they will all know what he has found. Soon the name of *P.e.* will be on everyone's lips as the link that is missing no longer from Darwin's chain of evolution.

Dubois is uncharacteristically relaxed, happier and more at peace than he has been for years. The major tasks before him are packing up his fossils, making sure their documentation is complete, and planning his trip to India. Best of all, Prentice's long year at Geboegan is drawing to a close and he will soon be back at Ngrodjo, where Dubois can see him often. He, Boyd, and Prentice will be able to sit together again on the veranda and try to bring order to the universe. The prospect fills his heart with joy.

In late October, Prentice receives word at long last that his application to develop waste land as a plantation is likely to be granted. He will be given a plot south of Kediri, near the village of Ngawiluwih, not far from Mringin. He writes to Dubois in a state of high excitement; his ambition to have his own plantation is at last coming to pass.

Mringin   October 20, 1894

My Dear Doctor,

Thanks for your kind letter of the 17th inst. Mr. Boyd is very sorry indeed that you should have misunderstood him especially as he had before asked you to take nothing amiss from him as he meant it all well with you and has always considered you a good friend. Nor can he even in the least recall the circumstances that seem to have made a painful impression in your mind. The weather, too, unfortunately for next year's coffee crop, is and has been miserably wet at Mringin and misty so that visitors cannot enjoy the climate much, but all the same any friend who does not mind these things is highly welcome to come, especially an old frequenter like yourself.

I am intending to begin a new estate close to here & the commission will be here shortly to inspect the pillars & etc. I can get no permission to begin planting until all these matters are arranged. Thanks for your hearty good wishes for my success in this new venture which I anticipate will turn out well.

The "Climates" book remains long in suspense, but I trust the work will be published soon in English. What is the general impression now in Germany & among the learned respecting the views you express in this book?

Mr. Van Velthoven and Mr. Goernat are coming to Mringin tomorrow for one day. Messrs. Henderson & Budd, friends of mine, are coming here today to stay two days. They are going to inspect the forest here with me. I have cut a narrow foot path through it to facilitate inspection but I trust it won't be wet tomorrow, as it is today.

My child is in magnificent health in Malang.

When are you going to Calcutta?

Come up please to Mringin at any time that suits you, with Mrs. Dubois and the children, and receive the kind regards of Mr. Boyd and

Yours Sincerely,
Adam Prentice

P.S. I often look back with pleasure to our excursion to Trinil. Whenever I pass [the turn-off to Trinil at] Kedoeng Gale (I rather forget the name) my recollection dwells upon the rustic holiday you provided for us. Indeed, I would like to make another visit with you. A.P.

Dubois' answer to this letter is a silent gift, a copy of the monograph itself. There is no greater treasure he could give to anyone, nor no person in the world whom he values more highly than Prentice. Because it comes from Prentice, Dubois decides, too, to accept the secondhand apology from Boyd, who showed an appalling lack of enthusiasm for the monograph when Dubois took it to him. How could he react so offhandedly to receiving the greatest scientific work of the century? Perhaps, Dubois concedes grudgingly, the Old Warrior was simply a bit under the weather or preoccupied. He will let the matter pass now.

Prentice soon writes again.

Toeloeng Agoeng   November 3, 1894

My Dear Doctor,

I was agreeably surprised this morning to be presented with a handsome volume relating full particulars of your wonderful discovery of the missing link (or one of them). I shall look through the volume with great interest.

I am going today at 12 o'clock to Blitar. Perhaps I may be back again on Tuesday and if so I shall endeavor to give you a call in the evening when I hope to find you in good health and spirits.

Meanwhile, with best thanks, and with kind regards to you all (in which Mr. Boyd who is down here changing money heartily joins).

Believe me,

Sincerely yours,

Adam Prentice

But Prentice does not come. Dubois waits out Tuesday afternoon and evening in a pleasant state of expectation; Anna, too, is looking forward to the promised visit, for she has come to enjoy Prentice's good-humored company greatly. The hours pass, but neither Prentice nor word from him arrives. When a second day and evening pass without word, Dubois becomes alarmed. He pens a hasty note and sends a boy to Prentice's lodgings, inquiring as to his well-being. Is his friend down with fever, perhaps, in need of Dubois' medical attention? He would willingly drop everything and go to him. The truth is better than he fears.

Toeloeng Agoeng   November 7, 1894

My Dear Doctor,

I have been 3 days in house with a swollen throat inside, in consequence, I think, of going with wet shoes 2 days.

However, with taking care of myself, I am now almost better, and tomor-row I am going again up to Mringin. If you are receiving no visitors this evening and have nothing particular to do, I should be glad to call round at your house this evening at any hour that best suits your convenience.

When your messenger called last night, I had just retired to my room to escape the night air.

With kind regards,

Believe me,

Sincerely yours,

Adam Prentice

P.S. I write this letter on my knee to be out of all drafts.

Dubois is relieved that Prentice is only mildly ill. As the time of his extended journey to India approaches, he finds himself worrying more obsessively about the health of those he holds dear. Little Anna's death still haunts him and casts a pall over his relationship with his wife. He thinks it would be an unbearable load of guilt if another of his loved ones fell ill and died for lack of his attention. And then he reminds himself, there was noth-ing he could have done to save the baby's life. All his learning and skill and knowledge, all the medicines and scientific principles in his mental armory, were useless.

CHAPTER 32 TO INDIA

At last the end of the year approaches and Dubois' longed-for trip to India is about to begin. It will be his last scientific gift to himself before leaving the Orient. His departure is fixed for two days after Christmas. He leaves in a state of elation, carrying with him a journal in which to record his observa-tions and notes and a special parcel of letters from Prentice.

Mringin   December 21, 1894

My Dear Doctor,

Herewith the two letters of introduction (for Dr. Galloway & Mr. Lyon Sr.).

In addition, I shall write Mr. Lyon in the course of a day or two so that he knows in advance that you're coming. If you lodged with him or with Dr. Galloway it would be much pleasanter than in a hotel surrounded by strangers, but I do not know how either gentleman is at present situated as regards household arrangements. It is therefore better I write Mr. Lyon in advance.

I trust you will have a prosperous voyage, and profit greatly from your visit to Calcutta, so that you may always be able to look back upon it with pleasure.

The severance from your family is of course a great drawback, but it is only temporary, and it is unavoidable; and the time of your return will at length arrive. Till then adieu, and believe me to be

Sincerely yours,
Adam Prentice

Prentice encloses two nearly identical letters to Lyon and Galloway, introducing Dubois as a dear friend and asking them to show him every hospitality.

Dubois bids good-bye to his wife and family and departs by train for Batavia, arriving in Semarang early on the morning of the twenty-ninth. The steamer leaves for Singapore at nine on the morning of January 4. The day is fine and the voyage seems like a reward for his years of hard work. He starts a journal, to record his impressions of the trip.

> DECEMBER 29, 1894. *I met on board a very pleasant, though small, company: 1, Mijnheer Vermeulen, head of the customs in North Java, from the Noordbrabant province at home, thick and good-natured, witty and cheerful; 2, Harmsen, a globetrotter, making a trip around the world, at present coming from Siberia, who knows a lot of stories from his experiences there, in Japan, Africa, and America; 3, a midshipman, named Heus, who comes from the island of Lombok.*

> JANUARY 4, 1895. *Till yesterday afternoon we had excellent weather. Then, however, we passed the north point of Bangka Island and we got heavy wind and swell from the Chinese sea, which made more or less all of us seasick. The motion of water and wind, and hence of our ship, kept on toward morning and it was difficult not to wake up again and again because of the way I rolled to and fro in the berth. Still I slept the sleep of the righteous despite the uncooperativeness of the elements. From Batavia thus far we have passed thousands of smaller islands, which are placed like dark-green bouquets in the light green sea of the straits of Bangka and Malakka. To*

the right is the mountainous Billiton Island and the larger but
flatter Bangka Island, to the left is the long, low coastal strip of the
Palembangsche region of southern Sumatra.

Toward morning, on our right side, we passed the islands of the
Lingga Archipelago and then Riau, covered in forest and alang-
alang grass in patches and small clusters of coconut trees. To the left
was the splendid coast of eastern Sumatra—Siak and Djambi—
with its richly indented bays and the land covered in original forest.
It gave me inconceivable nostalgia for the pleasant times in
Sumatra.

JANUARY 7—SINGAPORE. I arrived here toward 4 p.m. yesterday,
still somewhat seasick, so perhaps the beautiful scene of the greatest
and most gorgeous harbor of these regions had a diminished effect
on me. But still I know that it was magnificent. First some small
islands, then the coast of the large island on which Singapore is situated, mostly green but relieved with pink-brown rocks of granite
and a cathedral and a great number of other large buildings and a
forest of masts of ships and funnels of steamers and all sorts of other
vessels. The rocky, capricious coastline tells me I am outside of the
Dutch colonies, as we do not possess such a harbor either on Java or
on Sumatra, nor on Kalimantan or any other island, for the surrounding land there in the Indies is always flat. The boat is hardly
tied up than the Indians arrive on board and go ashore with their
turbans and stately beards and their ridiculous thin statures. Going
ashore myself and seeing the hackney carriages and bullock carts
and rickshaws (which are pulled by the Chinese), I cannot help but
realize that I am in an entirely different country than the Indies.

Having placed our luggage on a bullock cart under the supervision of a porter from the Hôtel de l'Europe, Mijnheer Vermeulen and
I got into a hackney carriage and rode into the city. There are many
large, new, or nearly new buildings in the monumental style, both
private and public buildings, with excellently kept up roads and
beautiful lawns (where I see at this very moment, from my window
in the Hotel, a number of young Englishmen in white trousers and
shirt-sleeves playing lawn tennis). . . .

Dinner does not differ a lot from the Dutch dinner, but it is
followed by a sort of small rijsttafel with many savory dishes. Apart
from that, another difference from our ways is the division of the

day. In the morning at 7 o'clock a small teapot with its appurte-
nances and a thin slice of bread (with a little butter) is brought into
the room by the Chinese room-boy (all the servants, even the wait-
ers, are Chinese). After that, I have to wait until 9 o'clock when the
"brunch" starts. This is, however, a very substantial meal: fish with
bread, omelette with bread, drumstick with bread, cheese with
bread, then a cup of coffee. . . .

We drove yesterday evening to the waterworks, a reservoir that
provides the city with water, and afterward to the Botanical Garden.
Because of the way it is situated, its outlooks and views, and
because of its beautiful layout, the Botanical Garden here exceeds
the one at Bogor in Java, though it has much less scientific value. In
the garden are 3 mammals and 7 birds which purport to be a zoo.
The museum that I visited today is a beautiful building with a few
badly stuffed animals and other specimens, but also a large library
of all sorts of works, including belles-lettres.

Dubois finds the mixture of English, Malay, and Chinese cultures
immensely interesting in comparison to the Dutch-Indies hybrid with
which he is so familiar. The English are stiffer, more remote from the
Chinese and Malays than is common in Java. There are no mixed marriages
in evidence here, no crowds of pale brown children with curly hair, no
comely Chinese or Malay women dressed in European clothes standing
demurely by their English husbands' sides. The English women—not that
there are very many—maintain European dress.

Prentice's friends Galloway and Lyon adopt Dubois immediately as a
friend. They show him the sights of the town, take him to the museums and
introduce him to curators and other scientific men, and generally make
sure he is amused and welcomed. The few days in Singapore pass most
pleasantly. On the ninth, Dubois sets sail for Penang. It is an idyllic, comfortable voyage, he records in his journal.

JANUARY 10, 1895 IN THE "PUNDUA," IN STRAITS OF MALAKKA
BETWEEN SINGAPORE AND PENANG. *Yesterday afternoon at half*
past four, the vessel—a big beautifully furnished ship of 2100 tons
(3x as large as the packet service ferry), set sail. Luckily I received a
cabin alone, one that is more spacious than the ones on the Dutch
boats and furnished with a commodious sofa. The party of travelers
is very mixed, there is even a Chinese; most of those traveling alone

are the English, 1 English-speaking German (from Jena), further an
Arabian Jew, an American, and a lady with her good-looking
daughter. . . .

    The island of Penang stands out clearly, with its mountain slopes
dappled in dark green and yellow, the sky turning light blue with
little white cloudlets floating above the land. A prau, an Indies out-
rigger, with 2 brown batwing sails drifts past our boat. The moon is
at the point of sinking beneath the sea. A silver fish jumps out of the
water and dives back in again several times. The sea is very calm
and no wind blows.

The shipboard days pass in lazy comfort. For once, Dubois has no ambitions to pursue, no work to claim his attention: there is little to do except
wait for the accolades of his European colleagues when they receive his
book. Dubois passes the time in writing in his journal and observing his fellow passengers. He tells them amusing and exciting stories of his life in the
Indies and hardly mentions science or the missing link, which gives him a
delightful sense of traveling incognito. At Penang, Dubois changes ships,
taking the *Basnura* to Rangoon; from there he catches the *Abalda,* arriving
in Calcutta on January 20. This extended and ever-changing sea voyage is a
rare and happy time in his life. There are even a few pretty women who are
charmed by this handsome Dutchman, who tells such interesting stories
and speaks such good English. They do not guess what he is waiting for: the
recognition that he has made the most important discovery of the century.

CHAPTER 33 CALCUTTA

As the ship docks, Dubois wonders whether Calcutta will live up to its nickname as the City of Palaces. Certainly the white spires and saffron-colored
domes of the government buildings he can see, set amidst the palm trees
and banyans, gleam attractively in the sunlight. He can just see the green of
the enormous Maidan, a huge open park that stretches some two kilometers along the river's edge from the racecourse in the south to the Esplanade
in the north. He travels into the city by carriage, observing, to his immense
amusement, that the Maidan has a thriving population of marabou storks,
just like the Adjutant they kept in the garden at Toeloeng Agoeng for a while.
Dominating the Maidan is Fort William, a large, solid octagonal construction built in the late 1700s to house all the town's European citizens in case

of attack by the natives. Dubois can see that the British, like the Dutch in the Indies, have not always been welcome rulers. The Maidan owes its very existence to the fort; the space was initially cleared of tracts of forest to give the soldiers a long, clear line of fire. Now it is a valuable public park, heavily used for entertainments and endless rounds of sports and games.

The Maidan's eastern edge will become familiar territory to Dubois, for it is traversed by the broad avenue known as Chowringhee, the home of the famous Indian Museum, where he will study the Siwalik collections. Said to be the largest museum in Asia, the Indian Museum boasts collections of zoological, archaeological, and geological objects, as well as an art gallery, all housed in a vast two-story building completed in 1875. It is an impressive accomplishment, this elegant stone building with its double galleries decorated with stone arches. The whole embraces a central courtyard, with formal plantings and a fountain that casts a continuous spray into its octagonal pool. Dubois is a little embarrassed to compare this stately edifice with the shabby quarters allotted to the natural history museum in Batavia.

Farther still to the east, past the Indian Museum, is the wretched collection of huts and shacks of the bazaar, congested with the poor and full of every manner of food and merchandise, smelling of sweat and spices and fumes from cooking fires. It is still very like the native markets of the Indies, Dubois thinks as he explores its margins. The people are still small and poor, dark-skinned and brightly clad—the women in saris, not sarongs, the men in loose pajamas—and gabbling away in a hundred unintelligible languages. He ventures into the market only far enough to procure a fine turquoise-blue scarf for Anna, embroidered with gold thread in an intricate design. He does not suppose she would wear such a showy garment, but perhaps she will enjoy it. He purchases a few trinkets for the children as well. They are little different from what is available in Java, but they will show his family that he is thinking of them even when he is far from home.

What preoccupies Dubois more than sightseeing and shopping is settling into the Grand Hotel, an impressive building with an elaborate white-washed façade with balconies and balustrades everywhere. He has been told this is the best hotel in town; certainly, the room is spacious after his neat cabin on the *Abalda*. But he is soon disenchanted with his accommodations and the other guests.

JANUARY 22, 1895. *This morning it is again so really wintery and cold, with a draft on my face, that I am surprised that the shower is not as cold as ice; for that matter, it is cold enough. Still, it gives me a certain refreshment after the terrible all-night coughing of the man in the neighboring room. In addition to that, he makes the most filthy sounds and swears under his breath and moves roughly, crashing into things; he always seems to be drunk from the large whiskies that I hear him prepare in the evening before going to bed. Accordingly I cannot stay here a night longer. Some way, somehow, I have to seek shelter here or in another hotel. To stay here for four weeks, without getting a decent rest—even for someone accustomed to traveling and marching like me—would be anything but refreshing.*

At the museum, things go much better. The curators—Alcock, Finn, Holland, and Moll—are delighted to have a knowledgeable visitor. They are busy with their own work, of course, but they are hospitable and welcoming. Over tea and at lunch, he tells them of his discoveries in the Indies. They are fascinated to learn of his work and look forward to hearing his opinion of the resemblances between the Siwalik and Indies faunas. By the end of the day, he feels himself particularly drawn to Alcock, the friendly, intelligent head of the department of zoology and archaeology, which

In 1895, Dubois photographs his friends in India: Frank Finn (left, standing) and A. Alcock (right, standing), both of the Indian Museum; and two Dutch acquaintances, Carl Thieme (seated left) and Miss Thieme (seated right).

houses Lydekker's fossils. Alcock generously solves Dubois' accommodation problem.

> JANUARY 23, 1895. *Alcock is a good fellow, though unfortunately for me he is interested in lower marine animals and therefore has few points of scientific contact with me. In the afternoon he invited me to take tea with him and proposed that I stay with him, which invitation I hesitatingly accepted.*

It proves a good decision, for the two men get along congenially. Alcock's household is large and comfortable and well run, a bachelor establishment where Dubois soon feels content.

Most important of all are the fossils. Going through the collection is for him like walking blindfolded across unknown terrain, never knowing whether he will sink into a sticky, smelly morass or stumble into a fragrant mass of orchids. Has Lydekker found something like his missing link? Is there some specimen here that will disprove all Dubois' ideas? He never knows what each new drawer and cabinet will reveal; until he has surveyed the entire collection, he can only wonder what specimens are there, what animals are represented in this sizable fossil fauna from ancient India. He is soon reassured, however, and able to write of his preliminary studies with satisfaction.

> JANUARY 23, 1895. *Yesterday I studied in the museum some zoological and paleontological specimens, mainly the maxilla or upper jaw of* Anthropopithecus sivalensis. *According to me, Lydekker has described and pictured it wrongly, because he has inaccurately joined some of the fragments. The Siwalik fossil collection is in many respects much richer than mine, certainly so because a large part of it is in London and I cannot see it, but the condition and the preservation of the fossils disappoints me. In a number of aspects, my collection is richer. Buffalo, deer, and antelope—I have better and more beautiful ones, also many more elephant molars, and in general molars from all kinds of mammals. Nothing like my P.e. is to be found in the Siwalik collection. . . .*
>
> *Taken all together, I believe now that my collection is even more significant than I assumed earlier. In* Nature *(January 3, p. 230) is a short note about my* Pithecanthropus *description, to which my friend and host Alcock drew my attention. It is not much more than*

an announcement, but it makes clear that <u>my ape-man will not</u>
<u>walk into the world unnoticed.</u>

As Dubois does not have his own copy of *Nature,* and is reluctant to deface another's, he carefully copies the announcement out into his notebook.

The significant name "Pithecanthropus erectus" *is proposed by Dr. Eug. Dubois, of the Netherlands-Indies Army Service, for some fossil remains recently discovered in the andesitic tuffs of Java, as indicating the former existence in that island of an intermediate form between man and the anthropoid apes. The bones, which consist of an upper part of a skull, a very perfect femur, and an upper molar tooth, are elaborately described and figured in a quarto memoir recently published in Batavia.*

It is not much, but it is a beginning. He likes the phrase "elaborately described and figured." It shows that someone appreciates the great care he took over the monograph.

Dubois finds his colleagues at the Indian Museum extraordinarily congenial. The museum staff soon feels like a company of friends, exchanging ideas while at work and often meeting in their off-hours. They arrange a temporary membership for Dubois at their club, so he can join them for socializing and sport. They make a point of introducing him to C. L. Griesbach, the director of the Geological Survey of India, who knows the Siwalik Hills well and has much sound advice to offer. Dubois enjoys himself. On the day before his thirty-seventh birthday, he writes:

JANUARY 27, 1895. *This lady, the second one I talked to, was obviously a lovely and lively woman, different from what I had imagined English ladies to be: full of spirit and cheerful, and possessed of a calm charm that did me good, as if she were a pleasant scent or a beautiful flower.*

Even the weekends offer Dubois new entertainments. An opportunity too good to miss arises when he learns that a group of natives from one of the remote islands east of the southern tip of India and south of Burma is to be exhibited nearby.

FEBRUARY 2, 1895. *Yesterday instead of going to the zoological garden, I went in the afternoon with Finn to see a troop of natives from the Andaman Islands—remarkable, small, black humans of an*

Australian type, of whom there are only 500 in existence. They are somewhat lighter in color than negroes, but they are black enough, with faces that resemble those of Australian natives; their hair is crimped and plaited in bunches. The doctor (whose name I have forgotten) who collected them and will bring them to London asked me, as "the discoverer of the missing link"—that is how I am generally known here on account of the note in Nature—*if I took them for direct descendants of my missing link. English scientific men are so old-fashioned in their thinking! Meanwhile I am happy to have had a chance to see such a primitive and rare human race.*

The English, Dubois realizes, differ utterly from the Dutch in their exuberant emphasis on physical fitness and sports. As he has always prided himself on his strength and fitness, he finds this national characteristic admirable.

FEBRUARY 3, 1895. *I see with great regret that we Dutch are so far behind our English counterparts. Among the officers there is not a single thick, impossible belly, no stiff physique without resilience, no drunken soldiers with wretched, mean faces. Even old officers (who, anyway, must retire from the army at 55) look supple and yet strong and still possess (as do the ladies) something youthful, something I am not as a rule accustomed to finding among the Dutch. . . . Their interest in sport and games is extraordinary; I realize now that this is where the strength and power of the English nation originates. Even ladies take part in the hunt, so that the daughter of the Inspector of Forestry actually shot a tiger some days ago, in the neighborhood of Dehra Doon, where she is at present traveling. . . .*

Truly, it saddens me to think back on all those thick Dutch bellies in Java, as I must do whether I will or not. The great English public of ladies and gentlemen that surrounds me includes not one person (or perhaps only a single one) who resembles a degenerated lump of fat, as is so often the rule in Java. That some among us Dutch, like myself, differ in this regard is considered more a disadvantage than an advantage. . . . Here, without exaggeration, everyone keeps his body strong and supple and healthy through moderate exercise and is in his work as good and as thorough. No heaviness of body and mind—as is with us nearly the rule—but also there are none who lack seriousness, no clowns or acrobats. Even the making of music by the moderately talented seems to me to be the outcome of the correct equilibrium of English society.

Dubois spends the last few days of his time in Calcutta trying to organize his trip upcountry, with a lot of help from Griesbach in particular, who writes out pages of advice. Dubois feels it is essential to see Lydekker's sites for himself and to collect additional fossils from them if possible. It would be foolish to come all this way and not return with a comparative collection of the Siwalik fauna. But the Siwaliks are a long way from Calcutta: he must traverse some fifteen hundred kilometers of northern India. Although traveling on the Grand Trunk Road has an attractively romantic sound to it, Dubois cannot imagine that voyaging on this ancient trade corridor, stretching from Calcutta in the east to Peshawar in the North-West Frontier Province, will be either comfortable or swift. He is far better off relying on the excellent railway system that spreads across the Indian subcontinent like a spider web. In a first-class compartment, he can depart from Calcutta on the night of February 8 and arrive, fresh and rested, in Ambala on the tenth.

Alcock kindly sees Dubois off at the station, even though the departure is late. The previous day, they dispatched a large quantity of camping and collecting gear by goods train, to await Dubois at the other end. He carries with him only a small case with clothing, toiletries, a camera, notebooks, and writing instruments.

The station is an exercise in chaos. Travelers wend their way through a maze of humanity to buy their tickets and arrive at the correct platform. Boys and men carry luggage; children and women sell newspapers, flowers, candies, hard-boiled eggs, and every sort of food that can be eaten with the fingers; itinerant magicians perform tricks to coax the unwary out of their rupees; sadhus—holy men—meditate or hold out their begging bowls; Muslims spread their prayer mats and pray to Mecca; Hindu women squat to tend small fires over which they cook their family's meager meals. All this life and color and squalor is incongruously enclosed within the confines of the magnificent, soaring railway station the British have built. It is a wedding cake of a building, large, echoing, cool, smelling of people and curry and steam, with a front façade so white it almost burns the eyes. The whole scene is very like, and yet terribly unlike, the Indies, Dubois thinks.

He finds his comfortable first-class compartment, puts his case under the seat, and throws open the window to shout his last good-byes and thanks to Alcock, who is waiting on the platform.

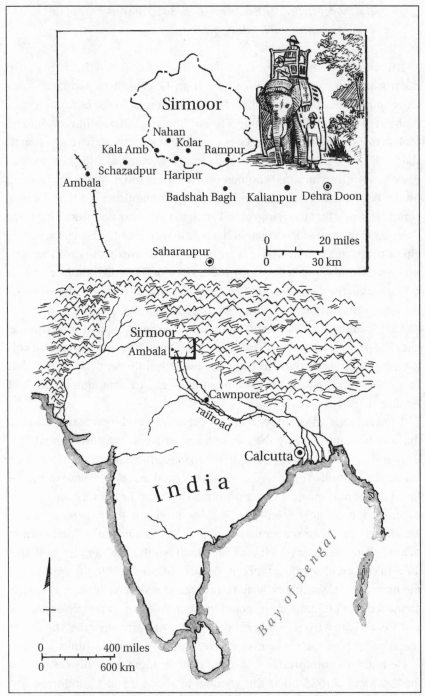

Dubois travels from Calcutta by rail to Ambala and collects fossils in the princely state of Sirmoor and neighboring districts.

"Have a good trip, Dubois!" Alcock shouts merrily. "I'll see you when you get back. You can tell me all about your discoveries then. Have you got Gamble's address at the Forestry School in Dehra Doon? You must look him up, now."

"Thank you, Alcock," Dubois replies, waving. "You've been a tremendous help. Say good-bye to the others at the museum for me. I'll make contact with Gamble at the Forestry School for sure."

Like a mother hen, Alcock cannot resist clucking over his visitor. "Now, you've got all the other addresses and papers and the receipts for your goods that went yesterday?"

Dubois smiles; this concern is so like Alcock. "Yes, yes," he says, pulling out a leather case in which he keeps his papers: receipts, addresses, advice on prices and how to organize his trip and his men. "Here are the ones for the small crates, and the one for the excavating gear, and . . . Ach! Where is the one for the tents?" As the train starts to move, he shuffles through the chits and receipts. Leaning out of the window, he calls back to his friend, "Alcock, I haven't got the one for the tents after all. I must have left it in my pocket yesterday. Can you check in my things at your place?"

"Righty-ho, Dubois," responds Alcock, trotting along beside the train. "I'll check for it as soon as I get home. Don't worry! I'll send it along after you, to Ambala."

Dubois settles back into his seat, disturbed. It is not like him to lose track of an important piece of paper; he is both embarrassed and concerned. All this traveling and packing and unpacking and depositing things here and there and the other place to be called for . . . it is all very complicated, especially in an unfamiliar culture.

He is pleased to find that the compartment is not fully booked. For the first leg of the trip, he is in company with two Army officers making their way to a new posting. They make polite conversation for a while, but all turn in to sleep on the bunks (precisely made up with starched white sheets and a blanket and pillow) before long. Looking around the compartment, Dubois thoroughly approves of the good mahogany woodwork, the folding berths that can be tucked up and out of the way in the morning, and the neat little sink that folds down from the wall in the small toilet and washroom.

Alcock's cook has packed him up a tiffin box with some tea, fruit, bread and butter, a few hard-boiled eggs, and a teaspoonful of salt wrapped in a twist of paper, in case he gets hungry before breakfast is served. For now, he is happy to read a little and then doze off, lulled by the rhythmic percussion of the train. In the morning he awakens early and is happy to watch the remarkable landscape out the window as he sips his tea. The tea is cold by now, of course, but it is better than nothing.

There is no mistaking this part of India for the Indies, Dubois decides. India is flat and hot and full of rivers from the swamps of Calcutta right across Bihar to the great Ganges River. But there are no cool green rice paddies, no groves of bananas and coconut trees, and no sheer, green-cloaked volcanoes suddenly arising out of the paddies, as in the Indies. Despite the rivers, this part of India seems a much drier country, brown and dusty rather than lush like Java. Before long, the train arrives at Mokameh, along the banks of the Ganges, where breakfast is served. The system is most impressive. Meals take place at preordained stations at preordained times. Once the train pulls in, an elaborate restaurant service springs into action. Each first-class compartment containing ongoing passengers is served by a particular cart, the responsibility of two restaurant-wallahs, as they are known in Anglo-Indian slang. The restaurant-wallahs position their cart outside "their" first-class carriage, handing up silver-covered dishes of food, linens, cutlery, and drinks in a matter of moments. The meal is eaten and cleared away in less than half an hour, so the train can continue on schedule, as it invariably does. Dubois is amazed at the efficiency and predictability of it all. He dines well and peacefully and soon the train is clacking down the rails again, following the curving course of the Ganges for some miles.

Two hundred kilometers later, they arrive in Baxar, where lunch is served in the station waiting room while the compartments are brushed and cleaned. Another few hundred kilometers onward, they have tea in Allahabad, served again from carts manned by restaurant-wallahs, and at 18:25 precisely the train pulls into Cawnpore for dinner. Once again, the linens are spotless, starched and ironed; the trains run exactly on time; the serving dishes are polished to a high gleam and the food is plentiful. The British have accomplished so much in India, have almost civilized the place; it seems so polished and well organized compared with Java.

At 8:04 on Sunday morning, two days after leaving Calcutta, Dubois arrives in Ambala. The city lies on the broad, fertile river plains north of Delhi. To the north he can see the impressive foothills and behind them the mountain ranges of the highlands, the Simla Hills, Mussoorrie Range, and the Siwaliks themselves. It is a dramatic landscape, enhanced by the high proportion of Sikhs who live there. The Sikhs, Dubois has been told, are mainstays of the British Indian Army. Tall, well-built, proud, and fiercely loyal warriors, the Sikh regiments are renowned for their courage, as well as for their magnificent beards and mustaches.

A letter from Alcock is waiting for Dubois at Lumley's Hotel in Ambala.

Indian Museum   February 9, 1895
I have been through all your papers and have also looked in the almi-
rah [armoire] among your clothes, but I cannot find the Railway Receipt
for your tents.
I cannot find it—my God!!
I hope you have had a good journey, not too hot and not too cold.
Mind you communicate with Mr. Gamble. I hear from Dr. King that Mr.
Gamble is most pleased to hear that a geologist is at last coming to re-
explore the Siwaliks, and that he will do all he can to help you. So mind
you write to him, if you cannot call on him.

He doesn't need the tents right away, but he can foresee that not having the receipts may be a problem. He sends a servant from the hotel over to try to collect them, along with the rest of his luggage. The other baggage and the small crates of equipment are all turned over cheerfully, but the luggage clerk obstinately refuses to relinquish the tents. He waggles his head and shakes his finger at the hotel servant, who dutifully repeats the entire performance for Dubois: "Ah, no, you must tell the Sahib, I must have the correct chit, for I must file my papers and mark my forms properly or there will be com-plaints!" Here the luggage clerk tapped his register book vigorously with his index finger, as if to emphasize how all-powerful the book is, and the proper recording of all documents. "Yes, I am deeply sorrowful," he continued, "but I cannot give up the tents without the chit. Nothing may leave my domain without its passport," he chortled gleefully, "no matter what the reason!" Dubois sets the matter aside to deal with later, for he has a great many things to arrange while he is in Ambala and this seems the least of them.

That night, relaxing in his room, Dubois unfolds Griesbach's sheets of scribbled advice and reads them over once again. Running an expedition in India is a little different from running one in the Indies, he thinks to himself. He will start tomorrow in trying to round up the permits, men, animals, and supplies he will need. He plans to be in the area fossil-hunting for over a month, perhaps six weeks all told, before heading back to Calcutta. Griesbach writes:

Monday. Call on Deputy Commissioner at Katcherry. I have written to
him about 10 camels, with 2 ordered to be ready on the 7th inst.
See that camel men have their ropes, they are obliged to supply one rope
per camel, which is included in their wages. The wages (so much per

camel per month) include also the services of the drivers, usually 1 for 3 or 4 camels. It is usually advantageous to carry ropes (hair-ropes) as reserve. Camel men expect to get at least ½ month's advances of wages.

Norton & Co. is a good shop for supply of stores etc. etc. Expensive but very good.

Get your servants; tell the hotel people, Norton & Co., or anybody else what men you require. Perhaps the Deputy Commissioner will know of men or give orders to his Tehsildar [the local tax-collector] to find suitable men.

You require:

wages about 12–15 rupees	*1 cook or Khitmagar who knows a little cooking*
7–8 rupees a month	*1 Bhistie (water carrier)*
6–7 rupees a month	*1 Syce (for horse)*
5 rupees a month	*1 Grasscutter (for horse)*
7 rupees a month each man	*2 or 3 Khalassis or Chuprassis for general service in camp & pitching tents. For the latter not less than about 8–10 men are required, but your Khalassis will hunt up coolies for help. It would be better however to engage 4 Khalassis.*

All these men will expect "warm clothing." I think they will be satisfied if you give them 6/– each man for that purpose. They also will want an advance of wages, generally a month's pay each man. The Cook will also want about 50/– advance for making arrangements about your food etc. He will have to buy all supplies at Ambala, such as rice, flour (Delhi flour in tins), salt, sugar, etc. etc.

Your man also will require all sorts of things such as "charans" (cloth for cleaning dishes etc.), bootblacking, Momroggan & yellow soap for brown boots & saddle etc. etc. all which is only a few rupees.

It sounds a daunting task, finding three or four camel men, plus their camels and the ropes and other equipment, and then another eight to ten laborers to organize the camp, fetch water and firewood, and so on. The key seems to be finding the right khitmagar, who is a combination of a butler and a mandur or foreman in the Indies. With the right khitmagar, all the rest will fall into place, so he will choose carefully. He wants an intelligent man, with good English and good knowledge of the country, and an honest, practical sort of fellow good at commanding men. He will have to ask around.

Then there is the matter of his horse. Dubois has no intention of struggling up the Siwalik Hills and down again on foot when wiry native horses are readily available at reasonable prices. Griesbach has opinions about that, too.

For horse get about a manud (84 lbs) of grain; it is always good to keep some grain ready, as it cannot always be got in villages.

Tell the hotel servant that you want a horse, & get the dealers to bring round their animals to the hotel. You ought to find many ponies in Ambala at this time & ought not to have to pay more than from 100–180 rupees for one. The horse will require clothing (a so-called Jul), curry-comb & brush, bucket (for water), headstall and hair-ropes (also one or two cotton ropes). The syce will make all these arrangements and will cheat a little over it, but not much. It will cost about 12–15 rupees.

You might ask Capt. Mardall of the 17th Bengal Cavalry about purchase of a pony, if he is in Ambala. I give you a letter to Capt. Mardall of the 17th B.C. who is a friend of mine.

Also a letter to Major F. Drummond, in case he is in Ambala.

Call also on Mrs. Melliss, who is a great friend of mine, if she is in Ambala, where she generally lives during the winter. Wife of Colonel Melliss C.B. the Inspector-General of the Imperial Service Troops.

Appended here is a list of the regiments stationed in the Ambala Cantonment, which must be a large one, Dubois supposes, for there seem to be a great many regiments: 18th Hussars, two batteries of the Royal Horse Artillery, two Mountain Batteries of the Royal Artillery, battalions from Norfolk, Somerset, and Derbyshire. There are, in addition, two native regiments commanded by British: the 17th Bengal Cavalry and the Depot 32nd Pioneers. While the military men may be of great help to Dubois, he is intrigued by the mention of Mrs. Melliss. From the notes, she seems to be a closer friend to Griesbach than her husband is. Dubois cannot help but wonder how close the friendship might be. Perhaps the stories about romances between lonely officers' wives and unmarried men upcountry are true. He had not previously suspected Griesbach of being a ladies' man, a . . . he searches his memory for the colorful English expression . . . a . . . poodle-faker, that is it! Perhaps he ought to call on this Mrs. Melliss indeed, if she is such a charmer. Certainly the Englishwomen are very attractive and, from the sound of things, he may have some time to kill in Ambala before he is ready to set out in search of fossils.

Griesbach's notes are a fount of information. Dubois reads on, trying to absorb the myriad pieces of advice. The most important concerns his permits and purwanas.

> *The Punjab Government has been informed about your traveling in the Punjab Siwaliks & also the Governor of the NorthWest Provinces, which begin on this side (east) of the Jumna River. You will get* purwanas *(sort of orders to local officials) from these Governments through the Deputy Commissioners of Ambala & perhaps Saharanpur. These purwanas are useful in case one gets into any difficulty about supplies or coolies & one gives usually one of the chuprassis the power to collect what one requires & armed with a stick and a parwana he manages generally to get what one requires. In the district you will get grain (for horse) perhaps, otherwise probably nothing but fowls, eggs, perhaps a sheep, milk now & then (best take a milchgoat with you) and practically nothing else.*

Good heavens, Dubois thinks: not only camels and ponies, but a goat! Whatever will Griesbach advise him to obtain next? Indeed, absurd as it sounds, Dubois has been advised to ask the Maharajah of Sirmoor to lend him an elephant for travel in the Siwaliks. Even after all his years in the Indies, British India seems colorful and exotic.

> *Let me know when you get into difficulties, and always let me know your next address. Make a rough sort of plan of operations & find out from the postmaster [at] Ambala where your next postal stations will be. Have your things addressed there & keep a man going backwards & forwards. You must give the man a letter to the Postmaster, asking him to hand over your letters etc. to your man, otherwise you won't get them. . . .*
>
> *Do not forget to take with you a lot of wooden tent-pegs for your tent. Your men will get them made in Ambala. You will require a good lot more than is actually necessary, as they continually break & get lost. Of course you can renew them in bigger villages from time to time.*

Griesbach seems to have thought of everything, from Dubois' social life to the payment of his servants and the stockpiling of tent pegs. Dubois is amused by his concern for minutiae.

Two tasks occupy Dubois the next morning, February 13. The first is applying to the Collector and Magistrate of the District of Saharanpur for a purwana. This is easily accomplished via the efficiency of the Raj postal system. Dubois writes a suitably officious letter—how these English love their administrative ceremonies!—and sends it off.

The second priority is rescuing his tents from the railway station, which he decides warrants a personal visit to the luggage clerk. Unfortunately nothing Dubois says makes the slightest impact on the man, who is determined to follow the exact protocol he has been taught. Exasperated and not a little insulted that this native clerk would question his word, Dubois goes directly to the stationmaster, a dignified Sikh with a massive beard and the largest turban Dubois has ever seen.

"Ah," says the man gravely, nodding as he listens to Dubois' explanation. "Wait here, Sahib. I will deal with this." He disappears into another office. Through the door, Dubois can hear raised voices in a language he does not understand. The Sikh stationmaster reappears in a minute or two, flashing a fine set of white teeth in a smile. "Now, Sahib, if you go outside, all of your things will be put into a bullock cart for you. There is a carriage waiting also. I have instructed the driver to take you and your belongings directly to the hotel."

"Thank you very much," says Dubois, wondering exactly what has transpired.

"Not at all," comes the answer, with a smile and an ambiguous shake of the head. Dubois slips a coin into the man's hand and departs to his waiting transport.

Now what remains to be done while he waits for word from the Collector and Magistrate of Saharanpur is to put together his expedition: men, camels, ponies, and a vast quantity of supplies. Once this caravan is assembled, he will take off for Nahan, in the foothills of the Siwaliks, to meet with the Maharajah. Griesbach has warned him that the Maharajah is a prickly, awkward fellow, an old-fashioned absolute ruler with modern ideas and a heart full of pride. The Maharajah's princely state, Sirmoor, contains many of India's best fossil sites, so his permission and assistance are essential. On the way to Nahan, Dubois can visit a few areas of which he has read in Lydekker's papers.

To Dubois' surprise, the arrangements seem to be relatively simple, thanks to Griesbach's tips. By the afternoon of his first day in Ambala, he has contacted Captain Mardall, Griesbach's friend of the 17th Bengal Cavalry, who has in turn sent word to the horse dealer he does most business with to find Dr. Dubois a good pony. And Norton & Company are organizing the khitmagar and the other men he needs, subject to his approval. What seemed a daunting task has suddenly shrunk to manageable size. Inwardly, he blesses Griesbach and his wonderful advice several times over the course of the next few days. He anticipates this part of his travels with a certain glee: it will be like the old days in the Indies, hunting for fossils alone with some natives to look after him, but even better. For one thing, he need have no concern that he will be plagued by the enervating, sultry heat and threat of fever that made his work so difficult in the Indies. Here in the highlands of the Siwalik Hills—really, the beginning of the mighty Himalayas—the temperatures at night are cool enough to leave hoarfrost on the ground. During the day, the sun shines brightly and warms the air enough to make for comfortable working conditions. No wonder that the entire British government moves to Simla, the prettiest hill station in the Siwaliks, to escape the stifling heat on the plains during April, May, and June. It seems a healthy, pleasant climate after the exhausting business of hunting for fossils in the Indies.

This expedition has another enormous advantage over his previous endeavors. Now Dubois is not looking for sites; he is visiting fossil localities that have already been found. Of course he will do some new prospecting as he travels; it would be foolish not to. But there will be no tedious scrabbling up mountainsides and slogging along swampy riverbanks to find nothing, as happened so often in the Indies. With Lydekker's maps and local guides, he should be able to find fossils with relatively little trouble most of the time.

He writes a quick letter to Griesbach to advise him of his safe arrival in Ambala, making an amusing story of the little inconvenience of reclaiming his tents. As he settles down to eat his second dinner at the Hotel Lumley, he thinks that all is well, everything is going remarkably smoothly. Only one matter worries him: he still has heard nothing from Anna since arriving in India. He sends a projected itinerary to Alcock, who will forward his mail from Calcutta if any should arrive. At home, the end of the rainy season is nearing: a bad time for fever. In a day or two, a child can sicken and die; in a week or two, his whole family may have perished, one after another, in dreadful agony. After Anna Jeanette's death, it is easy to imagine the worst.

He cannot understand why else Anna's letters would have grown so infrequent. At the beginning of his trip, she sent regular, twice-weekly missives. They were filled with trivialities and gossip, but at least they reassured him of the continuing well-being of his family. This recent, inexplicable silence seems ominous. What can she be thinking of? Has she perhaps relapsed into the despondency that threatened to consume her after the baby's death? He looks back through her letters, reading each one carefully for a hint of oncoming catastrophe. All she says in her last letter is that Prentice has stopped in from time to time to visit and share a meal or to play with the boys. The sentences leap off the page at him, filled with new meaning. Could it be . . . ? No, Dubois cannot believe such a thing, not of Prentice. There must be something else, something terrible. Perhaps that thing in the garden, the pontianak, has started crying again. But she does not mention it.

By the time a few more days have passed, Dubois is deeply concerned about his wife and family. He adds a note to his letter to Alcock, asking him to check with the Dutch consul, in case there is mail for him that has been put aside somehow or delayed. Should that yield nothing, would Alcock please send the following telegram immediately to Prentice: "Sir, Please explain visits to my wife in my absence. —Dubois."

CHAPTER 34 SIRMOOR STATE

On the tenth, Dubois departs Ambala at the head of a long caravan of porters and camels, astride a nicely moving little chestnut mare with good strong legs. She is called Tez, which means "bright" in Urdu. They are headed to Schazadpur some twenty or twenty-five miles to the northeast, across the Dangri Lake. They rest there overnight, Dubois' first experience of a dak bungalow. These small bungalows are sprinkled throughout the country, put up and maintained by locals for the use of visiting officials. By virtue of his association with the Indian Museum and the India Geological Survey, Dubois is fully entitled to use them—and he is grateful for it, too, as many areas have no European hotels at all. It is the first dak bungalow of many he will stay in: not elegant, but clean and sound and convenient. Dubois judges this amenity an excellent one and makes a note to himself to suggest it to the Resident when he returns to Java.

In the morning, he marches east and north with his men, crossing another lake and a river, arriving at the small village of Kala Amb in the afternoon. Now they are at the edge of Sirmoor State, with only nine miles to travel in a northeasterly direction to reach Nahan. Nahan proves to be a rather charming hill station, with well-maintained tree-lined streets laid out in a design based on a series of circles—a plan thought up, it is rumored,

by one of the previous maharajahs. Dubois has written in advance request-
ing an audience; when he arrives, he finds a response from the Maharajah's
secretary, whose name is a spiky drizzle of ink like the mark made by a spider that has fallen into an inkwell. The therefore anonymous secretary gives
him an appointment to see the Maharajah at noon. It is the beginning of a
prolonged wrangle to obtain the necessary permissions and purwanas.

Precisely at noon, Dubois presents himself at the palace. It is an extraordinary building, full of Moorish arches; courtyards enclosing fountains,
pools, and gardens; statues, rugs, and intricate screens and decorations.
The Maharajah receives him in a room awash in carvings, gold leaf, and
intricate designs. The decor is far too exotic and fussy for Dubois' taste—
every square inch of every surface is carved or gilded or painted or inlaid
with precious stones—but it certainly conveys the Maharajah's wealth and
power. The Maharajah himself, a small dark man with an enormous mustache, wears an ackhan, a long coat with high collar made of richly embroidered stuff, over tight-fitting trousers. On his head is a magnificent turban,
fastened in front with a large, gaudy pearl surrounded by gems. His fingers
are studded with jewels set in gold rings. Dubois' high-necked tutup jacket
and crisply ironed trousers are nondescript by comparison, but he fingers
his own luxuriant, ginger-colored mustache with a certain satisfaction.

The Maharajah is dignified and educated, with a good command of
English. Dubois has been warned not to come to the point too rapidly lest
the Maharajah think him rude. The Maharajah knows very well the point of
Dubois' visit; it remains the Maharajah's responsibility to turn the conversation in that direction. He is apparently interested in the rajahs and princes
of Java. Are their palaces like this one, so small and ordinary? he asks with a
deprecating gesture that begs Dubois to praise the extravagant surroundings. Dubois answers tactfully that he has seen nothing to compare with the
Maharajah's palace in the Indies, though there are many fine palaces there
also, especially in the region of Solo.

Pleased, the Maharajah begins to speak to Dubois of his great scheme for
modernizing his people. He has set up a school to provide higher education
in English for intelligent young men of good families in Sirmoor State. They
will acquire the knowledge and customs that have made the European
nations great, and will be his state's ambassadors to the outside world. The
Maharajah asks whether in Java there are facilities for the higher education
of the noble families in English. Truthfully, Dubois cannot boast that there
are. There are few schools of any kind for natives, although the odd

princeling may be sent to Batavia for an education or may even be privately schooled by a Dutch tutor. No, the only Javanese favored with a truly European education are the legally recognized children of Dutch fathers and Javanese mothers, and most of them are sent home to Europe for the purpose. He is duly admiring of the Maharajah's farsighted plans, while silently wondering whether education alone can possibly transform the ignorant, illiterate peasants he sees along the roadside into the equals of educated Europeans.

"Ah," says the Maharajah shaking his head sagely. "But I have forgotten. Your princes are no longer heads of state in their own country, so they can do nothing for their people. They are merely—what is the phrase?—*younger brothers*, to be commanded and led by their Dutch elders. Hmm." He pauses here and looks at Dubois accusingly, as if he is to blame for the difficulties of colonial rule. "So it is a very awkward situation. No prince wants to be ordered about in his own home," he says firmly. "No one does."

Dubois does not know how to reply and so says nothing.

The Maharajah leaves the room soon after that, rather abruptly. Dubois is bewildered. Has he insulted the Maharajah? He does not think so. Surely he has said nothing but the most polite platitudes. After some uncomfortable minutes alone in the room, Dubois is relieved when an aide enters and introduces himself. He questions the aide strongly, pressing him for a clarification of the situation. Has the Maharajah refused his request by leaving without discussing the matter? No, no, Sahib, certainly not. The Maharajah is a busy man, you know; things to attend to. Be flattered that he agreed to see you himself. Is he certain that the Maharajah has not somehow forgotten the very reason for Dubois' call? Oh no, Sahib, comes the reply. The Maharajah is the most intelligent of men. He never forgets a thing. No, he has a formidable memory, as befits a ruler. Then has the Maharajah's assistance been tacitly granted? The aide cannot say; such things are the Maharajah's business and he has not been instructed.

The aide will make a note of the things that Dubois will need while traveling in Sirmoor. Perhaps Dubois could draw up a list, and an itinerary? He can present it to the Maharajah tomorrow, perhaps, or the next day. In the meantime, a guesthouse will be put at Dubois' disposal. Perhaps the doctor would care to buy a fine pony, to make his own journey more comfortable? The Maharajah's Keeper of the Horse can find some prospects for him. Ah, the doctor has already purchased a pony. A pity, that; the horses are so much better here than in Ambala. And men; surely Dubois will need men. If he will simply let the aide know how many are required, he will send along some men needing employment. Because of the Maharajah's education scheme, it is possible to hire men with very good English indeed. No? Dubois has already hired all the men he needs for his entire expedition? Ah, what a shame. Yes, it would have been better to take some local men; they have more experience in the Siwaliks. Just as the Doctor wishes, then.

Dubois spends the rest of the day walking around Nahan and resisting all urging to buy more equipment, hire more men, and purchase additional ponies. The next day, gratifyingly, he receives a letter from the Collector and Magistrate of Saharanpur granting him the desired purwana for Saharanpur District. That decides him. If the Collector and Magistrate of Saharanpur can issue a purwana so easily, so can the Maharajah. Clearly the Maharajah is delaying, perhaps resenting the orders from the Raj to cooperate. The inactivity makes Dubois restless, the Byzantine politics impatient. When the silence from the palace continues the next day, Dubois decides to set out from Sirmoor, so that he can get something done while the Maharajah procrastinates. Soon he is trotting Tez along briskly in the early morning light, heading toward Kolar, fifteen kilometers to the east. From there, he sends a telegram to Griesbach complaining of the Maharajah's behavior. The man was cordial, to be sure, but he was in no way helpful. Can Griesbach somehow prod the Maharajah into action?

Dubois and his men travel eastward and then turn north, staying at the tiny village of Rampur on the Jumna River before ascending into the Siwalik Hills proper. Unbeknownst to Dubois, Griesbach has cranked the vast machinery of the Raj into action. Letters are written and notes filed, a flurry of telegrams is sent, until finally a telegram is dispatched to the government of the Punjab repeating the request to order the Maharajah to give Dr. Dubois every assistance. In the meantime, Dubois and his expedition march on, arriving at Kalianpur on the twenty-second. There they will spend a few days collecting fossils. At last a letter from Alcock comes by runner, sent on from Nahan.

Indian Museum   February 17, 1895
I was glad to hear of your safe arrival, and I hope you are now
unearthing the bones of Pleistocene ladies and gentlemen in plenty.
    The Dutch consul sent me a very thick envelope, which I took to be a let-
ter from your wife: I therefore <u>registered</u> it and sent it on. I hope it reaches
you safely. I have not sent the telegram yet, and shall not do so until I hear

*again from you. It is a great mistake—a mistake which I often make
myself—to translate one's imagination into action: the imagination upsets
no one but oneself, the action may upset other people.*

Dubois is livid. What audacity, not to send Dubois' telegram, and he took
Alcock for a friend, a colleague. He believed Alcock when he said how welcome Dubois was, what a pleasure it was to meet a fellow scientist doing
such interesting work. Yes, but not enough of a pleasure to carry out a simple request of the utmost importance to Dubois. Really, the man is unspeak-
ably rude and presumptuous. And as for Anna—well, time will tell what the
envelope holds and what story she has concocted to explain her actions.

All that day, and the next, Dubois drives himself and the men up and
down the gullies and hills in a sort of frenzy. At least it is cool enough. Some
of the men know something about collecting, so Dubois does not have to
teach them everything. They can recognize fossils and he asks them simply
to call for him when they find a good or relatively complete old bone. They
are not to touch it themselves until he has seen it, decided whether it is
worth collecting, and made a note of its location in situ. By the end of two
days, the accumulation of fossils is substantial and his anger has begun to
wane. They leave for Dehra Doon and the Forestry School station, where
Gamble and the envelope sent via the Dutch counsel await them.

Dehra Doon is a haven of civilization after camping in the remote settle-
ment at Kalianpur. The Forestry School is well organized and has an impressive library of information about the area, from wildlife to geology to
vegetation. A wiry, short man, little bigger than a native, with tanned,
weatherbeaten face, Gamble regales Dubois with accounts of his work and
studies: the relationship of the density of deodar trees to altitude; annual
rainfall figures going back many years; surveys of the forest animals and
their habitats. He has annotated copies of all the topographic maps from
which Lydekker worked when he was there. While not everyone would find
these matters entertaining, Gamble's enthusiasm and intensity are contagious. Besides he, like Dubois, is something of a scientific polymath, interested in almost everything in this tiny corner of the world. Only after Dubois
has had a cool drink, a bath, a tour of the station, and a hot supper does
Gamble produce the long-awaited envelope, saying rather casually, "Oh,
some things came for you from Cal the other day; a telegram and this envelope. It looks as if it must be from the Indies. Letters from home, eh Doctor?"

Dubois can hardly prevent himself from snatching the packet out of

Gamble's hands. Of course, the man has no way of knowing how important this might be, Dubois thinks generously. "Ah, thank you," he says a little stiffly. "I'm rather tired and I have been awaiting this package for some time. I think I will retire now and look it over. I'll see you in the morning, then?"

"Yes." Gamble nods. "If you need something before then, just ring for the boy. Mohammed will see to anything you want." He gestures to the slender young man in a spotless white uniform standing discreetly at the other side of the room.

Comfortably ensconced in his room, Dubois tears open the packet forwarded by the consul. It contains two letters from Anna, but no explanation. It is as if she is completely unaware of the anxiety she has caused him by her silence. She writes of the weather, the children's games and health— "you will be glad to see how well Eugénie is doing on her pony, and Jean seems to grow taller every day, you will hardly recognize him when you return." Then there is a small problem with the servants, described in boring detail, and an account of the friends she has seen and the gossip she has heard. Oh, and that charming Mr. Prentice has been by again on his way back from Malang. He brought her some of the most lovely potted plants that he bought there—a really brilliant hibiscus that stands on the front veranda, and a huge, fragrant frangipani that she has had the tukang kebun plant near the wall, in the far corner. The tukang kebun does not remember to water unless she nags him, she suspects, but so far it seems to be doing well because the small rains still come every other day or so.

The next letter is much the same, reassuring in that Anna sounds ordinary. There is no word of illness or fever, no hint of hysteria in her tone. But Dubois notices that Prentice has visited yet again—how many times is that since he has been gone?—and he wonders. Of course, Prentice is a charming man, and handsome: who should know that better than Dubois, his closest friend? And he has not remarried, though he has taken up with a lovely nyai from a nearby village. He must still be a little lonely for civilized company, especially with Dubois gone.

In the back of his mind, Dubois can feel the irritating wriggle of that small worm of suspicion, jealousy, and doubt. He thinks that Prentice ought not call on Anna so often. It is probably just friendly concern—Prentice knew how worried Dubois was at leaving Anna alone—and yet, and yet . . . The tropics move respectable people to disreputable impulses. He knows this is true; he has seen it many, many times. No threat of scandal, no possible disruption of the family's peace is enough to prevent it. And in the

Indies, where attitudes are more lax, the sinners need not even fear social ostracism, except for a brief period. All is forgivable in the Indies.

He determines to write Anna tonight, and Prentice. He will ask Anna to explain the sudden dearth of letters, to see what she says, and he will simply write Prentice an ordinary letter, telling him of his travels and discoveries. He need only mention at the end that he understands Prentice has been calling on Anna rather often in his absence. Just a hint, nothing more: that is all it will take.

CHAPTER 35 SIWALIK ADVENTURES

For the next week, Dubois spends most of his days traveling and hunting for fossils, with considerable success. His itinerary sounds like a travelogue of the Saharanpur District: Dehra Doon, Khagnaur, Kalawala Pass, Kerwapani, Badshah Bagh, Kalesar, and Kolar. What he most wants to do is spend a few weeks camping at the remote village of Haripur, in Sirmoor, where the best fossil sites are reputed to be. But he cannot work there without the Maharajah's permission, and so far there has been no word from the palace. The man's obstinacy is infuriating! Before leaving Dehra Doon, Dubois sends another telegram asking once again for the purwanas, guide, and elephant he needs. A reassuring reply comes the next day, but no guide, no elephant, and no purwanas appear.

Dubois travels on, exploring the areas where he is permitted to work. The scenery is spectacular, like nothing in the Indies. The plains and foothills are brown, dusty, dry; then, as he climbs up into the hills themselves, there are magnificent vistas of deodar trees, pine-covered slopes, much more vegetation of all sorts. His pony Tez proves herself a sure-footed, tireless, good-tempered beast; he slowly winnows his entourage of coolies, porters, and guides and trains them into a fairly effective force. The daily routine of setting up the camp, producing hot meals, settling in for the night, and then eating and packing up again in the morning is now well polished. Dubois has been fortunate in his choice of khitmagar.

When still more days pass with neither the guide nor further communication from the Maharajah, Dubois again telegraphs Griesbach, complaining in even stronger language about the lack of cooperation. Griesbach investigates and identifies what seems to be the heart of the problem: "Obstruction by Rajah Nahan owing it is stated to orders having been issued to Rajah through Extra Assistant Commissioner Ambala." There is another round of official orders, which seem to do nothing.

Events conspire to worsen Dubois' mood. Fossils seem scarce. Runners sent back and forth to various postmasters find no letters from Anna. Dubois' conviction that Prentice is responsible grows daily more certain. Of course, she would come to admire and respect his friend even more in Dubois' own absence. Of course; of course. Prentice winds perceptibly through her infrequent letters. Early in March, he hears from Alcock in Calcutta once again.

> *Indian Museum   March 3, 1895*
>
> *I am glad to learn that you are safely started at last and I hope you will have good luck. I got your box of photographic apparatus landed, and sent it off by post last week. I fancy it will have reached you by this time. No letters have come for you for some time.*
>
> *I am sorry that you misunderstood my remarks about bottling up one's anxieties and apprehensions. I did not mean that I was annoyed because you asked me to send the telegram; but I meant to suggest that your own friends might be needlessly alarmed by a telegram.*

He sets the letter aside—what does Alcock know of this, anyway?—and tries to concentrate on his work. At least the English he has met so far find his work interesting, and there has already been a brief notice in *Nature* about his monograph. Soon there will be more, he thinks: additional comments, reviews perhaps, or accounts of scientific meetings. If only he were not at the ends of the earth, where news travels so slowly. He has never been a patient man, and he is doubly frustrated now. When the box of photographic equipment arrives, he opens it immediately. To his surprise, the box also contains a copy of an article in the German journal *Naturwissenschaftliche Wochenschrift* (Natural Science Weekly) about his monograph. How kind of Alcock to send it along, knowing he would want to see it as soon as possible.

As Dubois skims the article, his pleasant mood evaporates in seconds. The piece is written by Paul Matschie, a zoologist, and his opinion of the monograph is scathing. Dubois has utterly failed to persuade Matschie of the most fundamental fact: that the tooth, skullcap, and femur belong to one individual animal. Without that basis, the rest of the work is a futile exercise in analysis of something that does not exist. "My God!" exclaims Dubois under his breath, as he reads. He is a man who rarely uses bad lan-

guage, but this is shocking, shocking. "Am I never to be freed of this absurd-
ity? Is the idiotic opinion of that fool in Batavia, Homo Erectus, to follow me
the rest of my days?" He throws the article down onto the table impatiently,
angrily.

"Sahib?" asks one of his men softly. "You need something, Sahib?"

"Yes, yes," he replies irritably, snatching up the article again and reading
the damning words over once more. "Bring me some tea, some strong tea."

"Yes, Sahib, right away, Sahib," the man replies, making the respectful
gesture of namaste—two hands pressed together, raised in front of the face,
as in the Indies—before leaving swiftly and silently.

Dubois has already written Gamble, asking for news of the reception of
his monograph, and now he writes again. Matschie cannot be the only one
to have noticed his work. There will be other, more intelligent assessments,
surely. He awaits the arrival of the post morosely, the triumphant mood he
has enjoyed for weeks now turned dismal. There is no word from Java, from
Europe, or from the Maharajah. He sends yet another letter to the latter,
hoping it will produce the desired assistance. Perhaps telegrams are too
terse, too curt; perhaps the Maharajah needs sweetness, coaxing. His letter
of March 4, 1895, is answered the same day by one asking him to pay for the
coolies he has requested and for fodder for the elephant. At last! So the
problem is that the Maharajah wants his people to be paid, not simply con-
scripted as in the Dutch East Indies. Dubois replies immediately, agreeing
to the projected costs. By letter the next morning, the Maharajah confirms
that a guide and elephants are on their way to meet Dubois at Haripur.

The guide is Sukh Chain Sinha, the son of one of the Maharajah's tehsil-
dars. He is an educated young man—a product of the Maharajah's own
schools, he proudly informs Dubois—with good English. Dubois finds him
a little oily and officious in his presence, yet arrogant with the coolies.
Heretofore the khitmagar has handled his responsibilities ably without fuss
and the expedition ran smoothly. Now Sukh Chain Sinha is here and there
are conflicting orders, disagreements and resentments. Dubois realizes he
is burdened with this young fool, the Maharajah's chosen one (perhaps his
spy?), for the duration of his stay in Sirmoor. He cannot dismiss the youth,
for his employment is clearly part of the price Dubois must pay for the
Maharajah's cooperation, in addition to the ridiculous two rupees a day for
the elephant's fodder and the four rupees a day for coolies. How could a
man with a gold- and gem-encrusted palace niggle over two rupees a day
for elephant fodder?

But the elephant is as helpful as it is exotic, for the thickly vegetated terrain is difficult to navigate on foot or horseback. The view from the elephant's back proves excellent for spotting areas where there are exposures; Dubois' keen eyes can even pick out individual fossils sometimes. Dubois and the expedition settle into a good campsite in Haripur and begin finding fine mammal specimens, similar to those that make up Lydekker's fauna. Dubois discovers that the Maharajah has even ordered his representative in Haripur to hand over to him "some 5 fossil bones" in their possession. Five bones! To add to his hundreds . . . He must remember to express his thanks to the Maharajah.

Dubois is further cheered when he receives a letter from Prentice a few days later. Now he can put his suspicions to rest, he thinks; now he can see if there is anything between his best friend and his wife. It is unthinkable, impossible, a completely unfair suspicion. No, Prentice cannot have . . . But the suspicion squirms and grows in his mind, try as he might to kill it. This new letter will settle things. He has had so many from Prentice, they have corresponded so often, surely he will know instantly if anything is amiss.

Dubois borrows this elephant from the Maharajah of Sirmoor to use in his explorations.

Toeloeng Agoeng   March 7, 1895

My dear Doctor,

I would have written you long ago but have been very busy at my work with the new coffee estate on the Willis and generally very wearied after my day's work or not much disposed for letter writing.

Things are going on after a fashion, but not very briskly. I have rather few workers and bad weather with difficult muddy roads & etc. & etc. Today I am down in town with Mr. Boyd as I require to see the Resident & controllers in connection with my coffee estate. I thus profit by the opportunity to write you a few lines but you will excuse brevity on my part as I have not much time and less news. Indeed there is nothing here worthy of mention. Everything is going along in its customary course. Dr. Van Buren of Djombang is come to Kediri on 1st April & Dr. v/d Veldt is going to Europe. Mr. Van der Woude is not in the best of health & is at present with the doctor at Blitar, I hear.

I suppose you have not seen much of northern British India? I trust you met with a kindly reception there and are pleased with your visit. You will no doubt have acquired fluency in speaking English and I sincerely hope your visit to Calcutta has been a profitable one for you from a scientific standpoint.

The time is nearing when your return to Java may be looked for as not far distant, & I trust you will find everything here in the best order and have no reason to regret the journey you undertook to British India.

When you return to Toeloeng Agoeng we shall be sure to meet you once more and hear of your travels.

Wishing you a safe return to Java, believe me to be, with kind regards in which Mr. Boyd heartily joins me,

Yours Sincerely,
Adam Prentice

The letter sounds just like Prentice, no different from any previous letter. The man is honest to the core, Dubois thinks; he should never have doubted him. The bad weather must be making it difficult for Prentice to plant his coffee trees and build his roads and warehouses. And the lack of able workers: this is a problem Dubois knows well indeed. But Prentice is very good with the natives, very fair; he will manage.

. . . Odd that Prentice should write from Toeloeng Agoeng and make no mention of seeing Anna or the children. Isn't it? If he is innocent?

Some nights later, he is again in the grip of a black mood. All his troubles over the last few years crowd in on him: the deaths; the endless difficulties of trying to find *P.e.;* the fevers that threaten his life and sanity; the lazy, insolent laborers who cheat at every opportunity; the villagers who will not tell him where fossils or caves are; the days of scrambling up mountains in that awful, awful heat. Now most of these tortures are repeated all over again in an Indian fashion, with that obstinate Maharajah, his odious guide, and the endless troubles with food for ponies, goats, camels, and the elephant. And the review by that useless German scoundrel Matschie, who knows nothing about those fossils, nothing! Dubois should be hailed by his success at finding *P.e.,* honored by every natural historian in Europe, not criticized. Somehow his courage seems a little weak tonight. Perhaps he has a touch of fever coming on. Uncharacteristically, he tries to share his mood in a letter to Anna: "The case of my missing link evokes more general interest than I had thought. If I possess enough cheerful fortitude to introduce her in Europe personally, I do not doubt she will be wonderfully victorious. That would shed a little light on my poor life, which has been so marked by sadness in the last years."

A warm and sympathetic letter from Gamble in Dehra Doon arrives the next day, March 15. Of the attacks by Matschie and others on Dubois' monograph, Gamble writes, "I hope you will succeed in bringing them all round eventually & possibly in making them stronger believers even than if they had accepted your demonstrations straight off. I suppose it is only human nature when a new thing is brought out, for the authorities to try to find all the arguments they can against it."

Another missive comes, this time from Griesbach, with a troubling remark about a review of Dubois' monograph. "Did you read Lydekker's critique of your essay in *Nature*?" Griesbach asks. "He is convinced that *Pithecanthropus* is no transitional form!"

These words haunt Dubois. What, exactly, did Lydekker say? What objection could he have to *P.e.*'s transitional status? How could any anatomist consider *P.e.* otherwise, with its apelike skull and humanlike femur? Where has he failed to convince?

In the meantime, the fossil-collecting in Haripur goes well—very well, and Dubois will have an excellent collection to ship home to Holland—but there is nothing to compare to *P.e.* She is truly the greatest find of the nineteenth century, Dubois thinks, never mind what Lydekker says or Matschie or anyone. On the twenty-fourth, he calls an end to his adventures and sets out for Barara, where he will catch the train back to Ambala and then back toward Calcutta. Though he tries to suppress them, his concerns color his next letter to Anna, written from Delhi.

> *Griesbach writes about the critique of Lydekker in* Nature. *I regret that I have not yet read this critique; but I conclude it must be of little merit because Lydekker suffers from impossibly anti-Darwinian tendencies. After becoming better acquainted with his writings of the last few years during my visit here, I see him as a productive writer but a scientific man of only moderate value. He is an Indian scientist. Meanwhile, I desire very much to make the acquaintance of the piece in question.*

He cannot help pressing the pen more strongly into the page for that last sentence. Soon he will be back in Calcutta, catching a steamer to Rangoon, then on to Toeloeng Agoeng in late April or early May. His thoughts cycle over and over again, like a maddening tune that will not quit the brain. He convinces himself of the worst, of the ruin of both his career and his marriage. There is nothing he can do from this distance about Anna, but he sends a desperate telegram to Griesbach pleading for more information. No letters come to ease his mind, no copies of articles, no hopeful reviews from Europe: simply silence. It is worse than hearing bad news, this endless, mysterious, infuriating silence.

On the way back to Calcutta, he visits the Taj Mahal. Who could be in India and not see this exquisite tribute of Shah Jahan to his late, beloved wife, Arjumand Banu Begam? The visit leaves him colder than ever. He can see the elegance in the building's design, its rounded minaret, its tranquil reflecting pools, its beautifully kept formal gardens. He especially admires the warmth of the rosy marble chosen and carefully transported from Makrana in Rajasthan. But the experience raises only bitterness in his soul, by forcing the comparison with his own fate. Where are the admiring crowds for his monument to science and evolution? Where is the praise, the glory, the appreciation of the magnificent intellectual edifice *he* has constructed? And where, where indeed, is the loving, faithful wife who should be waiting for him? The train journey to Calcutta seems much longer and more arduous than the trip out. What once charmed now irritates; what fascinated bores.

CHAPTER 36 LEAVING INDIA

Calcutta seems dirty, ugly, and noisy to him now, after the cool green quiet of the hill country. He wonders if his eyes were deceived when he first saw Howrah Station, thinking it colorful and cheerful. His huge piles of luggage have been joined by crates of fossils from the Siwaliks that will need to be dispatched to Holland. At least he finds a warm welcome at Alcock's house, and he has forgotten his irritation with the man.

"Doctor!" cries Alcock happily, coming out of the front door when he sees the carriage pull up. "So you are back at last! And was your trip successful? Have you found many fine fossils? You must tell me all about it! And what are these—boxes of fossils, eh? Jolly good, jolly good." Dubois can hardly get a word in edgewise, so effusive is his friend at his return. Perhaps it is for the best, for Dubois is feeling travel-stained and exhausted.

"Come in, come in," urges Alcock, while the servants take care of the mountains of luggage. "You must be in need of a wash-up and a cool drink. Go on up to your old room, I'll send Mohammed up to you." Dubois is a good deal refreshed by his bath and his tumbler of cool lime juice. By the time he comes down to greet Alcock properly, it is nearly time for lunch. He can see Alcock has laid on a special meal for him. Such a thoughtful friend, Alcock. They relax with cool drinks in the drawing room for half an hour or so while the preparations for the meal are completed.

"Thank you, Alcock, for making me welcome once more. The trip was exhausting, I don't mind telling you. I did get a lot of good fossils, though I never found any more of Lydekker's *Anthropopithecus*," says Dubois.

"Ah, what a shame! Still, the beast must be very rare, I suppose," replies Alcock. "Griesbach has sent over a copy of that *Nature* note by Lydekker— unfortunate, that, really a little tough on you, but I suppose he has a right to his opinion. And some mail has just come for you, too."

Dubois cannot keep his weary eyes from skimming the *Nature* article, despite Alcock's and Griesbach's warnings. Lydekker's article starts nicely enough, but only, Dubois thinks, to make his later criticisms more pointed.

> *Review of Dubois'* Pithecanthropus erectus, eine menschenaehn-liche Uebergangsform aus Java.
>
> *Java, from its geographical situation, being just one of those countries where the remains of a connecting form between man & the higher apes would be extremely likely to occur . . .*

"Ha!" exclaims Dubois and Alcock looks up from his newspaper, sympathetically curious. "At least the man has understood my fundamental reasoning about why the missing link would be found in the Indies," Dubois explains to Alcock. "I spelled it all out in my 1888 article and it is clear that Lydekker accepts my argument. How widely disbelieved that notion once was—'Don't be misled by Darwin's crazy book,' they said to me—and now it is nothing more than a commonplace observation in an article in a learned journal by one of England's best-known paleontologists."

"Ah yes," murmurs Alcock, knowing there is worse to come.

Dubois returns to his reading.

> ... zoologists have naturally been attracted to the title of the work before as it proclaims in no uncertain terms that such a missing link has actually been discovered. A feeling of disappointment will, however, probably come over the student, when he finds how imperfect are the remains on the evidence of which this startling announcement is made and when he has submitted them to a critical examination he will probably have little difficulty in concluding that they do not belong to a wild animal at all.

What audacity! That Lydekker, who named *Anthropopithecus sivalensis* on the basis of a scrappy jaw fragment, should criticize Dubois for working with "imperfect remains." Indignation burns hotly in his mind. It is small consolation that Lydekker announces himself "content" to accept Dubois' assertion that all three fossil specimens are derived from a single animal, for he finds that animal to be human. The size of the braincase—huge for an ape, deficient for a man—he attributes to the skull owner's being "a microcephalic idiot, of an unusually elongated type." It is the same old jibe that was made about the Neanderthal skullcap: can't be primitive or old, must be pathological. Lydekker's prejudice against Darwinism shows clearly in this, Dubois thinks, just as Virchow's did when he tried to discredit the Neanderthal remains.

Lydekker's conclusions are flatly dismissive:

> Haeckel's "Pithecanthropus" *may, therefore, be relegated to the position of an hypothetical unknown creature for which it was originally proposed; while the specific name* "erectus" *must become a synonym of the frequently misapplied* "sapiens."

The piece leaves a bitter taste in Dubois' mouth and a conviction that Lydekker is neither anatomist nor scientist at heart.

Dubois never anticipated a problem of acceptance once the fossils were found. To have failed in his great quest would have been difficult to bear, for so much was at risk. But now he sees that the pain of outright failure would be much less than the anguish caused by success achieved but unjustly ignored. He has found the missing link, and still it is not enough. What he reaps for his years of backbreaking, mind-wrenching, courageous work is bitter betrayal and criticism from armchair experts. Pah! He has no respect for these men, but their knives are sharp and cut deeply. And he bleeds and bleeds, with no one to stanch his wounds.

He turns to the second envelope Alcock has handed him, seeing at once that it is neither from Anna nor from Prentice, as he had hoped and feared. It contains a begging letter from the Maharajah's spy, Sukh Chain Sinha, asking for a testimonial letter. It is as oleaginous as the man himself, and Dubois has no intention of recommending the youth to another innocent European. He folds the letter tightly and puts it away. After a pause, Dubois asks Alcock hesitantly, "I don't suppose that there were any other letters for me, letters from Java?"

"No." Alcock shakes his head sadly, reading Dubois' eagerness. "Nothing has come. I've checked with the Dutch consul, too. Perhaps your wife feared to miss you, that you'd already gotten on the ship for Rangoon."

"*Ja*," remarks Dubois, dubiously, sadly. "That must be it." He knows full well how scatterbrained his wife is. Would she even consider that the letter might miss him?

"Oh, but there *was* an envelope from America, though—came some time ago," Alcock remembers suddenly. "Mohammed!" The servant comes running. "Mohammed, where did we put the doctor's letter from America?"

"In the desk drawer, Sahib, in your office," Mohammed answers. "I will get it."

In a moment, the envelope is in Dubois' hands. "Maybe it's about your monograph," Alcock guesses.

And so it proves to be. It is from the paleontologist Othniel Marsh, one of the few Americans to whom Dubois sent his monograph. Dubois looks through the article quickly, and is pleased. "Here, now, Alcock, at least someone appreciates my work!" he says cheerfully, brandishing the article. "This is from Marsh, at Yale you know, in the States. He's written a piece about my monograph in the *American Journal of Science*. He calls the unearthing of *P.e.* one of the most important discoveries since the finding of

Neanderthal man. And listen to this: 'It is only justice to Dr. Dubois and his admirable memoir to say here, that he has proved to science the existence of a new prehistoric anthropoid form, not human indeed, but in size, brain-power and erect posture, much nearer man than any animal hitherto discovered, living or extinct.'"

Alcock listens attentively and then bursts out, "Jolly good! That's more like it. I think we ought to celebrate that. Shall we have some beer at lunch?" The two go in to eat, happily gossiping and exchanging news.

The next day Dubois calls on Griesbach, who has helped him so much, to discuss his findings and the various articles about *P.e.* Of all he has yet received, only Marsh's article has so far supported him. He is dismayed to learn from Griesbach that Marsh's reputation is somewhat tarnished.

"Y'know, Dubois," Griesbach says confidingly, "that Marsh chap seems a damned fine paleontologist, but there is talk about some bad business between him and Matthew Cope. I don't remember all the details—something about stealing fossils, or what the Americans quaintly call claim-jumping, I think. They're both after those dinosaur bones that the American West seems to be full of, and there's been some dirty dealing there. But Marsh still has a good reputation as a paleontologist, so his endorsement should help you out."

Dubois visits the Indian Museum to see his other friends and to show them his fossils, taking some hours to compare his new specimens carefully with Lydekker's in the museum's collections. Now that the man has come out against him, Dubois would hate to make a mistake in identifying a Siwalik species and leave himself vulnerable to Lydekker's further scorn. For the first time Dubois notices undercurrents of disagreement, quarrel, and jealousy among his colleagues at the museum. Was he blind to them before? Or had he simply been too much a stranger for them to talk frankly in front of him? Whichever the case, he is sadly disconcerted by the pettiness and edgy atmosphere he finds among the curators. They seem self-absorbed, all-consumed by their own territorial squabbles. He had hoped for more concern for his successes and interest in his opinions.

That evening Dubois writes to Anna, mostly focusing on his worries about the reception his precious monograph and fossil are receiving abroad. Lydekker and Matschie are definitely on the "con" side; only Marsh has so far enlisted "pro" *P.e.* If it is to be a battle, he is as yet outnumbered. "Marsh sent me a reprint from an American magazine about *P.e.*, in which figures are reproduced, pointing out its great importance," he writes Anna. "He praised my work. But he has done a lot that has given him a bad name despite all his fame, because he has been involved in some American humbug."

In the days that follow, there is no mail from Anna, nor from Prentice. Dubois' patience, limited at the best of times, snaps. There is no longer any doubt in his mind what has occurred. By their silence, his wife and his dearest friend have told him more plainly than any words that they do not care for him. Their pledges of constancy are rendered as cold and worthless as the ashes from a fire that once warmed and lighted the darkness. It is a bitter, sour-faced man who crates up his Siwalik fossils a few days later to ship them off to Holland. With each passing day, he is more and more discouraged, more irritated by the eccentric ways of the British in India and by the squabbles that buzz around the Indian Museum like a vicious swarm of insects.

On April 14, he is more relieved than saddened to bid farewell to India, the Raj, and her people. He sails for home on the S.S. *Lindula,* calling first at Rangoon. With every passing mile, his anger grows blacker and his imagined reproaches more savage. He has spent four months and a great deal of money on this trip, and what has he to show for it? A few crates of fossils, miles upon miles of dusty, uncomfortable travel with useless, lazy natives, a few scenic views from the famous Siwalik Hills, and the public ruin of his private life. He has no doubt that the affair between Anna and Prentice is the talk of Java. Servants always know these things, and they spread the word farther and faster than even the highly efficient European gossips do. In return for his eight long, hard years of work and perseverance, he has been served up dishes of acidic criticism by stay-at-home strangers and bitter faithlessness by those closest to him. The scientific skepticism burns like the hottest Indies pepper, the betrayal repels him like the noxious smell of the durian fruit so beloved of Javanese natives. He cannot imagine any combination of events that could pain him more acutely than these. He writes to Anna:

> *April 14, 1895*
>
> *So I have now left the ground of India, and I am happy to have done so because, taking everything together, nothing—not the stay here, not the traveling, nor the country itself—has pleased me. Even the friendships that I have enjoyed here are spoiled because those who gave me friendship hated each other. . . . The amount I have profited scientifically has certainly not been worth the labor and costs.*

He is delayed in Rangoon several days on account of some confusion at the booking office, but sets off again by steamer for Penang on the twenty-third.

With a good connection between ships there, it takes him just a few more days to go from Penang to Singapore, where he calls in on Prentice's friends Galloway and Lyons. Welcoming as they are, they have no news from the Indies to report. No letters, no explanations; nothing at all from Anna or Prentice. It is true, it is all true. He knows it by the almost tangible pain in his stomach, as if he has been poisoned. But there is no doctor's medicine that can heal him, no art that will diminish his anguish.

He arrives at Tanjung Priok in Batavia on May 2. The exhausted and nearly broken man who has returned is in shocking contrast to the hale, optimistic Dubois who left months before. His acquaintances in Batavia wonder what happened to him in India, whether he has caught another fever perhaps. Rather than struggle to be sociable, he pleads travel fatigue and declines all invitations to dine that night or to stay on a few days. The next day, he sets forth by train to Toeloeng Agoeng. It is a two-day journey that seems much longer. Each passing mile brings him closer and closer to the ruin of his personal life and the confrontation with those who have been faithless. There is no avoiding it. He will have to survive whatever occurs and leave for Europe as soon as possible. He does not know how he can bear to live with Anna any longer. For that matter, if his missing link is totally rejected by the scientific world, he does not even know how he will find the courage to live at all.

CHAPTER 37 TOELOENG AGOENG

As Dubois steps off the train in Toeloeng Agoeng, he sees that everyone has come to meet him. Anna is standing there, in her prettiest European hat and dress, pale blue trimmed with a darker border that shows off her lustrous hair and fair skin to advantage. And Prentice is there, too, standing by her side, tall and sun-browned and handsome as ever. When he first spies Prentice, his heart leaps up with the anticipated joy of sharing with his friend all his discoveries and experiences in India—and then falls, as he remembers those damning letters. Prentice. Prentice and Anna. He had not anticipated the confrontation with them coming so soon, or so publicly. His face darkens with choler as he appreciates their cunning bravado. Oh, *ja,* come to the train together to meet him, all innocent and aboveboard! He looks around involuntarily, as if seeking an escape route, and instead sees the children and Babu waiting in a delman, all dressed in their best finery as if for a celebration. "Papa, Papa!" the children cry, waving frantically at the father they have not seen in more than four months. Babu lowers her eyes respectfully and waves shyly also. Behind their delman stands another, presumably to carry Anna, Dubois, and Prentice back home, and behind that, an oxcart waits to receive his luggage. How horribly well-organized it all is, this nightmare.

Dubois has not succeeded where everyone else has failed by avoiding the difficult. "Recte et fortiter," he thinks to himself, squaring his shoulders and putting on his hat; recte et fortiter. He will not be shamed or defeated, not here in public on the platform in full view of all of Toeloeng Agoeng. He walks forward boldly, carrying only his small case and leaving the rest for the coolies to fetch. "Anna, my dear," he says upon reaching the couple, his wife and his best friend. She takes a step forward, holding up her arms for an embrace. He deflects this, kissing her a little coldly on the cheek. He can see in her eyes that she is puzzled and a little hurt.

"Dubois, my dear doctor!" exclaims Prentice. There is no avoiding his firm hug of welcome, though Dubois stiffens in his arms. He can bring himself neither to wholeheartedly return his friend's affection nor to openly reject it.

"Prentice," he murmurs over his friend's shoulder. "I had not thought to see you here."

"No," says Prentice, stepping back and grabbing his hand for a warm shake, "Mrs. Dubois and I planned it as a surprise for you. We knew you'd think she'd be waiting at home, just as usual, but we found out what train you were on. Welcome home, old friend, welcome home!"

Dubois is overwhelmed by the apparent genuineness of the man's affection. Could his awful suspicions be wrong? Could he have hardened his heart against Prentice for nothing? His thoughts flip back and forth wildly, like a leaf in a high wind. No one could feign such fondness for Dubois were he seducing Dubois' wife, no one. Surely . . . But there are those letters, and those weeks and weeks of no letters, and all the visits to Anna . . . What a terrible blackguard the man must be, to practice such deception while seeming open and honest. Prentice must be truly a snake in disguise, a viper that has wound its way into his heart while intending only to impart the poisonous bite. Dubois does not know how to react, and so grows stiff and taciturn.

"And here are the children, Eugène," Anna says gently as Babu brings them over. Dubois kneels down to greet his brood, tidy and neat for once and behaving properly. He envelops them, all three at once, in a bear hug, and kisses their heads.

"And how have you been, my little ones? Eugénie? Jean? Victor? Are you all well? Have you obeyed your mama and been good children?" They nod solemnly, round-eyed at so much attention from their father, who seems like a stranger to them.

"Papa, I have been learning how to ride my pony," says Eugénie proudly, hoping for his approval. "I could go out for rides with you. We can look for fossils." Dubois chuckles and pats her curly head.

"And Papa, Papa, I can read lots of new words, now, and books, big books from home," brags Jean, pulling on his father's sleeve. He does not like to be outdone by his big sister. He expands in his father's smile like a flower in the sun.

"I know my letters, Papa," whispers Victor, a little anxiously. "And the cat, the little tabby that lives in the kampong, has kittens and Mama says I can have one. Can I, Papa? Please?"

"Yes, Victor, all right, we will pick out a kitten for you," says Dubois gently. He looks at Anna over the children's heads, while he talks to them. "That's very good about your lessons, Jean. You shall read to me when we get home. And you can show me how well you can ride, Eugénie." Anna cannot interpret the meaningful glance Dubois has cast at her; she has no idea of his suspicions. She sees he is behaving oddly, but she does not understand why.

"You must be very weary, Dubois," surmises Prentice, looking at Dubois' masklike expression, "after such a long journey. We'll get you home quickly." As he hustles Dubois' family down the platform toward the waiting delmans, he tries to fill the stony silence with conversation. He cannot imagine what is wrong with Dubois; he seems not to be himself. "And was the journey successful, then? Did you find lots of fossils in the Siwaliks? Mrs. Dubois showed me some of your letters; she used to send me a note up at Mringin when one arrived, so that I could come down and learn the latest news," Prentice continues.

"*Ja, ja*," says Dubois, nodding. "It was a difficult journey in many ways, but I learned a great deal." He pauses and then continues, heavily, meaningfully, "I even learned some things I never expected to. Sometimes you only understand the true nature of a thing when you are far from it." With these words, he stares directly at Prentice, his eyes cold and penetrating.

Prentice has no idea what to make of this remark. He can feel the hostility emanating from Dubois, but he can think of nothing he has done to warrant it. Is Dubois perhaps angry that Prentice has joined in a family event? But he has always been so welcome at Dubois' house, so often a part of family occasions. All he can say by way of reply is, "Ah, yes, Doctor. Well, you must tell me all about it when you have had time to rest. I am most anxious to hear about your discoveries and adventures."

Both Prentice and Anna fall into an uneasy silence on the brief ride back to the house, cowed by Dubois' manner. Something has gone very wrong with their welcoming party. When they get back to the house, Anna shoos the children and Babu into the back garden and leads the men onto the back veranda for cool drinks and a light meal she has had Kokkie prepare. Dubois excuses himself to go wash his face and change into more comfortable clothes while the others wait on the veranda.

"He is very tired, Mrs. Dubois," Prentice offers, settling into a rocking chair. "That must be it."

"Of course," she agrees uncertainly, taking the seat closest to him. She places her hand on his arm, pleadingly, hoping for understanding. "But I thought it would please him so, to have us all there, you, me, the children. I thought it would show him how much he was missed while he was away. Why was he not pleased?"

"I don't know, Mrs. Dubois," replies Prentice, patting her hand comfortingly. "I don't know. He seems almost angry about something. Perhaps . . ."

Dubois comes back to the low murmur of their voices and the intimacy of their posture. He loses his temper in a flash. He strides over to them, keeping his voice down with effort, his face distorted with fury. "Can't you even stay apart from each other for a few moments while I leave the room?" he hisses. "Am I to have this going on under my very nose?"

Anna springs back, hearing at first only the anger and not the specific accusation. Dubois turns to Prentice, who has stopped his rocking and sits, utterly still, staring at Dubois in shock. "And *you*, my dear friend," Dubois says bitterly, "*you* have so kindly looked in on my wife while I am gone, time after time, visit after visit, staying so long and so often that everyone in Toeloeng Agoeng must be talking of it. What *kind* concern for the well-being of my family," he continues sarcastically, "what an exhibition of true friendship. How could you treat me so?"

"Just what do you think I have been doing?" asks Prentice in a quiet, deadly tone of voice. He stands up to look his dear friend directly in the eye, for they are almost the same height. "What exactly do you mean by those remarks?"

"Eugène—" Anna breaks in breathlessly, pulling at Dubois' arm. "Oh, no, no, you don't think—" She is so stunned that she is almost gabbling. The men are too absorbed in their confrontation with each other to pay much attention to anything she is saying. "But, I mean, what do you think—do you think—with the *children* here? My children?"

"Prentice," answers Dubois in a voice full of the sadness of all eternity. "You know exactly what I mean." He tries to control the quaver in his voice, but he cannot, and carries on regardless. "You have been seducing my wife while I was in India. It is perfectly clear from the correspondence I received. I am only appalled that you have the treachery in your heart to come with her to the station to greet me."

"*I did not!*" roars Prentice, flailing with his strong right arm as if to sweep away the accusation, and instead knocking over a table and potted palm. "I have done no such thing!" He reaches out with both hands and grips Dubois' shoulders, the two of them only a foot apart and looking like two versions of the same man. "I did not," he repeats in a stony whisper, staring into his beloved friend's eyes.

The servants have been squatting in the shadow of the veranda. Now they scurry over to right the table and sweep up the mess from the plant pot. As soon as possible, they back away quickly, crouching and averting their faces from the scene in front of them. They return to their stations, deaf and mute to the unfolding conflict. They want only to be unseen, unnoticed, forgotten. They do not want the Tuan and Njonja to be angry with them later for what they have witnessed.

Dubois and Prentice stand like a tableau, face to face. There is hurt on both sides, and anger: the measure of the deep bond of affection and understanding that they have shared. Either their intimacy will break under the strain of this accusation or it will survive forever. They need, both of them, to become sure of what they see in each other's eyes, each other's heart.

Anna interrupts, fussing and trying to smooth over the awkward situation. "Mr. Prentice," she says a little shrilly, like a parody of an imperious njonja. "Please *do* sit down again. I shall have the boy bring you another drink. Adik!" she calls over one shoulder.

"Yes, Njonja?" inquires the djongas softly, not daring to look up or to approach too closely.

"Another drink for Tuan Prentice," she says, never looking at the servant, who sidles silently into the house on bare feet. "Now, Eugène, we'll have no more of this," she continues in an artificial voice, as if scolding a fractious child. "You are tired and hungry and it makes you bad-tempered. Prolonged trips are always a strain on your constitution, and the last few months have been difficult for us all. After you've had your lunch and a rest, I'm sure you'll think better of this."

Something has resolved itself between the two men, who have been ignoring Anna completely. Prentice releases Dubois from his grasp—his embrace?—but not from his gaze. Dubois sits down in his chair, first looking at the floor thoughtfully and then lifting his head to return Prentice's look.

"And Mr. Prentice," she says, turning to the visitor and eyeing him down into his rocking chair. "I'm sure no one is more grateful than the doctor for the thoughtful way you called in on me and the children from time to time, to hear the news of the doctor's success in India and inquire after our well-being. It was most . . . civil of you, most kind. The children missed their father so much, I'm sure it was good for them when you had time to talk with them a little." Some of this is lies, little social white lies, but she does not stop to think of that. Her objective is to reestablish normal social behavior, and in that she succeeds. Abruptly, she runs out of courage and inventiveness and falls silent herself. At least she has prevented anything worse happening. There are no blows, no further shouting, only the echo of that terrible, terrible, suspicion in Eugène's voice. She is still shocked at his words. How could he think . . . ? Whatever gave him the idea that she . . . that she and Mr. Prentice . . . that . . . She cannot even formulate the words in her mind, for she is still a prim little totok from Amsterdam at heart.

Prentice makes his excuses and rises to leave a few minutes later. Dubois stands up to see him out while Anna waits, frozen in fear, on the back veranda. She prays that nothing will happen between them, not at the front, in the open, where everyone will see. She can hear no raised voices, so perhaps everything is all right.

Prentice turns to go down the steps to his horse and then turns back to face his friend. He cannot leave without a last word. "Doctor," he says quietly, so no one else can hear, "my dear doctor." He grips Dubois' hand firmly in both of his. "You misjudge me. I am your true friend. You have had a bad time of it in India, I can see, or you would never think such a thing of me. Please come up to Mringin, as in the old days. You can tell me of your discoveries in India. Please." His face is intense with sincerity and affection, as he thrusts aside his own indignation in order to reach his friend's heart.

"I . . ." Dubois starts to reply, but he cannot control his voice. He places a hand over Prentice's, but he is not sure yet what he thinks, now that he has seen his friend. The entire confrontation has gone differently from the scene he anticipated. His suspicions have been dislodged from the rock-hard certainties to which they once clung limpetlike. He cannot look at this man and think evil of him. He finds it hard to speak. "I . . . perhaps, Prentice, perhaps. I shall come if I can," he finally stammers.

Prentice nods, presses Dubois' hand once again, and walks across the yard to mount his horse. The syce stands silently holding it, trying to be invisible. "Mringin," Prentice calls in farewell, raising one arm in salute.

"Mringin," replies Dubois softly, as if he is repeating a sacred pledge.

Prentice wheels his horse and rides off, up the mountain, never to see Dubois again.

CHAPTER 38 DEPARTURE

The remaining weeks of May and early June are spent in arranging transportation to the Netherlands for fossils and family. The Duboises will sail on a French mail steamer departing on June 27 from Batavia for Marseilles, where they will catch a train to Paris and then another on to Amsterdam. In the meantime, there are possessions to be sorted—these to be sent home; these to be auctioned to clear their debts and provide a cushion of money to see them through the first few months at home. The servants are given notice and Anna approaches her friends and acquaintances to find positions for the better ones, the ones who have been really loyal and faithful. Each servant is given a written reference, which Anna knows will be produced proudly for the next newly arrived totok.

Dubois tries to put in order all of his notes and maps and scientific writings, making as sure as he can that every last detail that needs checking in Java has been verified. He organizes the correspondence and field reports from De Winter and Kriele, placing the most extremely important papers in waterproof boxes in case of disaster on the voyage home. He remembers the trials of Alfred Russel Wallace, the great naturalist, who lost almost his entire collection of specimens and his notes—indeed, and nearly his life, too—when the ship on which he was returning from the Amazon sank within sight of land.

Dubois has a sort of wooden suitcase made to hold the two lovely wooden boxes that house the precious *P.e.* fossils, so that he will have only one case to look after as he brings them home, to Europe and their acceptance. He postpones the half-promised trip to Mringin to see Prentice, distracting himself with the many tasks that must be completed before leaving. "Next week I shall go up," he promises himself, "next week." But in the end, he cannot muster the courage to go and see Prentice again. What could he say to this man, the man who kept him from going mad when his father died, who went to Trinil with him? What could he possibly say to him now?

The criticisms from Europe continue to pile up mercilessly. The Dutch anthropologist Herman ten Kate joins the attack, writing in the *Nederlandsch Koloniaal Centraalblad* (Colonial Dutch Journal) that he doubts the association of the skullcap, femur, and tooth. In a cruel wordplay on Dubois' assertion that it would be foolish *(thöricht)* to presume that the specimens come from different individuals, Ten Kate remarks he would rather be regarded as *thöricht* than leap to such an unsupported conclusion. He regards Dubois' method of calculating the cranial capacity from measurements of the skullcap as unproven.

And Rudolf Martin, the Swiss anthropologist, expresses his serious reservations, too, in an article in *Globus*. He too refuses to accept the association of the three fossils and is skeptical of Dubois' estimated cranial capacity for *P.e.* Most of all, he faults Dubois for failing to compare his fossil with the Neanderthal fossils, a point that particularly annoys Dubois.

It is not as if I was sitting in Europe, Dubois fumes, needing only to stretch out my hand to obtain a cast of a Neanderthal skull. Besides, Neanderthals are much younger and really human. The *point* of my monograph was to compare *P.e.* to apes and humans, not to place it within the spectrum of the races of man. *P.e.* is no human, that much is certain.

Dubois begins to feel like a sinner tied to the stake, the flames licking at his feet. Instead of praise for his astonishing achievement, he receives only accusations of foolishness and incompetency. What crime has he committed? Finding the missing link when no one else could? That is, perhaps, the truth. He is the Man Who Found the Missing Link, and now they resent him for it.

The next is Daniel Cunningham, a self-important anatomist in Dublin who makes snide remarks in *Nature* and then has the audacity to give a lecture on *P.e.* on February 23 that is published in the *Journal of Anatomy and Physiology*. Dubois is outraged. What does this man know of *P.e.*? He can only have just barely read the monograph before giving his lecture. Cunningham, too, chastises Dubois for omitting comparisons to Neanderthals. Cunningham maintains that, since Neanderthals have cranial capacities of about 1,200 cc, they must lie on the direct lineage between *P.e.* and man: "By a series of easy and nearly equal gradations we are led from the fossil form up through the Neanderthal and Spy forms to the modern cranial arch." Feeling brain size is the more important issue,

Cunningham denies the apelike shape of the *P.e.* cranium and concludes, magisterially,

> *The fossil cranium described by Dubois is unquestionably to be*
> *regarded as human. It is the lowest human cranium which has yet*
> *been described. . . . The so-called* Pithecanthropus *is in the direct*
> *human line, although it occupies a place on this considerably lower*
> *than any human form at present known.*

Dubois snorts incredulously as he reads these words. But of course—how convenient for Cunningham and the others, how simple. If the skullcap is a primitive human, coupling it with the humanlike femur of *P.e.* eliminates all the awkwardness of assessing a transitional form. All evolution, in fact, neatly disappears, and there is no need for Cunningham to grapple with the uncomfortable truth that there is, and was, a missing link between apes and man.

Dubois keeps tally, dividing those who favor his views from those who oppose them, noting also who believes the fossils derive from two or three separate animals, who thinks the skull and femur human, who thinks the skull apelike and the femur human, and so on. Opinions are so varied it is difficult to keep track. He will go back to Europe and show his specimens to every important man of science; surely that will convince them all. For some reason they do not understand his monograph properly. Perhaps, at thirty-nine pages, the work is too brief, but it is the best he could do in the Indies. Or perhaps the ideas are simply too new to be readily accepted. Darwin, he reminds himself, was hotly criticized too.

The nearly endless details of packing and leave-taking finished, the Dubois family leaves Toeloeng Agoeng forever on June 23, accompanied by Janet Boyd, the ten-year-old granddaughter of the Old Warrior. Janet will help with the Dubois children on the trip and then attend school in the Netherlands. Anna and Dubois make one last pilgrimage to the tiny grave in the Javanese cemetery across the road, laying flowers there and planting a frangipani. They instruct the servants to look after the grave and water the young tree, but they have little faith their wishes will be carried out for long. One more dead totok baby is nothing special to the Javanese. They have seen so many Europeans come and die that they are stoic. Anna weeps at the thought of leaving that poor baby behind and all alone in the vastness of Java. Little Anna Jeannette would be almost two years old by now if she had lived. There is nothing to be done, they must go home again, and Anna accepts that, with difficulty. The grave of Anna Jeannette who might have been will be shaded by the frangipani as it grows, protected from the torrential rains and the burning sun. Perhaps in time the tree will drop its fragrant, thick-petaled, translucent flowers on her grave. While Babu sits with the children in the delman, they walk around their house once more, admiring its fine spacious proportions and cool marble floors, its lush garden and cool verandas. They will never have such a grand house again. Then they close the doors for the last time and go off to the station with Janet, and the children, and the mountains of luggage.

When they arrive in Batavia, a letter from Prentice is waiting for Dubois.

Mringin   June 23, 1895

My Dear Doctor,

You will now be sitting in the train on the way to Batavia. I should indeed have liked to see you at your departure. Your promise of old, however, was to come up to Mringin for a day or two before you left. As you did not find time to do so, and I heard of your stay at Toeloeng Agoeng, I thought it better, <u>even for your own sake,</u> not to go down to take farewell of you & your family, & I therefore commissioned Mr. Boyd to bid you adieu for me.

The reason is this—You know you have taken me up wrongly on more than one occasion. You know what you once suspected me of in connection with Mrs. Dubois—a wholly groundless suspicion of course. Well, since I heard nothing from or about you just before your departure, I considered it best to keep away. An unguarded word, or even look, however innocent in itself, might revive some torturing suspicion in your mind & cast a cloud over your happiness. That is really why I kept away (as well as being very busy here) and I think it was better to keep away. At all events it obviated the smallest chance of any misunderstanding or wrong impression arising by any possibility in your mind.

However, from the bottom of my heart I wish you & yours a safe voyage, & good arrival in Europe, as well as health, happiness, & prosperity there. I hope to hear from you afterwards, and if I am ever rich enough to return to Europe I shall be sure to look you up in Holland.

Meanwhile believe me to be

Very Sincerely Yours

Adam Prentice

/In haste/ (I predict you will appreciate Java more once you are settled in Europe?)

A.P.

Dubois reads it through twice, slowly, as if memorizing the words. Then he folds the letter carefully and places it inside the wooden suitcase that holds the *P.e.* fossils for safekeeping. It seems only right that Prentice should keep company with the missing link. They are the best of Java, Dubois thinks.

CHAPTER 39 EUROPE

The ship to Marseilles becomes a little society, a tiny world unto itself, and the Duboises settle into a lazy rhythm of eating, reading in deck chairs, gossiping with fellow passengers, and napping. Restive from anxiety and lack of exercise, Dubois embarks on a regular program of brisk striding around the deck, increasing the number of circuits every morning and evening like clockwork. He has Janet see to it that his children participate. They shall not grow fat and lazy and sluglike, now or ever. He has no hope of overcoming Anna's basic nature or the indolence that has become ingrained in her during their eight years in the Indies.

Dubois and Anna never really reconciled after the death of the baby and the matter of Prentice has only made relations more difficult. Dubois has not raised the latter subject with Anna since his homecoming and she does not dare disturb the tense civility between them by probing. She does not know if he has ever resolved his suspicions. She thinks perhaps he has, for there was that letter awaiting them in Batavia, but of course he does not disclose its contents to her. The growing distance between them offers Anna a shield from Dubois' rigid, demanding nature. She grows more frivolous still, less systematic, less orderly, in a kind of instinctive, passive rebellion against his hardening nature. She never guesses that he is beset with worries all the way home, battling a terrible presentiment of failure. He cannot admit the possibility to her. After eight years of work—eight years of death and fever and despair—he actually found his fossils. Is he now to be denied the honor and acknowledgment he has earned? The cold fear settles deeper into his soul with every passing mile. The closer they come to Europe, and to judgment, the more Dubois armors himself against hurt and rejection.

Somewhere in the long crossing of the Indian Ocean, nature chooses to reflect Dubois' inner turmoil by producing a ferocious storm. It begins with a sudden blackening of the sky at midday and a tearing, rising wind. Before an hour passes, the waves have been whipped into a frenzy and the steamer tosses and heaves badly. A cold, gray, pelting rain drives the passengers into their cabins, where seasickness inevitably awaits them. At first the captain hopes to ride out the storm, but it is stronger and more extensive than he first estimated. The ship is pounded, dragged off course, covered in spume and rain that at least cleanse the decks of the vomit left by passengers trying, and failing, to make it to the rail before becoming sick. There seems to be no end to the surging waves and wind. The rain strikes like a hail of bullets.

Finally, fearing for the safety of the ship, the captain orders everyone to leave the cabins and get into the lifeboats. He has not yet decided to abandon ship, but he wants his passengers organized and ready, just in case. Ill, cold, wet, and frightened, the passengers huddle miserably in the small boats, not daring to look at the tumultuous dull green sea beneath them. The men fall silent; the women weep helplessly or clench their teeth and straighten their backs, according to their dispositions; the children wail and snivel miserably. The Dubois party is no different from the rest, though Anna shows surprising courage lest she frighten the children further. She whispers to them, "Papa will keep us safe, you'll see." Overhearing this fairy tale, Dubois wonders what exactly he is supposed to do. Banish the storm? Chastise the waves for being too boisterous? For all his physical courage and strength, Dubois is daunted by the look of those tossing seas. These little, overcrowded boats cannot last long if they are put to sea in such a storm, he fears. But if Anna's reassurance quiets the children, he supposes the story is a good one. Things will not be helped by hysteria.

And then he remembers the worst. He has left *P.e.* back in the cabin. If the ship goes down, the fossils will be lost forever. If he escapes drowning, he will arrive in Europe empty-handed, with no proof to show for his years of work. More than the fear of his own death, this fear of being mocked galvanizes him. "Anna," he says, climbing out of the lifeboat, "stay here. I will be right back." And he runs across the heaving decks, slippery with rain, back to the cabin. In minutes that seem like hours to his frightened family, Dubois returns, the precious wooden suitcase strapped to his chest and protected by one large, muscular arm. He clambers back into the lifeboat.

"Anna, listen to me," he says firmly. "If something happens, you must look after the children." She nods dutifully as she tries to anchor the bedraggled strands of hair that have fallen loose. She cannot imagine what he is talking about. "Anna, you take care of the children, *ja?* If the lifeboat is lowered, you see to the little ones, for I shall have to look after this." Dubois places his hand across the flat surface of the suitcase.

His meaning dawns on Anna and she turns her face away from him, so that he does not see her expression. *Ja, ja,* she thinks resentfully, wiping the water from her children's faces with a sodden handkerchief. *Ja,* he will look after the fossils and I am to save the children, all three of mine and Janet too. The fossils were always more precious to him than we were. Always. That is why he was always leaving us. The great swimmer, Dr. Dubois, will save his fossils from the waves, not his children.

There is no more energy to waste in talk; the shivering family sits in the lifeboat, passively awaiting their fate. Their terror does not wane but exhausts itself; the children fall into a soggy, miserable sleep, for which Anna is grateful. After an interminable time, the storm begins to wane. It feels like a miracle. Soon the captain orders the passengers out of the lifeboats, back to their cabins. The ship has done it, she has ridden out the terrible storm, and they are safe. As they slowly climb out of the boats, stretching cramped limbs and frozen hands, they can see the sky is lightening, the rain lessening, the waves calming. Dry clothes, warm blankets, and hot-water bottles await them in their cabins. The captain orders the cook to serve up hot soup and bread as soon as possible to passengers and crew alike. Later that night at dinner, when everyone has recovered from the ordeal, there is a frantically gay atmosphere and plenty of wine. Dubois, with the others, raises his glass to toast their captain for his skill, but under his chair is the wooden suitcase. From now on, only when he is exercising will he let it far out of his grasp. He has almost lost everything and he will not be caught unawares again.

After the colorful and exotic harbors he has seen in the East, Dubois is not much impressed with Marseilles. It seems dirty, rough, and dingy; the sailors look like scoundrels and the only women who present themselves to the eyes of the arrivals are less than respectable. Getting the family and the luggage through French customs and onto the boat-train seems to take forever, but at last they settle into a comfortable sleeping compartment, the ever-present wooden suitcase tucked neatly under Dubois' seat. By the next day they are in Paris, looking at the fashionable European people and places that they have not seen in eight long years. The children are agog at the strangeness of it all. "Where are all the natives, Mama?" Eugénie asks, querulously looking around the hotel. "Why don't we have proper servants? Where is Babu?"

"Shhhh," says her mother. "We are in Paris now, in France. It is a different country from the Indies and natives don't live here. We won't have a babu anymore."

"But we always have a babu, Mama," Eugénie persists, logically. "Can

Janet be our babu? And Mama, why do all the houses look funny, and the gardens? Why doesn't anyone wear sarongs and pretty clothes here? And it is so cold, Mama, like in the mountains at Mringin."

Anna bends down to hold a quiet conversation with her daughter, this child who remembers only the sunlight and colors and dark-skinned natives of the Indies. The boys are quieter, but just as confused as their big sister. They have never seen a place like Paris before, and they are not sure they like it. There is something indefinably . . . wrong . . . about Europe. They have been told all their lives that Europe is "home," but it does not seem like home to them. Home is warm and sunny and full of flowers and loving babus; home smells of incense and spices and woodsmoke. This cold, gray place, with its large stone buildings, gilded imperial statues, and ugly cement sidewalks is not home. This is some dreadful, dull, cold place. All the rest of their lives, the three Dubois children will remember Java as a paradise from which they were taken without explanation.

They lunch in a café near the station before catching the next train for Amsterdam, where they will change to get the branch line that will take them to Eijsden, to "Groetma's house." They have deposited most of the luggage in the station. Only the wooden case still accompanies them, tucked securely under the table, where Dubois can feel it with his ankle. The children do not like the food and beg for nasi goreng, the fried rice dish of the Indies, and fruit. Oblivious to their complaints, Dubois stares fixedly at one of the waiters, while Anna tries to coax the children into eating.

"Eugène," she says finally, in a soft tone of voice, "could you please explain to the children why they cannot have Indies food here?"

"What?" replies Dubois, startled. "Oh, *ja*, of course. Now, children, we are in Paris, in Europe. We are not in Java anymore. And so here we eat the food that Parisians eat, and in Eijsden we will eat good Dutch food. We cannot get Indies food here; the people do not know how to cook it." His explanation, or perhaps simply his greater moral authority, quiets the children and they begin to eat, after first dubiously examining every mouthful.

"What *are* you staring at, Eugène?" Anna asks him.

"The waiter, Anna," Dubois replies, as if it is obvious. The man has a high, prominent forehead, an aquiline nose, and a lantern jaw. "Look at that skull! What I would give to have one like it for my collection!"

"*Ja*, I should have guessed," Anna says with a sigh. "Always your skulls and your bones. Don't you ever think about people?" But her husband does not hear her.

They send a telegram to Dubois' mother; late the next night, they finally arrive at Eijsden. It is early August; the trip home has taken them just over six weeks. Dubois' mother has sent some carriages to the station to collect them, and various relatives have turned out to greet the long-absent Dubois family. It is a good thing so many have come, for there is so much luggage, and the children are so weary that they must be carried. It is nearly midnight by the time they arrive in the Breuestraat in Eijsden, where a whole crowd of villagers has turned out to catch a glimpse of the professor they remember as a young boy. He's made some great discovery, they say to one another. A neighbor stands by the doorway, holding a lighted lantern aloft, while others unload the cases, trunks, boxes, and crates from the carriages. Someone carries the sleeping children in and Anna puts them straight to bed upstairs, while Dubois stays below to supervise the unloading of the mountains of luggage. Only a few notice the heavy, wooden case that he hands over so reluctantly, saying, "Take care with that! Don't drop it! It is the ape-man!"

Once all the things are safely stowed inside the house, and each helper has been thanked, the crowd disperses. Anna and Dubois settle gratefully down to share a cup of tea and a slice of bread and butter with Dubois' mother. They feel disoriented, what with the lateness of the hour, the length of the journey, and the confusion of all of those people, most only half-remembered. It is a real pleasure to sit quietly and eat and drink a little.

After a while, Dubois feels somewhat recovered and decides it is time to make his announcement. It is no surprise, of course, but his sense of occasion demands that this be done formally. "Thank you, Mama, for all the help, and for welcoming us so late at night," he begins.

"Of course, of course," his mother says, wearily but with a smile. "My son and his family are always welcome here. I am only sorry your father is not with us to see you, it would have pleased him so."

"*Ja*, Mama," Dubois replies, saddened. "I miss him, too." Brightening a little, and squaring his broad shoulders, Dubois stands and walks into the next room, extracting the precious case from the chaotic pile of luggage. Taking out the skull box, he walks back into the room where his mother and his wife are seated. "But here, Mama, here it is. Let me show you. Here is the missing link, the thing that took me so far from home and took so many years to find." With a flourish, he opens the box to reveal the brown, shiny skullcap, nestled safely in its bed of velvet. He holds it out for his mother to see.

"So this is it?" she says dubiously, prodding the strange object with a gnarled forefinger.

Dubois nods proudly. "*Ja*, Mama, that is the skull. That is *Pithecanthropus erectus.*" His mother looks up at him and he sees how much she has aged in eight years. "*Ja*, Mama, this is it," he repeats softly, gently.

"But, boy"—she sighs heavily, looking bewildered at his treasure—"what use is it?"

A terrible pain is visible on Dubois' face for only an instant before he arranges his expression into a mask of neutrality. Not even Anna notices; she is too tired herself to watch carefully, and she does not understand the enormous significance of the gesture of showing *Pe.* to Dubois' mother. This was to be the moment of triumph, the justification for the years and the risk and the distance from home. Dubois' mother understands none of it.

When the Dubois family returns to the Netherlands, they stay with Trinette (seated, left) in Eijsden. Anna (seated, right) helps Jean (standing), Victor (seated, left), and Eugénie (seated, right) adjust to their new life while Dubois works harder than ever.

Dubois does not answer his mother's question, cannot answer. There is no answer. If even now she cannot see the importance of his discovery—even now, with the fossil in front of her—how can he possibly explain its value? It *has* no value to her, none at all. All he has sacrificed, all he has endured, all the trials through which he has persevered by sheer determination and force of character: these are as nothing to her. He was so sure she and his father would finally understand when he came home in triumph. But now he is back, and his father is dead, and his mother sees nothing. The prodigal son has returned, but there is no fatted calf.

What use is it, boy? What use is it? The words might as well be carved with a knifepoint into the beating muscle of his heart, so deeply is he wounded. No use, Mama, he thinks painfully, no use at all to you. Papa is still dead. I was still gone for eight years. Your grandchildren are strangers to you, and one lies dead in Toeloeng Agoeng. So the fossil is of no use at all.

He says nothing. He closes the box carefully, precisely, securing its lid with the fine brass clasp, and walks quickly out of the room. He replaces it in its wooden suitcase and puts the suitcase in a safe, out-of-the-way spot. He goes back and suggests that he and his wife retire for the night. "It is time we got to bed, Anna. We are all very, very tired."

CHAPTER 40 THE BATTLEFIELD

In the morning, Dubois determines to ignore his pain. As if he is wrapping a fragile fossil in teak leaves, he packs determination, perseverance, and pride around his delicate dream of accomplishment. I know this, he thinks to himself, I have learned that, I have found the missing link, I have, I have, I was right all along. By an effort of will, he summons up his natural confidence and acts upon it as if he has not been bitterly disappointed. Very well, neither his mother nor his late father will ever understand that he has made the greatest discovery of the century; very well. His mother is not a scientist, any more than Anna or the children. He must not mourn over those who cannot understand; he must not let them undermine his convictions. All he has risked is worthwhile, for he *has* found what he sought. It may be tragic to be unappreciated, but it is not fatal. Perhaps in time he can resurrect his hope.

Dubois settles his family in a house in The Hague, where ex-colonials cluster nostalgically to talk of the tempo doeloe, the good old days in the Indies. Then he embarks on an almost holy crusade that lasts for the next few years of his life. He plans a campaign to convince not his family—he has given up on them—but those who matter: his scientific colleagues. He must catch up on all the journals, visit De Vries, Place, Fürbringer, and the others, find out what the talk is in scientific circles. He must prepare himself for conferences and lectures. He must publish more. The world may be skeptical now—it *is,* he reminds himself sternly, thinking of the articles he has seen—but it will soon come to agree with him. Once men of science have seen his beautiful *P.e.,* they will agree that there is nothing to compare with her anywhere. He will introduce her to scientific society, he thinks whimsically, as if she were a debutante at a coming-out ball. She will dazzle them with her sheer beauty; she will win the day, if only he can make them *look* at her.

And somehow, he must secure a position suitable for one of his learning and accomplishments, where he can settle and study his fossils. True, he is

still technically a military surgeon, albeit on the unattached list with no active duties, so the Ministry for the Colonies pays him a paltry salary. The Ministry for the Interior is responsible for the costs of housing his collections and arranging for assistants, but they are not generous. Dubois' first priority is to make people pay attention to his fossils, to his missing link, but after that, he must attend to mundane financial matters, too.

His first salvo is fired in Leiden, at the Third International Congress of Zoology, held over six days (the sixteenth to the twenty-first) in September. It is a golden opportunity, for many of

Rudolf Virchow, the pathologist from Berlin, fiercely criticizes Dubois' interpretation of P.e. *This cartoon appeared in* Vanity Fair *on May 25, 1893.*

the elite of science are there: W. H. Flower, director of the British Museum of Natural History; A. Milne Edwards; Othniel Marsh, from Yale; and none other than the elderly pasha of German science himself, Rudolf Virchow. If he can change Virchow's mind, Dubois knows he will have won the battle of scientific opinion. Virchow is his fiercest opponent, and Virchow has not mellowed with age. His passionate rejection of Darwin's evolutionary theory and of all claims of human evolution have only grown stronger and louder. His opinions have petrified over time, like bones turning into fossils deep within the earth's sediments. Virchow's control over German science has been only marginally diminished by his public battles with Ernst Haeckel over matters of evolution. Yes, Virchow is the enemy, the target, the one whose stubborn convictions must be overturned if Dubois' wonderful find is to gain acceptance.

Virchow has already expressed himself unfavorably on the subject of *P.e.*, in a discussion following a skeptical lecture given by Wilhelm Krause on January 19 and in a later lecture of his own, both in Berlin. Like so many others, Virchow argued that Dubois' cursory description of the conditions under which he found the three fossil fragments was simply insufficient to support the claim that these were three pieces of a single individual. Cruelly echoing Dubois' assertion, Virchow declared it would be truly "foolish"— *thöricht* indeed—to fail to make a critical inquiry into this unproven point. The skullcap he regarded as coming from a giant gibbon of some kind. As for the femur, it was human. Who but a human, helped by others, could have survived such a terrible injury to the thigh long enough for the clot to ossify? At best, he conceded, such a femur could just possibly have come from a bipedal gibbonoid form. The molar he completely ignored. His summary of Dubois' interpretation of *P.e.* was devastating: "Here the fantasy passes beyond all experience."

Dubois is of two minds when he finds out that his presentation of *P.e.* will be presided over by Virchow himself. On the one hand, Virchow's presence and attention are assured; on the other, Dubois may be exposed to harsh criticism in a most public forum. Well, so be it. Dubois has the fossils and he knows what they are. He is certain of his conclusions. He vows to confront the old tiger in his lair, risking a painful clawing if that is what it will take to change Virchow's mind. He even invites Virchow and a select group of scientists to view the fossils before the public presentation, so they can examine and handle the specimens for themselves. It is a courageous move that clears doubt from several minds but does not sway Virchow.

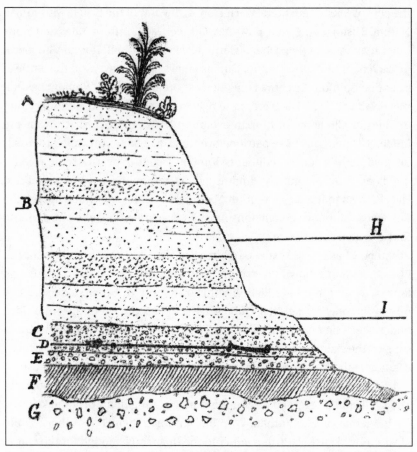

At the Third International Congress of Zoology in 1895, Dubois shows the find-spots of the P.e. specimens on this geological section of the deposits at Trinil. "H" indicates the level of the river in the rainy season; "I" shows its level in the dry season.

Having heard the criticisms of his monograph, Dubois takes this opportunity to counter them. He squares his shoulders, stands up even straighter than usual, and prepares to fight for the minds of those who listen to him. His anxiety makes him stiffer, more formal than ever; he knows this is not a good thing, but he cannot relax when so much is at stake. Now, for the first time, he presents the details of the discovery. He displays diagrams of the excavations and shows a schematic section of the geological formation at Trinil, including the all-important layer from which the fossils came. He indicates where in the geological sequence the fossils were found; he asserts that the layer is intact and the fossils show no signs of having been transported from anywhere else, as anyone present can verify for himself. The striking chocolate-brown color of the specimens and their patently obvious heaviness, showing that they are thoroughly fossilized, bespeak identical geological histories for the skullcap and femur. Also, for the first time, he speaks about the fauna, the abundant remains of other fossilized mammal bones, and the detailed and meticulous comparisons that lead him to conclude the entire collection is of late Pliocene or early Pleistocene age.

Then he turns to the fossils themselves. If the femur is so similar to that of a human, it cannot also possibly belong to a giant gibbon, as Virchow and others have previously suggested. Yet, Dubois argues, if bipedal walking evolved at the very beginning of human evolution—contrary to the predominant expectation among the scientific community—then just such a strongly human femur, with a few apelike characteristics, would be expected. The skullcap, too, reveals its transitional nature even in the criticisms offered of his interpretation. Why does one scholar consider it fundamentally apelike while another claims it is human? *Because it is in truth transitional, between ape and man,* Dubois tells them. Only that interpretation accounts for the diverse anatomical features of the skullcap they see before them.

It is painful, humiliating even, to have to restate publicly the criticisms of his monograph, but he must not dwell on that. The others have simply misunderstood, out of ignorance and surprise. He must show them how to reach the conclusions that he struggled so long and hard to reach, for he is right. He must be patient. He must be clear. This is not the time to express his feelings.

Next, he addresses the complaints about his estimate of cranial capacity. There is no accepted technique for making such estimates and, indeed, only one other scientist has tried to estimate *P.e.*'s cranial capacity for himself: Léonce-Pierre Manouvrier. Since 1885, this dapper Frenchman has been the professor of physical anthropology at the prestigious École d'Anthropologie in Paris; more to the point, he is a leading authority in the new "measuring

Léonce-Pierre Manouvrier of the École d'Anthropologie in Paris defends Dubois and collaborates with him on a study of the variation in human femurs.

school" of anthropology. Measuring, calculating, quantifying, and comparing anatomical features in consistent, reproducible ways are Manouvrier's passions. He is well-respected as a meticulous scientist, a man not given to fanciful interpretations. What is stunning is that Manouvrier's wholly independent estimate of the cranial capacity of *P.e.* matches Dubois' own: 1,000 cc. This is more than a telling coincidence! It is verification. While microcephalics and members of some races of humans have brains as small as this, Dubois concedes, these unusually small-brained types also have small bodies, which *P.e.* demonstrably does not. Thus, these estimates support the idea that the braincase is small for a human, but very large for any known ape: transitional, in fact.

Dubois finds himself breathing slightly rapidly from the effort of speaking in English; he is nearly fluent, but the finer points of meaning and usage sometimes escape him. And he must get his words right now. Trying to decipher and rebut every criticism in a foreign language is a formidable task. It would be better if people simply read and understood his monograph, much better, for in print he can write and polish his words carefully. He pauses for a moment and looks around the room, gauging the effect of his words on this learned audience. He can see some nods of agreement, some eyebrows raised in surprise and concession. But he knows in his heart that his delivery is still too stilted, too reserved; it is the old problem his anatomy students always complained about. He does not capture the imagination. Even though he passionately believes every word he speaks, the issues are so important that he freezes when he speaks of them. He cannot help holding himself aloof. It is a deeply ingrained reaction from a man too often misunderstood, who has been too often deeply wounded by the rejection of his novel ideas.

And Virchow, the worst of them all, sits there looking more like an ancient tortoise than an aging tiger. Even a tortoise can administer a deadly bite, however. He is the master of the sarcastic rejoinder, the subtle twist of words that leaves his enemies bleeding. He is wrinkled and wizened, thin-haired and old: a small, bent man who is physically unimpressive, especially in contrast to Dubois' own square, muscular physique. Virchow is seventy-four, Dubois still a handsome thirty-seven. Dubois has youth and energy and the future on his side, not to mention an intellectual power probably the equal of Virchow's own. But behind his gold-rimmed oval glasses, Virchow's eyes are like black bullets, revealing the essential nature of the man: cold, proud, unyielding. Dubois has not yet reached Virchow, he can see that. Virchow's eyes show only deep contempt for this obdurate young Dutchman.

Dubois takes a deep breath and starts on one of his two last points. He dislikes public speaking and resents having to defend his ideas so vigorously, but it must be done. In his monograph, he did not compare *P.e.* with the Neanderthal skulls from Europe; this was an error. Now it is time to do so. He points out some similarities between the Neanderthal skulls and the skullcap from Trinil, like the strong brow ridges and the elongated cranium. But, he emphasizes, there are also important differences between the skulls. The Neanderthal skull is much more humanlike than the skullcap of *P.e.*, for the former has a much higher and more domed cranial vault. The important difference is that, in overall capacity, Neanderthal skulls are enormous. The estimated cranial capacity of the original Neanderthal skull is 1,230 cc, and the Spy skulls are comparably capacious—as big as many modern European skulls—while *P.e.* at about 1,000 cc matches only the smallest-brained human races or microcephalics. Still, the brain size of *P.e.* greatly exceeds that of modern apes, which never surpasses 600 cc. At this point, Dubois is happy to see some signs of agreement. He is filled with hope, for a moment; perhaps no one will ever again say his beautiful *P.e.* is just another Neanderthal.

There, he thinks, that's a tricky point made. Now for the finale, the closing. If I can hold their attention to the very end, perhaps I can change some minds.

And with that, he moves on to discuss the molar and, stunningly, to produce a new find. There is another molar, he tells his surprised listeners, also uncovered in October 1892, that unfortunately escaped his attention until recently. It was among a large number of other mammalian teeth found only three meters from the skullcap. The new tooth is more worn than the original one, he points out, which is to be expected since the original is a third or last molar and the new one occupies the position just in front of it. Since second molars erupt through the gum and come into wear well before third molars do, a discrepancy between teeth from one individual is normal.

In conclusion, despite many questions and criticisms raised in response to his brief monograph on *P.e.*, Dubois can see no reason whatsoever to revise his original interpretations and opinions. *Pithecanthropus erectus* is a transitional form, a missing link between apes and man, and he has found her. Now that they have been properly introduced to her, the distinguished members of the audience can surely see the wisdom of this judgment. As soon as he has expressed the sentiment, he realizes it sounds pompous when he has meant to be lighthearted.

Dubois leaves the podium to mixed applause and murmurs of confusion and disagreement. Virchow opens the discussion, as merciless as ever. He is unconvinced that Dubois has found anything more than the remains of a giant gibbon—very nice, but no missing link—and reiterates his earlier criticisms. It is clear that the elderly pasha of German science prepared his remarks in advance; there was never any chance of changing his mind. Dubois is grateful when the American Marsh comes forward to defend him. Marsh has observed many similar healed injuries in the femurs of apes' skeletons, he tells the listeners, so Virchow's assertion that only a human could survive such a severe injury cannot be correct. The healed injury itself is no grounds for assigning this femur to the human species, rather than to an ape-man. This is another telling point in Dubois' favor. Other speakers mention the illuminating presentation of the geology of Trinil. Showing the stratigraphic section and describing the layers was an excellent strategy, and Dubois vows to keep repeating this information wherever he speaks on the subject.

Now that opinion seems to be turning his way, Dubois stands up to publicly invite the chemist J. M. van Bemmelen to carry out age tests on several of the mammal bones from the same layers as the *P.e.* fossils. It has been accepted for some time that the fluorine content in fossilized bones increases in proportion to their antiquity, though this technique cannot provide an exact age for fossils. (Although some believe that Virchow has scored a complete victory over the foolish Dutchman, van Bemmelen's findings, which he publishes later, support Dubois' suggestion that these fossils are from the Pleistocene.)

Dubois returns home from the congress, pleased with the reception of his paper and determined to fight on until he has completely won over all the most influential scientists of the day. But almost immediately, Virchow starts making trouble. Though Dubois allowed this arch-skeptic special access to the fossils at the meeting in Leiden, Virchow modified his criticisms not one whit as a result. And now Virchow complains to F. A. Jentink, the director of the National Museum of Natural History in Leiden, that he cannot continue with a more detailed examination of the fossils. Jentink informs Dubois of this via postcard:

September 17, 1895
Professor Virchow complains that your pieces of evidence of P. *erectus that were exhibited in Leiden are not here to look at, possibly to be studied by him or other scientists.*

Really, Dubois thinks to himself, this is outrageous. The man had an hour to look at the specimens in Leiden, yet wrote complaining to Jentink the very next day. Virchow has received a copy of the monograph and he heard my illustrated talk. What more does he want to know, before I have finished publishing my analyses of my fossils myself? It is obvious that there is much, much more to do, now that I am back in Europe with proper libraries and museum resources. Does he want me to hand my baby over to him, lifting her out of her very cradle? Shall I entrust her to *him,* the man who above all others has doubted my word and impugned my scholarship? Virchow is too used to having his own way in everything in Germany, that is what is wrong with him. He can be rude and insulting to German scholars and they still kowtow to him and defer to his opinion. Well, I am under no such obligation to him, and I shall not do it.

Dubois formulates his reply immediately.

September 17, 1895

Judging from your recent postcard, it seems Professor Virchow does not know that the parts of Pithecanthropus erectus *have not yet been fully described by me in such a manner that I can place them at the disposal of others as museum objects. The description that I have given already is, self-evidently, tentative and incomplete, because in Java I lacked the necessary comparative material and sufficient literature. . . . Meanwhile, it is obvious that the tentative description I have already published is so incomplete that there has been much misunderstanding. . . .*

In October, he travels to Brussels to speak before a meeting of the Société Belge de Géologie. Next on his itinerary is Liège, where he studies the Neanderthal fossils from Spy for himself, for the first time. It is a revelation. He is surprised by their appearance, for all that he read the descriptions and studied the published photographs closely. There is indeed a similarity in overall skull shape to his *P.e.,* a compelling one. And he can now see, quite clearly, that Virchow's dismissal of these specimens as pathological was entirely incorrect. Dubois is indefatigable. In the next few weeks, he travels to Paris to the Ecole d'Anthropologie, striking up a close collaboration with Manouvrier. At their first meeting, they review some four hundred human femurs, many of which Manouvrier has already checked at Dubois' request. They compile a case for the distinctions between this large sample and the femur of *P.e.* The information they gather is complex and voluminous; they work late into the night and adjourn to a nearby café to dine, talking all the while of the meaning and significance of their observations. With pencils and notebooks spread on the table next to their plates, the two men scribble notes and sketch anatomical details of femurs they have measured, comparing and sharpening their conclusions. The by now battered suitcase, the one that has carried *P.e.* from Java to Holland and now halfway across Europe, sits beneath the table at their feet, like a naughty child who has crept in to eavesdrop on the adults' dinner conversation. So engrossed are the scientists in their discussion that the restaurant empties without their noticing; finally a waiter comes and asks them to leave, as the restaurant is closing. They fold up their notebooks, tuck them and their pencils away, and rise to walk back to Dubois' hotel.

But they never stop talking, arguing, reasoning. Deep in the scientific intricacies of the form and function of the femur, neither man notices that the abandoned suitcase sits beneath the table where they dined. They walk some blocks before Dubois senses the emptiness of his left hand.

"Mon dieu, Manouvrier!" he cries out. *"Où est* Pithécanthropus?"—"My God, Manouvrier! Where is *Pithecanthropus*?" And with that, he dashes across the street, running panic-stricken back to the café. Manouvrier, older and less fit, follows at a more dignified pace, but one that reflects no less concern. They arrive at the restaurant as the proprietor is locking the doors.

"Où est Pithécanthropus?" Dubois blurts out, in an appallingly bad accent. He is flushed and breathing heavily, he is frightened, and his command of French deserts him. "Pithécanthropus!" he shouts at the uncomprehending man. "Pithécanthropus! *Les fossiles!*" He grabs the man by the shoulders and shakes him, trying to make him understand. Manouvrier hurries up, suave as ever, to extract the proprietor from Dubois' frantic grasp. He soon smooths over the situation. He explains that the professor has left some valuable specimens in a suitcase under a table at the restaurant—that one over there, by the window, surely the good proprietor remembers how long they sat there talking. The proprietor is puzzled; he remembers them, *oui, bien sûr,* but not any suitcase.

Then their waiter approaches, drawn no doubt by the shouting and excitement. *"Ah, oui,"* he admits agreeably. *"J'ai trouvé une valise; j'ai cru que vous retournerez pour elle."* (I found a suitcase; I thought you would return for it.) The words are like a blessing to Dubois. He closes his eyes a moment in silent thanks. The waiter takes the keys from the proprietor, unlocks the door, and leads them inside, to the cupboard where he has placed the suitcase. *"Voilà!"* the waiter cries cheerfully, producing it like a magician extracting a rabbit from a hat. *"C'est à vous?"*—"It's yours?" He gestures toward Dubois.

"Ah, oui, oui, merci mille fois," Dubois replies in a tremulous voice, opening the case to check that his fossils are there and then closing it with shaking hands. He clasps the suitcase to his chest. *"Merci; vous êtes très gentil; merci,"* he mutters fervently, bowing and offering his hand for the waiter to shake. (Thank you; you are very kind; thank you.) Then he presses a bill into the waiter's palm, not looking to check its denomination. Whatever it is, it is worth less than *P.e.* Manouvrier also has a few words with the man and then they leave, walking back toward Dubois' hotel.

Dubois is completely distracted now, unable to carry on their conversation. A terrible disaster has been averted, he thinks. If I cannot persuade my skeptics with the fossils in my hand, how much harder would it be if I had lost them, through carelessness? What would I have done? What would I have done? He cannot hear Manouvrier's words, the question echoes so loudly in his head. What would I have done? The words are repeated in the hopeless sound his feet make on the pavement: What would I have done? What would I have done?

After a short while, Manouvrier realizes his friend is not responding to his attempts at conversation and falls silent. Outside the hotel, Manouvrier bids Dubois good night, promising to call for him in the morning so they can continue their work. With a twinkle in his eye, he suggests Dubois sleep with the case under his pillow. If it were only possible to sleep thus, Dubois would do just that.

*   *   *

Some months later, Manouvrier publishes their comparisons and observations, and, with Dubois' full permission, also publishes his own reconstruction of the entire skull of *P.e.*, including the missing face and jaw, which graphically reveals his view of the skull as intermediate between apes and man. Dubois returns to Paris the next year in June, to attend the Quatorzième Conférence Annuelle Transformiste de la Société d'Anthropologie. His presentation there and previously at the Ecole d'Anthropologie, combined with Manouvrier's influential endorsement, soon wins over other French scientists. The anatomist August Pettit, the well-known prehistorian Gabriel de Mortillet, and the leading anthropologist René Verneau all openly accept Dubois' interpretation. He feels triumphant.

In mid-November 1895, Dubois travels to Edinburgh to lecture. Edinburgh is the home of the eminent anatomist and skeptic Sir William Turner, who discussed and rejected Dubois' interpretation soon after receiving the original monograph. Edinburgh is also the home of the archaeologist Robert Munro, president of the British Association for the Advancement of Science. Munro declares himself more favorably inclined toward Dubois' theories. But then, as he himself says in a letter, "Of course, not being a special anatomist my views have little weight. But I hold that your discovery is a practical illustration of theories propounded previously and of course this should be a strong argument in favour of the correctness of your opinion of *Pithecanthropus erectus.*" Hoping to bring Turner over to his point of view, and believing that Turner's support will be crucial, Dubois makes sure that he has a cast of the skullcap and molars. After seeing the fossils and listening to Dubois' lecture, even the rather pompous Turner must openly admit that the skullcap is more apelike than Dubois' exquisitely illustrated monograph had led him to think, though he confesses to some lingering reservations. Still, he concedes, *if* the deposits in which the bones were found can be shown to be contemporaneous with the Quaternary deposits in Great Britain, Turner argues, then Dubois' discovery is very ancient and "the most important hitherto recorded."

Dubois is pleased. *Ja*, he thinks, Turner is beginning to see the truth. Now we have only a short distance to travel to our rightful home, *P.e.* and I.

But Turner will not budge on the matter of the femur, which he persists in thinking fully human. At a later meeting of the Royal Society, Munro argues against Turner. Of course the femur of an animal that has achieved an erect stance will resemble a human femur, for the two species' bones share the same function. This is a subtle argument that Dubois likes; he makes a mental note of it, for future presentations. Munro goes one step further: "After the erect position was attained, another evolution commenced, viz. the development of the brain, and this was facilitated by the setting free of the upper limbs."

Dubois speaks the next week at an evening meeting of the Royal Dublin Society. Once again, he is facing critics, for this is the very group before which, some ten months before, the anatomist Daniel Cunningham openly criticized Dubois' monograph. Cunningham felt that Dubois' failure to compare his skullcap with the Neanderthal skulls was a fatal error. He argued for an anatomical continuum from *P.e.*, through the Neanderthals, to modern humans, and came to the bizarre conclusion that *P.e.* was merely the "lowest human cranium yet to be described," a view he later defended in an address to the Anatomical Society of Great Britain and Ireland.

Knowing the predisposition of his audience, Dubois is handsomely honest from the outset. He acknowledges that his interpretation of his finds is controversial: "Professors Sir W. Turner, Cunningham, A. Keith, Lydekker, Paul Matschie, Rudolf Martin, and A. Pettit, held that the thigh bone and calvaria were human. Only Professor Manouvrier of Paris, and Professor Marsh in America, admit to the *possibility* of the remains belonging to a transitional form between man and the apes," he concedes in his talk. And he confesses openly to having had mistaken views of Neanderthals, which led to his erroneous omission of them from his monograph. "I am now wholly convinced," he allows, "that they are not at all pathological, and was much struck by the great resemblance with the cranium of *Pithecanthropus.*" He repeats his geological information, presents his arguments about the anatomical features, explains how only a transitional status will accommodate such a creature.

The question in Dubois' mind is, How will the discussion go? He feels the talk has gone well, that he has changed some minds, but he keeps his expression guarded until he is sure how things are going. Several of the most important men present—Sir William Flower, John Lubbock, Professors Thomson and Thane, Dr. Garson, and Sir William Turner—unite in congratulating Dubois upon his remarkable and interesting discovery and express their gratitude at being able to examine the specimens firsthand.

Flower has a particularly graceful way of expressing his opinion. "It is unfortunate that the fragmentary condition of the remains of *Pithecanthropus* was such as to leave much of its real nature open to conjecture." Undecided, thinks Dubois, or unwilling to commit to a position publicly.

When Turner stands up to speak, Dubois prepares himself for a lengthy discourse, for he knows by now that Turner is enamored of his own opinions. As expected, Turner does not discuss, he proclaims.

> *The opportunity which Dr. Dubois has given us of seeing his very interesting specimens and the fuller description of the conditions under which they were found, have enabled us to realize their characters and antiquity much more clearly than was possible from a perusal of his memoir published last year. . . .*
>
> *As regards the thigh bone, the opportunity of carefully examining it, both last week in Edinburgh, and now at this meeting, does not lead me to alter the opinion which I expressed in my published criticism of the original memoir, that there is nothing in its form and appearance which would lead one to say that it possessed char-*

acters specifically or generically distinct from those of a human thigh bone. . . .

As regards the skull, now that one has seen it, there is more difficulty in coming to a conclusion. If, however, the thigh bone and the calvaria belong to the same skeleton, and Dr. Dubois, from his personal examination of the locality, has no doubt on this point, the establishment of the human character of the femur would require us to regard the calvaria as also human. . . . The calvaria is less distinctively human than the Neanderthal skullcap which everyone now admits to be human. In the latter there is a forehead with rounded frontal eminences, but in the Java specimen the frontal bone is flattened and slopes abruptly backwards in a manner such as approximates it much more to the shape in the ape than to a human skull, even as low as the Neanderthal. . . .

In conclusion, may I express the thanks of the anthropologists in this country to Dr. Dubois for his courtesy in bringing the specimens for our inspection, and the further hope that the government of the Netherlands may continue the search for additional remains in the same locality.

Wants more evidence, surmises Dubois, but is coming around. He sees the dual nature of the skullcap, that much is certain, even if he won't accept the femur as being from an ape-man. It is an absurd proposition, for what creature has an ape-man's head and a man's thigh?

Next is Dr. John George Garson, who, like Turner, emphasizes the value of seeing the specimens for himself. Says Garson,

I have studied Dr. Dubois' memoir on Pithecanthropus erectus *very carefully, and also the various criticisms of it which had been published. I am therefore extremely glad to see the specimens themselves, as they showed many morphological features of which the plates and diagrams gave but an imperfect idea; the paper which Dr. Dubois had read that evening shed further light on the specimens. In the first instance, I was very uncertain as to the geological epoch to which they should be referred, but from additional information Dr. Dubois has just given regarding the mammalian fauna found in the same formation with them, I am satisfied as to their being Pliocene.*

Dubois nods unconsciously, satisfied that his discussion of the geology and fauna has been useful. He should have put more geology in the monograph. Garson continues, and some of the elaborate phrases simply wash over Dubois like a voiceless wind.

> *The femur is extremely human-like, and if taken alone would undoubtedly be said to be that of* Homo. . . . *The calvaria, on the other hand, is very different and much more gibbon-like than one would imagine from the drawings of it. . . . The characters of the calvaria . . . if taken alone, might be ascribed to a large extinct* Hylobates, *although it should be remembered that in these large extinct forms the brain is proportionately smaller than in the recent, whereas in this specimen the capacity of the calvaria indicates . . . a proportionately larger brain. . . . Considering the strong opinion Dr. Dubois has formed from examination of the strata and other mammalian remains therein contained, it is most reasonable to conclude that the specimens are probably parts of the skeleton of one animal, and that it belonged to one of those extinct species of primates more or less related to* Homo sapiens. . . .

Professor John Arthur Thomson, speaking later, echoes Garson's remarks. "Yes," Thomson observes,

> *What strikes me most forcibly is the very different complexion put upon the case, now that I have had an opportunity of examining the specimens. This only went to prove how difficult it is to form any correct opinion on such a matter by mere perusal of a monograph, however good. For my part, I feel justified in saying that the calvaria is undoubtedly ape-like in all its characters, except in regards to capacity: on the other hand the femur displays all the features of a well-developed human thigh bone.*

Aha! rejoices Dubois to himself at this juncture. They are beginning to see it, as I do. Neither ape nor human; both together; transitional. That is it; *ja,* that is it.

After some further pronouncements, Thomson concludes,

> *As to whether or no the calvaria and thigh bone belonged to the same individual is a matter of vital importance. Unfortunately, the evidence advanced is not conclusive, and the only course left open*

at present is to reserve one's judgment. This, however, does not detract from the remarkable value of the discovery of this skull, which I regard as by far the most important contribution to our knowledge of an intermediate form between man and the known apes.

Arthur Keith, a young London anatomist, deftly summarizes the problem about P.e.: "The chief question to be settled is whether the skull is human or not."

The only scientist to come close to wholeheartedly endorsing Dubois' interpretation is a young surgeon, Arthur Keith, then newly appointed senior demonstrator in anatomy at London Hospital. An ambitious man at the start of his academic career, Keith finds the opportunity ripe for expressing an intelligent opinion in front of this august body of scientists. He is a little hesitant to engage in public debate, on account of his strong country Scottish accent and a minor speech impediment that makes him self-conscious in formal settings But speaking in public, declaiming on evidence presented, is an essential skill for a British academic, and Keith aims to be one of note. Although at first Keith was skeptical of Dubois' assertion that the three original specimens are derived from a single species, he now generously reverses himself, while confessing to doubts about whether only a single individual is represented.

"The chief question to be settled," Keith declares, "is whether the skull is human or not. What is the criterion of a human skull? What is the criterion of an ape's skull? How are they to be distinguished?"

How indeed, Dubois thinks. This young man has stated the problem admirably, neatly, for all that he is not a facile speaker.

To my mind there are only two differences between the skulls of men and apes, and they are differences, not in kind, but in degree. The first difference is the large excess of cranial capacity of the human

skull: in the extent of its cranial capacity, the skull before us this
evening merits to be called human. The second difference between
the skulls of apes and men lies in the large development of muscu-
lar ridges and processes for the fixation of the masticatory appara-
tus, for chewing; the development is extensive in apes; it is slight in
men. In the extent of this development, also, the calvaria in ques-
tion is distinctly human.

Keith makes clear that he agrees "thoroughly" with Dr. Dubois as to the genealogical position of his fossils; they represent the human race during the late Tertiary period. He would prefer to call them Pliocene Man rather than *Pithecanthropus erectus,* but this is merely a matter of nomenclature.

In all, Dubois' heart is lightened by the number of scientists who have begun to echo his observations. First they must observe, then they will conclude, he thinks to himself contentedly. At the close of the meeting, by popular acclaim, the participants elect Dr. Eugène Dubois an honorary fellow of the Anthropological Institute of Great Britain and Ireland, in grateful recognition of his important and interesting discoveries. The report of the meeting in the journal *Nature,* some weeks later, is largely favorable.

A friend sends Dubois a copy of *The Evening Telegraph* from Dublin, which takes a still lighter view of the debate.

Saturday November 23, 1895 BONES OF CONTENTION

The lecture of Professor Dubois in Dublin on the Pithecanthropus
erectus *is a matter of extreme interest to intellectual people. There is*
nothing I revel more in than being translated into the dim ages
when our forebears made large use of their toe-nails for climbing
purposes, and of their teeth and foreclaws in order to vanquish their
enemies. It is necessary to counteract the swelling pride and impor-
tance of people that they should be reminded occasionally by a
Dubois or a Virchow from what very nasty creatures we have
descended. The modern professor exists apparently to degrade his
species as much as he can for fear it would think too much of itself
and in order to show it that it cannot think too much of the profes-
sor. The Pithecanthropus erectus *is one of the effects of the craze for*
distinction in this way. . . .

These fossils in the opinion of the Professor constitute the
Darwinian missing link. The skull must have been bigger than that

of an ape, while the thigh-bone is bigger than a man's, and the tooth is something between the two. From the shape of the thigh-bone it must have been upright—hence the word Erectus; *from the shape of the same bone, Professor Cunningham thinks the owner must not have learned the art of sitting down. Hence we arrive at a point when we can say that we have the fossil remains of an animal who could stand but who couldn't squat. Fill up the rest of the animal from your imagination and you have the* Pithecanthropus erectus, *who some time between the end of the tertiary and the beginning of the quaternary period went about like a roaring lion seeking he who pretended to better femoral development than him in order that he might kick him. Now, was the* P.E. *aforesaid a man or was he an ape? Judging from the femur, as man walks erect, so did the* P.E.; *ergo, he was a man. But then a man has evidently some more brains than* P.E., *and an ape has less; and the question is whether the* P.E. *is to belong to either class, and if so, which. Neither upon the thigh-bone nor upon the question of brains can the professors agree, and some of them even contend that the thigh-bone has nothing to do with the skull, and that the tooth is not connected with either. In fact these osseous fragments are regular bones of contention between the professorial pundits. Virchow says one thing and Dubois says another. Dubois thinks that Virchow did not examine things enough, and if Virchow's opinion of Dubois were put in plain language, it would probably express itself as* tête du bois. . . .*

    Now for my opinion of this find. This world is full of irregularities. . . . Now physically this* Pithecanthropus erectus *was in my opinion either a fool of a man with a big leg or an ape with a big head, or we might have in these three specimens the leg of a man, the head of an ape, and the tooth of the tiger that ate them both. The really important thing for mankind, however, seems to be that it does not matter which is the real state of the case.*

The article, signed "O'Mulligan," closes with numerous stanzas of doggerel.

Dubois has no time to savor his apparent fame, for he is due to speak in Berlin shortly. This battle is in deadly earnest, for all that Irish journalists may find it comical. He fights for his reputation, for *P.e.*, for the missing link itself. Berlin is Virchow's stronghold, so he can expect few converts there. Rudolf Martin, once a ferocious critic of Dubois, markedly softens his point of view, but Virchow and his followers Wilhelm Krause, A.A.W. Hubrecht, Herman Klaatsch, Johannes Bemuller, and others hold fast to the belief that *P.e.* is nothing but a giant gibbon. "According to all the rules of classification," Virchow intones seriously, "this creature was an animal, to wit, an ape," and he will not be moved.

By the time of his lecture in Berlin, Dubois has managed to remove nearly all of the matrix from the inside of the skullcap. For this nerveracking endeavor, he has invented a new technique using a heavy, footpowered dental drill. Hours of experimentation on rocks, using different drill bits, lead Dubois to decide that the most satisfactory bit is a diamond bur, with which he becomes quite proficient. He learns how to remove tiny bits of rock at a time, working with the utmost precision and delicacy. Only then does he cautiously apply this tool to the inside of his skullcap, skimming off the thin layer of matrix, in places only a tenth of a millimeter thick. It takes weeks of concentrated work of the very sort to which Dubois' impatient temperament is poorly suited. But he is good with his hands and, besides, he would never trust another person to clean his precious *P.e.* During the process, he is noticeably short-tempered and irritable with his family and servants, so fearful is he of making an irreparable mistake. Every interruption is seen as a threat to *P.e.*'s physical integrity.

Now Dubois has freed the fossilized bone almost completely from its encasing matrix, revealing the impression of a riverbed of blood vessels and irregular, rounded depressions into which the lobes of the brain once fitted. He prepares an endocast, an impression that shows the inside of the skull and of the brain that once nestled there. He hopes the new anatomical information about *P.e.*'s brain will prove persuasive, and he discusses it at length. As in humans, he tells his skeptical German audience, the foramen magnum—the bony hole through which the spinal cord exits the skull—is far forward. This position is typical of animals with an upright posture. There is thus an anatomical consistency between femur and skullcap that reinforces the geological information suggesting the two come from a single animal. Too, some of the sulci and gyri—the ditches and hillocks—of the long-decayed brain are still evident on the inside of the skullcap, indicating a more human arrangement of brain tissue than in apes.

He even proposes a phylogenetic tree, a graphical description of *P.e.*'s evolutionary position, though he knows Virchow's deep resistance to evolutionary ideas.

"I am well aware," he admits modestly, "of the exceptional mortality of

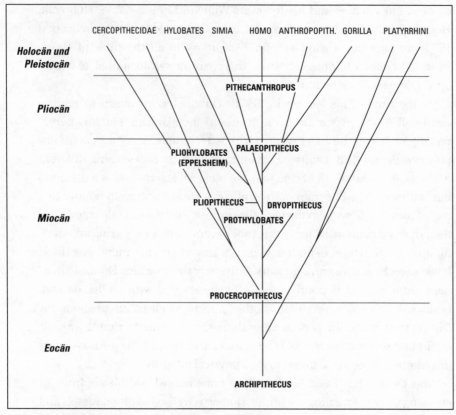

CERCOPITHECIDAE HYLOBATES SIMIA HOMO ANTHROPOPITH. GORILLA PLATYRRHINI

Holocän und Pleistocän

PITHECANTHROPUS

Pliocän

PLIOHYLOBATES (EPPELSHEIM)   PALAEOPITHECUS

Miocän

PLIOPITHECUS — DRYOPITHECUS
PROTHYLOBATES

PROCERCOPITHECUS

Eocän

ARCHIPITHECUS

During his lecture in Berlin, Dubois presents this phylogenetic tree showing the place of P.e. *in the human family, between* Paleopithecus (*Lydekker's ancient* Anthropopithecus) *and man* (Homo).

such trees, but I also know that parts of them at least often survive, from which new life emerges. One has to try to visualize the kinship relations of the forms presently known, and I know of no better means to this end than the form of a phylogenetic tree." The strategy fails utterly. Dubois' indicated willingness to modify his ideas in the future—the mark of a dedicated scientist—earns him no respect from the dogmatic Virchow, who is always certain of his opinion. The statement simply makes Dubois appear weak in his eyes. Virchow is adamant: this skullcap is a gibbon and nothing more.

Yet by the end of 1895, a few significant cracks have appeared in the edifice of German skepticism. The paleontologists Wilhelm Branco and William Dames come forth with strong endorsements of Dubois' view, Branco even saying that no one familiar with paleontological research could doubt that the three fossil specimens of *P.e.* belonged together. Dubois' last engagement in 1895 is at Jena, the home institution of Ernst Haeckel, a longtime foe of Virchow's and one of Dubois' staunchest supporters. Haeckel makes sure Dubois is duly honored during his visit.

It has been six months since Dubois arrived home with his precious *P.e.*, on that dark, disappointing night in Eijsden, and he is exhausted, emotionally and physically. There is nowhere he can unpack his enormous collection (414 crates' worth) and, even if he had the space, he is too weary with fighting. He has done his best, he thinks as he returns to The Hague for the holidays. He has done what he said he would do and now they do not believe him. But he will make them believe him. Next year, when he has rested: next year, he will fight again.

CHAPTER 41 MORE SKIRMISHES

In the early part of 1896, Dubois stays at home more. In February, after much correspondence, Haeckel lets him know that his attempts to persuade the University of Jena to give Dubois an honorary doctorate have been in vain. Haeckel has argued and presented opinions from many fine scholars but, in the end, anything to do with human evolution is simply too controversial. And although his lair is in Berlin, Virchow wields too much influence in biology for the University of Jena to oppose him in this way.

Despite this frustration, in May 1896 Dubois receives a friendly letter from Gustav Schwalbe, an anatomist at the University of Strasbourg. Like Haeckel, Schwalbe is one of Germany's few well known and outspoken supporters of evolution. It is a fateful letter, for Schwalbe does Dubois an invaluable service.

May 12, 1896

I delivered my lecture on your P.e. last Friday and expressed my conviction that, in any case, the skull cannot belong to a human and not to a monkey. . . . Concerning the femur, I am convinced that you are right, when you think that it belongs to the skull and to the molar, and I have expressed my opinion. . . .

Might you be interested to learn that we have here in the preparation room obtained a femur of an actor aged 52 years, with an exostosis resembling that on the Pithecanthropus *femur?*

Dubois most certainly is interested; what a wonderful thing it would be, to have a matching specimen of known history! In a few days' time, the actor's femur is on its way to him in The Hague. The resemblance is indeed striking; from that moment on, the actor's femur lives in the box with *P.e.*'s own femur, for handy comparison. It is especially useful, for the actor was said not to be lame following the injury; he made a full recovery, as did, presumably, *P.e.*

Though Dubois has already been in every European capital that matters, he continues to press his case over the next few years. The scientific community *will* listen to him; his colleagues *will* look at his precious *P.e.* and see her for what she is. He thrusts the fossils under their noses. By sheer force of personality and conviction he refuses to allow them to dismiss his ideas casually, without serious consideration. And all the while, he is constantly researching and revising, adding new information and new diagrams to his lectures, answering the criticisms of his talks and the resultant publications with an unceasing flow of argument and logic.

In June 1896, he returns to Paris, this time to be awarded the Prix Broca for outstanding achievements in anthropology. It is a moment to be savored. He is surrounded, for once, by supporters: Manouvrier, Verneau, Pettit, De Mortillet, and others. He is fêted and praised and honored at dinners. The only awkwardness is that Anna has begged to join him in Paris this time, especially to attend the elegant banquet held in Dubois' honor at the Ecole d'Anthropologie, and he has agreed. Anna spends all afternoon constructing an elaborate hairdo, which does not suit her, and wears her most fashionable dress. She is determined to be charming, gay, and much admired by all these learned men. Her manner instead strikes Dubois as

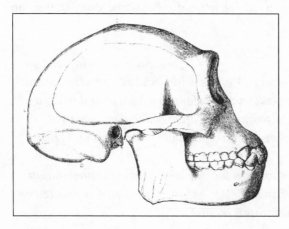

In 1896, Dubois reconstructs the shape of the complete skull of P.e.

boorish, uneducated, even lingeringly colonial. Anna has no conversation, no wit, no appreciation of the scientific accomplishments of the men who politely speak with her. She laughs too shrilly and too loudly; she flirts like an unmarried coquette. In all, her behavior is utterly unsuitable for the wife of a great scientific man, Dubois thinks; it is an occasion he longs to forget.

From time to time, his old enemy resurfaces and he is flattened with fever again. He did too much in 1895, he realizes now: he went to India, climbed the Siwaliks and searched for fossils, returned home to Java and those awful suspicions about Anna and Prentice, sailed home to Holland, and, barely settled in, left to lecture everywhere in Europe he could wangle an invitation. And all the time, every day, even now, he is thinking, learning, fending off criticisms, persuading skeptics, carrying out new research and new analysis until he is ready to drop. He vows to guard his health a little more, to plan a less demanding schedule, even as he learned in the Indies that to spend weeks camping and trekking through jungles was likely to kill him. The academy is a different sort of jungle, but the dangers are not to be underestimated. And things are not entirely happy at home, though he has little time and energy left over to deal with Anna and the children.

Dubois desperately needs a fixed academic post, an institution and a title to shelter him and his fossils. On January 8, 1897, the University of Amsterdam awards him an honorary doctorate in botany and zoology, but not a professorship and not a salary. That same year, he is appointed curator of paleontology (and of the Dubois Collection) at the Teyler Museum in Haarlem, by an act of parliament. Dubois hopes it will finally be possible to unpack and arrange a suitable resting place for his large and important collection, which has so far been unceremoniously stored in unheated attics and cramped coach houses. Though the curatorship carries only a modest salary, it is something. He moves his family from The Hague to Haarlem, taking a house convenient to the Teyler. Dubois feels that he is making progress in the battle for acceptance of *P.e.*, inch by inch, just as in excavating the sediments at Trinil. His missing link is coming into her own. He is finally beginning to assume the role that he has earned of leading Dutch scientist.

Early in 1898, the professorship of anatomy at the University of Amsterdam becomes vacant. It is the very position—Fürbringer's old professorship—that Dubois turned his back on when he left Amsterdam to find the missing link. On February 5, Thomas Place, Dubois' old friend and mentor in physiology, writes to ask Dubois if he would like to be nominated for the post. His competition is Otto Seydel, a German who has held the readership of anatomy, a position just below that of the chair, and the anatomist Louis Bolk, whom Dubois has never liked.

Dubois replies to Place's kind letter gratefully, but he stresses that the position is surely Seydel's by right; Seydel has doubtless worked hard for it. A few days later he receives another letter about the matter, this time from Pierre Schrijnen, an apothecary, who knows Dubois' brother Victor. Victor is now a well-regarded physician, and Schrijnen is a man with some influence at the university. Like Place, Schrijnen is concerned that the best man be appointed to such a key position. His letter to Dubois is blunt and direct. "Among the students," he writes, "there is a rumor that you will not accept the position if you are nominated. Is this true?"

Dubois writes to Place, hoping he can clear up the confusion with Schrijnen and the students.

February 18, 1897

In my view, there is a regrettable agitation concerning the undertaking of this professorship in anatomy. This has given me the idea that perhaps my opinion concerning the nomination (that it should go to Seydel) should become more widely known or perhaps there is another way in which I can promote Dr. Seydel's nomination.

Place replies that the students dislike Seydel and are protesting his possible appointment. He has only a slim chance of being accepted. But if Seydel is rejected and another candidate does not appear, the position will go to Dubois' old enemy Louis Bolk.

Oh, the students! Dubois sighs as he remembers his own trials with student popularity. They care nothing for the quality of science or of intellect, only for the amusement a lecturer offers. Bolk may be popular, but his research is unsound: Dubois and Place agree in this. Dubois would be a much better choice, as he is a much better anatomist and he is now a famous man in Europe, even if he has not been at the University of Amsterdam for many years. But Dubois' problem is twofold: the students do not know him personally, and what they do know of him is the ominous rumor that he would not accept the position if offered it. But that is a misunderstanding based on Dubois' noble effort to leave the field clear for Seydel. Dubois decides to give in to his friends' urging to put himself forward. Someone must repair the damage done by the false rumor, however, if Dubois is to have a chance. In the meantime, these complex machinations must remain confidential. Dubois sends Schrijnen a telegram via his brother Victor, for privacy's sake. He assures Schrijnen that he is willing to vie for the post, but he must be assured that he will not be hemmed in by the position. He must be free to carry on his research, whether in geology, fossils, brain weight and body size, or anything else that takes his fancy. After all, a professor is a little king in a university and he must have the right to move freely within his kingdom, as Fürbringer always did. In the telegram, it amuses Dubois to refer to Pierre Schrijnen as "Piet," the Dutch equivalent of Pierre, and Place as "Plaats," the Dutch translation of his name. As for himself, Dubois is "Piet's friend." It is like playing at espionage.

> *Dr. Victor Dubois, Venlo   20 February 1898*
> *Plaats wishes Piet's friend success which is not impossible if nobody believes him unwilling to accept the post. Will write to Plaats to do his best and communicate this confidentially to the examiners. Tell Piet that his friend always kept this possibility seriously in mind but in the first place his friend did not wish to be troublesome. His head and hands are full of other things which he will in no case give up. Maybe a combination is possible. Can Piet put this forward cleverly? If not, better to be silent. All this confidential.*
> *Eugène.*

Dubois and his supporters must sway opinion subtly for this plan to succeed. In Dubois' favor, he was trained at Amsterdam and would have been likely to assume the professorship had he never left; he is also internationally recognized. His main handicap is that many at the university remember how he cast aside his brilliant future to go fossil-hunting, and no one likes to be rejected. Another problem is that Dubois has never been a skilled lecturer and the students are already up in arms. Too, Dubois' theories, though well-known, are certainly controversial.

In the middle of the plotting and planning, Dubois receives an infuriating letter from Victor. Somehow it is always Victor who sticks the knife in Dubois' side, reviving painfully Dubois' worst fears. Dubois thinks Victor acts like the head of the family instead of the younger brother. In the letter, Victor chastises Dubois for doggedly pursuing his research, come what may, reminding him that he has a family to support and an aging mother who is sick with worry about her elder son's lack of a permanent position. Victor closes his letter with an admonition: "I believe it is your duty still to do everything that is possible to be nominated for this position. . . . It will need a lot of willpower to conquer your aversion to this, but (1) the final goal is not ignoble, and (2) it leads certainly to the goal."

Dubois is enraged and deeply hurt. The goal? Who is Victor to speak to him of a goal, as if burrowing into a job, no matter what job at no matter what cost, ought to be Dubois' primary ambition? His lectures, his research, his writing, are as compelling as they are exhausting. Doesn't Victor know that in two and a half years Dubois has lectured at no fewer than nine European cities and institutions, refuting the criticisms of his monograph? He cannot abandon *P.e.* now.

And has Victor overlooked Dubois' own publications? There were two in 1895, the year after the monograph; another eleven in 1896; three in 1897; and perhaps another two to be out before the end of 1898. Some are brilliant, innovative; none is mundane. Dubois is working harder than he ever did as an anatomy professor, back in his early days. Simply because he is not very well paid, Victor—his younger brother Victor, the one who never believed in working hard for any reason—thinks he has the right to tell him what to do. It is insupportable, humiliating. Will his family never understand him? He writes his reply in a fury, his pen nearly tearing into the page.

February 22, 1898

Your letter of yesterday did not please me, because of the advice you gave me in it but above all because of the way in which you gave it. . . . That you give me advice, certainly wrong advice that might even lead to my misfortune, is not your fault, because you do not know the situation well enough. . . .

But what grieves me even more is that you place me in a position of inferiority, which you have no right to do. You must admit that with my "impractical mind," which you ascribe to me as well as to our mother, I have managed to achieve a few things. Since I returned to the Netherlands, I have for example acquired two things: the opportunity to work and money for the publication of my description. . . .

In the Indies, I achieved literally everything that I wanted. Abroad, my name is perhaps better known than that of any other living Dutchman. I have even been nominated for a professorship, even though I had explained explicitly that I did not desire it. . . .

You cannot ascribe all this to luck, because I have desired and effected all of this. . . . I have fixed my eyes upon a still further future. What I really desire is a professorship not in anatomy . . . but in paleontology, and . . . I am confident of reaching that goal.

Have you ever considered that to execute a function well, one must also

be suitable? Otherwise, to accept such a position is to walk into misfor-
tune.... What life would await me if I accepted such a position and then
did not, as I am afraid I would not, meet the expectations[?] ... Certainly I
could eventually find another post, but in the meantime it would be
dreadful, especially under the current circumstances of agitation.

When, Dubois wonders, when are his trials to be over? Would he never stop
needing to fight the skeptics? He is too proud to tell Victor that he is already
doing everything he can to obtain the post. It would be too humiliating to
admit he desires it, and then be rejected.

Indeed, the scheming and plotting of Place and Schrijnen on Dubois'
behalf are all in vain. For whatever reason, the examiners award the post to
Louis Bolk, a man Dubois never liked before this and now likes even less.
Dubois is deeply disappointed, even though he once fled halfway around
the world to avoid the prospect of taking up just such a job. Some days he
looks back on his forty years of life and sees only a long line of naysayers:
there was Fürbringer, Weber, and De Vries; then the Secretary-General of
the colonies; the entire military establishment; and, worst of all, his family.
They should have shown him respect and support, but they have been trai-
tors, offering only scorn and condemnation. And now the university examiners prefer that idiot Bolk to him!

And yet, Dubois reminds himself, he has met with kindness, too. There
were Groeneveldt and Kroesen, supporting his expeditions in the Indies; De
Winter and Kriele, faithfully supervising the coolies and never questioning
Dubois' word; there was Weber, come around at last and sending that chimpanzee skull; there was Sluiter, who sent the Wadjak skull and drew him to
Java; now there are Manouvrier and De Mortillet in France, Marsh and
Osborn in America, Haeckel, Branco, and Dames in Germany, W.H.L.
Duckworth and William Sollas in England. Yes, he has his supporters and
defenders, and they are fine men. But above all, he realizes, his greatest supporter, more important than any of these, was Prentice, with his intelligent
thoughtfulness, his understanding. It was Prentice who helped him through
the most trying days, those days of discovery, and fever, and desperation.
Even now that they are separated by so many miles, Prentice still writes him
from time to time, still applauds his triumphs, stoutly denies the legitimacy
of his critics. Dubois smiles gently at his fond reminiscences of days spent
with Prentice, of the man's honesty and loyalty.

Suddenly he is spurred out of his reverie into action. Where is that last letter Prentice sent? He wants to read it through again, to hear Prentice's voice once more. Dubois rummages through his desk and finds it, slowly opening the folded pages that bring back those happy days in Java once more. Is it his imagination, or is there a faint, spicy, indefinably tropical smell clinging to the pages?

Toeloeng Agoeng   15 April 1896

My Dear Doctor,

Your pamphlets about the missing link reached us safely two days ago and both Mr. Boyd and I sat down diligently that evening to read the work through. I had just gone over to Mringin to see the Old Warrior. We were naturally greatly interested in your article and most pleased to see you were causing a stir in learned circles in Europe where you must now of necessity be well known. I shall not enter into any discussion on the matter of which you treat, after so many learned & qualified authorities having investigated the questions and being unable to come to any agreement as to the fossils being human or ape. It would be idle for a mere layman like myself to offer any opinion as the same must be a matter of pure guess as anything else. After reading your work, however, and pondering over the views entertained by the many eminent men who have investigated the matter, it would seem to me that the bones are from one and the same being, that he must have been a very low man or a very high ape, thus probably between both, seeing no such highly developed ape has yet been found and no human remains have ever been found of such antiquity as you calculate for that of the formation in which you found the relics. If ape, then it is a species between any known ape and man, and if Homo, *then it is of a race nearer the ape than any fossil* Homo *yet exhumed. Thus if not a missing link is discovered, you have at least discovered not only the lower but the most ancient type of man yet met with, and your discovery has in that case the merit of proving that man existed in an age in which it was till now deemed he had not been. I cannot help thinking what a pity it is that you were not allowed to prosecute your excavations further at Trinil. I did not then realize so fully as I do now the vast importance of your discovery there or I should have urged & urged, & urged you to continue & continue it. Of course your health prevented you greatly and your insecure position in gov't employment also unsettled your mind for continuous and laborious work of a mental kind. These things the world cannot know, but it is a pity all the same as further excavation on that lucky spot might have brought to light similar fossils. I notice from your*

pamphlet that it will take a year or two yet to complete your report on
your collection. I hope you find it in every way agreeable and that both
yourself and your family are all well and happy.

Mr. Boyd is as usual & is glad Janet is at such a nice school and under
such nice teachers.

With me things are going along slowly. My plantation does not stand as
well as I expected as my crop has not turned out well. This gives me much
mental worry. I have been reading about nothing but coffee of late. I have
lots to do & lots of little worries. I have not the leisure now that I had at
Tempoersarie or Geboegan for scientific or learned reading, which I greatly
regret. My boy is still at Malang & is growing well. With kindest regards &
best wishes to you all.

> *Believe me,*
> *Ever Yours Sincerely,*
> *Adam Prentice*

N.B. Mr. Boyd sends his thanks and kind wishes. A.P.

Yes, there is the man who understands him best, the true brother of his
soul. There is the man, a true scientist by reading, not education, and yet a
deep thinker. Prentice still believes in him; Prentice always did. Has it been
nearly two years, then, since he had a letter from Prentice? He must make
the time to write to Prentice, see how he is faring, learn the fate of his last
crop. He will draw his strength from Prentice's faith.

CHAPTER 42 USING HIS BRAINS

Dubois' next big battle is in the summer of 1898, when he attends the Fourth
International Congress of Zoology in Cambridge, England. It is an enormous affair, with all of the British and many international figures of science
in attendance: John Lubbock (Lord Avebury), W. H. Flower, Adam Sedgwick,
E. Ray Lankester, William Turner, D. J. Cunningham, Grafton Elliot Smith,
John Evans, Arthur Keith, Richard Lydekker, John Forsyth Major, William
Pycraft, D'Arcy Thompson, and Arthur Smith Woodward. The participants
list is also sprinkled liberally with military men, Fellows of the Royal Society,
nobles, professors, and men of the cloth. Just before Dubois speaks, Ernst
Haeckel himself addresses the assembly.

Dubois' ally Ernst Haeckel holds the skullcap of Pithecanthropus *in this portrait, painted by Gabriel Max in 1896.*

Haeckel's stirring presentation is masterful. He is a handsome, vivacious man, now silver-haired and silver-bearded, but still vigorous and strong as ever. His resonant baritone literally shakes the crystals on the chandeliers from time to time, when Haeckel emphasizes a particularly telling point. He is mesmerizing. What a command of language Haeckel has! What marvelous clarity in explaining such vast topics! Watching Haeckel speak, Dubois thinks to himself that it is good that his enemy is the dry and pedantic Virchow, with his slow and crackling speech, rather than Haeckel, with his ability to sway an audience.

Haeckel summarizes the evidence for human evolution—indeed, for the evolution of all life on earth—brilliantly, alluding to the enormous body of evidence now amassed by comparative anatomists in support of Darwin. Then Haeckel mounts a vigorous defense of Dubois' interpretation. As Dubois listens, admiringly, he realizes he could not have asked for a more favorable summary of his own embattled ideas.

> *The next question now is, What has paleontology to say regarding these important results of comparative anatomy and their application to the system of the primates and to phylogeny? For it is the petrifactions that are the true "footprints of the Creator," the immediate testimonials of the historical succession of the numerous groups of forms which have peopled this earthly ball for so many millions of years. Do petrifactions of the primates give us any determinate points of support[?] ... The most important and interesting of these petrifactions of the primates is the renowned* Pithecanthropus erectus, *which Eugène Dubois found in Java in 1894. As this Pliocene ape man brought out a lively discussion at the last zoologi-*

cal congress held three years ago at Leiden, I may be permitted to
say a few words. . . .

From the proceedings of the congress at Leiden (at which I was
not present), I learn that the most distinguished anatomists and
zoologists expressed different views as to the nature of this remarkable Pithecanthropus. Its remains, a skullcap, a femur, and some
teeth, were so incomplete that it was not possible to arrive at a conclusive judgment regarding them. The final result of the long and
spirited debate held on this subject was that among twelve distinguished authorities three declared the fossil remains to be those of a
man, three that they were those of an ape. Six or more zoologists, on
the contrary, stated what I believe to be the real fact, that they are
the fossil remains of a form intermediate between ape and man. . . .
The Pithecanthropus erectus of Dubois is in fact a relic of that
extinct group intermediate between man and ape to which as long
ago as 1886 I gave the name Pithecanthropus. He is the long-sought
"missing link" in the chain of the highest primates.

The able discoverer of Pithecanthropus, Eugène Dubois, has not
only convincingly pointed out his high significance as a "missing
link," but has also shown in a very acute manner the relations
which this intermediate form has on the one side to the lower races
of mankind, on the other hand to the various known races of
anthropoid apes. . . .

For forming a correct judgment concerning this important
Pithecanthropus and its immediate position between the anthropoids and man, two features are especially valuable: first, the close
resemblance of the femur to that of man, and second, the relative size
of the brain. Among the few anthropoid apes yet living the gibbons
appear to be the lowest and oldest . . . they are also the most generalized and appear especially adapted to illustrate the "transformation of apes into man." The gibbons more than the other anthropoids
have the habit of voluntarily assuming the upright position, whereby
they walk upon the entire sole of the foot. . . . The other modern
apes . . . seek the upright position, and when they use it do not tread
upon the entire sole but upon the outer edge of the foot. . . . It is
thus explained why it is that it is exactly the femur, in the gibbon
Hylobates and Pithecanthropus, that is much more human in form
than that of the gorilla, the orang, and the chimpanzee.

> *But also the skull, that "mysterious vessel" of the organ of the*
> *soul, approaches nearest the human proportions both in*
> Pithecanthropus *and in the gibbon in important particulars—the*
> *rough, bony crests which the skulls of other anthropoids show are*
> *wanting. . . . The capacity of the skull of* Pithecanthropus *is from*
> *900 to 1,000 cc, therefore about two-thirds of the capacity of an*
> *average human skull. On the other hand, the largest living anthro-*
> *poids show a capacity half as high as this—500 cc. So the capacity*
> *of the skull and consequently the size of the brain is in* Pithecan-
> thropus *exactly midway between that of the anthropoid apes and*
> *the lower races of mankind. . . .*

The swell of Haeckel's rhetoric is spellbinding. Dubois listens as if he has never heard these arguments before, as if they were all fresh and new, instead of being his own offspring. How can anyone doubt such wisdom! And yet it *was* doubted, even scorned. But now, as the contradiction raises questions in the audience's mind, Haeckel moves skillfully on to demolish the opposition:

> *To this momentous interpretation, which is now accepted by nearly*
> *all naturalists, the renowned pathologist of Berlin, Rudolf Virchow,*
> *set up the most obstinate opposition. He went to Leiden for the spe-*
> *cial purpose of contradicting the idea that the* Pithecanthropus *is a*
> *transitional form, but met with little success. His contention that*
> *the skull and the femur of* Pithecanthropus *could not have*
> *belonged together, that the first belonged to an ape and the second*
> *to a man, was rejected at once by the expert paleontologists present,*
> *who declared unanimously that, in view of the extremely careful*
> *and conscientious account of the discovery, "there could not exist*
> *the slightest doubt that the remains belonged to one and the same*
> *individual."*

Dubois wishes that Haeckel's felicitous reading of the events at that congress were accurate. What trouble would have been saved if only everyone had accepted at once that the fossils belonged together! How many gibes and cruel jokes would he have been spared!

Haeckel continues,

> *Virchow further asserted that a pathological exostosis in the femur*
> *of* Pithecanthropus *likewise attested to its human characters, for*

only by the most careful attention by human hands can such disor-
ders be cured. Immediately thereupon the famous paleontologist
Marsh showed a number of similar exostoses upon the leg bones of
wild apes, who had no "nursing care" and yet recovered. . . . Finally,
Virchow asserted that the deep notch between the orbital edges and
the low skullcap of Pithecanthropus—*a sign of a very deep confor-*
mation of the temporal fossa—were decisive for the ape-like char-
acter of the skull, and that such a formation never occurs in man. A
few weeks later, Nehring . . . showed that exactly the same formation
was presented by a human skull from Santos in Brazil.

Virchow formerly had the same want of success with his "patho-
logical significance of the skulls of the lower races of man." The
famous skulls of Neanderthal, of Spy, of Moulin Quignon, of La
Naulette, etc.—which taken together are the interesting isolated
remains of an extinct lower race of man standing between
Pithecanthropus *and the races of the present day—these were all*
declared by Virchow to be pathological products; indeed the saga-
cious pathologist at last made the incredible assertion that "all
organic variations are pathological"; that they are produced only
through disease. According to this all our noblest cultivated prod-
ucts, our hunting hounds and our horses, our noble grains and our
fine table fruit, are, alas! diseased natural objects that have arisen
by pathological change from the wild original forms that alone are
"healthy."

. . . It must be remembered that for more than thirty years,
Virchow has regarded it as his especial duty as a scientist to oppose
the Darwinian theory and the doctrine of evolution necessarily con-
nected with it. . . . The most important conclusion from the latter,
the "descent of man from the ape," Virchow is well known to attack
with zeal and energy. "It is quite certain that man did not descend
from the apes." This assertion of the Berlin pathologist has been for
twenty years past repeated innumerable times in religious and
other periodicals—cited as the decisive judgment of the very highest
authority—not caring in the least that now almost all experts of
good judgment hold the opposite conviction. According to Virchow,
the ape-man is a mere "figment of a dream"; the petrified remains of
Pithecanthropus *are the palpable contradiction of such an*
unfounded theoretical assumption.

Haeckel's talk continues for some time, but Dubois can no longer absorb his words. Haeckel has so boldly supported him, so ably defended him that Dubois finds himself in a sort of deaf halo of pleasure.

This is his best opportunity, he now realizes. He knows many of the participants personally, having lectured at their institutions and sometimes stayed in their homes. His ideas have been endorsed by one of the great men of German science, and his worst doubter derided. And now he must complete the job, persuade the audience, shatter their preconceived notions of what a transitional form will be like—because he has the goods, he has *P.e.*, and nothing else in the world can compete with her. He once risked his life to find her, and he will not abandon her now. No, he will fight for her to the end, until she is seen for what she truly is: the missing link.

Dubois has invented an utterly new tactic for this conference. During the previous year, 1897, he became fascinated with the idea that there might be a fixed and predictable relationship between brain size and body size among different types of animals. The idea of such a fixed ratio is not obvious, yet the concept is a magnificent one that reveals a crucial element in the design of being a mammal. Dubois and a few other anatomists—Manouvrier, Otto Snell, Fürbringer, and Lapicque—pursue this notion, which will eventually become a fundamental area of biological study.

Dubois' insight was born long before the Cambridge congress. It has been evident to him for a long time that *P.e.*'s large braincase cannot belong to an ape, for the body attached to such a brain would be enormous in an apelike creature, while although the femur's possessor was tall, the bone is not big enough to have supported a *huge* body. But what exactly *is* the size required for an apelike animal with a brain of 1,000 cc? The question niggles at him and haunts him. He intuits that there is some consistent mathematical relationship between the two, brain and body, if only he can derive it. He believes that theoretically, brain size is determined by two factors, the first being the animal's total body weight. But the relationship is not a simple one.

For example, anatomists have long noticed that small animals tend to have relatively bigger brains—more brain per unit of body weight—than large ones. The second factor at work is the developmental level of the animals' nervous system, its *cephalization*. Clearly some types of animals are more advanced or brainer than others. So there must be some sort of sliding scale of brain size to body weight, with lower organisms having a lower ratio and higher ones (apes and man) having a higher ratio.

Fürbringer, Dubois' old mentor, suggested that the explanation lay in the relatively larger body surface of small animals, which would cause them to lose body heat faster. Maybe, Fürbringer postulated, the "extra" brain in smaller animals is occupied by an extensive heat center. And as early as 1892, Snell proposed that the surface area of the body, which he symbolizes by the letter "P," is equal to the body weight taken to the ⅔ power, or $P^{0.66}$. To demonstrate this, Snell considered pairs of closely related animals with similar degrees of nervous development, like a lion and a cat. Within such a pair, the brain weight is a simple function of body weight. Thus, the ratio of the lion's *body* weight (P_1) to the cat's body weight (P_2) would be identical to the ratio of the lion's *brain* weight ($P_1^{0.66}$) to the cat's brain weight ($P_2^{0.66}$)

Dubois admires Snell's attempt, but thinks that he and Fürbringer have gone astray in attributing the constancy of the relationship to a metabolic function. For him, the issue is one of the structure and function of the nervous system. Dubois' starting point is the observation that the nervous system comprises two main types of structures: *sensory nerves,* which receive information from the world through smell, taste, touch, vision, and hearing; and *motor nerves,* which produce actions by triggering muscle activity. Higher animals, Dubois thinks, have more of each type of nerve and more complicated connections among them than lower animals do. And the relatively larger surface area of small animals means that they will have relatively more sensory nerves per unit of body weight; they are more extensively innervated, in effect. If there are more sensory nerves, then there are more connections to the motor nerves and a higher overall cephalization. By analyzing closely related pairs of species with similar cephalization, Dubois derives the exponent that expresses the relationship between surface area and body size. His result is 0.56, not Snell's 0.66.

"It is actually," he wrote in his first paper on the subject, in 1897, "the size of the perceptive surface of the sense organs that determines the quantity of brains in animals of equally high organization." In the same paper, he ranked various species according to the extent to which their brain size deviated from that expected for their body size. For each species, he calculated a cephalization value, *c.* If $c = 1$, then the species had exactly the predicted brain size for its body weight. If *c* was smaller than 1, the species was relatively small-brained; if *c* was greater than 1, the species was large-brained.

Although Dubois is very proud of this ingenious article, he knows few of the scientists at the Cambridge congress will have read it, for he wrote in

Dutch. And now is the time to introduce them to the ideas and their application to *P.e.* If brain size has a predictable relationship to body size, then this relationship proves *P.e.* to be the perfect transition between apes and man.

First, he explains his new method of estimating cranial capacity for *P.e.* The last traces of stony matrix have been removed from the braincase, so he can estimate her cranial capacity more precisely than ever. Since the

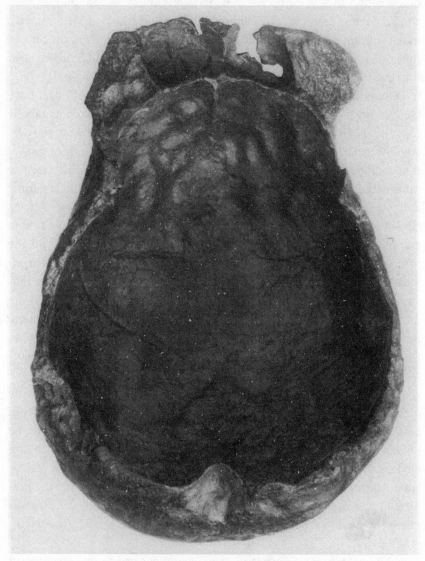

Dubois removes the last matrix from the inside of P.e.'s skullcap, revealing new information about its brain which he presents at the Fourth International Congress of Zoology in 1898.

incomplete skullcap has a volume of 570 cc, he tells his audience, the total brain was obviously much larger. If the relationship between the volume of the skullcap (enclosing the upper part of the brain) and of the entire brain matches that in humans, then the cranial capacity of *P.e.* is approximately 798 cc. If the appropriate ratio is that found in apes, which have a greater part of their brain housed in the missing lower section of skull, then the cranial capacity of *P.e.* is 861 cc. As a compromise value, he accepts 855 cc as the correct value.

Next, he shows that the cephalization coefficient—the amount of "extra brain"—varies little in modern humans, since even the smallest human brains are about 90 percent the size of the largest. If *P.e.* were as cephalized as a human, a brain size of about 800 cc would accompany a body weight of 19 kilograms, or 41.8 pounds. This is a patently absurd weight for a creature standing about 5'7" tall with a strong and robust femur. Dubois himself is two inches taller and weighs much more than 42 pounds. If *P.e.* were as cephalized as an ape, then the resultant weight would be 230 kilograms, or just over 500 pounds: another ludicrous result. These calculations demonstrate how much the ratio of brain size to body weight varies in different types of animals. As both answers yield absurd body weights, it is obvious that *Pithecanthropus* is neither an ape nor a man.

Rather than estimating body size from brain size, Dubois prefers to estimate each, independently, directly from the bones themselves. The brain he has already shown to be about 855 cc; the femur is constructed to carry an animal weighing about 70 to 75 kilograms, or 154 to 165 pounds. These values give *P.e.* an entirely novel ratio of brain size to body weight, *intermediate* between apes and man. And that, Dubois emphasizes triumphantly, reveals clearly the true phylogenetic position of *Pithecanthropus*. His is an elegant mathematical argument, based firmly on anatomical structures and biological laws.

His conclusion is, he hopes, irrefutable. "From all these considerations," he declares boldly, "it follows that *Pithecanthropus erectus* undoubtedly is an intermediate form between Man and the Apes." He does not need to add, "as I have said all along." His meaning is as brilliant and hard as a diamond. Yet once again, the august body before which he speaks fails to completely embrace his gems of wisdom.

It is ridiculous, Dubois thinks. What better evidence could they expect? Are these men fools? Are they so jealous of my success that they would ignore the truth? He shakes his head in disbelief. What poor vision so many of these scientists have, that they cannot see the facts he lays before them so plainly. Almost as an aside, a minor point in what he feels has been his most compelling lecture on *P.e.*, he reveals that the Dutch Indies government has undertaken another dry season of excavation at Trinil, pursuant to the digging of a large irrigation canal. A valuable supplementary collection of fossils has been made, including a second, left premolar tooth from the lower jaw of *P.e.* Although there is nothing in this specimen to alter his conclusions, the new tooth has enhanced knowledge of the complete skeleton of the creature. Unfortunately, the excavations of 1897 are the last the Dutch Indies government intends to undertake, so no more fossils of *P.e.* are likely to be forthcoming.

When he finishes his presentation, Dubois is satisfied. If he has not completely won over his audience, he feels he has stirred their interest in *P.e.* once again. They will, in time, come to understand the complex business about body size and brain size, he is convinced.

Participants in the Fourth International Congress of Zoology in Cambridge include (left to right, back row): Arthur Keith, Grafton Elliot Smith, Eugène Dubois, T. H. Gurney, J. F. Gemmil; and (left to right, front row): G. Swainson, Dr. Stokvis, W.H.L. Duckworth, Judge Peepers, K. Newstead.

Though not everyone at the conference understands Dubois' innovative techniques or agrees with his conclusions, there is no doubt that his is among the most interesting and important of lectures. His fossils are clearly priceless as scientific objects. Word of the closure of the Trinil excavations— although further digging is just what is needed to resolve some of the debates—spreads from scientist to scientist. "You know," one man confides to a friend, "whatever that fossil is Dubois has found, it is dashed important! You'd think they could spare a few coolies to keep excavating at the most important fossil site anyone has ever found."

In another corner of the meeting, the American Othniel Marsh is making much the same assertion to his British colleague Alexander Macalister. "I like what this Dubois has done," says Marsh. "A good man, I think, very clever. But what we need, to be sure of his conclusions, is more fossils of *Pithecanthropus*. And we've got to know exactly where they come from. It's a real shame the colonial government has stopped the excavations. Not that I blame Dubois for leaving the Indies; the fever nearly killed him, I hear, and more than once! But the work ought to go on."

"Ummm, yes, of course. Do you think," Macalister wonders, "that a resolution from this congress—an international resolution—might influence the Dutch colonial government? Maybe we could propose something tomorrow? An endorsement of Dubois' work, or at least a statement of its extreme importance and a request that excavation be resumed."

"Good idea!" replies Marsh, striking his hands together enthusiastically. "Let's do it! I'll just have a quick confab with Dubois, make sure he approves. Then you draw it up and I'll second it."

Dubois is immensely pleased at the idea. While he isn't optimistic that more specimens of *P.e.* will be forthcoming, one never knows. Anyway, Kriele and De Winter and a team of laborers could be kept working until the fossils give out. The next day Macalister introduces a resolution that is quickly adopted by the congress.

> *IVth International Congress of Zoology*
> *Cambridge: August 28, 1898.*
>
> That, *in the opinion of this Meeting of the members of the IVth International Congress of Zoology, the Dutch Indian Government, by ordering the exploration of Trinil, Java, leading to that most remarkable (among many) discovery of* Pithecanthropus erectus, *have laid the Zoological World under a most weighty obligation;*

and that *the aforesaid members of the IVth International Congress of Zoology hereby desire to express their fervent hope that these investigations may be continued in the future with the same thoroughness as in the Past.*

The above Resolution, *having been adopted by the above mentioned meeting is herewith presented for signature by Members of the IVth International Congress of Zoology:*

Alex. Macalister (Proposer)
O. C. Marsh (Seconder)

Amazingly, the resolution has the desired effect. In 1898, excavations in the Trinil area resume, and they continue until 1900. No new fossils of *Pithecanthropus* are found.

CHAPTER 43 BETRAYAL AND RESURRECTION

In the new year, 1899, things grow much, much worse.

One of Dubois' handful of supporters in Germany since he returned home with *P.e.* has been the anatomist Gustav Schwalbe, a large, bluff man with silver-white hair and beard and an enthusiastic, even boisterous personality. It was he who had the actor's femur with the exostosis sent to Dubois, and he who boldly defended Dubois' ideas in print against Virchow's slashing attacks. Moved by Schwalbe's kindness, Dubois sent

Excavations continue at Trinil through 1900, when Kriele takes this photograph from Standpunkt II (marked on map, page 141).

Schwalbe a cast of the skullcap when very few were available, and in 1896, even granted Schwalbe's request to spend some days in Haarlem, studying the original fossils.

It proved an uncomfortable interlude. In person, Dubois found that Schwalbe made him uneasy. It was as if Schwalbe were appropriating the fossils; ownership was somehow implied in the way Schwalbe pointed out anatomical features to Dubois and lectured him upon the significance of *his* finds. Dubois sensed a too-sharp ambition and a hint of ruthlessness underlying the jovial exterior. Like Haeckel, Schwalbe seemed to be looking for something with which to combat Virchow. Dubois has the evidence, the tangible evidence of human evolution. But it is his to present, not Schwalbe's.

The days soon passed and Schwalbe left, but Dubois' suspicions did not die. Instead, they grew and flourished, torturing him with visions of his prize discovery being usurped. At the end of 1897, Dubois decided he must write to Schwalbe, spelling out his own plans for the further publication of his fossils and making clear what possible openings would be left for other scholars after Dubois has finished his own work. Schwalbe's reply was less than reassuring.

December 20, 1897

I shall be glad to leave the femur to you for some time before publishing. Certainly we are competitors. . . . I have almost finished a manuscript on the skull and another one on the femur. . . . We will both have to accept the fact that one or the other of us will finish sooner.

Competitors? But the fossils are *mine,* Dubois raged. Schwalbe has no earthly claim on them. Schwalbe did not find them, did not sacrifice eight years of his life in the grueling search for them. He did not pick the spot for excavation or teach the men how to work; he did not lie in that wretched tent on the banks of the Bengawan Solo, sweating and close to death from fever, hoping only to live long enough to announce his finds to the world. No, I showed these fossils to Schwalbe as a professional courtesy; I never granted him the right to publish on them. He tried to calm himself with the thought that no editor of a reputable journal would accept such an article from Schwalbe when Dubois' intentions to publish more lengthy studies are well-known in Europe.

But between the end of 1897 and the beginning of 1899, Schwalbe has founded his own scientific journal, designed in no small measure to loosen Virchow's stranglehold on matters of anatomy and evolution in Germany.

Schwalbe is editor and chief reviewer of the *Zeitschrift für Morphologie und Anthropologie* (Journal of Morphology and Anthropology), and of course he needs something extraordinary for his inaugural issue, so he opens it with an essay explaining that the purpose of the journal is

> to contribute to the important question of the origin of the human kind by careful comparative anatomical and developmental investigations, and to investigate the relationships between the human races, their bodily development, and their higher or lower position via a purely morphological approach. . . .
>
> A zoology of mammals without paleontology is an extremely deficient science that can provide only highly incomplete information on the evolutionary history of the entire mammalian group with its individual members. . . .
>
> The apparently wide chasm which separates Man and the simians can be bridged by a consideration of the fossil forms which have only recently become known. A thorough study of the fossil remains of the order of primates . . . is an absolute prerequisite for the foundations of zoological anthropology. . . .
>
> In this way, paleontology becomes the principal guide for our fascinating field of research and thus, at the end of . . . our investigation of the aims and methods, we arrive at Cope's dictum "The ancestry of man is a question to be solved by paleontology."

That is all very well, a welcome endorsement of Dubois' own views on the importance of fossils in evolutionary studies. What follows is an absolute betrayal. Dubois cannot believe his eyes: most of the *Zeitschrift*'s first issue is taken up with several hundred pages of description and analysis of P.e.'s skull. The sharpened blade of the weapon Schwalbe wields is Dubois' own fossil. It is the first installment in a series of articles in which Schwalbe measures the P.e. skullcap and compares it with everything: gibbons, other apes, Neanderthals, humans. He addresses Virchow's attribution of the skullcap to a gibbon in a long, meticulous discussion, pointing out detail after detail in which P.e. differs from a gibbon. Schwalbe emphasizes and reemphasizes how the anatomy of P.e. reveals not only its way of life but also its phylogenetic position. He proposes a new evolutionary tree, starting from Dubois' ape-man, leading through Neanderthals to modern humans. The pathway of evolution seems self-evident, obvious; Schwalbe's mastery of the material unquestionable. In thoroughness, complexity of analysis, and most of all in sheer length, Schwalbe's effort outshines Dubois' thirty-nine-page monograph. Somehow it does not seem to matter that Dubois' work was written under the most difficult of circumstances during a period when he was isolated at the ends of the earth, far from libraries, comparative collections, or colleagues.

Dubois can dispute none of Schwalbe's observations, only his overall interpretation. More than that, he is bitterly wounded. Schwalbe has stolen Dubois' fossils, as surely as if he had absconded from Haarlem with them tucked under his arm. Not one new word would he have written on *Pithecanthropus,* but for Dubois' generosity. The impact of his work does not stop there. Next, Schwalbe reexamines the Neanderthals, which he now finds to be less than human and a distinct species, *Homo primigenius.* This is not a novel suggestion, but Schwalbe takes it as his battle stance for the rest of his scientific life. It is *P.e.* that pushes Schwalbe to create a new phylogeny, to launch a new journal, to fly into the bright light of scientific celebrity like a moth to a flame.

But it is Dubois' flame in which he shines: Dubois' by right, by the sweat of his brow, by the genius of his intuitions, by the daughter he left behind in a small cemetery in Java. It is his, and Schwalbe has stolen it. Dubois will never forgive him, nor ever trust another so naïvely. He feels that he has been ruined, all his efforts and perseverance cast aside.

Although in 1899 Dubois is offered—and accepts—a position at the University of Amsterdam as Professor Extraordinarius of Crystallography, Mineralogy, Geology, and Palaeontology, he finds no peace. His rate of publication has been slowing since 1896 and continues to drop off sharply; he can find little more to say about *P.e.* that he has not already said repeatedly. He has told his colleagues, and shown them, and argued with them. His fossils—*his* fossils—have been discussed at enormous length by Schwalbe, not to mention the more than eighty publications by various scholars before the end of 1899. They are the most important discovery of the nineteenth century, without a doubt, but their discoverer is bored.

He is also bone-weary of the battle. He has no further patience—indeed, he has never had much—for the frivolous criticisms and alternative interpretations offered by those who do not know the fossils or who will not understand them. He has shown the fossil to everyone, everywhere, even allowed access to one whom he clearly ought not to have trusted. Dubois has diagrammed and explained the geological setting at Trinil until even a child could understand it; he has analyzed and calculated and estimated and compared until his brain is sore, and still they will not understand. And now he is repaid by Schwalbe's blatant, despicable theft.

Dubois feels like an old, old man with no physical or mental power left at all. What can await him now except death? What matters in his life is surely finished. He cannot remember feeling this bad except during attacks of malaria. What is done is done. Schwalbe has laid claim to Dubois' ideas and Dubois' discoveries, and there is no erasing all those detailed pages of meticulous German science. The new security offered by the position at Amsterdam gives Dubois some respite from financial worry, but there is no relief from the outrage that seems to be eating him from the inside out. For once, he has no energy, no ideas, no recourse. He wants to hide himself and his fossils in some dark corner where no one will ever find them again.

Ironically, just when he wants to withdraw he is invited into the spotlight. The government wishes him to supply the fossils of *P.e.*, or good casts of them, for the Exposition Universelle, to be held in Paris in 1900. Along with exhibits from many nations, the Dutch are planning a special display in the Pavillon des Indes Néerlandaises, devoted to objects from the Dutch East Indies. It is to be a celebration of national pride. National pride! Dubois, beleaguered and wounded, cannot imagine that the general public will have any interest in the bones of his missing link. After all, the scientific community still cannot understand the most basic truths about *P.e.*, so how could the uneducated public make sense of her? He is too drained even to write his refusal. The official letter sits for weeks, unattended-to and gathering dust, on his desk. For weeks Dubois does no work.

And then he has an idea, a crazy, brilliant idea. He will produce a sculpture of *P.e.* as she looked in the flesh—or rather, he will do a life-sized statue of a male *Pithecanthropus*, a hide draped around its waist for decency. Even the public will appreciate this, the image of a male *Pithecanthropus* as he might have looked in the jungles of Java so many years ago.

Dubois is an excellent artist with pen and pencil, and skilled with his hands; he has no doubts that he can produce a credible sculpture. He finds a disused attic, in an old two-story building, where he can set up his model, his clay, and the metal frame to support the sculpture. He will need to work without interruption. His plan is to make a clay figure that can be cast in plaster and painted in lifelike colors, with brown skin and orang-utan red hair. He needs, urgently, a suitable model. It is an awkward thing, to approach a man of his acquaintance and ask him to pose naked in an unheated attic in winter for a statue of a primitive ape-man. He turns the problem over in his mind for some days, until at a crucial moment, his elder son, Jean, comes in from a day of skating. Flushed, disheveled, and bright-eyed from cold and exertion, Jean at eleven seems the image of a little savage, a primitive man. "My son," Dubois says, addressing his offspring with unusual fondness, "you'll do. You'll do very nicely." The boy looks up, confused. Do? Do what? Seeing his puzzled expression, Dubois explains: "You'll make a nice ape-man. You are to pose for me."

For the rest of his winter holidays, Jean poses, miserably naked, cold, and cramped, while his father sculpts. Dubois has a precise posture in mind. He directs the boy to stand very still, and just so: his knees a little bent, feet spread apart and pointed slightly inward, eyes fixed on a deer antler that he holds in his right hand. The antler, like the many fossilized antlers found at Trinil, is to provide just a suggestion of possible tool use, which Dubois finds plausible but unprovable. The boy's other hand, his left, is to be just in front on his thigh, palm forward and fingers slightly open.

Jean soon finds that holding still in the bitter cold is intensely difficult. He begs for frequent breaks, to warm himself in a blanket and relax his cramped muscles, and these are granted. Jean obeys his father, in this as in nearly everything else, for Dubois is a stern paterfamilias and his word is not to be contradicted. The statue proceeds quite rapidly, given that the sculptor is a novice and the model a restless eleven-year-old boy who would rather be reading or skating or playing ball with his brother. When it is completed, Jean is quite impressed and brags to his fellows at school about his modeling job. "Only my body was being used," he adds quickly, to forestall the teasing he can see coming, "*not* my face!" Because of

Jean reluctantly poses for this sculpture of Pithecanthropus, *nicknamed Piet, for the Paris Exhibition of 1900.*

P.e.'s transitional status, Dubois gives the statue elongated apelike fingers and toes, not to mention a big toe that diverges from the others on the foot.

Before being shipped to Paris, the finished statue—nicknamed Piet, for *Pithecanthropus*—goes on exhibit in the front hall of the Colonial Building of the Industrial Arts Museum in Haarlem. Dubois takes his entire family along for the opening, to observe the public's reaction. Jean is both embarrassed and proud to see his likeness displayed in public and attracting such attention. There is no doubt that Piet catches the eye, he is so large and naked and odd-looking.

Jean stands near his statue for a long time, watching as people approach and study it. His attention is drawn to an elderly, countrified couple who walk in, so busily gazing around themselves that they come quite close to the statue before they seem to see it. They halt, bewildered expressions on their faces, in front of Piet.

"And who," the woman asks her husband, querulously, "is that?" Before the man has a chance to reply, Jean takes a step toward them and blurts proudly, "That is my father!" He means, of course, that it is his father who has found the being and reproduced it as a sculpture. The couple look at him, confused, and then hastily move away.

Later, in private, Jean confesses his outburst to his father. Dubois breaks into a low chuckle and smiles at the boy, patting his shoulder warmly. Jean is confused, but grateful. It is perhaps the only time in Jean's short life that he has been so irreverent about his father and not been punished for it.

CHAPTER 44 FAMILY

During those first five trying years back in Holland, Dubois finds no solace in his family. His father is dead; his mother is not proud of him; his brother misunderstands and chastises him. Nor can he warm himself with his children's affection or his wife's sympathy, for these are sadly strained by the abrupt and disturbing changes in their world. His children were in urgent need of his attention when they returned to Holland in 1895, Eugénie eight, Jean seven, and Victor four. Removed from the casual, languorous Indies, the children's shortcomings in manner and habit became all too evident to their father. They ran wild in the Indies, playing only with native children and half-castes, exploring the jungle, chasing after birds and butterflies and anything that moved. They did not learn to sit and read, or even to listen or think carefully. Their babu spoiled them—babus always do—but their mother has not exercised any counterbalancing influence. Anna is hopeless, Dubois realizes when he compares her with the wives of his colleagues in Europe. In company, she knows how to mask her weaknesses for a short time. But she always likes to join in the conversation, and her empty-headedness is soon apparent. She is no longer a decorative companion at a banquet, nor a good housekeeper, nor even a firm mother. And Dubois' mother has only encouraged the children's harum-scarum ways since they returned. This will not do.

Dubois decides that he can yet inculcate the boys with his values, mold their personalities. He can make the boys into sons that he can be proud of, uproot that Indies indolence and inject some Dutch vigor. Eugénie does not worry him; she will never have to earn a living and she is her father's daughter, strong-willed and intelligent. He resolves to exercise much more control over the boys' daily lives and characters, before it is too late. As his discouragement and weariness wean him from an intensive scientific schedule, Dubois starts to play a larger role in the boys' upbringing.

He establishes a firm routine for the boys. He makes sure that they rise early, well before six, and have a cold bath every morning. Before breakfast, they do fifteen minutes of exercises with an apparatus that Dubois devises himself, to build the strength in their arms and chests. He attaches a pair of handgrips to the wall of the house. The handgrips are connected by thick elastic cords to a pair of adjustable cables that move over pulleys, to give resistance. Moving rhythmically, each boy pulls against the cables in a pattern Dubois has

Dubois tries to shape the characters of his children (left to right, Jean, Eugénie, Victor) but only Eugénie has inherited his intelligence and drive.

devised to build their muscles. Sometimes he commands them to repeat the entire set of exercises, if they are too reluctant. They go inside for breakfast (two slices of rye or wheat bread, a boiled egg, and a glass of milk) and go promptly to school. If there is extra time before school, they are sent off for a vigorous walk, regardless of the weather. "I will not have weak sons," Dubois tells them all too frequently, "and I will not have lazy ones."

When the boys return from school, they are to wash and change. They are not to appear in front of their father dirty or disheveled. They are expected to sit quietly through dinner, which is followed by an evening walk with their parents and Eugénie. Dubois feels the walk is a good time to instruct them in natural history, to see if they have learned anything about the animals, birds, and plants that surround them. He tries desperately to teach them to observe and think. He might say suddenly, "Jean, what bird is that calling now? Yes, very good, it is the magpie. And what does it look like? Yes, black and white, with a long tail. Does it migrate, or does it stay here through the winter? Hmm, don't know? You'd better refresh your memory when we get back." And a little later, "Now, Victor, do you see the nest up in that tree? You must learn to look with your eyes, to see what is around you. There it is, there! Hurry up, boy, climb up and look in the nest. Come on! Don't touch it, just tell me: Are there any eggs in it? How many? What do they look like? Whose nest do you think this is?"

After the walk, the boys do their homework. There is no playing or reading for enjoyment until their work is done. They also often look up information about the questions that their father posed during the family's walk. There is no point in hoping that he will forget to ask them the very same question the next day. Failing to answer the same question twice is the cause of considerable paternal disapproval and wrath; they have never dared to find out what failing to answer three times would provoke. Dubois quizzes them weekly on the subjects they are taught in school, to make sure that they are learning and that the teachers are not filling their minds with rubbish. At first, their ignorance is appalling, but with Dubois' daily intervention, their memories and habits improve. Finally, under his guidance, they learn to study diligently, to conduct themselves properly in the house, to be seen and not heard.

These boys bear Dubois' name, a proud name, and he is determined that they will also bear his character, if it can be instilled in them. Even when he is working hard in his study, Dubois listens for their return from school, to see if they slam the door and thump up the stairs or if they remember to walk quietly like civilized people. Slowly, they begin to behave with dignity and calm, as befits their father's sons. But even after months of his tutelage, he is saddened to receive complaints about their conduct at school. At school, so their teachers say, the boys are rude, noisy, and sometimes insolent. They interrupt classes, do poorly on tests, and befriend the very worst students. It is as if they have only so many hours of good behavior in them, which are expended at home. He tries to teach them, by example and explanation, the importance of truthfulness, hard work, and discipline. In this family, a detected lie—and Dubois always knows when the children are lying—merits a much sterner punishment than an admitted wrongdoing.

After some years of this regimen, Dubois realizes he cannot alter their basic natures further. Despite all his efforts to be a diligent father, Jean and Victor do not do well in school and do not qualify to go to university. They are not intelligent enough to do well without working, and they are not hardworking enough to do well without being smart. It is a bitter disappointment. All his sons want to do is return to the Indies, to the land of sunshine and freedom, where they were indulged and petted. They talk of it as a paradise lost.

Surprisingly, Jean and Victor have no trouble obtaining jobs once they leave school. They take the chance to fulfill their dreams and travel. In 1906, when the boys leave for the colonies for the first time, the entire family has been for a few days at De Bedelaar, an estate in Limburg that Dubois has bought as a country retreat. He already has plans to make it into a sort of nature park, to restore the ancient habitat and vegetation to this peaceful corner of Holland. He is already busy inventorying the plants, studying the two fens on the property, and learning what animals and birds dwell there.

When Jean and Victor are ready to go to the station, Dubois is swimming naked in the lake. He has tried not to cast shadows over their bright plans, not to make his skepticism about their futures discouragingly plain. As so often before, he takes refuge in fierce physical activity to quiet his mind. He cannot even bear to get out of the lake and dress to accompany them to the station, for fear of sending them off with his disapproval ringing in their ears, as his own father did so many years before. Better to let their mother take them, for she will send them off with a bright smile and a dream in their heads. If, as he believes, their disgrace is impending, there is little to be gained from saying so now.

The situation is a poignant replay of his own father's despair when Dubois quit his job at the university and left for the Indies. He understands now how his father felt—though of course, his father was wrong, while he, in this instance, is right. He went off with a purpose, a holy mission of science, and he succeeded. His sons leave with no education, no purpose, no real skills: not as he did. They go off to play in the warm sunshine, like children. Their guileless, smiling faces are young, so young. They probably think that the flowers are always in bloom in the tropics and the girls always beautiful; they dream they will be rich and happy in a few years' time. Jean heads for Dutch Guiana, to work as a planter for several years; he ends up in Java, using his knowledge of Malay to manage plantations, first rubber and coffee, then tea. Victor follows a similarly wandering track through the Dutch colonies.

They dream of wealth, but Dubois knows better. Statistically, they are more likely to be dead than rich after a few years in the colonies. And he knows they will find the Indies a different world now that they are adults; things are not so easy when there is work to do and responsibilities to fulfill. The tropical climate is hotter than they remember, and it requires more fortitude to get real work done. His sons don't understand that the natives have to be taught everything, and reminded and watched all the time to see that they do the work carefully, or at all. Failure always lurks just around the corner on plantations; disease and death are the constant companions of Europeans in the tropics. At every turn, they will have to guard against slacking and cheating and lying and fever. It will be a big change for them, to be, for once, the ones who have to maintain standards and hold others to them. He says good-bye to his sons, shakes their hands, and then plunges back into the cold lake water as soon as they have left, as if to drown his feelings.

For several years after their departure, Jean and Victor earn a good living in the colonies. They do well enough managing their workers, better than most young men newly out from Holland. Their employers seem happy with them, for one of the plagues of plantations is the problem of constantly firing men—or having them simply wander off—and then taking on untrained newcomers; Jean and Victor have the knack of keeping the workforce stable and happy. They write home from time to time, bright and cheerful letters mostly addressed to Anna, who misses them keenly. To their father, they write only of crop yields and profit, with occasional anecdotes about unusual animals they have seen or shot. They send home the skulls and notes of the body weights of unusual animals they have killed, for his studies of cephalization and body weight. This pleases him. But when Jean writes home dramatically describing how he has killed a Sumatran royal tiger with a single shot, his father has no praise for his son's prowess. His only question is "Did you save the skull?"

CHAPTER 45 THE NEW CENTURY

With the beginning of the new century, things change for Dubois. It is as if the dawning of the long-forecast future has brought a change of heart—or, more accurately, a petrification of heart. Dubois is beaten and bruised from his years on the lecture circuit, fighting, always fighting, with words and principles and theories and new techniques. Nothing ever brings him victory.

Worse yet (he hardly dares admit it), he is restless. He has studied, analyzed, compared, and reconstructed *P.e.* until he has nothing more to say. Now that I have found the answer for myself, he thinks, I have no more interest. There is no more to do with those few, wonderful fossils, and he knows from bitter experience that if he allows someone else to study them, that man will attempt to steal his glory. That will not happen again, he vows: never. He puts the bones away in their special cases and locks them away in their own special cabinet at the Teyler Museum. He rarely takes them out, only sometimes in the afternoons if he is feeling melancholy. Then he extracts them and holds them up to the light and he is once again filled with wonder at the remarkable objects he found.

What days those were, in Java! What dreadful conditions, nearly fatal: fever and heat, in a country where even the vegetation and the rivers seemed to be determined to kill him; lazy, ignorant coolies who couldn't be bothered to understand the work; and mountains so steep a man's legs ached for days. What obstacles he overcame during those long lonely weeks of anguish when only Prentice understood him, when only Prentice's faith in his find and his intellect kept him going. And here is the proof, his *P.e.*, his missing link. Whatever they say, those skeptical scientists of Europe, they cannot change the truth. He is the man who found the missing link.

From time to time, he gets an idea about the bones and carries them home to work on. As matter-of-factly as he closes and locks the door to the street, Dubois rearranges the china in one of the glass-fronted cabinets in the dining room, places the bones within on a bed of fabric, and carefully pastes newspaper over the glass in case a fellow scientist should call. In the end, inspiration always fails him and eventually he takes the fossils back to the Teyler, to live once again in their special case. He has found his missing link; he knows the truth; and it is done. He has nothing more to say.

He publishes nothing on *P.e.* in 1900 save a brief pamphlet to be given out at the exhibition in Paris, explaining how the statue "Piet" was made. The rest—the other five publications that year—focus on geology, the age of the

By 1902, Dubois is focusing on discovering the biological laws that govern the relationship of brain size to body size in mammals.

earth, the circulation of carbonate of lime. He is, after all, a professor of crystallography, mineralogy, and geology as well as of paleontology. In a few years, the university adds geography to his purview as well, as if he does not have enough to do. He pays little attention to the thousands of mammalian fossils he collected and brought back so painstakingly from Java, Sumatra, and India. They are his capital, his investments in his future, and yet he has not been able to obtain suitable quarters for storing and working on such a vast collection. Frankly, his interest in them, too, has waned. He knows enough of those fossils anyway; he saw enough of them back in the days when every crate sent by Kriele and De Winter was a treasure-trove eagerly explored.

The years roll by, 1901, 1902, 1903, and no papers appear on the fossils from Trinil even though Dubois continues to draw a salary as curator of the Dubois Collection. His friends grow anxious. Karl Martin, the geologist who first described Raden Saleh's fossils from Java, starts to take more interest in Dubois' neglected collection. There is a move afoot—is it Martin's scheme?—to turn the entire Dubois Collection over to the Geological Museum in Leiden, where it would come under Martin's control. Dubois fights off this proposal, claiming that most of the specimens have not yet been individually labeled and he is the only person capable of overseeing the basic labeling and registration. He leads the Ministry of the Colonies to believe that one-third of the text of a larger description of the fossils is already at the printers, with many plates published, but the work never appears. Dubois' friends begin to worry that he will lose control of his collection to Martin. Jan Lorié, the geologist, writes,

April 30, 1903

How is the situation, yet, with the Indonesian bones? Have you totally forgotten them? Do be careful with the Colonies and our "mutual friend" of Leiden! I fear for a small catastrophe.

Dubois does not want to hear these fears, and buries himself in research on groundwater and geology. But Lorié persists, warning him more explicitly a few days later,

May 4, 1903

... You yourself informed me how much Martin preyed upon the Indonesian bones, which I think is very understandable. Further I know, from cases with Wichmann and Molengraaf, that he does not shrink from tricks and prevarications to gain control over what attracts him. ... It is difficult for you not to attract attention with your lectures about the water supply of Amsterdam. But by now, people must find it strange that you, properly called a paleontologist, lose yourself successively in $CaCO_3$, $NaCl$, H_2O. ... while it is generally known that you receive a salary from the Colonial Office especially to work on the bones, and yet you do nothing about them, so people think. It is very obvious that the Colonial Office ultimately want to have value for their money and it appears entirely likely to me that Martin, who often comes to the ministry, will use this to grab your bones. ... Therefore, I offer you a friendly warning. If I have everything wrong or exaggerated, then good, but I do not believe that.

Lorié's forecast of doom echoes in Dubois' ears, accompanied by an ominous announcement from the Dutch Minister of the Interior that the final description and publication of the fossils *will* be completed within three years. Does that mean his curatorial salary will be stopped in three years' time? What will happen to the bones themselves?

Dubois broods, but still he does not publish anything about the Trinil fossils. The Dutch East Indies Government continues excavation in the Trinil area in 1900 and then ceases. Now Dubois has no longer any hope of receiving a crate with additional fossils of *Pe.* in it, one that might move him to new insights and energy. In 1905, Dubois is appalled to be recalled to active duty as an army surgeon. He is almost fifty years old and still suffers from occasional bouts of fever! It is an absurd idea. He cannot find out who is behind this plot, but he knows it cannot be a coincidence. He suspects Martin, carrying out some devious plan to get the fossils away from him, but he cannot work out how Martin managed to influence the Army. In any case, Dubois calls for an immediate medical examination and is promptly declared unfit for service, with an honorable discharge.

The vultures are moving in. Where he once worried only about Martin, Dubois soon comes under attack from another source. For some years, Emil Selenka, a German zoologist especially interested in apes, has been laying plans to reopen Dubois' excavations at Trinil. Now Selenka has actually won the financial support of the Prussian Academy of Science. Would that I had been so blessed, thinks Dubois when he reads of this miraculous news. Through Dubois—who cannot refuse, for the request comes through government channels—Selenka even contacts Kriele and De Winter and obtains Dubois' old site plans. Disheartened, bored, and unwilling to return and risk his own life again, Dubois cooperates, albeit grudgingly. Steal my site, steal my men, Dubois thinks bitterly. Go ahead! See what you find. I wish you well of it!

Before the expedition can be mounted, both Selenka and Kriele die. Dubois thinks this nightmare is over, but he is wrong. Selenka's ambitious widow, Margarethe Lenore, decides to head up the expedition, though she has no particular academic qualifications. She puts the German geologist Dr. Johannes Elbert in charge of the geological survey; the Dutch mining engineer Fritz Oppenoorth will handle the technical side of the excavation. She is undaunted, insufferable. Reviving the organization and raising additional money takes time, so it is not until 1907 that Oppenoorth leaves for Java, to be followed shortly by the others.

Dubois knows all about this expedition, of course. His reaction to it is to start publishing on the Trinil fauna; the first paper appears in 1907 and opens with the following remarks:

> The excavations which at present will be carried out at Trinil, under the auspices of the Prussian Academy of Science and the widowed Frau Selenka, with the support of the Dutch government, have once again called attention to the extinct mammalian world of Java, known as the Kendeng or Trinil fauna, from which one form, the much-discussed Pithecanthropus erectus, has become generally renowned. It is this species, Pithecanthropus, in particular that has given rise to the German expedition.

There is little about P.e. in this article—he does not write anything new of any length about P.e. until many years later—only the reaffirmation of belief by a bitter, tired man: "In short, I consider Pithecanthropus to be a descendant of less-specialized (less long-armed) ancestors of the Gibbons . . . a descendant which has assumed the erect posture."

Although the mere existence of Frau Selenka's expedition offends Dubois, he has no fear that they will find what his nine seasons of excavation at Trinil have failed to: more evidence of P.e. Still, he does not want her expedition to claim priority for describing the various new species of mammals that he has already retrieved at such cost. A second article on the fauna follows in 1908. These establish his priority, so his names for the new mammalian species he discovered at Trinil will endure.

The new expedition is an open insult to Dubois. How could his work be superseded by that of a woman! And such a woman, too! The sheer naïveté, the simpering idiocy, of her answers to a newspaper interviewer in Batavia—of course, she is seeking fame and glory before she has accomplished a single thing!—provoke his ire.

April 25, 1907

BATAVIAASCH NIEUWSBLAD: *"And do you not shrink from the loneliness and isolation at Trinil?"*

WIDOW SELENKA: *"Oh, that problem is so very small in comparison with those I have already overcome."*

Dubois places a large exclamation point next to this answer. Oh yes, survival at Trinil is only a matter of facing loneliness, a trivial thing, he fumes. She knows nothing of what she faces, nothing. As he reads further, he loses his temper completely.

BATAVIAASCH NIEUWSBLAD: *"We ask again, is the purpose of your excavation exclusively to trace the fossil remains of* Pithecanthropus erectus?*"*

WIDOW SELENKA: *"What else?" came the astonished counter-question.*

BATAVIAASCH NIEUWSBLAD: *"For all that, the main thing will be also, if possible, to establish the age of the geological strata in which the skeletal remains are found, in which Dubois made his find?"*

WIDOW SELENKA: *"Oh, yes, certainly, but the one issue encompasses the other. Certainly the age of those strata is a very important point. . . ."*

BATAVIAASCH NIEUWSBLAD: *"And how are you disposed with regard to the outcome of your investigations? Do you think you will find what you are looking for?"*

WIDOW SELENKA: *"It is surely possible, isn't it?"*

Ja, ja, Dubois mutters to himself. No doubt you, the Widow Selenka, will succeed where my excavations have yielded only a few teeth, one skullcap, and a femur. *Ja,* it is so simple to establish the geology and antiquity of those beds, if only I had been more observant during the long years I worked there. Pah! I wish you luck, you black widow, I wish you luck.

By 1908, the Widow Selenka is triumphantly showing around Europe a few teeth she has found. She believes them to belong to *Pithecanthropus,* of course. Dubois is scathing about her finds, in print.

> *From the two teeth, which the Widow Selenka showed me as her most important finds, one was quite whole, white-looking and recent: a human lower molar to the rootless underside of which was stuck sand similar to that from Trinil, although it was "not found at Trinil." The other tooth, which really was excavated at Trinil, is the upper premolar of a pig.*

Selenka responds vigorously to this accusation of "forgery," as she terms it. On February 9, 1909, she rushes into print with an article on the fauna from Trinil, including expert opinions about the tooth from Dr. Schlosser and Professor Dr. Walkhoff. Schlosser asserts that the tooth is "much darker than [is the case] among recent human teeth, indicating real fossilization" and points out that the dentine has been completely removed by natural geological processes, a characteristic of some fossils. Walkhoff asserts that the tooth cannot be a forgery, for he has studied it with X rays and microphotography.

Sensing a lively debate, the journal invites Dubois' immediate reply. He agrees that the removal of the dentine is natural, but not that it is a sign of fossilization, and that the color and condition of the tooth are completely incompatible with the color of the sand stuck to it. As for the question of forgery, Dubois denies he has ever suggested this is not a genuine human tooth, only that it "imitated" something it was not. Dubois' considered opinion is that the tooth is subfossil—too modern to be fully fossilized—and may originate from a relatively recent grave dug into the sediments near to Trinil at Sondé, where the tooth was found. It is nothing to do with *P.e.*

In the end, despite the flag-flying and trumpet-blowing in the news-papers, Selenka's two-year expedition comes to naught. She and her col-leagues confirm the antiquity of the fossil-bearing strata—no surprise to

Dubois there—and find mammalian fossils, lots of them, as Dubois knew that they would. What they do not find is *P.e.* He is more than a little smug about this. They cannot find *P.e.* because he has already found her. She is not there, waiting for some silly woman to mount an expedition. *P.e.* is his, his idea, his find, his life.

During the years of Selenka's expedition and afterward, Dubois is not silent in the scientific world. He writes prolifically. In addition to the three paltry articles about *P.e.* and the Trinil fauna, between 1901 and 1910 he publishes sixty-eight about other things: climate, geology, groundwater, drinking water supplies, paleoglaciers, minerals. Although in time even his friends feel he has become reclusive, Dubois cannot be accused of being unproductive. What he can be—and is—faulted for is abandoning his *P.e.*, leaving her alone in the cupboard. None of the numerous topics he writes about in these years ever engage his heart, only his cold, crystalline intellect.

Though Dubois is effectively absent, much happens in the study of human origins in the early years of the new century. More Neanderthal remains are found, most spectacularly burials at Le Moustier and La Chapelle-aux-Saints in France, another at La Quina, and a mixed, broken collection of many Neanderthals' skeletons from Krapina in Croatia. Some important monographs appear, too. Gustav Schwalbe, propelled perhaps by his underhanded analysis of *P.e.*, produces yet another detailed reexamination of all known Neanderthal remains and declares them to be a separate species, *Homo primigenius*. In 1906, a Croatian scientist, Dragutin Gorjanović-Kramberger, describes the new Neanderthal fossils from Krapina that he has discovered. Gorjanović's work remains obscure, despite being published in German, for the French and English scientists are reluctant to read German even if they can. Gorjanović also prejudices his case by making a sensational claim: that there is evidence of cannibalism in the smashed, charred bones at Krapina. Still, however sensational, Gorjanović's is a lengthy and thorough treatment, 218 pages long with fourteen plates of photographs. Dubois, who reads German effortlessly, cannot help comparing Gorjanović's impressive monograph to his own poor effort of thirty-nine pages. He despairs once again, overlooking the fact that Gorjanović has had access to all the libraries and museums of Europe in writing his work.

Even more stunning is the main scientific event of 1911: the publication of the first installment of a massive monograph on the La Chapelle-aux-Saints Neanderthal skeleton by Marcellin Boule. Although Boule has headed the laboratory of paleontology at the Musée National d'Histoire Naturelle in Paris since 1902, this work is a venture into new territory. His predecessor and mentor, Albert Gaudry, deliberately avoided writing about or working on human evolution, for it was still a dangerously controversial subject in his day. Boule's ambition is to use this monograph to establish human paleontology as a respectable discipline in France, and himself as the leader of it. The subject is good, for the fossil is splendid: a remarkably complete skeleton of an aged male Neanderthal, intentionally buried in a cave, discovered with the remains of an animal's leg and a collection of stone tools, dressed flints, and flakes. Boule's thoroughly scientific analysis of these remains is designed to demonstrate how systematic and scientific the study of human fossil remains can be. In this intention, his ideas accord well with those of Manouvrier, Dubois' great friend at the École d'Anthropologie. Where the two Frenchmen part company violently is over the subject of phylogeny. Manouvrier, the scientific liberal, firmly defends the idea that *Pithecanthropus* is the ancestor of Neanderthals, who are in turn ancestral to modern humans, while Boule's work is underpinned by a conservative conviction that no such savage and apelike a creature as a Neanderthal—much less *P.e.*—can lie anywhere in the direct ancestry of modern man.

Dubois is ignorant of the politics of French science and cannot make himself care about such matters. All he sees when he reads the full 279 pages of Boule's work is the deficiencies of his own monograph. He forgets that his was the first, the groundbreaking monograph on an early human ancestor, the article whose shortcomings saved later authors, like Gorjanović and Boule, the same painful treatment Dubois experienced. All he can do is reproach himself for not producing such a lengthy, detailed monograph. He does not like to admit it, but once he lifted the corner of the mystery of *P.e.* and saw what lay underneath, that was enough. He is an impatient man with a fertile, restless mind, ill-suited to prolonged, meticulous work. He knows *P.e.* now, knows her inside and out. He cannot understand why, after all he has done, others do not believe him, and he cannot bear any longer their academic jibes and idiotic theories. They doubt him and accuse him of lying or foolishness; the wounds are too deep to heal. The only thing is to closet *P.e.* away, where she—and he—will be safe from further torture.

CHAPTER 46 DIVERSIONS

If he is to do no more with *Pe.*, Dubois must find something else to occupy his brain. Boredom is agony; he cannot endure it, he will not sit mentally idle. In August, work starts on a large, comfortable house to be built at De Bedelaar, his country retreat in Haelen. He inventories the plants and trees and wildlife on the thirty-eight-hectare property, intending to attempt to restore the late Pliocene habitat of the region, which is known as the Tegelen Clays after a nearby fossil site that preserves abundant plant and seed remains. He occupies his mind for months with schemes to lower the water level in the fens at De Bedelaar. This is but the first step in creating a sort of prehistoric nature park, where plants and animals surviving from the past can live at liberty. The next step would be to fertilize the water and soil so that he can carry out a massive planting scheme to restore the original forest vegetation. He plants tulip trees and Chinese rubber trees, sequoias and swamp cypress, nine species of oaks, and numerous conifers; he reintroduces tench, rudd, and other fish to the fens and constructs thickets to attract nesting birds. After a few years, he erects bat towers of wood and brick in hopes of encouraging a colony to roost there and control the mosquitoes.

Dubois' vision of re-creating a natural past habitat is an extraordinary one that few understand. Locals generally think him demented and make jokes about this Amsterdam professor who wants to bring the past back to life. Maybe he wants to provide a home for his missing link? Their skepticism disturbs him only a little; he is used to pursuing, through sheer strength of will and intellect, goals that others do not understand. Still, he treasures the support he finds among the members of the newly established Netherlands Society for Nature Conservancy. In 1912, he is appointed a member of its board and of the board of the international society with which it is affiliated.

De Bedelaar is a welcome outlet for Dubois' energy and intelligence. As soon as the house is finished, he finds it a peaceful retreat from Amsterdam and the university. He spends more and more time there, as much as his teaching duties will permit. He realizes, too, that De Bedelaar provides him with an acceptable way to get away from Anna. There is no longer any overlap between their interests. The countryside is boring and cold, Limburg too foreign for Anna's tastes. She prefers Amsterdam comforts and society, is utterly indifferent to Dubois' plans for a prehistoric nature park, and cannot muster any curiosity about fens or drainage or fertilizers or trees. The whole

idea strikes her as just another one of her husband's expensive and inconvenient dreams. When they are together, he lectures and commands, she
takes refuge in misunderstanding: "This is all too much for me, Eugène; I am
not educated and clever like you!" When she is feeling greatly frustrated, she
slyly undermines him. She no longer has any faith in his obsessions; she no
longer thinks him a great man, only demanding, irritable, and invariably
stubborn, while he finds her more trivial and thoughtless than ever.

But De Bedelaar is not enough to keep Dubois busy. Nature is a slow
companion and Dubois is impatient. He decides to fill his time by resurrect-
ing the Wadjak fossils from his collections, to set about cleaning and reconstructing them properly, as he has never before found time to do. It proves
more interesting than he expected. Now that he looks at them again,
cleaned up, he sees that the two partial skulls—the one sent him by C. P.
Sluiter so many years ago and the other found when Dubois first started
work on Java—have considerably more to tell him. In a state of high excitement, he settles down to write to Sluiter.

December 17, 1910

*And it is now evident that everything is so important that I feel the need
to share this information with you, without delay. If I had thought twice,
soon after writing you in 1889, I would have realized that I am not dealing
with a Papuan type but with an Australian type. I learn now, on closer
acquaintance with the jaws, in particular, that this type is of a very primitive nature and in many ways links up with the Neanderthals.*

*Actually it is good that the task has lain fallow for so long, because in
the meantime the Australians and also the Neanderthals have become
better known, and because our early Javanese is in that line of descent,
which on the basis of that knowledge is believed to continue further back
into the past.*

Typically, Dubois' interest in the Wadjak skulls wanes after some months
and he soon sets aside the unfinished manuscript on their anatomy and
significance.

The numerous fossil discoveries of recent years foster enormous public
interest in the evolution of man. In 1910, Alfred Haddon of Cambridge publishes the first-ever history of anthropology, expending an entire chapter on
"The Unfolding of the Antiquity of Man." In it, Haddon summarizes the various fossil finds thought to pertain to human evolution, starting with the
famous *Homo diluvii testis,* found in 1726 and now under Dubois' care in the
Teyler Museum. In reality, this specimen is a fossilized skeleton of a giant

salamander that was mistakenly hailed at its first discovery as "Man, witness of the Flood," a "rare relic of the accursed race of the primitive world," and the "melancholy skeleton of an old sinner." Haddon then surveys the rest of the known fossils, but they "fade into relative insignificance compared with the sensation caused by the discovery made by Dr. Dubois in Java in 1891."

A most satisfactory assessment, Dubois thinks, reading Haddon's words. Alas for Dubois' temper, Haddon alludes to every nuance of the debates, challenges, and arguments that ensued over *P.e.*, its dating, and the association of its parts into a single individual. Haddon writes,

> *Dubois published his account in Java in 1894, and since that date a vast amount of literature has accumulated round the subject, representing three antagonistic points of view. Some, like Virchow, Krause, Waldeyer, Ranke, Bümüller, Hamann, and Ten Kate, claim a simian origin for the remains; Turner, Cunningham, Keith, Lydekker, Rudolf Martin, and Topinard believe them to be human; while Dubois, Manouvrier, Marsh, Haeckel, Nehring, Verneau, Schwalbe, Klaatsch, and Duckworth ascribe them to an intermediate form. The last-mentioned sums up the evidence in these words: "I believe that in* Pithecanthropus erectus *we possess the nearest likeness yet found of a human ancestor, at a stage immediately antecedent to the definitely human phase, and yet at the same time in advance of the simian stage."*
>
> *The English, as Dr. Dubois somewhat slyly noted, claimed the remains as human; while the Germans declared them to be simian; he himself as a Dutchman, assigned them to a mixture of both. . . .*
>
> *The discovery of these human remains has had a very noticeable effect on anthropometry. Most of them are imperfect, some very much so; as in the cases, for example, of the partial calvaria of* Pithecanthropus *and of the Neanderthal specimen. The remains are of such intense interest that they stimulated anatomists to a more careful analysis and comparison with other human skulls and with those of anthropoids. . . . New ways of looking at problems suggested themselves, which led to the employment of more elaborate methods of measurement or description.*

Very fine, Dubois nods to himself. Haddon understands what I have done, how I have changed evolutionary studies forever.

Though he does not travel to examine the new fossil specimens that seem to be announced every few months, Dubois keeps careful track of them. Neanderthals and more-modern skeletons seem to be unearthed in every cave in France; other types are discovered in England, Italy, and Austria; in Germany, a massively built fossil jaw from Heidelberg (*Homo heidelbergensis,* some call it) is found that might go with a skull somewhat like *P.e.*'s. Each new fossil produces a new round of inquiries about Dubois' specimens, both from up-and-coming young scholars and from the old giants to whom he showed *P.e.* back in those early years.

In 1911, Dubois' old ally Arthur Keith follows Haddon's lead and publishes a little book, *Ancient Types of Man,* summarizing the most current information on human evolution. Now curator of the Hunterian Museum at the Royal College of Surgeons in London, Keith has decided to mount an exhibition on early man in England to attract attention. At forty-five, Keith is settled in a relatively unimportant post that he finds unsatisfactory. He is a tense, anxious-eyed man with fair, frizzled hair and a pronounced Scots accent that reminds Dubois a little of his old friends Prentice and Boyd. Keith and Dubois first met years ago, when Dubois spoke to the Anthropological Institute of Great Britain and Ireland in 1895. On that first occasion, Keith was one of the few who rose to defend Dubois' ideas. They renewed their acquaintance three years later at the zoological congress in Cambridge. Dubois is interested to see what Keith has to say about *P.e.* in his new book.

Keith fills a complete ten-page chapter with an account of the discovery and interpretation of *P.e.,* Dubois is pleased to see. But when he reads the chapter, his pleasure turns to exasperation. Dubois himself is described as "now Professor of Geology in the University of Amsterdam . . . trained under that veteran Dutch zoologist, Max Weber." Weber! Trained by his colleague, Weber! Dubois is appalled. What of Fürbringer, the great anatomist who taught him so much? Keith is a little sloppy with his facts, he thinks; I would not have thought him so careless. I hope he is better about the fossils. But as Dubois reads on, the text becomes worse. "Went out to Java in 1889 as a military surgeon. At the request of the Governor-General of Java, he explored the fossil bed of Trinil, a native hamlet in the Province of Madiun, near the center of Java."

Eighteen eighty-nine? What about the years he spent on Sumatra? Dubois cannot help fuming; he hurls the small red-backed book across the room in disgust. So, I only carried out orders from the Governor, eh? What about all the working and writing to convince my superiors to relieve me of medical duties so I might look for fossils? What of the publication in the *Tijdschrift*? As for the Governor-General . . . well, *ja*, the man was helpful, very helpful at a time when few others would offer their support, but going to Trinil was hardly the Governor's suggestion. It was mine, my insight, my genius to look for open-air sites along riverbanks, and the Governor had nothing to do with it!

After a minute of indulging his temper, he gets up from his chair and stumps across the room to fetch the book. He might as well know the worst. But the remainder of the chapter is not so bad. There is a reasonably accurate account of the locality and its stratigraphy, and, although Keith wants to call the fossil *Homo javanensis*, he admits that Dubois' name is "justified." The most important thing is that Keith supports Dubois' interpretation of the significance of the fossil specimens' shapes.

> *Whatever the exact date may be . . . the characters of the femur leave no doubt, in spite of minor peculiar features, that the fossil man of Java was as completely adapted for erect posture and erect progression as the man of to-day. There are no features in it which suggest the slouching gait of Neanderthal man. . . . The modern human posture was attained long before the human brain reached its modern size. . . .*
>
> *The brain is the characteristic organ of man. Dubois estimated that of the fossil man of Java at 855 cc, but it is highly probable that the estimate is somewhat under the truth. . . . If we accept the Java specimen as representative of late Pliocene man, then we must admit that the human brain was then in its more primitive stages of development. . . .*
>
> *An analysis of the dimensions and form of the Trinil skull cap reveals all the characters of the Neanderthal type in a nascent or rudimentary form.*

The next year, another Englishman Dubois remembers from international conferences, W. H. L. Duckworth of Cambridge, follows Keith's lead and publishes a little book of his own called *Prehistoric Man*. Dubois is surprised; these books seem to be popping up everywhere, one after another, like tulips in spring. It is as if everyone finds it time to take stock of the evidence and compile a synthesis of what is known about human evolution. At least Duckworth seems to have listened carefully to Dubois' arguments about *P.e.* at the zoological congress in Cambridge in 1898:

Here we find a creature of Pliocene age, presenting a form so extraordinary as hardly to be considered human, placed so it seems between the human and simian tribes. It is Caliban, a missing link, —in fact a Pithecanthropus.

With the erect attitude and a stature surpassing that of many modern men were combined the heavy brows and narrow forehead of a flattened skull, containing little more than half of the weight of brain possessed by an average Englishman....

The arguments founded upon the joint consideration of the length of the thigh-bone and the capacity of the skull are of the highest interest.... The body-weight is asserted to be about 70 kg ... and the brain-weight about 750 gm, and the ratio of the two weights is approximately 1/94. The corresponding ratios for a large anthropoid ape (Orang-utan) and for man are given in the table following, thus:

Orang-utan	*1/183*
Pithecanthropus erectus	*1/94*
Man	*1/51*

The intermediate position of the Javanese fossil is clearly revealed.

Dubois finds the attention satisfying, especially as his colleagues seem to be coming around to his point of view. Late in 1912, only a week before Christmas, Arthur Smith Woodward, a fish paleontologist at the British Museum, announces an extraordinary new skull at a meeting of the Geological Society of London. Some credit is given to a solicitor, Charles Dawson, but it is quite clear that he is only the amateur antiquarian who stumbled upon the thing. Dubois does not know Dawson, but Smith Woodward he remembers well. Calling their new find *Eoanthropus dawsoni* (Dawson's Dawn Man), Dawson and Smith Woodward exhibit the back of a skull and a partial jaw with only a few teeth in place, excavated from a gravel pit near Piltdown, Sussex. Moreover, they confidently reconstruct its gently curving forehead, a surprisingly modern face, and a complete and some-what apelike dentition. It is a large-brained, ape-toothed wonder.

Dubois watches from the sidelines with some glee while debate ensues. Clearly the braincase of *Eoanthropus* is large, but exactly how large, and how it is to be reconstructed, become points of heated contention. It is like a replay of Dubois' trials. The face is generally agreed to be nearly vertical, quite modern in shape, with no trace of the heavy brow ridges that typify Neanderthals and *P.e.* But what puzzles the scientific community is the unexpected combination of a large brain and modern face with an apelike jaw. Grafton Elliot Smith seizes upon the fossil as proof of his theory that the growth of the brain came first in human evolution, but the shape of this fossil does not jibe well with the evidence of *P.e.* and Neanderthals. *Pithecanthropus* and Neanderthals clearly have humanlike teeth and ape-sized brains; Piltdown shows the opposite. Some paleontologists and anatomists question whether the bones come from one individual: mostly Americans take this approach, Dubois notices. It is just like what happened to me with *P.e.*, he thinks. Surprise them, and they say your evidence is faulty. Besides, each scientist wants to voice his own opinion, whether or not he has examined the fossils firsthand. An anthropologist at the Smithsonian with an unpronounceable Bohemian name, Aleš Hrdlička, even suggests the jaw comes from an ancient chimpanzee while the cranium is from an intrusive, modern burial. Hrdlička alone persists in saying plainly that this new Piltdown specimen cannot be right if all the bones of Neanderthals and *P.e.* are also right. The finding of an additional tooth, a canine, after further excavation seems to quiet the dissent, but murmurs of protest persist.

Keith now moves to the fore with a new reconstruction of Piltdown, challenging the work of Smith Woodward. Both parties churn out criticisms and corrections and reassessments of the course of human evolution in light of this new find. The controversy is so compelling that the debates soon enable Keith to establish himself as a major expert in human evolution. He forms a canny alliance with Marcellin Boule, the French paleontologist who tried to prove, in his monograph, that Neanderthals are not ancestral to man but belong on some primitive side branch of the human evolutionary tree.

Very interesting. Dubois reads and observes all this activity, but is not tempted back to his own fossils. He will not jump into another flaming fray like the one he barely survived over *P.e.* Indeed, he cannot understand why colleagues who have seen his specimens before—even Manouvrier and Haeckel—now write, asking to come and examine *P.e.* once again. Dubois is annoyed; his fossils have not changed, after all, and he dislikes having colleagues visit and interrupt his routine. Besides, he has never forgotten Schwalbe's perfidy; he would prefer that only the utterly trustworthy examine his fossils, and he cannot think of anyone he trusts so deeply. He receives letters and telegrams, as well, from some to whom he has more tenuous ties, such as Hrdlička. Even though he and Dubois have in common a warm friendship with Manouvrier, who trained Hrdlička at the École d'Anthropologie, Dubois will not make an exception.

Dubois' hesitancy notwithstanding, Hrdlička has determined to spend 1912 on a grand tour of Europe, examining all the ancient human fossils. Perhaps, he hopes, Dubois will change his mind. In Paris, Manouvrier warns him that Dubois has become something of a recluse in recent years. "I don't know," he says regretfully, "if my old friend will allow you to see his precious fossils. He has withdrawn from discussions of human evolution entirely. I no longer seem to have any influence with him."

Hrdlička resolves to try to see Dubois' fossils anyway. His determination is fueled by stories he hears in Liège, from Charles Fraipont, the son of the anatomist Julien Fraipont who described the Spy Neanderthals. Fraipont says that Dubois has become very strange indeed, his strongly held convictions turning and twisting inside his head into grotesque shapes. In fact, Fraipont fears Dubois may be going mad and might have destroyed the fossils. Hrdlička hears other wild speculations during his trip: that Dubois has been pressured by the Roman Catholic Church to put an end to "all this talk of descent from apes"; that Dubois has returned to Catholicism and is hiding away his own evidence because it challenges the biblical view of Creation. "His sister," one scientist says slyly to Hrdlička, with raised eyebrows, "is a nun."

Hrdlička does not believe half of what he is told, but even the remaining stories are alarming. It is terribly important that these fossils be saved and exhibited for study. He writes Dubois begging for permission to see them. He arrives in Amsterdam on June 20, 1912, and calls at the medical school of the university. Professor Dubois is not there, for he is a professor of geology, not anatomy.

The anthropologist Aleš Hrdlička is furious when he cannot see Dubois' fossils in 1912.

There is no point going around to geology, however, for Dubois is not at the university that day at all. He can be found at home, perhaps, in Haarlem. Hrdlička explains pleadingly that he has traveled all the way from America to see Dubois' fossils.

"I'll give you his address, *ja,*" his informant says, "but do not be angered if he will not see you. He is not in the habit of receiving visitors these days, especially those who want to see his fossils."

Hrdlička goes anyway, unable to believe that Dubois will turn a true colleague away. He finds the address, Zijlweg 77, Haarlem, without difficulty and rings the bell. When the maid, a rather young and pretty girl in a uniform, opens the door, she holds out a small silver plate for Hrdlička's calling card.

"Dr. Hrdlička to see Dr. Dubois,"Hrdlička says with dignity.

Without even looking at his card, the maid replies immediately, as if by rote, "Dr. Dubois is not at home."

"I am a fellow scientist," Hrdlička responds, with what he hopes is a reassuring smile. "I have come from America to see the professor and his famous fossils. I have written him of my plans to visit. Perhaps you would tell him I am here? I hope he will agree to see me."

"The doctor," repeats the maid more loudly, turning stony-faced, "is not at home." In an instant, her attitude has changed from friendly to dismissive. She stares at Hrdlička haughtily and he stares back, waiting. "Would you kindly remove yourself from the premises?" She flushes in embarrassment and then closes the door in Hrdlička's face.

At that very instant, Hrdlička's eye is drawn by a lacy white curtain twitching suddenly closed over the window on the floor above. Someone has been watching him. Hrdlička can now see a small round mirror there, carefully positioned so as to show a watcher on the second floor who is at the door. He is convinced that Dubois himself has observed the entire exchange. The arrangement, the deliberateness of mounting a mirror in such a place so that visitors can be turned away, staggers Hrdlička. Dubois must be mad indeed, he thinks. This is something out of a cheap novel! Unfamiliar as he is with the Netherlands, Hrdlička does not know that such mirrors are a common feature there. Rather than having a servant go "to see if the master or mistress is in," as is done elsewhere, the Dutch in Haarlem simply watch their own front doors and instruct the servants in advance.

Hrdlička does not know what to do. He stays for three hours, drinking coffee in a small café near the house that contains the man and the fossils he has come so far to see. He feels as if he is being watched, and it is true. From time to time, Dubois glances at him through the curtain to see what he is up to. Hrdlička stands and paces irritably on the sidewalk. From time to time, he stares hard at the very window behind which Dubois sits, concealed. It is as if Hrdlička is challenging Dubois, without words: "You know I am here; well, I know you are there, too." So, Dubois thinks, so. This is Hrdlička, the physical anthropologist at the Smithsonian, the man with the big, dark mustache, who questions the Piltdown fossils. What is he doing? Does he not understand the message the maid has given him? Dubois will not see him: that must be plain. Why does he remain?

For his part, Hrdlička is engaged in an active internal debate. He knows Dubois must be in the house; he is almost sure Dubois is watching from the upstairs window. Shall he ring again? Perhaps it would be too rude, too open a contradiction of the not-so-polite fiction that the doctor is not at home. But he wrote in advance, politely, respectfully, telling Dubois he would come today. How can he refuse to see him? Did the maid actually tell Dubois that Dr. Hrdlička is here? Perhaps it is a misunderstanding; perhaps he has been mistaken for someone else. Perhaps if he rings again, this time he will be admitted. Is Dubois testing his sincerity, his interest?

Frustrated, tired, and not a little annoyed, Hrdlička rings the bell again late in the afternoon. This time, it is answered by another woman, better-dressed, nice-looking, with dark hair streaked here and there with gray. It is Anna herself. "*Ja?*" she says neutrally, looking carefully at the stranger.

Hrdlička bows and removes his hat. "Do I have the honor of addressing Madame Dubois?" he asks hopefully. She nods uncertainly. He presents his card once again, holding it out to her. Before she can dismiss him, he declares in a rush, his words tumbling over one another, "I am Dr. Aleš Hrdlička, of the Smithsonian Institution in Washington, D.C., in America. I made an appointment with your husband, Dr. Dubois, to call today to see his famous *Pithecanthropus* fossils. I went first to the university, but they told me he was at home, so I came here. Your maid told me earlier that the doctor was out. Perhaps he will be able to see me now? I have come so far, and his specimens are so important."

"*Ja,* that may be so," concedes Anna with a small smile, "but my husband is not in a position to receive foreign visitors today."

"But—" Hrdlička interjects, desperately, pleadingly.

"Perhaps you come again tomorrow? Maybe he can see you then," Anna offers hesitantly, softening a little.

"Alas, no, I cannot. I must travel to Berlin tomorrow. I would be most grateful if the doctor could spare me a little time, I know he is a busy and important man. If he would just grant me an hour or so to examine the fossils . . ." Hrdlička replies earnestly.

"No, Dr. . . . . Urdluck?" answers Anna, looking at the card and mangling the Bohemian name. "No, I am sorry. The doctor cannot see you."

Hrdlička's temper flares. He is hungry and weary and offended. Really, he thinks, I am being treated like a tradesman with inferior goods to sell. He retrieves his calling card from Anna's hand and scribbles a note on the back, hurriedly: "The anthropologists of the world owe you a great deal, but it is a damned shame that it is not possible for scientific purposes to even glance at the specimens! Y.T., A.H."

He hands it back to Anna with the words, "Here, then. Please give this to Dr. Dubois, with my compliments!" Turning on his heel, he jams his hat angrily back into place and strides off.

Anna, shamefaced, closes the door. She carries the card up to Dubois and hands it to him without a word, embarrassed at the role she has been forced to play in all this. Dubois looks at the card, turns it over and reads the back, and then roars with laughter. "Ah, Anna, Anna!" he chokes out through the laughter. "Anna, look at this!"

"*Ja,*" she replies quietly. "I know. I read it. He was angry, very angry."

"Anna, you understand nothing, nothing," Dubois chastises her, his laughter dying away. "The man has passion, he has spirit. I like that. He cares about the fossils and not so much about himself."

"Then why not admit him and let him see the fossils?" asks Anna, sourly.

"I cannot, I cannot. But oh! I could like that man very much," Dubois replies regretfully. Anna leaves the room. Her husband's behavior makes no sense to her at all, and she no longer tries very hard to divine what is in his mind.

Had Hrdlička known of Dubois' response, it might have defused his anger. As it is, he is so furious at his treatment that he repeats the story over and over until it reaches the ears of a reporter at *Het Algemeen Indische Handelsblad,* a newspaper widely read in the Dutch Indies. Hrdlička has been treated scandalously over the matter of the Indies fossils, and he is pleased to tell the story to the newspaper. It would serve old Dubois right, Hrdlička thinks, if his behavior gets him in trouble. The printed story is flamboyantly critical of Dubois, part of whose salary is still paid by the Ministry of the Colonies.

Hrdlička is still indignant as he writes his monograph *The Most Ancient Skeletal Remains of Man,* which is published in 1914. He has seen nearly every fossil representative of an ancient human or human ancestor, except *Pithecanthropus.* All he can say of it is this:

> *On account of the peculiar circumstances an attempt to describe firsthand the important pieces under consideration met with serious difficulties. It would surely seem proper and desirable that specimens of such value to science should be freely accessible to well qualified investigators and that accurate casts be made available to scientific institutions, particularly after twenty years have elapsed since the discovery of the originals. Regrettably, however, all that has thus far been furnished to the scientific world is a cast of the skullcap, the commercial replicas of which yield measurements different from those reported taken off the original, and several not thoroughly satisfactory illustrations; no reproductions can be had of the femur and the teeth, and not only the study but even a view of the originals, which are still in the care of their discoverer, are denied to scientific men.*

Dubois reads this passage in Hrdlička's book, chuckles a little at the display of temper, then clucks his tongue and dismisses the incident from his mind. Ah, that Hrdlička is too excitable, he thinks, and he was unlucky. All these people reproach me for not showing *Pithecanthropus* to everyone. Dubois shakes his head wearily at the thought. Well, of course I did not run after people to show it to them. I had already had enough misery from those bones, when I carried them all over Europe.

That this is not quite the truth, Dubois does not care to remember. Like a man with scalded skin, he is hypersensitive, and it seems to him that the criticisms are gaining strength and volume. He does not have to hear the gossip firsthand to know what is being said.

They want his fossils; they all do, but they shall not have them.

CHAPTER 47 TRAGEDY

Despite his feeling of impending doom, Dubois cannot stir himself into action. His Indies fossils sit, unstudied, undescribed, gathering dust first in this basement, then in that warehouse, never in a place where scientific study is easy. Besides, his fascination has played itself out. Most of the fossils are neither numbered nor registered; there is no master list of specimens. While that tedious but essential task remains undone, he feels they are safe from the grasping hands of others. He hoards the fossils like gold bullion, but he does not involve himself with them.

As ever, his attention and brilliance are focused elsewhere. Some papers by the Frenchman Lapicque reawaken Dubois' interest in the question of brain size and body size. Dubois knows that something important is at stake, that some universal law or mathematical principle governs this aspect of mammalian anatomy. It is big enough to interest him, but too big to grasp yet. He knows it will bear on *P.e.* and her intermediate status, if indirectly. He begins gathering information with the same fervor he once applied to searching for fossils. He scours the literature, writes endless letters, begs specimens and information from museums, colleagues, acquaintances, and complete strangers around the world. Birds, reptiles, fishes, and mammals: he pursues their body weights and brain sizes like one possessed. He must know more, still more, before the grand pattern will become clear.

By 1914, he is ready to publish a significant article, one that reveals the grandeur of his vision. In it, he articulates the principle that becomes known as Dubois' Law:

> *In species of Vertebrates that are equal in organisation (systematically), in their modus of living and in shape, the weights of the brains are proportional to the 5/9 power of the weights of the bodies.*

It is a breathtakingly elegant proposition, a biological law that applies across all the vertebrate animals. If one compares animals of equal stages of evolutionary development—cats to tigers, for example—then there is an absolutely predictable mathematical relationship between brain size and body size. It is a magnificent, sweeping observation.

Within species, the story is somewhat different. Dubois' masses of data seem to suggest that the variability between individuals of one species obeys another law, based on an exponent of 2/9 of the body weight. But this, too, is fixed and predictable. It is a stunning concept: Nature works by laws, mathematical principles. There is reason behind the obvious diversity of animals in shape and size. The next question is, for Dubois, Why? Why should brain weight and body weight scale together in a regular fashion, both within and between species? What biological fact dictates that this be true?

He addresses this question in the second section of his paper. His idea is that this law, this regular relationship, actually reflects the number of "sensory-motor units" in the body. By "sensory-motor unit" he means a part of the body consisting of the sensory nerve fibers that perceive the outside world and the muscle or motor nerves that then prompt action. A sensory-motor unit is thus a single functional entity for perceiving and reacting to the outside world. While no one could possibly count the number of sensory-motor units in a body—the sheer task of dissection would require a microscopic precision far beyond what is possible—Dubois can measure something that he believes is closely related: the diameter of the eye. His logic goes like this. The eye is a sensory organ first and foremost; it is a supreme organ of perception that directs muscular or bodily responses to what is seen. The size of the eye thus reflects the size of the optic nerve that gathers and processes visual information. Dubois' insight is that the size of the eye also reflects the extent of the eye's perceptual and responsive power, and yet it can be conveniently measured.

As an animal increases in length (he uses the symbol "L" for this), its muscle mass must increase not by length but according to the cross-sectional area of those muscles (L^2), for everyone knows that a more powerful muscle may be longer or thicker or both. And, as the animal grows longer, its mass or weight must also increase—not in two dimensions (length times breadth) as an individual muscle does, but in three dimensions, symbolized by L^3. The longer an animal is, the greater its muscle mass and the greater its surface area, which means it must have more innervation and more sensory-motor units. More sensory-motor units means more brain, to process information and direct movement. How much more? Across species, as body size increases by length L, brain size increases by $L^{5/9}$. The chain of reasoning is intricate and complex, but also stunningly simple.

The third significant observation has to do with cephalization, the relative braininess of various species. In some species, brain size increases exactly as predicted by Dubois' law. In others, there seems to be an extra dollop of brain for the body size, a factor Dubois calls a cephalization quotient or coefficient. Why? Because some types of species have specializations and adaptations that require more brain per unit body size. As examples, he presents the highly sensitive trunk of the elephant; the prehensile tail of South American monkeys; and the unusually sharp hearing and large ears of hares. These specializations cause the elephant, the monkey, and the hare to have larger brains than would otherwise be expected. The elephant's trunk and the monkey's grasping tail are almost like extra limbs; these structures require more sensory-motor units to function—and hence more brain. The hare's sharper hearing, shown by its enlarged ears and quick reaction time, also requires a bigger brain.

Dubois is very proud of this remarkable new synthesis of ideas and information, but the reaction to the article is less enthusiastic than he hoped. Those scientists already intrigued by the possible relationship between brain size and body size notice and respond favorably, but few others seem to appreciate the magnitude of the problem or the beauty of the answer. Through all his calculations and computations, Dubois can glimpse the vague shape of something compelling, something terribly fundamental and important. No one else seems to care. Never mind, he consoles himself. Someday they will understand; someday they will all share my vision.

He carries out more research, collects more measurements, recalculates and rethinks. More and more articles come out, fully twenty-seven between 1911 and 1920, and over half of these concern brain size and body size. It is work he can do in the library or museum, despite the war that absorbs the attention of most of Europe between 1914 and 1918; though the Netherlands has declared itself a neutral country, the war rages all around it. Tired of his own battles, Dubois publishes almost nothing on human evolution in those years. Others are far more interested in Dubois' fossils, and far more concerned at his lack of visible research on them, than he is himself. The fossils are like a nagging injury, a dull ache that never quite goes away but that cannot be alleviated by any remedy he can imagine. He should do something; he cannot be bothered to do something; he cannot see what there is of real interest left to do.

His sense of uselessness is accentuated further when Eugénie, the one child of whom he is really proud, announces her intention to marry a man twenty years her senior, Carel Hooijer. Dubois has met Hooijer and does not think him worthy of his clever, beautiful Eugénie. Anna is concerned about the difference in their ages and perhaps suspects that her daughter is in some way trying to marry her own father. Their disapproval does not sway their daughter. When Anna and Dubois threaten to fail to attend the wedding, Eugénie picks a day on which there is likely to be little newspaper coverage of the event so she can avoid public embarrassment. She and Carel Hooijer marry quietly on March 11, 1915, but there is no lasting happiness. Before many years pass, Eugénie decides that her parents were right and Carel is not smart enough or energetic enough. Rather than divorce, she simply ignores Carel whenever possible for the rest of his life.

Events arising from the 1914 meeting of the British Association for the

Advancement of Science in Sydney, Australia, eventually revive Dubois' interest in anthropology. Dubois himself was not present; few European-based scientists were, the notable exception being Grafton Elliot Smith, a native-born Australian and then president of the Anthropological Section of the British Association for the Advancement of Science. Like the other participants, Smith was startled by the presentation of a virtually complete (if somewhat crushed) human skull of Pleistocene age, which had been found in Talgai, Queensland, as long ago as 1884. It is surely the oldest evidence of the existence of man on the Australian continent, so why had no one heard of this specimen? Questions abounded, answers were scarce. The newspapers picked up on the story and, before the meeting was over, the Talgai skull was famous worldwide. Smith's brother, Dr. Stewart Arthur Smith, was appointed to describe, investigate, and analyze the specimen. His monograph on the oldest Australian is published in 1918, the year the Great War ends. Smith finds the Talgai skull to be large and robust, like those of living Australian Aborigines. It has heavy brow ridges, a prognathous, forward-thrust face, and a broad palate with rather large canine teeth for a human, though the canines are more modest than is typical of even female apes. Despite its primitive facial features, the cranial capacity of the Talgai skull is estimated at 1,300 cc, notably larger than the average for modern Aborigines. Smith concludes that the skull represents an ancestor of the Australian Aborigines, showing a surprising decrease or deterioration in brain size in Aborigines over time.

Smith's publication on the Talgai skull gives Dubois something to say, for his Wadjak skulls are also primitive Australian types, according to his re-appraisal of some ten years before. Convinced that the Wadjak skulls make better ancestors to Aborigines than this Talgai fossil, Dubois reenters the anthropological realm to say so. He is back, once again touting one of his fossils as the most primitive. Before the end of 1920, Dubois gives a series of four presentations to the Royal Academy of Science in Amsterdam (published in both Dutch and English) on the Wadjak fossils and their significance. There is some irony in this entire chain of events, for the Wadjak skulls have been undescribed for nearly as long as the Talgai skull. Now Dubois gives his skulls a new name, *Homo wadjakensis,* and declares this type ancestral to both modern Tasmanians and modern Australian Aborigines.

As for the Talgai skull, in Dubois' view it is no closer to the common ancestor of modern mankind than Australian Aborigines of the present time. Talgai is simply a somewhat old and primitive Aborigine, and Aborigines are the most primitive and lowest race of man known. Even Wadjak is not "a distinctly lower type than the Australian of the present time," Dubois writes, "for this ancestor had reached the same stage in the evolutionary scale as the living race, at least almost." What, then, accounts for the anatomical differences between the modern skulls and those from Wadjak? Dubois proposes a new and interesting hypothesis.

> *The differences may nearly all be attributed to the more vigorous development and GREATER PERFECTION OF THE TYPE, in surroundings more favorable than those in which the Australian native finds, and has found for a long time, a scanty subsistence.* Homo wadjakensis *was an optimal form. In the present race the type is evidently in a state of decadence, as also* Homo neanderthalensis *is the less vigorous and less perfect descendant of* Homo heidelbergensis.

In other words, the skulls of Wadjak and modern Aborigines differ because they live (or lived) in different habitats. Like other creatures, he argues, humans evolve and adapt to their environments. He extends this hypothesis to account for the shape of Neanderthal skulls as well. Wadjak men survived by hunting and fishing, giving their jaws and teeth almost carnivore-like features, while Neanderthals ate tough vegetable food that they ground up with powerful teeth and jaws more like those of a gorilla than a carnivore. The resemblance to apes in Neanderthal skulls can only be explained as a functional analogy, Dubois suggests.

Talgai has reengaged Dubois with the fossil record of human evolution. He publishes on Neanderthals, more on Wadjak, and more on *P.e.* He is rejuvenated, alive again and full of ideas. He is ready to apply his brain-weight and body-size work to *P.e.,* to show she is the missing link he has always claimed.

He is distracted from this ambition, though, by bad news from the Indies. The plantations where Jean and Victor have been working have gone bankrupt, and the owners are ruined. Was the cause his sons' mismanagement? Dubois never knows. A lot of plantations close down at that time, for prices are very low and the world economy is depressed. Anna hopes that now her sons will return to Holland, settle down, and marry and produce grandchildren that she can visit and indulge. The "boys"—boys no longer, but grown men—decide to return to Holland via America to follow some other crazy dream. They are young and charming; they meet two American girls and marry. They settle down and begin to succeed modestly at this and that in the new country.

Jean initially works managing an export company in San Francisco. By the time a few years have passed, however, he is billing himself as a photographer of big game, a naturalist, and an explorer, giving lectures and entertaining talks in various cities around the States. Dubois is amazed that anyone would listen to him. What does Jean know of animals, natural history? He cannot tell a magpie from a mockingbird, he cannot remember the name of a plant from one day to the next. His father is disgusted with Jean, and even a little bitter. He tried so hard to teach his sons about natural history and they resisted at every turn. And now, now Jean thinks he is some kind of expert. It is nonsense.

Victor's choice of profession is even worse. Victor, the son who would never listen to his father, who never studied a bone in his life, who never learned an ounce of geology, has the audacity to give lectures about the missing link. Dubois is scandalized. Some of those foolish people who pay to hear Victor might even think that this is the discoverer of *Pe*. But Victor was only a few months old when the tooth and the skullcap were found, not yet two years old when the femur came to light. He cannot possibly remember anything about that time and place, or even the scientific principles by which his father interpreted the fossils. He cannot even remember the campaign his father waged, traveling all over Europe, to win scientific opinion to his side. It pains Dubois that his son trades on his good name and hard-won accomplishments this way. Why does he do it? All he can conclude is that Victor has no moral character.

Even in adulthood, Victor is the one whom Anna loves the best, the younger son still closest to her heart. She thinks she can forgive him anything, but she finds to her sorrow it is not so. What she cannot forgive him is his thoughtlessness in dying young. In 1922, at age thirty-one, Victor dies of tuberculosis, far away from home, in America.

Though she has recovered from some hard blows in her life, Anna's heart and mind are shattered irreparably by Victor's death, like a vase knocked to the marble floor by a boy's careless elbow. Even that bad, bad business in Java in 1893 was not so bad as this. Anna is confused, bewildered, furious. She even accuses Dubois of killing his son, of sending him off once more to the colonies to die. "You never loved that boy as you ought to," she says bitterly. "No wonder he ran away from you, to his death. He tried so hard to please you, but no one ever can. No one is good enough for you: not me, your son, your students, or your colleagues. You never loved him because he wasn't clever enough for you." Tears pour down her face and she cannot speak any longer, only sits, sobbing, in a chair by the window.

Dubois is at a loss. Maybe it is true, he thinks guiltily. Maybe I should have loved more, expected less. Did I drive Victor away, and Jean? I don't know. I suppose now Jean will never come home again, not now that he has married an American and settled there. But could I let my sons—*my sons*—live by lax standards? Could I let them float aimlessly through life, without any passion, any sense of conviction? It was my duty to try to instill some character in them, some backbone. "Recte et fortiter," that is our family motto. Was I wrong? He does not know.

Anna has found her voice again, and the accusations continue to flow. "You only use other people, you never value them. Everything must give way for you and your precious Science, even me, even the children. You have driven my children away and now you humiliate me with that hussy, that so-called housemaid from Limburg you have installed here." She looks up at him, her face filled with anger and grief. "I have seen how you look at Claartje, you dirty old man! I have seen it. Oh, yes, she is young and pretty and knows nothing except that you are the rich professor who employs her. Well, don't think you can carry on like some Indies colonial with a little nyai to keep his bed warm. I won't stand for it!"

Dubois is completely taken aback. He has admired the new housemaid's firm young figure and glossy hair, and he likes to hear the Limburg accent in her voice. He has always appreciated a lovely woman. It is true, he has entertained thoughts, from time to time . . . but he has done little, just a pat on her bottom from time to time, a simple touch. He has not seen his behavior as being like that of their Indisch friends, with their native concubines and half-caste children. Is Anna right? Does she see the sin in his soul, the sin he has been concealing even from himself? He can make no reply to her and he cannot examine his soul in front of her. He is too proud to defend himself, too honest to maintain his innocence. He leaves the room abruptly, without a word. Henceforth, they inhabit the house in chilly silence. No more than is necessary is said between them, barely enough to maintain a veil of civility, when he is in the Haarlem house. They have been on distant terms for years, but this is the end of everything. Whenever possible, he retreats to De Bedelaar.

He still has ideas, research to do, plans and dreams, but Victor's death makes him feel his age. He is no longer as quick as he once was, nor has he the stomach for the cruel scientific infighting of the past. He does not have so many brilliant ideas anymore, either. On damp days, when his body aches and creaks, Dubois mourns his lost youth, his lost family, his lost dreams. It is a sad business, this marrying and having children. They break your heart; they always break your heart. He would have been better off if he had stuck to science and never married, gone to the Indies alone. Dubois, the stern paterfamilias, the man who always knows where he is going, is wounded beyond all expectations. He feels weakened, as if the foundation has been swept out from beneath his feet. This is impossible, unbearable: a son dying before his father is surely against all the laws of nature. Yet the laws of nature are what Dubois clings to, as a drowning man hugs a life preserver. He has done so all his life. It is only the laws of nature that give him any chance of making sense of the vast, chaotic world. Now, in one wretched instant, those laws have been violated. He does not know what to do, so he does nothing.

CHAPTER 48 DANGEROUS TIMES

As Dubois has always feared, forces are gathering against him. The powerful men who would wrench *P.e.* from his grasp choose this moment to strike, sensing the time is right and he—now an old warrior—is weakened.

The attack is orchestrated by Henry Fairfield Osborn, an American vertebrate paleontologist who is about to assume leadership of one of the great museums of the world, the American Museum of Natural History in New York. In 1915, Osborn publishes his own definitive account of human evolution, a nearly six-hundred-page tome entitled *Men of the Old Stone Age*. Osborn is somewhat critical of

Henry Fairfield Osborn, of the American Museum of Natural History, mounts an international protest against Dubois for sequestering the P.e. fossils.

Dubois, referring to the "scattered and scanty materials [Dubois] collected" and emphasizing the resemblance of *P.e.* to a Neanderthal—a view with which Dubois disagrees completely. Osborn also uses the book to propagate his pet theory, that the origin of man and of many other mammalian species lies in Asia. For Osborn, "Asia is the mother of all continents." He garners support from Marcellin Boule, among others, for Boule is ever anxious to banish the brutish Neanderthals from direct human ancestry. They expect a bigger-brained, more human sort of ancestor evolved in the East and will be found there. Osborn declares boldly,

> *It is possible that within the next decade one or more of the Tertiary ancestors of man may be discovered in northern India among the foot-hills known as the Siwaliks. Such discoveries have been heralded, but none thus far been actually made. Yet Asia will probably prove to be the center of the human race. We have now discovered in southern Asia primitive representatives of relatives of the four existing types of anthropoid apes . . . and since the extinct Indian types are related to those of Africa and of Europe, it appears probable that southern Asia is near the center of the evolution of the higher primates and that we may look there for the ancestors not only of prehuman stages like the Trinil race but of the higher and truly human types.*

This argument differs only slightly in logic and evidence from the one Dubois published back in 1888, when he first argued that the Indies were the most likely home of the missing link. But Osborn dismisses *P.e.* as too apelike for his purposes. Osborn longs to find the first man, not an apeman.

Determined to put his theory to the test, Osborn persuades the American Museum of Natural History to mount the Central Asiatic Expedition starting in February 1921. The leader of the explorations is the dashing zoologist Roy Chapman Andrews, who with his handsome face and aristocratic pince-nez becomes a celebrity almost overnight. Dubbed the "Missing Link Expedition" by the press, this adventure is funded by donations from the public that pour into the museum. It is nothing like Dubois' save in intent. Dubois marched on foot, with a few horses, two civil engineers, and fifty convict laborers or coolies. Andrews has modern vehicles (five Dodge cars and two one-ton trucks), plus spare parts, gallons of gasoline and oil, and trained mechanics; he has the latest equipment for photography, surveying, and mapping, as well as eighteen tons of food, tents, sleeping bags, cooking gear, and camp furniture which are carried by seventy-five camels. The admitted cost of the expedition for five years is $250,000, not counting donations in kind; in the end, the bill is more nearly $600,000.

In 1922, at the start of the expedition, Dubois and Osborn exchange letters. The first concerns a privately printed paper on Pleistocene elephants by a Dr. Hay, which Dubois arranges to have sent to Osborn because of Osborn's interest in proboscideans. Osborn ungraciously returns the work to its author, on the grounds that it is not properly published. He explains to Dubois, haughtily,

August 1, 1922

A privately printed and distributed list [of species] is an innovation which, if imitated in other countries and in other languages, or in other localities in this country, will arrest the progress of Palaeontology at the present time and make the science absolutely impossible in the future.

American palaeontologists are now in a position of leadership wherein they must set an example in all matters of authorship and of procedure.

It is an arrogant, self-righteous response and Dubois is embarrassed on Hay's behalf. Dubois has no inkling that he is about to suffer still worse insults, triggered by an incident concerning J. H. McGregor, a former student of Osborn's, who wrote to Dubois in the summer of 1921 asking if he might come to Amsterdam to see the *Pithecanthropus* and Wadjak fossils. McGregor also wanted to obtain casts of *Pithecanthropus* for the forthcoming Age of Man Hall to be erected in the American Museum of Natural History. Dubois declined, saying that the time proposed for the visit was not convenient. Now, a year later, Osborn writes, asking again whether McGregor might come to Holland to study and make casts of *P.e.*, and Dubois refuses once again. Osborn waits a few months and writes another time, explaining that McGregor has already completed a series of sculptures of *P.e.* for him for his book and longs to do something more accurate based on a firsthand inspection of the fossils.

Dubois is seriously offended. First of all, the work has already been done, by him, for "Piet" was sculpted for the Paris Exhibition twenty-two years ago. Second, he has already refused access and it is rude to press him. Letting another anatomist make casts of his material is the same as giving the fossils away! How could any professional ask such a thing? Besides, Dubois is hard at work on the fossils now himself. Surely he has the God-given right to first access to his own fossils. Besides, he has no assistant or technician who might make casts for distribution to other scholars, even if he were so inclined. Further, Osborn has embarrassed Dubois by enlisting the aid of Professor Hendrik Antoon Lorentz, a Dutch Nobel laureate, to plead his case. Dubois telegraphs Osborn baldly, "As Professor Lorentz on your request wrote to you October in my name, remains of *Pithecanthropus* which I am now describing inaccessible in Haarlem. Please wire your intention."

Osborn takes the refusal very badly. He considers it insufferable pigheadedness on Dubois' part. The man has had those fossils out of the ground for over twenty years! How long does he think he can sequester them away? Something must be done. As president of the vast American Museum of Natural History, Osborn understands about politics and power. The way to influence this stubborn Dutchman is to bring pressure to bear from his source of support. Osborn finds it simple to rally international opinion, for Hrdlička's well-known and similar story has already done some of the work for him.

Osborn addresses a letter to Dr. Louis Bolk, the Secretary of the Royal Academy of Science of the Netherlands, not knowing that Bolk and Dubois have been at odds for years.

December 2, 1922

We address your honorable Academy on behalf of the naturalists, and especially the anthropologists, of America, to express a universal desire that the specimens of Pithecanthropus *found by Doctor Eugène Dubois be placed in some institution in Holland where they may be examined and studied by properly qualified experts and students in anthropology and in anatomy. These priceless objects are of world-wide interest. According to an immemorial custom of both investigators and learned societies, such objects are placed where they may be examined with all possible precautions for their protection and preservation.*

We write after having addressed Doctor Dubois himself on this subject and after having communicated with the Minister of the Legation. We have also had the pleasure of discussing this important matter with a distinguished visitor from the Netherlands to the United States, Professor Lorentz.

We trust that in considering this matter the Royal Academy of Science will take into consideration the fact that the American Museum of Natural History is extending to anthropologists and anatomists from all parts of

*the world the opportunity of examining and studying its unrivaled collec-
tions in anthropology, anatomy and vertebrate paleontology. We regard
these precious objects preserved by nature as the natural property of the
scientific world, to be easily accessible to all investigators.*

The first, flamboyant signature on this letter is that of its instigator, Henry
Fairfield Osborn. His name is followed by those of as many colleagues as he
can induce to cosign.

The letter from Osborn to Bolk starts a conflagration within Dutch science. On January 3, Bolk writes sternly to Dubois, like a headmaster calling
a naughty student into his study. As the Secretary of the Royal Academy of
Science, and a member of the board, Bolk would like to talk to Dubois about
this matter before it is brought up formally at the next meeting. He will call
on Dubois at the university. At the appointed time, Bolk clumps stiffly into
Dubois' office, using two canes. In 1918, he lost his right leg to cancer and his
prosthesis has never been entirely satisfactory. Walking and standing are
painful, so he has perfected the art of dignified slowness. He has dressed
with especial care, wearing his royal decoration in his lapel to remind
Dubois of the august body he represents. He finds Dubois not humble but

Dubois' old enemy Louis Bolk reprimands Dubois for his uncollegial behavior.

already resentful at being treated like a child. Worst of all, it is happening in front of his staff and students.

"Miss Schreuder." Dubois addresses his assistant formally. She does not immediately answer, for her eyes are big with curiosity at the sudden appearance of the Secretary of the Royal Academy of Science. He addresses her a trifle more sharply the second time. "Miss Schreuder!"

"*Ja*, Professor Dubois?"

"Would you and the others kindly leave us? I must speak with Dr. Bolk privately on matters of some importance."

"Oh, *ja*, Professor, of course," she says hurriedly, herding the others out of the room like a mother hen chasing her chicks. As the door closes behind them, Dubois can hear one of the students ask earnestly—he does not know who speaks—"Antje, what is going on? Why is Bolk here? Is the professor in some sort of trouble?"

Dubois goes rigid with anger when he hears this. His students and underlings are gossiping about him! He turns to Bolk, determined to fight for his honor.

Bolk's air of righteous superiority is infuriating. He feels that Dubois has altogether too high an opinion of himself, just because he found a few old bones, and his time has come. "Professor Dr. Dubois," he says to open the discussion, planting himself heavily in the most comfortable chair, the one behind Dubois' massive desk, "let us behave like gentlemen. I am forced to call on you like this because I have had a disturbing complaint about your behavior from scientific colleagues abroad."

"*Ja?*" queries Dubois, as innocently as he can with his jaw clenched. By sitting in Dubois' chair, Bolk has transformed Dubois into a visitor, a mere transient in his own office. "And what is that, Dr. Bolk?"

Bolk pulls a thick envelope from his inner pocket and taps it meaningfully with his spectacles. "They say," he replies, "that you are refusing colleagues access to the *P.e.* fossils."

"Who makes these charges?" challenges Dubois, reaching out his hand for the envelope. "May I see that letter, then?"

"No," answers Bolk, hastily moving it out of reach and looking as if he might want to slap Dubois' hand. "No, you may not. I do not think that is necessary."

Dubois is indignant. He may not see the charges against him? "But if the letter concerns me," he splutters, "I think I have a right, Dr. Bolk—"

Bolk cuts him off firmly: "You have no right to see my correspondence."

He replaces the envelope carefully in the inner pocket of his jacket and pats his chest as if to settle the document into place. "None at all. And you know we have heard this complaint before; there was that incident with that fellow from the Smithsonian some years ago."

"Hrdlička," supplies Dubois, widening his eyes, which have turned a dangerously cold icy shade.

"*Ja*, that is the name. His complaint about your behavior was in *Het Algemeen Indische Handelsblad* for all the world to read. So this"—Bolk pats his jacket pocket irritatingly—"this is not the first time."

"Dr. Bolk, what you have heard is exaggerated," replies Dubois proudly, straightening his broad, muscular shoulders and spreading his strong hands apart in a gesture of innocence. "Hrdlička was simply unlucky. I know he made a lot of noise about things, but it was only that he went to Amsterdam while I was in Haarlem. They sent word to me he was in Amsterdam, so I left Haarlem, but he left Amsterdam to come to Haarlem, so we missed each other."

"Such goings-on give Dutch science a bad name," scolds Bolk. "Something must be done." He leans forward and stares directly into Dubois' eyes, daring him to make an apology or offer a solution.

But Dubois has other ideas. "Dr. Bolk, who discovered these fossils?" Dubois frames the question sharply.

"You did," Bolk concedes.

"*Ja*, that is right. And who realized that the missing link had to be in the Indies in the first place?" demands Dubois. Bolk opens his mouth to answer, but Dubois continues relentlessly without waiting for a reply. "And who risked his life and his health to recover these fossils?" Bolks nods in Dubois' direction but, again, Dubois leaves him no room to speak. Dubois' voice is growing stronger and his eyes are piercing. Now he rises from his chair, standing above Bolk and staring down at him. "Who found *Pithecanthropus?* Who found four hundred and fourteen crates of mammalian fossils that weighed nearly thirty-seven thousand kilograms? *Ja*, Dr. Bolk, that is right. It was I." Dubois places his hand over his heart. "All of it was my work, mine alone. So now you tell me, sir, Mr. Secretary of the Royal Academy of Science, who do you suppose has the right to the fossils of the Dubois Collection?"

Bolk is speechless, flattened by the hammer of Dubois' anger and indignation. He can think of no reply to make in the face of Dubois' emotion.

"Then that is settled," concludes Dubois, dismissively. He turns on his heel to exit the room. Stopping at the doorway, he asks one last question: "But tell me this: why should I give my fossils away to men who did nothing for them?" And then he is gone.

Bolk stares after the silhouette of Dubois, who walks swiftly out of the door and down the corridor. The Secretary's fury grows larger by the second. The arrogance of the man is astounding. Why, he has walked out on him! Somehow this interview got totally out of control. Bolk had intended to reprimand Dubois firmly, to make him see that he must behave better, more professionally. Instead, Bolk feels as if he has been ground to dust beneath Dubois' sturdy heel, like some insect. Now he must struggle to his feet and make his slow way past the gawking students to leave the building. It is a sensation he does not care for.

A few days later, at the Academy's regularly scheduled meeting, Bolk circulates copies of Osborn's letter and raises insidious questions about it. "There are four issues in this matter which I think we must consider," he intones magisterially.

> First: Is the Academy competent to take note of this request officially and to deal with this matter?
>
> Second: Is the Academy the proper organization to bring this matter to a favorable conclusion, or must the petitioners be referred to the Dutch government?
>
> Third: Is the request of the petitioners considered to be reasonable?
>
> Fourth: In the case of an affirmative answer to the third question, how can a satisfactory answer be given to their request?

After brief discussion, the members of the board of the Academy are unanimous on the first three issues. They are the highest representatives of the scientific life of the Netherlands, so of course they are the competent and proper body to take action on this matter, which concerns the entire scientific world. Clearly the Dutch government technically owns the Dubois Collection, having paid Dubois and having financed the shipping home of the 414 crates of fossils and their storage since 1895. Still, the Academy can see no need to refer the petitioners who have signed this letter directly to the government, which would only engender hostility. They already know Dubois is possessive about his fossils, rather too much so, but publicly humiliating him will not help matters. They choose to remind him of his duty and to hope for his sensible cooperation.

> *The unwinding of this tangled affair is made easier by the fact that Mr. Dubois, under whose care the objects in question are to be found, is a member of our Academy and there is no need to invoke a government censure, which would be contrary to the notion of collegiality. The Board is grateful for Mr. Dubois' membership in our Academy and feels that a satisfactory resolution to the problem may be brought about with the cooperation of Mr. Dubois and the advice of this Board.*

There is to be no threat to Dubois' position as curator, so he has escaped the worst. However, the board must be satisfied that Dubois will make the fossils and casts of them accessible to fellow scientists within a reasonable period of time.

The day after he receives the report of the meeting, Dubois sends a reply to Bolk.

> *I ask the Board to clarify when the casts are to be placed at the disposal of scientists. It is my explicit wish that the casts not be made available to interested parties until I have had the opportunity to publish in the Academy's journal the description and complete figures of the fossils in question. I have promised to make as much haste as possible in writing these descriptions, so that the first part (dealing with the skullcap, the endocranium or inside of the skull, and maybe the teeth) may be presented at the May meeting of the Academy. The casts of those parts can be made available immediately after this description appears. I will undertake to present the description of the remaining objects (the femur and the teeth, if not dealt with before) before the end of the year, with the casts likewise being made available after the description appears.*

True to his word, Dubois begins to work like one possessed. During 1923, he publishes more detailed works on the skull, brain, and teeth of *P.e.*, as well as the fragmentary jaw found at Kedoeng Broebus which he now feels is also *Pithecanthropus*. His works are not long monographs, heavily illustrated, but they are something. He is able to write to the Royal Academy on June 29 that the cast of the outside of the skull of *P.e.* is made and now drying, so that it—along with the endocranial cast of the brain impressions on the inside of the skull—can be reproduced and distributed. Dubois is not easily satisfied, so many casts are destroyed, until shards of plaster litter his workroom, which is now in a house formerly occupied by a section of the

Ethnographical Museum. He will not let anything that is less than perfect out of his hands and into those of another scientist. Casts are duly made and sent to the Smithsonian and the American Museum of Natural History.

To pacify the Royal Academy, Dubois invites Hrdlička and a group of his students who are in Europe to inspect the original *P.e.* fossils. It is perhaps a sort of apology to Hrdlička, who has increasingly come to support Dubois' position on *P.e.* In a lecture at the American University in Washington, D.C., in 1920, Hrdlička describes *Pithecanthropus* as "a creature that stood at the threshold of humanity." He goes even further:

If science was to construct an advanced precursor of man it could hardly do better than what nature has given us in this specimen. . . . Its femur, though primitive in some important respects, indicates a perfect biped posture, and a fair human-like stature . . . the two molar teeth show primitive human or subhuman characters. The skullcap is in nearly every respect about midway between such a skull as that of a chimpanzee and a human. The brain, as seen from the cast of the cranial cavity, is in size intermediary between that of a high ape and human, but in conformation of the frontal lobes and some other parts approaches closer to the human. . . . Taking everything into consideration, the Pithecanthropus *is just about what its name indicates, namely an ape-man.*

In 1923, Hrdlička and the students come to Haarlem, to the Teyler Museum, where Dubois himself opens the safe he has just persuaded the trustees to buy and extracts the precious remains for his visitors. He spends a full hour explaining and exhibiting the fossils to his fascinated audience. It is the first time Dubois has permitted anyone to examine the originals in thirty years. Hrdlička is not insensible of the honor and he and Dubois like each other on this meeting. Having so widely spread the story of Dubois' refusal to see him in 1912, Hrdlička deliberately sets out to make amends. He speaks to the American newspapers, which publish statements like this: "*Pithecanthropus erectus,* the Java ape-man, the world's most famous prehistoric creature, has come out of retirement." More to the point, in Hrdlička's next book, *The Skeletal Remains of Early Man,* he includes a gracious passage about Dubois and his fossils.

We found Professor Dubois a big-bodied, big-hearted man who received us with cordial simplicity. He had all the specimens in his possession brought out from the strong boxes in which they are kept

and demonstrated them to us personally and then permitted
me to handle them to my satisfaction. . . . The examination was in
many ways a revelation. When Dr. Dubois publishes his detailed
study, which he tells me to expect before the end of the year,
Pithecanthropus erectus *will assume an even weightier place in*
science than it has held up till now. None of the published illustra-
tions or the casts now in various institutions are accurate.
Especially this is true of the teeth and the thigh bone. The new
braincast is very close to human. The femur is without question
human.

The next to visit is J. H. McGregor, Osborn's former student. Dubois is cordial but does not like the man much, for he seems full of questions and challenges. Dubois answers them easily, of course, for he has been thinking about *P.e.* for years upon years; there is nothing McGregor can come up with that he has not considered thoroughly, as quickly becomes evident. McGregor's attitude toward Dubois grows hourly more deferential as he realizes how badly rumor has slandered this brilliant man. But Dubois never warms to McGregor, for he cannot help but hold the younger man partially responsible for Osborn's move against him.

In 1924, Dubois publishes more about *P.e.* in both Dutch and in English, with superb illustrations of the fossils from all views, as well as complex, dense papers about brain size and body size in mammals. Unfortunately, it is not long before he hears from Osborn once again:

I am awaiting with great impatience the publication of your promised
account of Pithecanthropus *and the receipt of additional casts, etc., which*
you have so kindly offered to send us. Your Pithecanthropus *has been*
given an entire case of honor in our great HALL OF THE AGE OF MAN
which is nearing completion after ten years of assembling of materials
from all parts of the world. There is intense interest in this subject at pres-
ent in the United States and thousands of people are visiting our AGE OF
MAN HALL. I am therefore daily awaiting a copy of your Memoir so that
the Memoir itself can be placed in the case with the casts when they arrive.

As requested, Dubois painstakingly makes casts of the *Pithecanthropus* fossils for Osborn's exhibit and resentfully delivers them to the Royal Academy, that they may certify he has complied with Osborn's request. Soon Dubois receives a copy of a letter Osborn sends to Bolk.

I am very glad indeed to inform you, in reply to your letter of February thirteenth, which I have just received, that a week ago the precious casts given by Doctor Eugène Dubois arrived here in excellent condition, accompanied by the cards which Doctor Dubois has so kindly inscribed. . . .

I have never had a moment's doubt that the delay was not either on your part or on that of Doctor Dubois. Now that the specimens have safely arrived I trust we may both forget as soon as possible this regrettable incident. . . .

In acknowledgment of this gift from Doctor Dubois and of the courteous manner in which it has been made, as well as in recognition of the great importance of his scientific discovery, I shall propose to our Trustees at the coming meeting in the month of May that his name be placed in the roll of Honorary Fellows of this institution, the highest scientific honor in our power to bestow.

Will you be good enough to ascertain from Doctor Dubois if we may feel at liberty to use his materials for our own scientific description, for comparisons with other materials in our large collection of duplicates? Also, if one may purchase from him another set of these casts, very carefully colored after the originals at our expense by an artist working under Doctor Dubois' direction. This colored set will conform with all our other duplicates received from various parts of the Old World which are colored after the originals.

Dubois cannot believe his eyes. Not satisfied with a single set of the casts, now Osborn wants a second one, hand-painted to resemble the originals. Where does Osborn suppose Dubois will find the time and money to engage and supervise an artist to paint them? Obviously, the promise of an honorary fellowship is meant to soften the request, but Dubois has no illusions that this is anything but a spoonful of sugar to sweeten the sour taste of yet another command from the great Henry Fairfield Osborn. Acting upon advice from Grafton Elliot Smith, Dubois arranges for Messrs. R. F. Damon & Company in London to make and sell both plain and accurately colored casts of *Pithecanthropus*, in return for a royalty on each sale. Perhaps this will deflect these incessant requests for casts from Osborn. Thus Dubois is able to reply to Osborn in the most cooperative of tones.

You may make all the use you like of the materials for scientific description and comparison. I only request that publications on the femur should be deferred until 1926.

Having transferred the right to reproduce and sell casts of the Pithecan-thropus *fossils to Messrs. R. F. Damon & Co., 26, Conclurry Street, London, I cannot directly supply the colored casts you desire, but I have made arrangements that you may obtain a set, very carefully colored after the originals by an artist working under my directions.*

There, thinks Dubois, signing the letter with a flourish that nonetheless leaves his signature only half the size of Osborn's. He smiles to himself, not in the least displeased that he will doubtless become an Honorary Fellow of the American Museum of Natural History.

CHAPTER 49 A NEW SKULL

Now that the nasty business with Osborn is behind him, Dubois' interest and energy revive until his research chugs along like a well-stoked steam engine. He publishes on glaciers and the "paleothermal problem" and its bearing on stellar evolution; he writes about brain size in specialized mammalian genera; he describes and produces new illustrations of the brain, skull, teeth, and femur of *Pithecanthropus* and reassesses her place in evolutionary history; he describes a new pangolin, *Manis palaejavanica*, from Trinil. In this new frenzy of research and writing, he is too preoccupied in late 1925 to supervise yet another move of the Dubois Collection, this time to the unheated attic of a not-yet-completed university hospital. In his absence, the entire affair is rushed and disorganized. Labels and specimens are misplaced or transferred from box to box, and some specimens are broken.

On September 27, 1926, the Batavia news agency ANETA sends messages throughout the world announcing that a new and complete skull of *Pithecanthropus erectus* has been found near Trinil by Dr. C. E. J. Heberlein, district government physician at Soerabaja. Reporters from all over Europe contact Dubois, asking for his opinion of the sensational new find, but he is in the embarrassing position of knowing no more about it than they. On October 2, Heberlein issues a second statement amending the first: "It is not correct to speak of a complete *Pithecanthropus* skull, but only of the front part of a human skullcap, from the apex to the line of the orbital arches, which projected slightly more than in an ordinary human skull. Of the bone matter, only a thin layer has been preserved on the spongious rock mass of volcanic origin." Apparently, in ancient times, volcanic ash or tuff filled the empty cranial vault, adhering to the remaining bone of the skull, which then became fossilized.

That same day, Dubois cables Heberlein directly: "Accept my grateful homage and will you send me some tentative measurements and photographs from frontal lateral vertical and basal skull-aspect for Academy meeting of October 30?" Dubois' old friend Max Weber writes, and Grafton Elliot Smith from London, and, of course, the ever troublesome Osborn, who says,

> *The newspaper notices concerning the new cranium said to belong to* Pithecanthropus *naturally inspire the hope that the new form may be specifically identical with your original specimens and that it will definitely settle the problem of the form of the face.*
>
> *I trust we can count on your cooperation to secure us a cast of the new skull at the earliest possible date.*

Dubois is as amused as he is annoyed upon receipt of this letter. He has not even set eyes on this new find, much less confirmed that it is what Heberlein has announced it to be, and already Osborn is pestering him for casts! On October 7, Dubois gives a statement to *Het Algemeen Handelsblad* and sends a copy to the *Illustrated London News*.

> *According to the ANETA telegram from Batavia . . . we cannot yet speak of a new skull of* Pithecanthropus. *It is clear to me that the object found by Dr. Heberlein, government district physician at Soerabaja, is a mass of volcanic tuff, consolidated by calcareous impregnation and partially covered with calcareous concretions, which petrous mass fills and envelops a small remnant of a skull or of another fossil resembling a human skull more than a small rock. Probably the petrous mass was picked up from the ground. It could not be obtained by excavation for this cannot be done without the permission of the Dutch East Indies government, "Trinil" being a nature reserve.*
>
> *. . . With such a strong impregnation and encrusting as that of the "new* Pithecanthropus *skull," the elimination of the rock from the bone is a difficult and subtle work, which can only be done by the trained hand of an expert. It is to be hoped that this object found by Dr. Heberlein, which is in any case important, can soon be*

> *expected in Europe. Only then will we know exactly what is the*
> *nature of the object.*

He is vague because he can deduce little from Heberlein's description of the object. The Minister of Instruction, Arts, and Sciences formally asks that the new skull be forwarded to Leiden for complete study, where it will come into Dubois' hands.

Heberlein telegraphs Dubois on October 16, thanking him and promising to send photographs and to place the skull at the disposal of the Dutch government. He does not venture to give measurements of the skull, as it is deformed. Dubois is dancing with curiosity and anticipation; this new specimen is almost within his grasp. He telegraphs Heberlein urgently on November 2: "Thanks can you inform me when Government the Netherlands can expect skull which they will place in my hands. Not yet received photographs."

Heberlein replies on November 7 with a full account.

> *I found the* Pithecanthropus *skullcap on August 1 last, at Trinil, when I*
> *made an excursion there with a few other gentlemen just to see the classic*
> *place. Since before I studied medicine I was a geologist (more especially a*
> *paleontologist), it interested me highly, but in my 23 years of service in*
> *Indonesia I had not, until then, been able to pay the place a visit, because*
> *in the main (until 1922) I was an officer of health, almost always in the ter-*
> *ritories of the Dutch government outside Java and Madura. Now I was*
> *able to last to go to Trinil, it was an enormous stroke of luck to find some-*
> *thing interesting immediately. . . .*
>
> *We ourselves found on the dry banks of the Bengawan River a couple of*
> *bones, when I saw a skull fragment in the hands of one of the native boys*
> *who poke about there daily, which I immediately asked for. Thus the piece*
> *was found by this boy and the place where it was found lies at the right*
> *bank, practically directly beneath the* Pithecanthropus *monument. It was*
> *not found in situ in the strata, but grubbed loose and lying between the*
> *gravels or shingle.*
>
> *First, I spent two months studying the find a little by myself, but I had*
> *decided from the beginning that this find should be preserved for the*
> *Dutch Indies or the Netherlands. (I am German by birth.)*
>
> *The skull is a curved piece like the hull of a ship. The occiput is missing,*
> *likewise the left temporal and the largest part of the left parietal bone.*
> *Moreover, it is only a skullCAP, although the upper margin of the orbits is*

preserved. The actual fossil bone substance is reduced to a paper-thin lamella (probably by the action of sulfuric acid). I have the impression that the skull is also a little squashed by some force pushing from back to front, as is evident from the relatively large breadth and an isolated piece of the surface that sticks out, dislocated, at the back of the porous, lavalike rock. . . . Thus, you will understand that under these circumstances I, not being a professional anthropologist, could not give you measurements.

Moreover, so far as I can judge, I have the pretty clear impression that this skull-fragment is not from a Pithecanthropus, *but from a definite hominid or manlike form. I see this as increasing the interest of the case, given where it was found. I leave the rest with pleasure to those who are better qualified—in the first place, to you.*

Dubois is puzzled over some of Heberlein's points. There is no true shingle at Trinil, so Heberlein must be thinking of some of the rubble from old excavations, particularly Selenka's. If the find is seriously distorted, then Heberlein is probably not correct in thinking that this skull is more advanced and manlike than *Pithecanthropus.* Too, the mode of preservation, with the bone thinned to a paperlike width and an infilling of porous tuffaceous rock, seems very odd. Finally, on December 6, an envelope arrives from Java with photographs, which Dubois inspects immediately.

Oho! So that's what all the susa is about? The Malay word, susa, for a fuss or bother, flows into his mind as easily as if he were still in the Indies, so evocative is the look of the envelope. It is a terrible disappointment, but Heberlein has got it quite, quite wrong. The paper-thin bone that he took for the cranial vault itself is nothing more or less than the layer of articular bone of the rounded joint of some enormous species. What Heberlein thought was an infilling of the "skull" is actually the spongy, porous bone that typically underlies the articular surface on such joints. The object is almost certainly the head of the humerus, or upper-arm bone, of a *Stegodon,* a common elephant from Trinil. Yes, there is something evocative about the shape, something hauntingly skull-like, but it is not a skull at all. If Heberlein were better educated, he would have recognized that immediately.

Dubois writes to Weber immediately.

But for me this is not only a great disappointment (for, although I reasoned that I should not get my expectations up, I could not yet keep down a fearful hope), but also a painful one. My own careful consideration of the photographic images, and comparison of the object with what I saw like it in my collection, taught me that I cannot at all confirm Dr. Heberlein's opinion about the nature of the object.

For me there is no doubt: the object is neither a fragment of a skull of a human form, nor of a Pithecanthropus, *it is not even at all a skull fragment, but the largest part of a right head of a humerus of an elephant (probably* Stegodon, *which is common at Trinil).*

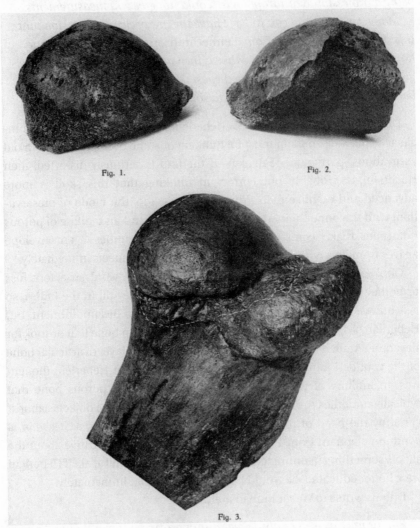

In 1926, the physician C. E. J. Heberlein mistakenly reports that he has found a new Pithecanthropus *skull at Trinil. Top: Two views of the "new skull." Bottom: A humerus or foreleg bone of a fossil elephant, with a dotted line indicating the placement of the "skull."*

Weber replies,

What a <u>sad</u> story that is. First of all because of the enormous disappointment in what the object really is. But it is especially an unpleasant and annoying story because of the great mistake made by Dr. Heberlein. Autosuggestion—wanting to see something and so, seeing it—is understandable, but what is unforgivable is this hurried and completely premature announcement which has, unfortunately, gone worldwide.

It is a bad affair for Dr. H., but it is, as you rightly remark, also embarrassing for our country, and also for the high colonial officials here and in the Indies. For that reason, it is best that you, a Dutchman, correct the record, as <u>quickly</u> as possible. . . . You can do it softly, pointing out the difficulties for someone who is not expert in such matters and who is not familiar with the special material from this particular site, and more of the like.

Where the newspapers have already published so much about this case, and have consulted you about this matter, like the Rotterdamsche Courant, *I should advise you to offer an article immediately. In it you may announce, without mentioning Dr. H., that the recently received photographs of the object lead you to conclude that there has been a mistake, understandable under the circumstances, etc. etc., and after that announce your conclusion.*

Dubois heeds Weber's advice. He is desperately concerned to quash any further rumors or stories about this false skull, lest the taint of mistaken identity spread to his own specimens or judgments. He sends a statement to several Dutch newspapers, giving his diagnosis of the "new *Pithecanthropus* find" based on his inspection of the photographs kindly supplied by the discoverer. "It is not even a skull," *Het Algemeen Handelsblad* quotes Dubois as saying, "but beyond doubt the greater part of the articular head of a humerus, the right humerus, of an elephant—probably *Stegodon,* of which genus many bones have been found at Trinil." Slyly, *Het Algemeen Handelsblad* also expresses an anonymous opinion that Dubois' rapid dismissal of the find may have been influenced by professional jealousy. ANETA cables word of Dubois' diagnosis to the Indies, where his opinion appears in the press on December 9. To resolve the disagreement, the fossil will be entrusted to two scientists in Java, Professor W. A. Mijsberg and Dr. H. J. T. Bijlmer, for study before it is sent to Amsterdam.

Dubois cables Smith and Osborn, advising them of the identity of the new "skull." Next he undertakes the delicate task of writing to Heberlein. It takes him several tries before he feels the letter is right.

Since I first heard the news of your Trinil find, it has—the matter speaks for itself—not been out of my mind. December 6, when I received from you the long-awaited photographs and letter, was an important day for me. I heartily acknowledge my gratitude for that and for the information you sent earlier by telegraph to me. I am also indebted to you for your enthusiastic interest in the important Pithecanthropus *problem, that caused you to preserve the object that you rightly thought resembled* Pithecanthropus *or a human form, for science and for the Netherlands.*

... The diagnosis I must announce to you here is: the object is not a skull, but indubitably the largest part of a caput humeri (of the right side) of an elephant and probably of the genus Stegodon, *from which many bones have been found at Trinil.... Of course, the object itself, as direct evidence, must still be studied; I hope that we may receive it here soon.*

I know you must feel a great disappointment in this diagnosis, as I do. It may be a comfort to you to realize that it needs a lot of experience to recognize such fragmented remains of mammals and that I, and others, are nonetheless grateful to you for taking care to preserve the supposed remains of a Pithecanthropus.

This difficult letter written, Dubois sets about preparing a presentation for the meeting of the Royal Academy of Science on December 18, "The So-called New *Pithecanthropus* Skull." He recounts the history of the find and its newsmaking progress. He shows the photographs sent by Heberlein, in conjunction with the humerus of a *Stegodon* for comparison, and no one in the audience has any doubt that his diagnosis is correct. Dubois makes a point of being gracious to Heberlein, lest the physician feel publicly humiliated.

On a photograph shown me by Professor BOLK, on which the spot at which the "skull" was found was accurately indicated, I saw clearly the rubble of an excavation, a sight well known to me. For the rest, according to my experience obtained during five years, nowhere in that area are bones to be found that have naturally come forth out of their layer. Obviously the find that has created such a sensation is nothing but a fragment left behind as worthless by ... recent excavators. There are more such fragments to be found there. They are sought by native boys loitering around, and exchanged for some money with the visitors at Trinil.

Mijsberg and Bijlmer in Java confirm Dubois' opinion of the specimen in a written report to the Royal Academy, but unfortunately neglect to credit Dubois with making the first correct identification. Nonetheless, he feels the situation has proceeded fairly satisfactorily, until he submits the written version of his December 18 Royal Academy presentation to Secretary Bolk for publication.

Bolk cannot resist needling Dubois. "Well, Dubois," he remarks with feigned innocence, "I really don't know about publishing your contribution. The report by Mijsberg and Bijlmer says much the same thing and was written three days before your lecture, on December fifteenth. Surely they have priority."

"But," protests Dubois, "my opinion was widely reported in the newspapers, on the eighth and ninth of December, and they knew of it."

"What the newspapers report, Dr. Dubois," replies Bolk smugly, looking over the tops of his glasses, "is hardly scientific presentation and rarely accurate. Gossiping to reporters cannot give you priority."

Dubois' broad, handsome face turns choleric. "You are certainly correct about that, Dr. Bolk," he answers, "as we both know. However, the contribution from Drs. Mijsberg and Bijlmer was received at the Royal Academy weeks after my formal lecture was presented to that learned body of scholars. It is not, after all, when one composes a manuscript or conceives of an idea, but when it reaches a scientific audience that counts."

"Ummm," replies Bolk maddeningly, continuing to peruse Dubois' manuscript idly. "I suppose so. . . ." He looks up slowly, a faint smile on his face. "If you really insist, Dubois," he adds carelessly, "I suppose we shall take it." He has no real intention of forestalling Dubois' publication, only of twisting the knife in his side. He knows, everyone knows, that there is no surer way to infuriate Dubois than to play on his fear of having his research discoveries usurped by others.

"I do insist, Bolk," declares Dubois firmly. "I do."

CHAPTER 50 RUMORS AND ISOLATION

Dubois' paper is not all that is published in 1927. Late that year, another popular book comes out called *Evolution*, which repeats one of those rumors about Dubois and *P.e.* that never seem to die:

Their discoverer, Dr. Dubois, after exhibiting them at the Inter-

national Zoological Congress of 1898, withdrew into retirement in Holland, and under the influence, it is stated, of the Roman Catholic Church, refused to allow the fossils to be examined by any other scientific man.

Dubois remains unaware of this accusation until well into the new year, which begins auspiciously. On January 28, 1928, he reaches his seventieth birthday. The university holds a special gathering to honor him. The Rector Magnificus, Professor H. Burger, presents Dubois with a fine portrait of himself painted by the artist Frans David Oerder. Dubois also receives a barrage of letters of congratulations and best wishes from his many friends and admirers worldwide, organized by Antje Schreuder. She is deeply loyal to her esteemed professor, although over the years she has suffered a too-close acquaintance with his difficult character. Shortly after this celebration, the Dutch government awards Dubois the Order of the Knights of the Netherlands Lion, in recognition of the international importance of his scientific work.

Retirement is normally mandatory at age seventy, and Dubois is looking forward to severing his connection with the university and its incessant demands. Now at last he can get on with studying his long-neglected

Dubois Collection, which he has managed to keep intact and under his control for all these years, without ever completing even the most basic tasks of labeling, identifying, and registering all the fossils. Still, he agrees reluctantly to remain at the university for one more semester, until a replacement can be arranged.

The honors he has received prove no shield against criticism. Not long after the birth-

For Dubois' seventieth birthday in 1928, the university commissions Frans Oerder to paint this portrait.

day celebration, a colleague in London sends Dubois a copy of a letter to the editor of *The Tablet*, written by a Catholic priest in England.

February 11, 1928

In a recent contribution to Messrs. Benn's Sixpenny Library called Evolution, *by Professor E. W. MacBride, a guarded suggestion is made that the reputation of the Catholic Church is in some way involved in a policy of secrecy with which Dr. Dubois is alleged to have chosen to veil the prehistoric remains known as* Pithecanthropus *discovered by him in Java. . . .*

Curious to know what misunderstanding, if any, lay behind a suggestion which a Catholic can only regard as Fantastic [sic], *I have made inquiries from friends well qualified to give me reliable information. The particulars elicited go to show that the suggestion is entirely unfounded. I need hardly say that the Catholic Church as such has in no way intervened to determine the attitude which Dr. Dubois may have seen fit to adopt towards his discovery. This is his own concern, a personal matter into which it would be unbecoming, for me at least, to enter. I am informed that so far are individual Catholics from endorsing an unscientific reserve with regard to the* Pithecanthropus *and similar discoveries that such an attitude was criticized by a Catholic priest, a Franciscan friar, in his thesis for the Doctorate of Science recently before a board of examiners over which Dr. Dubois presided. This Roman Catholic Church man pointedly maintained that such discoveries should be made easily accessible to scientists in public collections. The supposed modern Galileo did not, I gather, take up the challenge with the record to hand.*

Prof. MacBride does not inspire confidence in the source of his information on this particular point, seeing that Dr. Dubois made his discoveries under the auspices of the Dutch Colonial Government; that in 1898 he was appointed an assistant Professor at Amsterdam and in 1905 a full Professor, a public position which he still holds; that he has written on his Pithecanthropus *recently and as recently as 1924 in* Verslagen der Kon. Akademie voor Wetenschappen, *2nd section, Vol. 33, I, (where he expresses the view that the remains are not human). I must leave it to Professor MacBride's valuation whether all this can be implied in his term "withdrawn into retirement."*

Further, the Pithecanthropus *remains are kept in the Teyler Museum at Haarlem, an institution in which Dr. Dubois occupies an official position. During the Anthropological Congress, held last year in Amsterdam in May or June, Dr. Dubois conducted a party of scientists to the Museum and there read a paper on his discovery.*

As Professor MacBride's interesting little book is likely to have a large circulation, I hope that he will see his way to withdraw a suggestion which I am persuaded is quite unfounded and which is calculated, though not intended, to give offence not only to Catholics but also, I imagine, to a foreign scientist not yet withdrawn into retirement.

This letter might be amusing if it did not bring so clearly back to Dubois' mind the troubled times of a few years ago. At least, for once, someone else has defended him.

With age and the still-constant barrage of criticism of his finds and ideas, Dubois becomes more and more rigid, less and less tolerant. His temper flares easily and he develops a tendency to bellow over petty mistakes. He can no longer bear to be around Anna for any length of time, she contradicts him so often, and she regards her once-strong and handsome husband as nothing but a quarrelsome old tyrant with an eye for the ladies. One day, when he has demanded and ordered too much, she asks him to get out and leave her in peace. A small house up the street in Haarlem is vacant, so Dubois rents it and moves in with his books and his specimens. He is a solitary man, alone with himself and his science. He thinks that now, at last, he can organize his household to suit only himself. It will be a pleasure. He must, of course, have servants to look after him, so he places an advertisement in the newspaper in Limburg: "Wanted a servant-girl, not older than 25 years, for a gentleman living on his own." This advertisement does not attract the wholly respectable. He chooses the prettier applicants, especially if they seem to indicate by their look or their manner that they would not be averse to a little romance. Now that he is separated from Anna, he thinks, it is not so wrong to seek his comforts elsewhere. After all, he brings the girls from Limburg to the big city, Amsterdam, which is what they desire, and he gives them a comfortable home and a decent wage. Is it so terrible that they wish to show their gratitude? He knows he is no longer young, but he still thinks himself attractive: he is powerful, knowledgeable, and still handsome.

It is a debacle. One after another his housemaids cheat and rob him, playing on his weaknesses to get higher wages, fewer duties, or gifts, and then they quit. His servants change so frequently, and are so good-looking and lazy, that the neighbors begin to talk behind his back. No wonder his wife refuses to live with him any longer! Finally, he begs Antje Schreuder to help him find an honest housekeeper, which she does. Schreuder's embarrassment is acute when, after only a few weeks, the woman comes to her saying she must leave the post with Dr. Dubois. "It seems," the woman says, holding herself very upright and stiff, "that you do not know him very well."

"Whatever do you mean?" asks Schreuder. "I have worked with Professor Dubois for many years, as a close colleague and assistant. He is a brilliant scientist, a genius. Of course, sometimes his mind is preoccupied with higher things, so he is thoughtless, or even irritable, but he is a man of impeccable character and honesty."

"Perhaps at the university," concedes the woman grudgingly, "perhaps that is how he behaves. But not at home. I cannot stay with such a person." She will say nothing more specific, but suddenly Schreuder knows exactly what the problem has been. With a hollow feeling in her stomach, she recalls all the rumors about Dubois' appreciation of pretty women, rumors she dismissed as untrue. Now she sees that his moral rectitude is nothing but a hypocritical sham. And what has everyone thought of her, his "protégée"? She burns with shame at her naïveté, at what others may have thought of her. She has admired Dubois for a long time, forgiving his irritable temper and his selfishness as adjuncts to his greatness. She has been happy to pick up the tedious work he left undone—supervising students in the crystallography laboratory, teaching them how to use the polarizing microscope, preparing maps or slides for his lectures—because, she thought, he was too busy constructing lofty ideas. But she can never look at him the same way again now.

Schreuder withdraws from Dubois emotionally and in turn falls out of his favor. Something about this wounds Dubois deeply though he does not realize why Schreuder has changed. He remembers, too, the look of horror on the face of the last servant before Antje's find, the country girl with the lovely fair hair. One day he gave her a little squeeze and she recoiled in disgust. In one awful, stomach-churning moment, he realized that she saw him as old and decrepit, an aging lecher rather than a vigorous, still-virile man. All he longs for is to be indulged and made comfortable in his own home. Of course he prefers to have a pretty girl around; who would not? If he had stayed in the Indies, he would have a beautiful gentle nyai who would care for him lovingly as he aged. But not here, not these girls.

He realizes as painfully as if he has been struck with a rock that his best days, the days of his youth and courage and great discoveries, are over. In his mind, the long string of incidents begin to fuse together, into an awful mélange of indignation, condemnation, and repulsion. He is not what he once was, what he thought himself still to be; he is far, far less.

CHAPTER 51   BRAIN WORK

Dubois turns to his science for comfort, concentrating once again on his brain researches. He gathers more and more data and struggles with ideas about how brain size increased as species evolved. Learned article follows learned article, until he feels he has proven his point thoroughly. The brain evolves by a doubling of the neurons during the embryonic period, through a process of cell division. Thus, mice have 1/32 as many neurons as humans have, and the next level of evolutionary organization, represented by rabbits, has 1/16 as many as humans, twice as many as mice. Then the group of species on the next step upward, among them monkeys, herbivores and carnivores, have again twice as many neurons (1/8 as many neurons as humans have). The species closest to man, apes, have 1/4 as many brain cells as humans. It is a geometric progression based on an extra division of the brain cells during embryonic life. What is missing from the series is any species that occupies the level of 1/2 as many brain cells as humans. That level is occupied by *Pithecanthropus,* in this as in other ways a perfect transition between apes and man.

The *pattern* of evolution in this series differs from the gradual accumulation of small changes envisioned by Darwin. Here is the answer to the problem of missing links—the "absence of evidence of continuous development," Dubois calls it. Darwin attributed these gaps to the imperfection of the geological record and the paucity of fossils, but Dubois sees something far greater, far more interesting at work. Dubois sees evolution *per saltum,* by leaps. Ah, Dubois thinks, I am not dead yet. My mind still works, still sees what others miss. This is one of my most brilliant endeavors!

In his sometimes tortured English, he delivers his ideas first as a lecture to the Royal Academy, then as a published paper. He starts with some commonplace observations, but ends with an extraordinary conclusion.

> *Clearly paleontology bears evidence of the growth of life on Earth. . . .*
> *The minutest transitions between the members of the many parts*
> *of lines of descent with which we became acquainted, were so regu-*
> *larly absent, that this absence can no longer be attributed to the*

"imperfection of the geological record." Every member of a sequence is stepwise distinguished from the preceding and the following one. Again and again we find the pillars of the expected bridges, never arches. . . .

The members of these sequences are intermediate forms no "transitional forms," no "links." Indeed, repeatedly paleontology has removed from the hypothetical stock line its at first much made of transitional forms, referring them to side branches in the genealogical tree.

Strictly speaking gradual transition is a priori impossible, because every species living as an independent creature being adapted to particular circumstances of life, must be specialized in its peculiar way. . . .

That we again and again meet only with pieces of these ties, and that gradual transition between creatures living independently as different species is a priori impossible, leads to the conclusion that the real transitions, the missing pieces of the ties of the relationship, the arches of the bridges connecting the species, took place in the embryonal period of the individual life, *before the independent existence of the individual.*

Of this paleontology could not furnish any documents, because they never existed.

At this point his audience is confused. What can old Dubois be talking about? Transitional species are a priori impossible? Aren't transitional forms just what Darwin predicted? Isn't that what Dubois has always said *Pithecanthropus* was?

His point, however, is that evolution proceeds by a cellular process. Yes, a series can be constructed that reflects the course of evolution, with 1/32, then 1/16, then 1/8, then 1/4 the brain size of humans, each level being occupied by a different sort of animal. But the series proceeds by jumps, by doubling rather than by gradual increments. There is no 1/30 level, no 1/25 level, nor any at 1/12, 1/10, or 1/3. Among species at a common level, there are gradual transitions, adaptations, specializations: all the diversity and variation that Darwin observed. But the "real evolution," as Dubois calls it, the "progress of the degree of organization," depends on the multiplication of the units of the brain, via cell division. There is thus only abrupt change in form and structure at this level of evolution, not gradual evolution.

Phylogenetic progress in the brain, in terms of the size of the cerebrum and the complexity of its function—what Dubois calls the *psychoencephalon*—is determined by internal, autonomous factors, not external natural selection, as Darwin has proposed. He continues:

> *Besides, in the case of the psychoencephalon, phylogenetic growth means at the same time perfection of the organ, continual phylogenetic progress of its functions, for the higher [level of] organization is here immediately attained by the increase of the number of cells, which at once leads to more multiple combination, greater functional complexity, direct enlargement of the animal's outer world.*
>
> *Here is a law of evolution come forth out of the nature of the living being itself, not imposed by the surroundings. . . . It appears that there actually does exist . . . a law of phylogenesis, and that [law operates] with progression, with perfecting.*
>
> *It is self-evident that this perfecting, this steady progression cannot have been caused by factors outside the animal, to which darwinism* [sic] *ascribed phylogenesis. External factors can only have effected diversity of the animal forms and functions, by adaptation. It is inconceivable that they should have acted continually in one definite direction, that of perfection.*

Dubois is proposing a cellular mechanism behind large-scale evolution. Darwin's ideas of adaptation and survival of the fittest account perfectly well for speciation, for events like the gradual transformation and specialization of the cat family into diverse forms like the slender, swift, running cheetah and the massively powerful lion. But Dubois' mechanism accounts for the creation of whole new types of animals through the simple and well-known process of cell division. This synthesis of developmental and evolutionary ideas is so radical that decades will pass after Dubois' death before his approach is widely understood and embraced.

Dubois finally retires in 1929 to work on the Trinil fossils. How little he has accomplished since he shipped them home from Java! That he has done too little, either on them or on *Pe.*, has caused him much grief and trouble. But his temperament is ill-suited to the mindless tasks of labeling, sorting, cleaning, and registering fossils; even now he seeks every excuse to turn to more interesting work.

In December 1929, he finds the perfect reason, when the newspapers announce that Davidson Black, a young Canadian anatomist working at the Peking Union Medical College, has found the fossilized skull of early man in China. Dubois knows of Black, a former student of his old friend Grafton Elliot Smith in London, and he recalls hearing that in 1919 Black accepted a position in China in hopes of finding human ancestors. Though Black soon found that his sponsors did not approve of fossil-hunting, he persisted despite repeated warnings to stop this foolishness and attend to matters anatomical and medical.

Black's early years in China yielded few fossils of merit. In 1926, he successfully argued during a visit by the Swedish Crown Prince that two humanlike fossil teeth, found by Swedish geologists, were "striking confirmation" of the hypothesis that China would yield a "new Tertiary man or ancient Pleistocene man." Calling the teeth "one more link in the already strong chain of evidence supporting the hypothesis of the central Asiatic origin of the Hominidae," the zoological family to which mankind belongs, Black persuaded the Crown Prince to offer funding and support for further excavations. This in turn swayed the opinion of Black's sponsors, the Rockefeller Foundation, who in 1927 allowed him to begin systematic excavations at a cave site near Peking known as Chou Kou Tien, or Dragon Bone Hill. By the end of the first field season, Black's expedition has found an additional tooth, which he celebrates with a new name: *Sinanthropus pekinensis,* "the Chinese man from Peking." Dubois, like Black's other scientific colleagues, found this rather scanty evidence on which to name a new genus and species. I was criticized for *P.e.* when I had a skullcap, a femur, *and* a tooth, after all, Dubois thinks. But Black has made a canny move, which leads to a three-year leave of absence from his teaching duties and to more funding. The shape of Black's story is so intimately familiar to Dubois; it might as well be his own early career in the Indies.

And now, after two years of hard work at Chou Kou Tien, Black's ambitions pay off. On December 2, the Chinese geologist W. C. Pei is in charge of the field operations. Digging deep in the Chou Kou Tien cave, by candlelight, Pei spies a large rounded fossil, a skullcap. By December 4, Pei's telegram reaches Black: "Found skullcap—perfect—look [*sic*] like man's." On December 6, the skull, carefully encased in glue-soaked gauze and plaster, with two thick cotton quilts and two blankets for padding, arrives in Black's hands. Pei's field identification is correct; the fossil is a largely complete skull of a primitive hominid or human ancestor. Black calls a press conference in Peking on December 28 and the news is telegraphed around the world.

Dubois, like most of his colleagues, learns of the find first through newspaper accounts. He clips a lengthy article, "The Peking Man—An Undamaged Skull," from the December 30, 1929, edition of the *Manchester Guardian* and peruses it for details.

> *A new chapter in human pre-history opened with the discovery of "Peking Man"* (Sinanthropus pekinensis). *The fossil remains most recently unearthed are of the highest significance. . . . The Peking ape-man shows a striking mixture of anthropoid and hominid features. . . .*
>
> *The skull has been prepared down to its eye-sockets. The lower face is apparently missing, but the ear-hole and the back of the skull are present. Looking from below, one can see the massive jaw-socket, which suggests a biter of no mean power. The brain capacity is clearly larger (by perhaps one-quarter) than that of* Pithecanthropus, *the Java ape-man. At the same time it is still a skull of small type, definitely inferior in size to that of Neanderthal man—in fact, taken as a whole, the skull-cap very strongly suggests a very primitive Neanderthal type. The skull is that of a young adult. It bears out strikingly the "intermediate" characteristics shown by the teeth and jaws, the teeth being definitely of human type, the jaw being ape-like.*

Before long, Dubois receives a copy of the first publication on the find from Black himself, as a kind of tribute. *Sinanthropus* will surely be compared with *Pithecanthropus,* and Black hopes to make a friend and colleague of Dubois. Dubois is delighted and honored, and writes Black to say so.

February 11, 1930

> *It is with much pleasure that I received a copy of your "Preliminary Note on additional* Sinanthropus *Material Discovered in Chou Kou Tien during 1928." . . . Please accept my cordial thanks for your kindness of sending this very important and interesting paper to me.*
>
> *Your discovery of* Sinanthropus *I regard as a <u>great</u> one, possibly <u>the greatest</u> one ever made in paleoanthropology. Indeed this new human form strongly compels us, in my opinion, completely to review the current conceptions on human evolution.*
>
> *It is with keen and joyful anticipation that we look forward to your full description of the splendid material obtained.*

The reply arrives nearly two months later.

April 3, 1930

Thank you ever so much for your kind letters of last February which reached me two weeks ago. I cannot tell you how much I appreciate your cordial good wishes and it is indeed a pleasure to be able to send you in this letter copies of the first photographs to be made of the new Sinanthropus *skull after its exterior has been freed from travertine.... The report itself together with reproductions of these six photographs is now in press. I am also sending you under separate cover three reprints of preliminary papers on the Chou Kou Tien region and on* Sinanthropus. *From the latter you will see that now there can be no question of the geological age of the deposit. It will be two or three months yet before my final report on the skull can be prepared since its whole interior must be freed from travertine before the bones of the vault can be replaced in their exact natural relations. The more I study this specimen the more clearly it becomes evident that here we have a form sufficiently generalized in character to be not far removed from the type from which Neanderthal and modern types were both derived.*

Dubois underlines Black's final sentence about the position of *Sinanthropus*. If his *Sinanthropus* is the ancestor of both Neanderthals and modern humans, where does Black place *Pithecanthropus* in all this? It sounds as if he follows that old theory of Schwalbe's that *Pithecanthropus* is ancestral to Neanderthals. If so, then Black would draw a letter "Y," with *Pithecanthropus* at the foot of the stem, *Sinanthropus* at the fork, and Neanderthals and man at the ends of the branches. Interesting . . . Then he turns to the second page of the letter and sees a postscript.

Boule pointed out long ago that the morphology of the tympanic element in the La Chapelle skull recalls in certain respects the conditions obtaining in that region in the chimpanzee, presenting characters somewhat intermediate in type between the latter form and Homo. *The relations of these parts in the* Sinanthropus *specimen may thus with propriety be termed pre-Neanderthaloid, representing an evolutionary stage preceding the Neanderthal-like types. The morphological evidence so far available with respect to the position of* Sinanthropus *in the hominid scale would thus place the latter form not far removed from the type from which evolved both the extinct Neanderthaler and the modern* Homo sapiens.

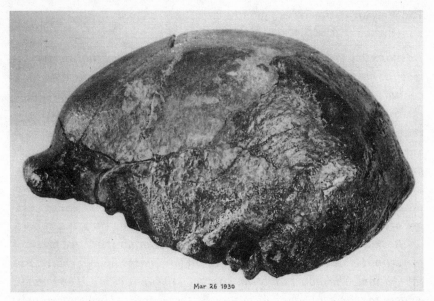

Mar 26 1930

Davidson Black believes that Sinanthropus pekinensis *is closely allied with* Pithecanthropus. *This fossil skull is found in 1929.*

Hmmm, Dubois hums slightly to himself, reading. "Boule pointed out long ago . . ." Yes, I suppose it was long ago, for a young man like Black, and Boule's monograph was published fifteen years after mine. My monograph was the one that started it all, that first compared a fossil form to living apes and man. Black clearly finds *Sinanthropus* a man, not an ape-man and not an ape. A pre-Neanderthaloid, more advanced than *Pithecanthropus:* yes. So my *P.e.* is still the true ape-man, the missing link.

Then he turns to the photographs Black has sent. What a glorious creature is there portrayed! The skullcap is more complete than that of *P.e.* and includes a large part of the base of the skull. A pity that most of the face is missing. Yes, there is a certain resemblance to *P.e.*, especially in side view, but also differences. What does Black say?

Since my last report before the Society, work has progressed on the Sinanthropus *skull specimen till now its whole external surface has been freed from travertine, with which however the interior is still filled. . . .*

The preliminary photographs show in unmistakable fashion certain of the major characters which serve sharply to distinguish Sinanthropus *from other hominid types ancient or modern.*

The next part, the technical description, Dubois follows closely, with one finger on the photographs. Black makes an important point about the shape of the skull in vertical cross section, a section to which Dubois never paid much attention because *P.e.* is missing the lower part of the skull and its base. In *Sinanthropus,* Black says, the maximum breadth of the skull lies just above the ear holes, even though the parietal eminences—the "corners" where the walls of the skull angle in to become a roof—are well-developed. Is this also true of *P.e.?* Hmmm, hard to tell . . .

The bones of the cranial vault in *Sinanthropus* vary in their thickness. They are reportedly thicker than in *Pithecanthropus* and, on the whole, much thicker than in a modern skull, especially low down at the back. There are some peculiar, primitive features about the bony ear as well—a region completely unknown in *P.e.*—where *Sinanthropus* appears more apelike even than Neanderthals, Black says. Black suggests, because of its delicate modeling, that this specimen of *Sinanthropus* comes from an adolescent female.

June 22, 1930

Accept again my cordial thanks for your great kindness of sending me . . . copies of the first photographs to be made of the new Sinanthropus *skull after its exterior has been freed from travertine, and of your interim report at the Annual Meeting of the Geological Society of China. Those documents concerning your great discovery and the reprints of preliminary papers on the Chou Kou Tien region and on* Sinanthropus *are indeed precious possessions to me.*

Now I take the liberty of writing you again, having studied with great care the photographs with your description, comparing them with Neanderthal Man, Homo sapiens, *and* Pithecanthropus. *The result is I quite agree with you in considering* Sinanthropus *a pre-Neanderthaloid form. I may now repeat my opinion that this is the most important discovery in human paleontology.*

As to the supposed nearer relations with Pithecanthropus, *I must avow not to find them.* Sinanthropus *is a very "primitive" type of man, but still a man, whereas* Pithecanthropus *is no man at all, in my opinion. . . .*

Please oblige me once more by communicating to me any data you should obtain on the size of the body of Sinanthropus, *and on the cranial capacity.*

The relationship between Dubois and Black continues on entirely friendly lines. Black always sends Dubois information, publications, and casts before anyone else; Dubois reciprocates. The finding of *Sinanthropus* seems to validate Dubois' claims for his Javan ape-man; the more abundant Chinese fossils ease the ultimate acceptance of *Pithecanthropus* as a very early human ancestor, even though Dubois always believes that his find is more primitive and older than Black's.

Despite the parallels in their lives, Black's pathway to success is much smoother than Dubois'. In 1932, with the sponsorship of his former adviser, Grafton Elliot Smith, Black is elected a Fellow of the Royal Society in England, even though his monograph on *Sinanthropus* has not yet been published. In 1933, Dubois initiates a move to heap yet another honor on the younger man:

March 28, 1933

Perhaps you do not know that this year again a Prix Hollandais is to be awarded by the Institut International d'Anthropologie, for the most prominent achievement in Anthropology, on which the author submits his works to the judgment of an international jury from the Institut.

Being a member of that jury, may it be permitted to me to ask you to send your works on the Sinanthropus *to the Secretary of the Institut International d'Anthropologie? . . . In my opinion yours is the most important achievement in Anthropology during the last cycle of the Prix Hollandais.*

The award of the Prix Hollandais to Black is the last occasion upon which his path and Dubois' will cross. Early one morning in 1934, Black dies at his desk, working on his beloved *Sinanthropus* fossils. He is only forty-nine years old.

CHAPTER 52 THE DILIGENT ASSISTANT

While the *Sinanthropus* fossils are new, Dubois spends all of his time poring over Black's letters, publications, and photographs, and none on the Trinil collection. The trustees of Leiden University begin to fear that Dubois will die before the collection is properly organized. Should this occur, most of the locality data, contained in Dubois' memory and semilegible notes, would be lost. This would be catastrophic, for it would make the vast collection nearly useless. To make sure this important work is done, in November 1930 they appoint an assistant.

Their choice is a curious one: Dr. J. J. A. Bernsen, a Jesuit priest who received a doctoral degree after studying the rhinoceroses from the Tegelen Clay deposits in Holland, working under Dubois himself. Dubois has mixed feelings about working so closely with a Roman Catholic priest, especially in light of the rumors that have circulated for years. Still, Bernsen is his three days a week, and Dubois surely needs assistance. Bernsen enters into the tedious work with enthusiasm. Constant exposure to each other leads to a growing intimacy. Bernsen and Dubois often fall into conversa-

Father J. J. A. Bernsen undertakes the enormous task of labeling, cataloguing, and ordering the Dubois Collection, but his relationship with Dubois is tortured.

tion as they work. Bernsen treasures these times, when Dubois seems to be confiding in him or at least speaking his true mind. Can the old man be lonely? Bernsen thinks it unlikely, though he knows Dubois has separated from his wife, for surely such a great man, a renowned scientist, has many friends and colleagues to spend time with. On November 10, Bernsen begins to keep careful track of these conversations in his diary, writing down what he can remember of Dubois' words each day.

> NOVEMBER 10, 1930. *The conversation comes around to belief and science. Dubois thinks that the naturalist has to keep belief and science absolutely separated. . . . Dubois tells me that he gets up in the morning at 5:30 and goes immediately to his study to think and read about evolutionary problems. "So," he says literally, "I think to best serve God."*
>
> *In the course of the conversation, Dubois said that the goal of many researchers is their own glory. Dubois is only looking for the truth. In an earlier conversation in October he said to me, "If they prove to me that Pithecanthropus was a donkey, I will be happy at last to know the truth. I am proud that I am still flexible enough to be able to change my opinion."*

Sometimes Dubois speaks to Bernsen of his frustrations.

NOVEMBER 12, 1930. *Dubois realizes that the study of the Dubois Collection should actually have been his life work. But it was not so. Other researches took up all his attention. . . . He said to me, "These days, with the moving of the collection, I have read once more what I said in 1894 about P.e. Now I am ashamed of it. I am happy that I waited until 1926 to write the definitive description of the 'Pitheek' remains. At that early time, my thoughts were not well-considered. Some thought that I hid 'Pitheek' intentionally, because of religious considerations. That is not true. Experience can also become a disadvantage. One can get fixed in one's own experience and become insensible to new impressions."*

Having something meaty to discuss makes the tedious work go more rapidly, for Bernsen's first major task is to supervise the moving of the collection one more time. Once again, the thousands and thousands of specimens are to be transferred, this time from the unheated attic room of the new university hospital to the building that once housed the pathological laboratory in the old plague hospital. The removal starts on November 3, 1930, proceeds through the bitterly cold winter, and is completed only on February 15, 1931. Then the boxes must be unpacked and the real work begun.

Once again, in the move, some specimens are damaged and some labels lost. This is inevitable, for neither Dubois nor Bernsen can possibly supervise every moment of the transfer; besides, Bernsen has never before seen the entire collection, so he is unfamiliar with the eccentric and sometimes haphazard way it is organized after so many moves. Yet Dubois is impatient with the confusion and breakage, unforgiving, and instructs Bernsen that he must do better. "Protecting this collection is everything," he declares coldly. "It will outlast us both. It must outlast us both. You will have to sacrifice more time and care in looking after it. These fossils are more important than you are."

It is such a ferocious dressing-down that Bernsen is cowed and depressed. During his studies on rhinoceroses, Dubois was impatient, certainly, with Bernsen's slowness, with his methodical nature, but never did he lose his temper like this. On November 23, after less than three weeks of working with Dubois, Bernsen worriedly writes in his diary:

I am very nervous. Feel myself overstrained. Worry about the distrust and small difficulties with Dubois. Can I never please him? He is a difficult man; his standards are so high.

Dubois says he has always gone through life honestly, not being diplomatic. Perhaps this is a sort of apology for his temper. "You are also not a diplomat," he says to me. . . . But he believes that diplomacy is incompatible with science. If you are a diplomat in life, avoiding the truth, then you are also one in science. . . .

He says that in the past he has been disturbed by the dishonesty of others and surprised that something like that could occur among scientists. He said this to Place, his teacher, and Place asked him how old he was.

Dubois answered, "Thirty-eight."

Place replied, "Do you still have to learn that even learned people have the ordinary human shortcomings?"

Some weeks later, Dubois returns to the subject of science and religion. Bernsen cannot but wonder how much the presence of a man of the cloth disturbs Dubois, raising questions he had long ago put aside. Bernsen hears in Dubois' voice the anxiety of one who still struggles with faith. "I myself speak in publications about brains, exclusively about brains—never about the psyche, let alone about the soul," Dubois declares defensively. "In the latter subjects I do not even want to lose myself, because I know that I cannot come any further toward the truth and because one blunders so easily on those subjects."

Bernsen is becoming a little obsessed with Dubois. Dubois is charismatic, brilliant, quick, impatient, brutally unkind, and in instantaneous reversal, breathtakingly considerate. Bernsen does not know what to make of him. He is learning much more about Dubois' mind and personality than he ever did as a student, when he was left largely on his own. In April, Dubois' former assistant Antje Schreuder comes to pay a visit. Dubois is visibly uneasy in her presence and makes an excuse to go off and do something else. Unperturbed, Schreuder stays and talks with Bernsen, asking how the work is going and explaining some of the procedures she used during her fourteen years as Dubois' assistant. She praises Bernsen for getting much more accomplished than she ever did, for she had great difficulty getting the professor to sit down and tell her what she needed to know about the specimens.

She asks how Bernsen finds working with Dubois and Bernsen hardly knows what to say. His thoughts of Dubois are intense and yet confused. "You know that Dubois is difficult, a demanding taskmaster," Bernsen confesses hesitantly. "But what an intellect! It is a privilege to work with him, truly." This confidence provokes a flood of words from Schreuder, one that both surprises and embarrasses Bernsen. Afterward, he records his impressions of the afternoon in his diary.

> APRIL 23, 1931. *Miss Dr. Schreuder from Amsterdam paid a visit.... Speaking about Professor Dr. Dubois, she said, "He is an intellect who rises miles and miles above the twelve or thirteen other professors we know in our field. Professor Ihle, de Beaufort, Boschma, etc. are very good people, but as professors just middling. Professor Dubois is someone who would bring honor to any university. But psychically he is a monster. His egotism knows no limits and he has brutal sensibilities." After this she told me, among other things, about his affair with Claartje, his personal maid. She feels his wife was right to run away from him because of his affair with Claartje, a servant girl from Limburg.... Miss Schreuder has a very critical mind. She can no longer stand Dubois as a human being, but judges him highly as a genius among scientists.*

The very next day the man who was Dubois' assistant before Schreuder, Professor Escher, comes to visit. Bernsen wonders at the coincidence. Are Dubois' former assistants trying to tell him something? He cannot decide. Escher actively avoids meeting Dubois, lest Dubois "suspect that something is behind" his visit, in the way that he always does. Escher, too, is eager to discuss Dubois with Bernsen. "He suffers from nerves," Escher confides, to Bernsen's surprise. Bernsen had not considered Dubois' behavior in this light before. Nerves? But the man is supremely self-confident, a man who can stand alone when no one else agrees with his point of view. And yet, Bernsen thinks, yes, there is a nervousness, the sign of an oversensitive temperament. He has that extreme touchiness, almost as if he expects betrayal. "He is always afraid," Escher continues, "that someone else will scoop up his honor and steal it."

But no one can steal the honor of an honorable man, Bernsen thinks. A priest's stock-in-trade is guilt, sin, and honor. So what is the real source of that fear? What is the guilt that makes Dubois look for punishment?

True to his nature, Dubois is deeply suspicious of these visits. It is not long before he denigrates Schreuder to Bernsen: "You mustn't believe the things she says, for she is always getting her facts wrong. It is typical of her; she is very confused."

"But I thought her thesis was a very good piece of work," replies Bernsen mildly. "Didn't you praise it quite highly for its precision?"

"Ja, ja." But Dubois then shakes his head. "I have said it was an important work, although I am sorry not to have told the exact truth. For I think it was not a significant work at all. It was real female-work. Females have no aptitude for science. That is reserved for males. I was wrong to encourage her and now she is angry, for she sees that I do not really respect her work."

He sounds to Bernsen like a man full of regrets, unable to face his own faults and consequently unforgiving of others'. Perhaps this is why he is so fearful that his place in science will be transient: in his heart, he does not believe he deserves honor and glory. It is curious to find signs of deep terror in a man so brilliant, Bernsen thinks. I am only now beginning to understand this man, with whom I have worked for so long. No wonder his other assistants have been so bitter and frustrated.

One day Dubois admits to a peculiar quirk, a dislike of walking alone to the train station at the end of the day. Bernsen is puzzled. Does Dubois fear that people will think he has no friends, if he walks alone? Does he care so much how random strangers might judge him? Out of kindness, Bernsen falls into the habit of walking with Dubois to the station every day. The walk provides another occasion to listen to Dubois' confidences and stories. One day Dubois offers an assessment of his life that seems almost to deny his professed agnosticism. "Yet I have never written about *Pithecanthropus* as Haeckel did, tendentiously, out of an aversion to faith. Never have I spoken about *Pithecanthropus* on these lines. Rightly, I have spoken about the missing link. But what yet is there against this view [that there is a missing link]? The loftiness, the greatness of life makes me believe in God."

Sensing a confessional mood, Bernsen remains silent, simply listening. Dubois' words pour from his mouth and his soul. "I believe that I, in leaving the Catholic Church, have done more for the Catholics, and in general for the Christian cause, than many advocates in the Catholic Church. Ah, you raise your eyebrows, Bernsen? You ask, 'How can that be?' I'll tell you. I have surely demonstrated the untenableness of Darwin's ideas of gradual evolution and survival of the fittest. My researches have clashed directly with social Darwinism, too. And I am convinced that Marx would never have written his book had Darwinism not preceded it."

It is a strange and despondent summary of a lifetime's work, Bernsen thinks. Does this man have no idea of his real contributions, his real worth? Working closely with Bernsen, trying to get his fossils in shape to be studied, Dubois seems weighed down by a deep melancholy, brought on by confronting the many years in which he has done no work on this vast collection. "You know, Bernsen," Dubois confides one day, "I never had the time to work on this material carefully, in all those years. The professorship in Amsterdam took up too much of my time; a professor must do research himself in the areas in which he is teaching, so I was drawn into those geological studies. And preparing all the lectures took so much time. You understand why I have done little on the collection, although I have always regretted it." Dubois sighs heavily as he thinks back.

"The difficulties that I have experienced spoiled my life," he adds. "Oh, maybe you don't think so, but the Trinil collection, too, caused me great difficulties. In the first place, in order to search in the Indies, I gave up the promise of a professorship in the near future. Also in the Indies I had many difficulties—at first, I had no opportunity to go fossil hunting. Later, after I had done all the work, Martin wanted to bring the collection under his control." He shakes his head at the perfidy, clucking his tongue. "After all I had done . . ." His voice trails off and he stares out the window for a few minutes. Then he continues, "Also *Pithecanthropus* caused me a lot of trouble. Among believers, the find caused conflict because of religious considerations; from nonbelievers, I received partly approval, for political reasons (I gave them an argument against faith), and partly condemnation, because they were jealous. . . . Now, finally, I am able to work on the collection, but my resilience is gone. If I meet with further difficulties, I will have to give up entirely. Not because I dislike debate, but because I cannot bear it any longer."

Dubois' depression lasts for weeks, as the two men toil over the boring but necessary tasks that face them. Unoccupied with scientific problems, Dubois turns to reflection and regret. "What a misery I have had from those fossils," he complains to Bernsen once, pushing aside a box full of specimens that he has been sorting and classifying. He stands up, stretches to relieve his aching back, and walks around the room as he speaks. "Sometimes I cannot sleep at night because of it. Now that I am old, I want to correct the omission, to get the work done. I hope to live long enough, until the collection is all right."

On another day, he continues the theme: "I get the impression sometimes that the others here at the museum regard my work as the making of my last will and testament. It is highly unpleasant for me, that they always look at me to see if I am dead yet." And death, thinks Bernsen, is when you will confront your Maker and his judgment on your life. Is that what really haunts you, the final accounting yet to come? Bernsen does not voice these thoughts; there would be no point. But he recognizes the anguish of the lapsed Catholic in his colleague.

In the spring of 1931, Dubois is pulled out of his gray mood by receiving Black's monograph on the Chou Kou Tien remains. There are some haunting photographs of a new specimen, found on July 30, 1931. It is another spectacular find: a second *Sinanthropus* skull, somewhat fragmentary, from a young adult male. This time the base is missing, and so is the face. It is almost a perfect match in preservation for *P.e.* Once again, Dubois has new fossils to think about, and he is full of ideas and energy. The Chou Kou Tien remains once again distract him from the mundane work on the collection, but Bernsen soldiers on, his frustration growing. And Dubois can't resist showing the photographs and monograph to Bernsen and talking with him about them, even though Bernsen's expertise (as Dubois well knows) lies in quite another area.

"From Black, I have received new photographs," he remarks to Bernsen, displaying them in his hand. "Every day I study them. In the open air, under the trees, I can think best about something like this."

"May I see the photos, Professor?" Bernsen asks tentatively, stretching out his hand.

"Oh, *ja, ja,*" replies Dubois, continuing to look at them himself, oblivious to Bernsen's waiting hand. He muses, "You know, I admire Black because of his bright mind, his excellent book. Now, you must know that, the longer I think about it, I see some resemblances to *Pithecanthropus*, although there are differences as well." He still does not relinquish the photos. Bernsen eventually gives up and withdraws his hand, knowing how reluctant Dubois is to share the photos and the information they hold. Dubois gestures, displaying the pictures but never quite handing them over, pointing out features to Bernsen as if the priest were as intimately familiar with the morphology of *P.e.* as he is himself.

"For example, here, the higher frontal bone, steeper than in *P.e.*, and the shape of the skull in horizontal section. Do you see? The *Sinanthropus* skull forms a long oval, like a Neanderthal or like many human skulls. But in *P.e.*"—Dubois puts down the photographs to gesture with two hands—"that outline is distinctly pear-shaped, narrow in the front and broad in the occiput." Bernsen nods agreeably, wishing he could remember the shape of the *P.e.* skullcap more clearly. For all his time working with Dubois, he has never been given opportunity or permission to examine *P.e.* at any length. He picks up the photos, hoping Dubois has relinquished them at last.

Dubois lectures on, unaware. "Yet, despite the resemblances, I feel that they do not belong to the same species." Bernsen nods again, not from personal conviction but from ignorance, scanning the photos greedily and trying to take in as much information as possible before Dubois reclaims them. Dubois continues, "Black gives a pleasant solution to this problem in his book. According to him, *Sinanthropus* is a general form—it has something of a human or Neanderthal, something of *Pithecanthropus,* but also very much something of monkeys—while *Pithecanthropus* is a very specialized form."

There is in Dubois' voice an uncharacteristic uncertainty, as if he is struggling with conflicting ideas. Does Dubois believe Black's pleasant solution? Bernsen cannot tell and, indeed, Dubois is himself unsure. Dubois wonders, in his heart, whether the differences between *Sinanthropus* and *P.e.* are too trivial for the weight they must carry. Are they enough to make these specimens truly separate genera and species? He will never express such doubts openly, not in front of a mere assistant and former student, in any case.

The distinction or lack of distinction between *P.e.* and *Sinanthropus* is a question upon which other anthropologists and anatomists are quick to publish their opinions. Grafton Elliot Smith immediately offers his support for Black's view that *Sinanthropus* is something entirely new. In Germany, Franz Weidenreich and Hans Weinert argue that *Sinanthropus* and *Pithecanthropus* are members of one group, which should be called *Pithecanthropus.* Weinert even declares the name *Sinanthropus* to be superfluous. Boule, in France, sides with the Germans, while Hrdlička in America feels Neanderthals are also somehow involved. All this debate echoes his own early trials and makes Dubois uncomfortable. He feels he must state his opinion and be prepared to defend it, but he knows what acrimonious exchanges may follow. He has little stomach for academic brawls. Besides, he cannot quite make up his own mind, for the evidence is maddeningly ambiguous. Some days he thinks one thing, another day, another. It is a most distasteful situation.

Dubois starts the conversation up with Bernsen about *P.e.* and *Sinanthropus* a week or so later, when the two scientists are once again working side by side, gluing the Trinil fossils back together and writing labels for them. "You know, Bernsen, the more I look at the smaller skull of *Sinanthropus* (of course, I mean of the photograph), the more he starts to resemble *Pithecanthropus.* I should say almost alarmingly. . . . I cannot make up my mind. I want to make a new reconstruction of the *Pithecanthropus* skull, to be able to compare it to the *Sinanthropus* remains, where they also have mandibles. When I told [a colleague] about this, he asked how I could do that, with so much missing. He asked me, 'But from the mandible, the jaw, you do not have so much, *ja*?' and I had to admit I do not. I have only that specimen from Kedoeng Broebus, the part of the symphysis at the front of the jaw up to and including the second premolar. Thus, it is not a small piece, but not so much as Black has."

Eagerly, Bernsen offers to assist Dubois with this proposed reconstruction, but Dubois brushes the offer away. "No, no, thank you, Father, but I can do it myself. It is better that you carry on with sorting and registering the Trinil bones. Only sometimes I like to talk these things over with another, you know?" Bernsen is growing a little resentful at being relegated to only the most tedious and mechanical of tasks, never being included in the research on new materials or new ideas. He hopes fervently that in time Dubois may trust him with more interesting work. Sometimes Dubois speaks of giving Bernsen whole groups of fossil mammals to work up and describe, while he, Dubois, will take others. That would be a wonderful opportunity for Bernsen, a challenging and exciting task. On other days, Dubois suggests they should work on them all together. As Bernsen comes to know Dubois better, he can interpret this suggestion more accurately: he, Bernsen, will probably do the routine work while Dubois polishes the ideas and interpretations. That, he would not like so much. But he must bide his time, work energetically, and show himself to be trustworthy. He knows well that Dubois is touchy and suspicious of other scientists, so he takes great care not to overstep the bounds of his assignment.

Two days later, still clearly troubled, Dubois stops to talk with Bernsen again. "I have been looking at those photographs of *Sinanthropus* again, Father," he offers. "The more I look at the pictures, the more differences I see."

Bernsen thinks to himself, a little tartly, that Dubois has changed his mind about the photos more times than there are specimens. He is too wise to put this thought into speech. "What differences are those, Professor?" he asks, trying to infuse his query with respectful tones.

"Here, Father." Dubois holds out a photograph of the endocranial cast of the brain of *Sinanthropus*, next to one of *P.e.*'s brain cast. "Look, see? The cerebrum of *Pithecanthropus* is more gibbonlike than humanlike, while that of *Sinanthropus* is humanlike." Bernsen nods dutifully, hoping Dubois will ask for his opinion, but it does not happen, now or ever. Dubois soon leaves the room, and the endless task of registration and organization, going back to more interesting dilemmas.

On their now-ritual walk to the station, Dubois raises the subject again. "I am still working on the problem of *Sinanthropus*," he confides to his assistant. "Although it certainly resembles *Pithecanthropus* in many respects it is, nevertheless, something entirely different. At home, I have compared the *Sinanthropus* skull to all sorts of gibbon skulls and among the gibbon skulls there are all sorts of great differences and variations."

Why don't you ask me? Bernsen cries silently. Don't you think that I have learned anything in all these years of work? But Dubois does not think about Bernsen and his state of knowledge at all. He writes, instead, to Black.

June 15, 1931

It was a great joy for me to receive, a few weeks ago, your splendid description of the adolescent skull of Sinanthropus pekinensis. *It is indeed a joy forever to one interested in phylogeny to possess such a work. . . .*

It is a source of particular satisfaction to me that your conception of the relation existing between Sinanthropus *and* Pithecanthropus *nearly agrees with that I have formed by the study of your preliminary descriptions and photographs. In recent weeks, at the behest of your perfect* Sinanthropus, *I thoroughly studied again* Pithecanthropus, *and I find the importance of the latter increased indeed in the way of "archaic specializations," as you express it.*

But I badly miss the principal data on the endocranial anatomy of Sinanthropus, *which I consider to be of the greatest importance in this respect, and probably gave you conclusive proof of the different character of* Sinanthropus *and* Pithecanthropus.

May we expect these data in the near future? You would highly oblige me once more, of course also with the complete casts.

Regardless of his stated willingness to alter his ideas in the light of new evidence, Dubois responds angrily any time Bernsen suggests a different viewpoint. This is a problem when, from time to time, he feels Dubois has made an error in his provisional identifications. The corrections ought to be made now, while formal registration is under way, but Bernsen can find no suitably tactful way to bring up the issue. One day, he commits a fatal error by suggesting in passing that a single, isolated tooth from the Tegelen Clays that Dubois has identified as belonging to a hippopotamus might actually be a pig's. Bernsen requests permission to examine the specimen, and Dubois becomes very agitated.

First of all, Bernsen is not employed to rummage among the Tegelen Clay fossils any further. Second, he, Dubois, has always been a little uncertain about this identification. He has always meant to double-check it. Bernsen knows Dubois too well to point out that twenty-five years have passed since the original identification, which has been repeated in scores of publications, and no such checking has yet occurred. For days afterward, Dubois hammers on the subject of his humiliation at Bernsen's hands, as if this small matter of misidentification had been broadcast to newspapers worldwide. From that day onward, Dubois' suspicions of interference by the Roman Catholic Church, in the person of J. J. A. Bernsen, are always active. All the resentment and regret Dubois has accumulated throughout his entire life seem to bubble to the surface like hot lava rising in a volcano.

"I have not published enough. How little I have done about *Pithecanthropus.*" Dubois mourns miserably one day early in March 1931, forgetting he has published a monograph and nearly forty papers on the subject so far. "I have too little ambition and was satisfied as soon as I knew it for myself. After finding the truth, my interest was gone."

Yet, Bernsen counters silently, you will not allow others to work on it, not even me, nor can I correct you in the smallest detail.

"Only after 1923 did I start to work on *Pithecanthropus* in earnest and to publish the results," Dubois continues morosely. "That will be of little account, that the discoverer says so little and so late about a famous find. And then Osborn was pressuring me through the Royal Academy that I should get the work finished and the publication done, so they will say that I would never have done it without him, and he will get the credit, not me. It has not been enough, what I have said about it. I should have written thick books, like the others who made famous discoveries. My work will be forgotten, overlooked."

When Dubois resumes this self-pitying monologue the next morning, March 3, 1932, Bernsen can remain quiet no longer. The frustrations that have built up over the months of working with this demanding and hypersensitive man explode into words. "You will be forgotten, Professor? You? And what about my work? What about the months and months of time I have put in on your collection? This will always be the Dubois Collection, and a few years from now, who will remember the priest, Father Bernsen, who brought order to the collection? I am allowed only to do the boring tasks, the tedious ones, and for that I get no thanks or recognition."

Dubois stares at Bernsen for a moment, as if he has suddenly sprouted tusks or wings. And then he understands, in an instant. "Oh, Father Bernsen, you must forgive me. I am too absorbed in my own work. I have too rarely taken the time to thank you for your labor on this long task, the sort of thing I do not have the temperament to do myself. It must seem to you as if I have never taken any interest in your progress, never appreciated the many long hours you have put in on my collection." Bernsen is not mollified, though he is glad to see Dubois feels some guilt. "You know, Father," Dubois says ingenuously, "if I have seemed uncaring, it is only because I know in what good hands my collection has been put. I never thought I needed to check up on your work or encourage you, but of course it is discouraging to work without any praise or recognition."

"It is indeed," replies Bernsen, tight-lipped.

"Ah, I understand the need for due recognition," Dubois answers soothingly. Bernsen nods, almost in acceptance of the tacit apology. "And perhaps now that we have cleared the air, we should speak of a more important thing. May I ask you a question? You may answer freely, with a clear conscience, but of course you don't have to answer at all. How do you imagine your future? Do you have aspirations for your career?" Dubois looks at him quizzically.

Bernsen is immediately suspicious. Why is Dubois asking him this? Is he going to hold out the prospect of describing various mammalian groups yet again, only to withdraw the offer the next day? He answers carefully, choosing his words with thought. "I aspire not after a particular post, Professor, but after time . . . time to carry out scientific work. Why do you ask me this?"

Dubois nods, furrowing his brow as if concentrating. "*Ja*, Father, *ja*. I see," he answers. "This is awkward. Initially I had you in mind for my successor at this museum. I have even spoken to you about it, *ja*? You remember, when we were talking about the saving of *Pithecanthropus* . . . But I must return to the subject at hand. Although I myself find you quite unprejudiced in your scientific work, the outside world should find it very strange indeed, if you, a Catholic priest, were appointed the guardian of *Pithecanthropus*. What would the Americans say, who have complained so much about access to the fossils? What about those who spread the false rumors that I was hiding them because of the dictates of the Catholic Church? They would surely see in your appointment a confirmation of those rumors. You can see the difficulty. So I think that while you might be in charge of the Dubois Collection, you can never be in custody of *P.e.*" He finishes strongly, but is unable to look Bernsen in the eye for some seconds.

Bernsen is stunned. So he is to carry out the saving of the collection but is not to touch its jewel. He raises his head and speaks clearly. "I must thank you, Professor, for speaking so frankly from the heart." Dubois still will not look at him. Bernsen knows, by this action, that his cause is completely and utterly lost. "I do not wish to criticize you," he continues evenly, "but I think perhaps it would have been fairer if you had said all this earlier. Now, after eighteen months, the collection is almost safe and the dull work is nearly completed. If that is all you think me suited for, then I shall finish up the registration and organization and then resign my post just as soon as the catalogue of specimens is completed." He will leave with dignity, if Dubois forces the issue.

"Do you mean this?" asks Dubois with transparent delight, raising his eyes once more to Bernsen's.

"Oh, *ja*, Professor, *ja*, I will resign once everything is in order." He cannot mistake the look on Dubois' face; the man for whom he has worked so hard is gleeful at the prospect of his resignation. That evening Bernsen drafts his letter of resignation and the next morning, submits it to the trustees of Leiden University who, after all, are technically his employers. Dubois has never personally had the power to hire or fire him.

Two days later, when Dubois next comes to Leiden to work on the collection, the situation is tense. He and Bernsen exchange only a few words, and Dubois leaves earlier than his accustomed hour. That night, Bernsen writes in his diary:

> MARCH 5, 1932. *No word of regret from Dubois about the sharpening of the case. On the contrary, my suspicion is confirmed by the letter I saw, saying that he (Dubois) is happy about the announcement of my resignation. Only I, the victim, must laugh sweetly as I am killed and be silent, that the outside world shall know nothing of my sacrifice.*

Unnerved by the hostility emanating from Dubois, Bernsen asks for a week of study leave. He does not even wish to speak to Dubois without witnesses present, for he fears that anything he says will be distorted and misinterpreted if it can be used against him. He wishes it to be known that he has done a good, thorough, and responsible job, working with the most difficult of taskmasters. Now he understands the import of those visits from Antje Schreuder and Professor Escher, who resigned their positions after similar quarrels. They were trying to tell him that Dubois would never share the credit for the work, that he uses his assistants and then casts them aside.

After a week of thinking about little else, Bernsen decides to go directly to the burgomaster, the man officially in charge of the Dubois Collection and of its director and his assistants. But when Bernsen arrives in his office to unburden himself of the awful tale of jealousy and quarrel, the burgomaster informs him that Dubois has already been there, complaining about Bernsen.

"What? Dubois has been here, telling you that I am difficult?" Bernsen is astonished. He had not suspected Dubois of dishonesty, only of selfishness.

"*Ja*, Father," the burgomaster answers curtly, watching Bernsen's reaction. The priest seems unsettled, but not guilty, so the burgomaster continues. "Professor Dubois came to me several days ago, telling me how difficult things were with the collection. He spoke for nearly an hour about his troubles. But I must tell you, I watched his eyes as he spoke and thought to myself, 'Man, you have lying eyes.' I do not believe he tells the strict truth."

"But—" Bernsen tries to break in, but the burgomaster carries on without pause.

"There is more, Father. As Dubois was getting up to leave, he told me that you, Father, are to be fired and that Dr. Van der Klaauw will be the future director of the collection. So only at the end did he state the real purpose of his visit. Until then, he only beat around the bush. It made a very unfavorable impression on me at the time," concludes the burgomaster seriously.

"What am I to do?" asks Bernsen.

"Leave the matter with me," replies the burgomaster. "But on no account should you resign your position voluntarily. I shall speak with Professor Dubois again."

The open enmity between Dubois and Bernsen is the talk of the museum. Bernsen hates the gossip, and although his colleagues mostly offer encouraging words, he is embarrassed at the looks he gets from some. It is all so tragic, so avoidable, he thinks. If only Dubois could treat me fairly, could give me the slightest room to breathe . . . Even Van der Klaauw, Dubois' proposed successor, urges him to stand firm and not to quit. "You have earned the directorship," Van der Klaauw tells him. "You now know the collection as well as anyone, except maybe Dubois himself, and he cannot live forever."

A week after Bernsen's interview with the burgomaster, Dubois comes again to work on the collection. Bernsen has thought hard about what attitude he will take toward this man, once so respected, now so feared. "Let us behave in a civilized fashion, Professor," Bernsen suggests. "We are professionals and we will carry out our work in a businesslike manner."

"*Ja,*" answers Dubois sourly. "That is easy for you to say, for you have not had a scolding from the burgomaster, as I have."

Bernsen makes no answer but continues silently to sort and classify and register the bones. He is now dealing with very fragmentary remains, so it takes some concentration to see what each might represent and check it against more complete specimens. Again and again, Dubois attempts to open a conversation about their disagreements, but Bernsen is determined not to start down the road to catastrophe.

"No, Professor," he answers. "We have agreed to be silent about it. It is the best way, or it will all start once again. I do not wish to argue with you, only to complete the work that needs to be done. Now, this specimen"—he changes the subject as he hands a fossil to Dubois—"do you think this is more likely to be a broken metatarsal or a metacarpal?" They discuss the fossil briefly, make a determination, start upon another one.

The day goes on, neither man comfortable, both painfully aware of the harsh words that have passed between them. Each feels put upon, beleaguered by the other. When Dubois rises to go, he turns to Bernsen and asks, "And shall you walk with me to the station, as usual?"

"No, Professor." Bernsen shakes his head, sadly. "I think it is perhaps better that I do not accompany you any longer." Dubois looks so gray and old and lonely as he walks down the street that Bernsen is tempted to run after him. No, he tells himself, it will not do. Dubois has chosen to cast me aside and he will have to live with the consequences of his unfair action.

Months pass, but the tension remains high, their interactions stilted. In May, their quarrel erupts again, for Dubois cannot leave the subject alone. He worries it like a terrier with a rat. "You know, Bernsen, we must talk once more about our relationship. This is all your fault, from the beginning. There is something hostile in you toward me, I have always noticed it. You have repeatedly humiliated me, corrected me, pointed out every error, criticized and questioned my judgments. Even as a small boy I was always treated with special respect. But no, not you, Father, you cannot respect me. You must humiliate me and bring me down out of jealousy at my high position. In recent months I have gone through so much sorrow. It has aged me. I have even wished for the release of death to end this misery. Oh, not that I would commit suicide," he adds quickly, knowing suicide to be among the worst sins a Catholic could commit, "for suicide is cowardly."

Bernsen cannot contain himself, he is so indignant at being accused of torturing Dubois with his criticisms. "Is not the most important thing that the collection be correct? Have you not said this, Professor? Now I see that you are hard and that everything must give way to your interests. I personally mean nothing to you, although for two years I have done the tedious work for the collection, day in and day out. Now I see you differently and my sympathy for you has cooled."

There is no denying the truth of the accusation. Dubois hangs his head for a moment, like a schoolboy. "*Ja,* Father, it is true. I am hard in that respect. I have always felt that everything must give way for the goal, everything must be arranged to serve the ends of science. So perhaps I have driven you too hard and given you only criticism, but it is for the collection, for science. I have driven myself as hard, sacrificed as much. Personally, I have always had compassion for you in this tedious work; I find you a good fellow, you know, Father." Dubois looks up hopefully, to see if these words have appeased Bernsen. But even this admission of his selfishness—the confession of his single-minded pursuit of scientific truth—is not enough to make peace between them. Bernsen thinks bitterly that surely science can be served without savaging others.

The weeks stretch on, full of barely contained ill-feeling. Soon the collection will be properly organized and registered; awareness of this fact heightens the anxiety. Both men know that something must be decided soon about the future of the collection and of Father Bernsen. The better specimens are all completed and registered by now; what remain are fragments, small, random bits and pieces many of which can be classified in only the most general way, by size and body part. Bernsen is scrupulous to the end, however, determined to make no mistakes or omissions for which he could be castigated. If leave he must, it will be with pride in a job well done, professionally done, even if it is unappreciated by the one whose opinion weighs most heavily.

Working systematically through some boxes of assorted rib fragments, Bernsen pulls out one that strikes him as particularly odd and troubles him. Both ends are broken off and it is rather thick. True, the ribs of large animals (such as elephants, buffalo, and giant deer) are just this thick; but the shape . . . the shape is wrong. He sets it aside to show to Dubois, as he does everything that strikes him as peculiar or as misidentified. Going through these oddities is always a trial, an occasion for possible explosions, but Bernsen will not neglect a single fossil. On June 1, 1932, he brings the specimen to Dubois' attention.

"Professor," he opens the conversation carefully, holding out the fossil for Dubois to see, "have you time to look at this specimen? It comes from

THE DILIGENT ASSISTANT   403

the box of rib fragments that I have been sorting through. I think it is peculiar, not like the others. Perhaps it is something important."

"Umm? Oh, *ja, ja,* I will take the time. Now, what is this? I see the problem, *ja.* It is different from the others." He wanders off with the specimen in hand, examining it inch by inch, opening drawers or boxes now and again to make comparisons. After about half an hour, he comes back to Bernsen with a smug look on his face.

"Father, you have indeed found something. After a great deal of inspection and many comparisons with other fossils, I think it is another piece of a femur of *P.e.* Imagine! After all these years, to find another piece. . . . This will be most important if it is true. You must look very carefully for the other broken parts. I'm sure there will be some additional pieces that glue onto this, that will make it more complete. I want you to set aside all other work, and get the technician Van der Steen to help you, too, until you have gone through every single fragment that might be part of this same bone."

Bernsen doubts that this fragment is a piece of *Pithecanthropus'* femur, but he does as Dubois asks. Over the next hours, he and Van der Steen turn up two more similar pieces out of the thousands of fragments. They bring these additional pieces to Dubois as they find them, noticing that the older man grows more—not less—dissatisfied, as if he hoped for failure. Before the end of the day, he and Bernsen once again have harsh words over Bernsen's cruel enjoyment (so Dubois thinks) in catching Dubois out and finding him wrong.

The next day, all three men arrive early to continue the search. Dubois announces, "After a close comparison with the original femur, I have arrived at the firm conclusion that the long bone you pointed out yesterday is indeed a part of the shaft of the femur of *Pithecanthropus.* You and Van der Steen must continue to search diligently among the boxes containing fossils from that same area, where this fragment was found."

Some hours later, elated, Bernsen comes again to Dubois, finding him lost in rapt inspection of the new femur. "Professor," he bursts out, "I have found it! I think I have found a piece that fits onto the one I found yesterday. May I try it, please?" And the two men, forgetting their disagreements, stare in wonder as the two pieces fit together so neatly that there is almost an audible click. "There," Bernsen comments. "It is clear. You are completely correct, Professor, it is the femur of *Pithecanthropus.*"

"*Ja,*" says Dubois heavily. "It is. Another left femur of *Pithecanthropus.*" He sits silently staring at the fossils for a moment. Bernsen is jubilant but

Dubois smells only disgrace. "This is a solemn moment," Dubois proclaims. "Until now, I stood alone. But now I have a second individual of *Pithecanthropus,* for this is surely a left femur just like the one we found in the first place. How they will look up and take notice in Berlin!"

Suddenly Bernsen realizes the implications of the find: not only has Dubois missed a bone belonging to *Pithecanthropus,* but now there are two individuals. The argument that the original skullcap, femur, and tooth must come from one individual, because there is no other primate in the vast collection, dissolves into nothingness. The harsh accusations and disbelief thrown at Dubois in those early years flood into both men's minds, especially Virchow's stormy insistence that Dubois had created a chimera out of several individuals and probably out of several different animals. What of all of the years and years of publications on brain weight and body size, based on that one skullcap and one femur? Are they, too, all swept away, drowned at sea? Yes, they must be, if the femur and the skullcap are not from the same individual. Two left legs: there could hardly be a more damning find. Bernsen can hardly look at Dubois' face, so clearly is the tragedy of this new scientific truth written there. Dubois' whole life's work is cruelly threatened.

"Still," Dubois remarks, trying to save the situation, "these pieces must certainly have come from the excavations of 1898 and 1900, after I had left the Indies. Otherwise I would have gone through the boxes more carefully and spotted these bones at the time. It is only that material that I have not examined thoroughly."

"In the box," points out Bernsen, "there was a note lying near these pieces that said B. 120." They both know that this is the standard code for the 120th box of the second, or B, shipment. That makes the specimen one excavated during Dubois' time in Java, not later. After a moment, Bernsen adds, kindly, "Of course, we both know that many of the notes were mislaid and mixed up during the moves. And it was you yourself who said to me, Professor, 'Pay careful attention to the small fragments, to see if there is not something yet of *Pithecanthropus* in there.' Of course, we know that to find a large piece of *Pithecanthropus* now is impossible, for those fragments you have checked very well."

Dubois does not answer, only sits staring at this new piece of left femur. Then he says with a sigh, "We must carry on. I shall soak the pieces in thin glue, to strengthen them, and then glue them together. You must continue to search for more pieces."

Although he is sympathetic to Dubois' dark mood, Bernsen can scarcely contain his exuberance as he walks out of the room, back to where Van der Steen is working. "We've got it!" he cries. "It fits!" Van der Steen is excited, but no more so than Bernsen himself. Bernsen feels as if his months of patient work have been vindicated, for he, too, has made a great discovery. More pieces are found, until there is a total of eight new fragments. Yet Bernsen worries over the long-term effects of these finds on his fragile and troubled relationship with Dubois.

> JUNE 4, 1932. *Dubois will surely find it unpleasant that I shall have a share in the discovery. If I had not taken out the first piece that was with the rib fragments, as a piece that struck me as peculiar (as I usually do, for I always show him such pieces), then it would have been maybe not for years, or ever, that someone would have turned to look at it. I am convinced that this will not soften his determination, restated only yesterday, to apply for my discharge. Even after the finding of the* Pithecanthropus *material, he was actually already speaking with others about how much longer it would take me to finish the work.*

It proves to take much less time than either of them expects. Hours after writing these words, Bernsen falls seriously ill. The next day, despite all the doctor can do, he dies of internal hemorrhaging, perhaps from a bleeding ulcer. No one had suspected the toll the incessant hostility was taking on Bernsen's health.

On June 25, Dubois presents the new material to a meeting of the Royal Academy of Science. The story he tells is compelling but not entirely honest. Dubois feels himself so vulnerable that he cannot bear to tell the whole truth. The only other man who knows the truth, Bernsen, is dead, after registering 10,411 fossils from Trinil.

> *Forty years ago the two principal skeletal remains, the skullcap and the femur, of* Pithecanthropus erectus *were excavated at Trinil, Java. It was then supposed by the author of this species, that both were remains of the same organism, species or even individual. The skullcap indeed so closely resembles that part of the body in the Anthropoid apes, especially the Gibbons, on one hand, and in Man, on the other, that the name* Pithecanthropus, *for the genus, is fully appropriate. The name* erectus *was given to the species on account of the strikingly human-like essential features of the femur, which*

imply erect attitude and gait. Together with those features, however, the Trinil femur, in the opinion of the author of the species, presented important differentiating characters, so that he found it possible, at least, to regard the skull and the femur as having been parts of one organism. . . .

The prevailing view on the Trinil femur, however, at present as well as in the past, is to consider it absolutely . . . human. . . .

It thus appears clearly that we have to regard the femur of the Pithecanthropus *species as the true key bone to his frame, a key admitting us to the knowledge of its organization. Now, as to know a species well one single individual is insufficient, what we have wanted, for forty years, are thigh bones of other individuals of the described species, to ascertain if the particular features seen in the femur of 1892 are essential characters of the species . . . or mere individual differences. . . .*

In these forty years, no remains of another "pithecanthrope" came to light . . . till the first days of this month other overlooked pieces appeared, three [partial] thigh bones of the described species.

On that day, at Leiden, in my Java collection, from a lot of inconsiderable fragments of ribs from different Trinil mammals, which I was minutely examining, were separated some dissimilar fragments, not belonging to ribs. Amongst them was a bone a foot long, still partially covered with rock, which my diligent assistant in the arranging of the Collection, Dr. BERNSEN, whose loss we now deplore, had put aside for my inspection, because he regarded it as a dubitable piece of deer's horn. To my great joy, I soon recognized it as the shaft of another Pithecanthropus *thigh bone. It presented some of the same characteristics which differentiate the Trinil femur of 1892 from a human femur. Then, searching further through the rib fragments for similar pieces, that might possibly fit, I found the defective upper extremity to that thigh bone shaft, which was broken off beneath the small trochanter, and 6 other pieces of different sizes, which enabled me to compose two more shafts of* Pithecanthropus *thigh bones.*

None of the three is by far comparable with the splendidly conserved femur of 1892, but they all unquestionably belong to the described species.

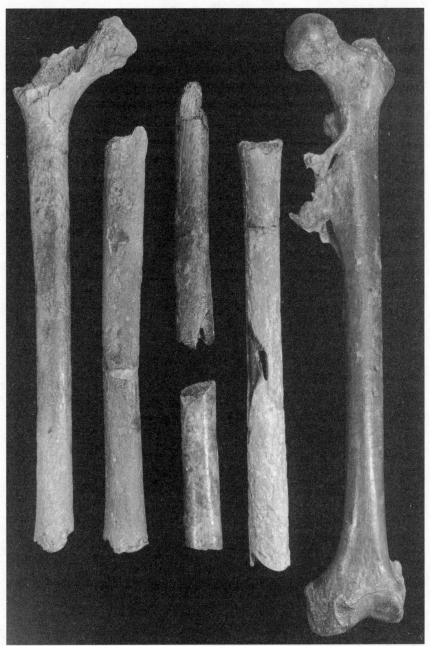

In 1932, Bernsen discovers the first of eight new fragments of Pithecanthropus *femur among the unlabeled fragments from Trinil. These finds disprove Dubois' long-held assertion that the original* Pithecanthropus *femur (at right) must come from the same individual as the molar and skullcap.*

Dubois attributes the new specimens to the excavations carried out under Kriele's direction in 1900, so that (although the exact site of the find was not noted at the time, as Kriele did not see their importance) they must have lain within sixteen and forty-eight meters of the skullcap, much farther apart than the original femur and the skullcap. For this reason, and because the new bones are much more heavily corroded than the original finds, Dubois argues that the new bones are from other individuals, now represented by two new but incomplete left femurs and two new but also incomplete right femurs. More to the point, all betray the very same features that initially convinced Dubois that the femur of *P.e.* was distinctly different from that of man. These anatomical differences, he hypothesizes, are due to a more tree-climbing habit in *P.e.*, although its primary means of movement was walking upright on the ground.

He concludes his lecture with a characteristically strong statement:

> *The morphological evidence acquired proves beyond a doubt that the skull and the femur which were excavated in 1891–1892, can have been associated in one and the same organism, a distinct species. . . .*
>
> *I still believe, now more firmly than ever, that the* Pithecanthropus *of Trinil is the real "missing link."*

The only mention of Bernsen's role in the entire discovery is the parenthetical remark about the "diligent assistant . . . whose loss we now deplore." Bernsen is replaced by Miss M. Sanders, who between October 1 and December 31, 1932, registers another 569 specimens. Dubois takes care that she never develops professional aspirations about the collection. Sanders is followed by her future husband, the herpetologist L. D. Brongersma, who brings the total number of registered specimens up to 11,284. Before long, Bernsen's heroic work on the collection is all but forgotten by most. An exception is, of course, Dubois himself, who can never seem to get the priest out of his mind.

CHAPTER 53 NEW SKULLS FROM JAVA

The importance of the new fragments of the femur of *Pithecanthropus* is overshadowed by other Javanese discoveries.

In August 1931, the Geological Survey of Java sends C. ter Haar to draw an accurate map of the Kendeng Hills in central Java. Ter Haar sets up his camp in the village of Ngandong on the Bengawan Solo some six miles north of Trinil. On August 27, he notices a series of three layers of gravel and sand that must be old river terraces, about sixty feet above the present bed of the river. One of them is full of fossils. Among the first specimens he pulls out is the remarkable skull of a giant water buffalo with a horn span of some seven feet.

When Ter Haar sends word back to survey headquarters in Bandung, the acting director, W. F. F. Oppenoorth, sends a team of trained Javanese to work systematically at Ngandong. Ever since participating in the Selenka expedition to Trinil in 1907, Oppenoorth has believed as an article of faith that there are great fossil discoveries yet to be made in central Java, but in his most ambitious dreams, he has not anticipated what is to come. Routinely, the men number all good specimens recovered from excavation at Ngandong. Number 29, found on September 15, is a large part of a fossilized human skull, although the Javanese collectors do not recognize this and label it a tiger skull. Another strange object—an "ape skull"—is found on September 30. In Bandung, Oppenoorth is startled to realize that both are archaic human skulls. He reaches Ngandong on October 21 to discover that yet another fragment of human skull was recovered eight days earlier. Oppenoorth brings along a young German paleontologist employed by the Geological Survey of Java to help: G. H. R. von Koenigswald, known as Ralph.

Oppenoorth and Von Koenigswald are stunned to find that their men have found an unprecedented three fossil hominid skulls. They institute a second numbering system, using Roman numerals, for the hominid skulls, which continue to be found over the subsequent months and years of work. It is a paleontological treasure trove. After the three finds in 1931, Skull IV is found on January 25, 1932; Skull V on March 17; and Skull VI on June 13. The skull that becomes number VII is found on May 24, before Skull VI, but is not recognized until later. Then there is a lull; no more are found until the last week in August 1933 (Skull VIII). Skulls IX and X are excavated on September 27 and the final skull, XI, is located on November 8, 1933. None of the eleven is complete; their faces are missing and the bases of the skulls are broken away. There are also thousands of animal bones.

Oppenoorth starts to write up the hominid skulls immediately, without waiting for the good luck to run out. Tactfully, he sends a copy of his first article on the new specimens to Dubois with a personal letter dated May 11, 1932, although they have not met before. Still, one aspect of the importance of the

Ngandong skulls is that they have been found so close to Trinil, the home of *Pithecanthropus,* and *Pithecanthropus* is Dubois' fossil. What Oppenoorth has written is a "provisional description" based on Skull I, nothing more. He compares the new skull to *Pithecanthropus, Sinanthropus,* various Neanderthals, the Wadjak skulls, and what he calls the "average" Australian Aborigine. He finds the new material unmatched by any previously found fossils and so proposes a new name, *Homo (Javanthropus) soloensis.* Dubois is fascinated and flattered by receiving this preliminary report. It is the beginning of a long correspondence, for Dubois writes to Oppenoorth with many questions about the brain size and endocast of the new skull. He is gratified to receive a prompt reply from Oppenoorth:

May 31, 1932

I have already received from Professor Black three casts of the Sinanthropus *material for comparison. Of your* Pithecanthropus *I have personally paid for a cast by Kranz in Bonn . . . but the* Pithecanthropus *cast is very bad. I hear that there are at present new casts of it, on which one can see the inside of the skull. If you are able to help our museum obtain such a cast, possibly also one of Wadjak man, then it will be certainly a pleasure for us to send you in due time also a cast of the Ngandong skull.*

The new skulls are, according to me, certainly not identical with the Wadjak type, but point to a more primitive human, something like the Neanderthal type. When you have seen my provisional description, you will probably agree with me.

I believe that with this I have answered all your questions, and I would like to make a few requests, namely if you can still help me to get reprints of your description in the Academy of the Pithecanthropus *and* Homo wadjakensis *(the description of 1894 I already own). You would greatly oblige me further by sending a few photographs of the Wadjak skulls.*

As you will perhaps remember, 25 years ago I went with the Selenka expedition, con amore, to Indonesia to continue your work at Trinil. After I went into government service, there was no longer any opportunity for such work, until some years ago when I became the leader of the newly established Geological Survey of Java, and with it came the possibility of bringing my earlier experience into play again.

Now, after 25 years, my first activities are crowned with success, and all those years I have had an interest in, and have kept myself apprised of

*further finds. You will understand that this completely unexpected find is
for me a great satisfaction, and you will certainly understand that I am
not inclined to let the work on the fossils out of my hands without protest.
I write this to you because I received a proposal to entrust the fossils to
others, which came from Professor Mijsberg in Batavia, who used your
name. [Mijsberg was one of the scientists to assess Heberlein's "skull" find.]
Hence, it is for me a great pleasure to hear from you personally on the matter. Of course I am completely willing to work together, and there are
themes, as for instance the endocranial anatomy, on which I will not
venture myself.*

Soon after receiving Oppenoorth's communication, Dubois publishes a
brief letter to the editor in the scientific journal *Nature* on the subject of the
new skulls. He endorses Oppenoorth's view that the new skull is a representative of a primitive human race, though it is less primitive than Wadjak
man in Dubois' view. Oppenoorth's error, Dubois suggests, springs from the
fact that the second, more fragmentary Wadjak skull has not yet been fully
described and is therefore less known to Oppenoorth. It is this second skull
that more closely approaches the new Ngandong skull, so he suggests that
they are "one identical type," a "proto-Australian." He reiterates this view in
a supportive yet tactful letter to Oppenoorth.

June 15, 1932

*I completely agree with you, that the fossil man of Ngandong is nothing
to do with the Neanderthal, but that it stands nearer the Australian
type.... But your description leaves little doubt in my mind, that Ngandong is a type identical to the Wadjak-man. The second (fragmentary)
skull from Wadjak that I have mentioned, but not further described, is
closer to the Ngandong skull than the other.... In my provisional opinion,
that is the greatest significance of this find. It is in any case a scientifically
valuable object.*

Soon afterward, Oppenoorth describes Ngandong IV and V in a June 1932
issue of *De Mijningingenieur* (The Mining Engineer). Skull V is the most
complete, missing only the lower part of the face and the base of the skull.
While all the skulls have remarkably thick vault bones and large brow
ridges, Skull V is gigantic. Oppenoorth estimates its cranial capacity at 1,300
cc, as large as many modern Europeans and larger than, for example, the
skulls of most Australian Aborigines.

Oppenoorth raises an interesting new issue as he tries to make some sense of the distribution through time and space of the various types of hominids that are now known from their skulls. In his eyes, the easiest way to align the fossils into ancestor-descendant sequences is to create two parallel but separate lineages, but how is this to be done? Initially, Oppenoorth proposed a geographic separation, making a European sequence (in which *Palaeanthropus heidelbergensis* → *Homo neanderthalensis* → *Homo sapiens fossilis*) and an Asian lineage (*Pithecanthropus* → *Sinanthropus* → *Homo soloensis* → *Homo wadjakensis*). Now he sees it is possible that the key factor is climate. Perhaps in temperate China and Europe *Sinanthropus* evolved into *Palaeanthropus heidelbergensis,* which evolved into *Homo neanderthalensis* and then into *Homo sapiens fossilis,* while in the tropics *Pithecanthropus* evolved into *Homo soloensis* and then into *Homo wadjakensis.* Dubois makes no response to Oppenoorth's suggestion, but the two men continue a cordial correspondence and Oppenoorth is welcomed by Dubois at the Teyler Museum where he examines *P.e.* and the Wadjak skulls, leaving a brain cast of the Ngandong I skull for Dubois' use. Thus, when the brain casts of the two *Sinanthropus* skulls arrive from China on February 8, 1933, Dubois can compare them with those from both Trinil and Ngandong. To Dubois, the differences are clear. *Sinanthropus* is apelike, though not so apelike as *Pithecanthropus; Homo soloensis* from Ngandong is simply a primitive human. The greater importance of the Chinese fossils is doubtless why Dubois nominates Black, not Oppenoorth, for the Prix Hollandais in 1934.

Dubois' next task is to prepare a detailed comparison of the brains of *Sinanthropus* and *Pithecanthropus,* research which he presents at the April 29, 1933, meeting of the Royal Academy. Because *Sinanthropus* and *Pithecanthropus* are much more apelike and much older (*P.e.* being the oldest) than the more humanlike Wadjak and Ngandong remains, it is the former skulls that record the earliest chapters of the evolution of the human brain. Ever the skeptic, Dubois takes the precaution of checking the external measurements of the endocast against those in Black's publications, to make sure the casts are accurate (which he concludes they are).

His first substantive assessment of the brains of *Sinanthropus* and *P.e.* startles the listeners at the Royal Academy meeting: "There is obviously little difference in size between the two brains, in *shape,* however, they are surprisingly unalike." Black's preliminary reports suggested that the brain of *Sinanthropus* was bigger than that of *P.e.,* by perhaps 25 percent. Is there, then, "little difference in size" as Dubois asserts?

The issue is that the most recently found *Sinanthropus* skull is smaller than the first. Dubois adjudges the new *Sinanthropus* braincase to be approximately 918 cc in volume, close to that of *P.e.* Although the new Chinese skull is of an adolescent, that fact cannot explain its unusually small brain.

> *Such a volume . . . is certainly a very low one for a human skull, as this* Sinanthropus *undoubtedly is. For at the age of this early adolescent human individual the volume of the brain is almost equal to that of the adult. . . . The shape and the major features of the* Sinanthropus *skull, on the contrary, are those of a full-grown male Neanderthaler. . . . We meet here with a contradiction of cranial form and cranial capacity, a contradiction emphasized by the other* Sinanthropus *skull, attributed by DAVIDSON BLACK to an adult woman. In contradistinction with the adolescent skull it, indeed, exhibits true female features. It is difficult to estimate the capacity of this very incomplete cranium; however, 1,150 cc will probably not be too high an estimate. In proportion to such a female capacity a normal adult male of the same race should have about 1,300 cc capacity. However the adolescent* Sinanthropus *exhibits adult morphology in combination with a brain volume very much smaller than the normal one of his age.*
>
> *This is a contrast which is perfectly unconceivable* [sic] *if we consider this* Sinanthropus *youth as a normal individual. . . . I may express my opinion that the adolescent* Sinanthropus *is a human male, belonging to the Neanderthal group of mankind . . . with an individually imperfect and hence abnormally small brain.*

Ironically, Dubois finds himself falling back upon Virchow's favorite excuse of old for any anatomical variation in ancient skulls: pathology. Later, when additional *Sinanthropus* skulls are found, this excuse must be abandoned, for even adults of *Sinanthropus* have brains close in size to that of *Pithecanthropus*. Black's original estimates of brain size were too high.

Dubois also emphasizes the difference in shape between the two endocasts, illustrating this point with a pair of photographs. Taken from the top and right side, these images unfortunately make the endocasts look remarkably similar to the untrained eye, notwithstanding Dubois' conclusion to the contrary. To Dubois' expert eye, the differences are evident and attest to these being related but quite different species, but his argument is unconvincing to most. The greatest differences in shape appear in side or

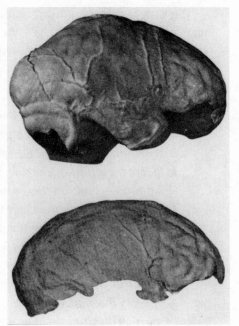

Dubois compares the brain casts of Sinanthropus *(top) and* Pithecan-thropus *(bottom), intending to high-light their differences but inadvertently emphasizing their resemblances.*

In lateral view, the brain casts of Sinanthropus *(top) and* Pithecanthropus *(bottom) differ moderately in shape.*

lateral view, but the *P.e.* endocast is so much less complete than the *Sinanthropus* skull that the comparison is difficult to make. However, Dubois is certain of his conclusions; he describes the "oblong and narrow" brain form of *Sinanthropus,* contrasting it with "the more rounded, broad form of *Pithecanthropus.*"

This is the beginning of a new trend in Dubois' research. The similarities between *Pithecanthropus* and *Sinanthropus* are beginning to trouble him, for his *P.e.* is losing some of her uniqueness. He tries steadfastly to maintain her "missing link" position, emphasizing the primitive, even apelike fea-tures of her anatomy. In 1935, Dubois publishes a paper entitled "On the Gibbonlike Appearance of *Pithecanthropus erectus,*" an astonishing move for the man who so vehemently fought Virchow's early suggestion that *P.e.* was naught but a big gibbon. But now things are different: he needs to emphasize the gibbonoid features of *P.e.* to make sure that *P.e.* remains dis-tinct from *Sinanthropus.* His fossil is apelike; Black's is humanlike. He adds his morphological observations to the results of his research into the proportions of brain weight and body weight, observing that *Pithecanthropus* occupies a position in cephalization that is perfectly intermediate between the apes and man.

> *It is therefore possible to arrange all the groups of Mammals existing at present in a geometrical series of progressive . . . cephalization, in which series there are no gaps (even when fossil groups are not taken into account), with one sole exception. . . .*
>
> *The only real void space in the series is between Man and the anthropomorphous apes (incl. Gibbons). This void marks the placement of* Pithecanthropus. . . .
>
> *The strongest evidence of the gibbonlike appearance of* Pithecanthropus, *however, is that given by the volume of the psychoencephalon, exactly doubled in relation to the body weight computed from the gibbonlike chief dimensions of the femur.*

Once again, the calculations reinforce Dubois' main conviction: *Pithecanthropus* is an ape-man, *Sinanthropus* is an early man, and Oppenoorth's *Homo soloensis* is modern *Homo sapiens,* part of a ring species that encircles the globe, varying somewhat from location to location. However, Dubois agrees with Oppenoorth that *Homo soloensis* is no Neanderthal, since the Ngandong tibias do not show the distinctive Neanderthal morphology.

Unfortunately, the harmony between Dubois and Oppenoorth does not carry over to Oppenoorth's successor at the Geological Survey, young Ralph von Koenigswald.

CHAPTER 54 A WORTHY OPPONENT

Von Koenigswald is a German, not tall but beefy of build, with somewhat coarse features and a blunt manner. He is also very smart, hardworking, and acutely ambitious to make his scientific mark in the Indies. His career at the Geological Survey of Java began back in 1931, when he was charged with establishing the stratigraphic and chronological framework of the island, using mammalian species as markers for deposits of different ages. Only eight months after his arrival in Java, he traveled to Ngandong on June 18 and 19, 1932, with Ter Haar, to witness the excavation of Skull VI. From the rural station of Paron, they rode in a two-wheeled horse cart to Ngawi and then walked the rest of the way through the steaming teak forests. When they finally arrived at Ngandong, Von Koenigswald was staggered by the scene that met his eyes.

> *Our excavation site lay a few hundred yards from the village houses. Our workmen had dug a pit 10 feet deep, the floor of which consisted of marly rock—a sign that the terrace gravel went down no deeper. One gravel bank had been only half dug away, and a few palm fronds stuck in the sand marked the spot where the skull lay buried. We removed the fronds and Ter Haar began to dig carefully with his hands, while I took photographs. Unfortunately, I was so excited that most of the shots were underexposed. After only a few minutes we came upon a large round object: this must be the skull. It proved to be the underside of a human skull; the cranium itself was still embedded in the gravel. In the case of every previous skull the underside had been in fragments; here at last it was intact. The foramen magnum was undamaged, but the place where the cranium had been knocked in was rather more to the front. In spite of a prolonged search we could find no trace of the facial section or the jaws. The skull itself was in an excellent state of preservation and undoubtedly the most perfect specimen discovered at Ngandong. We cleaned it as well as we could on the site and packed it carefully, for the final preparation could only be carried out at Bandung. This was the first skull at whose disinterment I had personally assisted.*

Von Koenigswald returns to the site twice more, witnessing the recovery of Skull VIII and Skull XI, but he longs to discover a new site, where more and more important fossil skulls will be found under his direction.

He begins prospecting in the Sangiran region, northeast of the central Javanese city of Solo, in 1934, near where Raden Saleh's fossils were found. The region's fertile soils are now prized by local farmers, who plant rice, maize, peanuts, papaws, and coconut palms. Von Koenigswald arrives with his native collector Atma, announcing to the local populace that he is interested in fossils. They obligingly produce fossil bones, teeth, and shells that they have found in the fields; Von Koenigswald pays small amounts of money for good specimens. Soon most of the local villagers are an avid collecting and excavating force. After some weeks, Von Koenigswald returns to Bandung, leaving Atma in charge of purchasing further finds.

Though he believes firmly that he has found an important site, perhaps as old as Dubois' Trinil site, Von Koenigswald's enthusiasm cannot counteract the realities of global forces. A disastrous economic crash, begun in 1929, has continued unchecked and the value of the Indies' main exports (rubber, sugar, tea, coffee, and tobacco) falls to a fraction of its former worth. Plantations fail and close monthly, their former workers returning to their villages and dire poverty. Too, the first groundswell of a movement for native independence begins. The colonial government responds decisively, disbanding native political parties and arresting their leaders. Salaries for government workers are slashed and all nonessential government employees—including Von Koenigswald—are fired.

Von Koenigswald asks Atma to stay at Sangiran and keep an eye out for good fossils. In 1935, Von Koenigswald and his family survive on his part-time work at the Geological Survey and his wife's salary as a teacher of German language and literature. Nonetheless, Von Koenigswald invites a French colleague to visit and inspect the Javanese fossils. Father Pierre Teilhard de Chardin is a perfect choice for colleague and possibly savior. He is an unusual character, a tall, thin, hawk-nosed Jesuit priest with a passionate interest in paleontology and archaeology. He has worked at Piltdown in England with Charles Dawson, in China with Davidson Black, and in India with the geologist Helmut de Terra. He is a well-known and internationally respected authority in the mammalian paleontology of Asia.

In 1936, a Javanese collector employed by G. H. R. von Koenigswald finds this fossil child's skull at Modjokerto. When Von Koenigswald identifies it as Pithecanthropus, *he starts a feud with Dubois.*

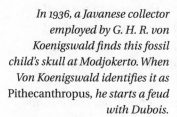

When Teilhard arrives in Batavia in January 1936, things are grim indeed. Von Koenigswald is still out of work and his wife, Luitgarde, has been stricken with typhoid. Despite the terrible strain, Von Koenigswald is eager to show Teilhard the fossils and discuss their interpretation. The two men leave for central Java so that Teilhard can inspect the localities himself; they forge a firm friendship during days of walking and nights of camping in tiny kampongs.

Teilhard leaves Java convinced of two things. One is the great importance of the material Von Koenigswald is finding; the other is the urgent need for institutional support for Von Koenigswald's work. He recommends that Von Koenigswald write to John C. Merriam of the Carnegie Institution of Washington, and the younger man complies, contrasting his complete lack of financial support with the bright promise of spectacular fossil finds. "I have found a new fossil locality here in Java," he writes to Merriam. "If *Pithecanthropus* is to be found anywhere, it will be here."

Von Koenigswald's fortunes begin to change. In late February 1936, one of his collectors, Andojo, finds a fossilized child's skull in the village of Modjokerto in East Java. Along with fossils of other species, the skull is sent to the Geological Survey in Bandung, where Von Koenigswald gleefully recognizes its importance. Here is the find he has been searching for. He announces it in a newspaper article, "*Pithecanthropus* Child's Skull," in *Het Algemeen Indische Dagblad* on March 28, 1936.

> *During the stratigraphic survey of the region SOERABAJA by the Geological Survey of Java, to determine the stratigraphy and age of the different strata, we also systematically searched for fossils. Only a few weeks ago was found a humanlike skull. It is a small skull, not more than 14 cm long! Consequently, this is absolutely the smallest skull found of a human up to this time. Unfortunately, it was found on the surface and is pretty well weathered. On the left side is mainly the skullcap with the upper rim of the orbits and the bone surrounding the left orbit. . . .*
>
> *Although [the skull is] more than 4 cm shorter than the skullcap found by Dubois, the height of the skull (measured perpendicular to the greatest length) is nonetheless the same in the two cases. This is one of the reasons I consider this small skull to be* Pithecanthropus. . . .
>
> *Geologically it is very important that our find from Modjokerto comes from a layer which, to judge by the fossil fauna in it, must be*

older than the layer in which the well-known fossils of Trinil are
found. In Trinil, there is underneath the fossiliferous layer a vol-
canic breccia, which also contains bones but fewer than in the
younger stratum above. This breccia is as old as the layer in which
our new find has been made. It seems possible that the Pithecanthropus *find from Trinil also originates from this lower level and*
that the place where it was found was a secondary berth, meaning
that Pithecanthropus *is older than we have supposed. Unfor-*
tunately Dubois has never published accurate statements about his
sites; at the same time, we know next to nothing about the fossils in
his very extensive collection.

Von Koenigswald sends a copy of the article directly to Dubois, who receives
it on April 15. Perhaps Von Koenigswald hopes for support; what he provokes instead is outrage.

Who is this upstart Von Koenigswald to criticize Dubois' work and suggest his geology is wrong? What has Von Koenigswald ever discovered? It is
certainly Von Koenigswald who is wrong, not Dubois. Dubois' rebuttal
appears in two Dutch newspapers on April 18, 1936.

On April 15 I got, as a cutting from De Preangerbode, *the article . . .*
by Dr. G. H. R. von Koenigswald in which appeared pictures of the
new skullcap of the "young Pithecanthropus*" from right and top*
views. . . . It is immediately clear to me that we are not dealing with
the skull of a Pithecanthropus *but of a REAL HUMAN CHILD,*
apparently of the Wadjak race. . . .

That this "new Pithecanthropus*" is a human, Dr. von*
Koenigswald might have known if he had noticed an article in the
Proceedings of the Royal Academy of Science in Amsterdam, *vol.*
36, (1933), titled "The Shape and the Size of the Brain in
Sinanthropus *and* Pithecanthropus*," and another article, in the*
same journal, "On the Gibbonlike Appearance of Pithecanthropus
erectus*" (vol. 38, 1935).*

The complete humanity of the "new Pithecanthropus*" cannot be*
doubted and it is entirely probable that this child skull did not
come from the geologic formations, in the region northeast of
Modjokerto, that contain "the remains of fossil vertebrates." The col-
lectors must have mixed up specimens collected on the surface and
those deeply buried, as may also have happened at Ngandong. . . .

Science requires urgently that research on such fossils not remain in the hands of the very few scientists present in Java, but instead be carried out in Europe.

To this it may be added that it is absolutely untrue what Von Koenigswald says, that "the original skull of Pithecanthropus *was found by an overseer and sent to Dubois, so we cannot be certain that it came from the Trinil horizon; it was probably washed in from the older Djetis stratum, after which it was found by the overseer."*

Both the Modjokerto skull and the Trinil skullcap were found by natives. If Javanese testimony about the location and manner of the find is not to be trusted, then the in situ placement of the Trinil fossil is more credibly documented, by the notes of Kriele and De Winter.

On May 5 and 7, Von Koenigswald replies hotly to the implication that he has made a geological as well as a paleontological error. If Dubois is right, then Von Koenigswald is unforgivably wrong, so the younger man counterattacks. It is the first skirmish in a long and bitter battle.

It is inconceivable how Professor Dubois from Holland, based on my first short publication in a newspaper, dares to condemn an entire case. I must repeat it once more: there exists no doubt that we deal here with a real fossil skull.

With regard to the circumstances of the find, Dr. J. Duyfjes, who surveyed this area, says, "I can give Professor Dubois the positive assurance that confusion over the finding of the new skull can be absolutely excluded. It was excavated by our collector Andojo from a pit one meter deep in a hard conglomerate sandstone, which certainly belongs to the fossil-bone layer of Modjokerto."

There is no change in my opinion of the identity of the skull, and before long I will publish on the subject copiously in a scientific periodical.

Professor Dubois' Wadjak man is a real primitive fossil of Homo sapiens, *according to a recent publication . . . and has nothing to do with either this new find nor with the Neanderthals of Ngandong. I can assure Professor Dubois that the eleven skulls from Ngandong are real fossils and belong to the Pleistocene fauna with which they have been found. . . .*

The publications Professor Dubois cites are known to me but have no direct bearing on this find.

So now it is clear how the matter stands with Professor Dubois.

Wouldn't it be better for this scientist to withold his judgment on the skills of paleontologists working in Indonesia, with respect to working on important finds made in this country, until the appearance of the definitive publication?

In the midst of this vigorous defense, Von Koenigswald slips an important change into his account. He initially said that the Modjokerto child's skull was found on the surface; now it is an excavated fossil, found in a one-meter-deep pit. Was he originally careless in questioning the collector Andojo about the circumstances of the find, or is he now recasting the truth for convenience? No one will ever know, except Andojo. Years later, scholars try to unravel the truth of the conflicting stories by asking Andojo to take them to the find-spot. Unfortunately, Andojo is by then elderly and somewhat confused about events long past. On different days, he takes different scholars to different places where, he always asserts, he found the Modjokerto skull.

Despite their vitriolic exchanges, Von Koenigswald sends Dubois additional pictures of the new skull, which Dubois acknowledges in a series of articles published in the Dutch newspaper *Nieuwe Rotterdamsche Courant* though he does not change his mind about the human identity of the skull.

Before long, Von Koenigswald begins to modify his position, dropping the idea that this skull is *Pithecanthropus* and calling it *Homo modjokertensis.* A newspaper report of a lecture in late August says,

> *Concerning the new find, the speaker thinks it is probable, but not proven, that this is a skull of a young* Pithecanthropus, *especially now that stone tools found in Java indicate that we must expect to find a more developed human type. The speaker thus thinks it is better to give the find a new name,* Homo modjokertensis, *since it will only be possible to decide whether this is a child of* Pithecanthropus *or a somewhat higher fossil human once new finds have been made.*

A temporary truce is established between Von Koenigswald and Dubois, though the former maintains steadfastly that the child's skull was not a surface find. The identification of the child's skull with *Pithecanthropus* becomes "a POSSIBILITY . . . although its appearance is also really different from the calvaria of Trinil." Years later, Von Koenigswald attributes this change of name to a desire to be courteous to the aging Dubois. At the time, it looks like a sign of uncertainty.

CHAPTER 55 TO THE BATTLEFRONT

The campaign against Dubois is soon joined by Pieter Vincent van Stein Callenfels. A striking and notorious individual, Stein is fully six feet tall and weighs over three hundred pounds, with an enormous beard and a thundering voice to match his size. He is an old Indies hand, having been a planter, an administrative officer, and an inspector for the Dutch Indies Archaeological Service. The newspaper *Handelsblad* refers to his "glorious disdain for all hierarchical traditions of the official world to which he belongs" and his "almost inconceivable influence on the Javanese." Stein imbibes enough for a family of drunkards and eats prodigiously. He is known to the Indonesians as Tuan Setan (Lord Satan) or Tuan Raksasa, after a giant of Hindu mythology with great canine teeth and a fearsome appearance. Stein is simply a force of nature, one of those eccentrics who sometimes flourish in colonial situations, learning multiple languages and dialects and memorizing innumerable local legends, folktales, and myths. He is also the undisputed world's expert on the archaeology of Southeast Asia.

In 1936, at a scientific congress in Oslo, he moves beyond his studies of stones, artifacts, and temples to challenge Dubois' view of human evolution in Java. Stein declares Von Koenigswald's Modjokerto skull to be *Pithecanthropus*, while denigrating the original *Pithecanthropus* as a human-ape chimera comprising a femur from one stratum and a skull from a much older level. It is a battle Dubois thought won long ago. He suspects that Von Koenigswald and Stein are colluding against him. His anxiety increases when he learns that Stein will lecture on the prehistory of the Indies to the Mijnbouwkundige te Delft (Mining Society of Delft) on October 28, 1936. Dubois dispatches his latest assistant, L. D. Brongersma, to the lecture to listen to Stein and report on his remarks.

Stein opens by announcing, disarmingly, that he has no understanding of paleontology, geology, and anthropology, but only about archaeology and the objects made by man. Thus the opinions he gives, Stein says, will be based on information given to him by experts, not on his own knowledge. (This is a clever way of denying responsibility if Stein is later proven wrong, Dubois thinks.) "In former times," Stein booms, "it was thought that the complete fossil vertebrate fauna of Java belonged to one single faunal complex of similar age. The researches of Von Koenigswald have demonstrated convincingly that this is incorrect. One can split the fauna in three (in comparison with the British Indies):

I. Young Pliocene
II. Middle Pleistocene
III. Young Pleistocene/Holocene."

The most recent period, he says, is adjudged to range from 50,000 to 30,000 years ago and contains *Homo soloensis* from Ngandong; the next, about 300,000 years old, contains *Pithecanthropus erectus* from Trinil; and the most ancient, about 750,000 years old, includes the child's skull from Modjokerto. The dates of these periods are not exact, for no precise means of dating ancient rocks has yet been developed. (Indeed, after such methods are perfected, many years later, the dates Stein gives prove to be wrong.) "*If,*" Stein adds slyly, "*Pithecanthropus* is something, then it is middle Pleistocene."

Next Stein accuses Dubois of documenting the finding of *Pithecanthropus* poorly.

> *The Chief of Mining, Zwierzycki, searched all existing reports of the excavations for data about the finding of P.e. Then it turned out that the reports of the two noncommissioned officers who supervised the excavations at Trinil were missing. Even if these reports are ever found, one cannot rely much on them, because these noncommissioned officers are not scientifically trained. One cannot doubt the place that the femur was found because Dubois himself was present when it was found.... Dubois was not present when the skullcap was found, and about that we have only the communications of the noncommissioned officers, which, as the speaker has already said, are of no value. It is not impossible that the femur and skullcap do not belong to the same species. Since the faunas of the Middle and Young Pleistocene are very similar, it is also possible that* Homo soloensis *could be found in the Middle Pleistocene and that the femur belongs to* Homo soloensis. *The skullcap, of which the origin is yet unknown, could perhaps be connected with the child skull from Modjokerto, found in the older Pleistocene strata.*

Dubois is angered by these words. Kriele and De Winter, though noncommissioned officers, wrote him frequent reports and were not such fools as Stein makes them out to be. They were civil engineers, fully trained, and well accustomed to dealing with geology and the placement of objects within the ground. Most unfortunately, Dubois was not present when any of the *P.e.* fossils were found, despite his dutiful monthly visits to the excavations.

Stein implies that Dubois has been negligent or careless, yet Dubois knows full well that nearly all excavations—including, probably, Stein's own—are carried out by unskilled workmen. The professional men who conceive of such work, analyze the finds, and write up the results do not shovel earth themselves, day after day. That lowly task requires little more than a dull mind and a strong back. No, the proper role for a paleontologist or archaeologist in such work is to select the excavation site and make sure the men understand how to work carefully. Then one appoints field supervisors and charges them with taking detailed information about important finds made between visits from the professional leader. This is how Dubois worked, and Davidson Black in China, and Von Koenigswald, too, in Java. Besides, if Dubois had spent all his time at Trinil he would surely have died of malaria long before anything of significance was found. Stein talks like one unacquainted with the realities of such work in the tropics, but he is not, so his motive is surely malicious. Finally, Stein asserts that he does not believe the missing link exists at all. The only solution, Stein announces boldly, is to fund Von Koenigswald to continue his researches.

While Stein speaks out on his behalf in Europe, Von Koenigswald and his wife have determined to leave Java. They sail on the S.S. *Baloeran* on November 18, planning to visit Berlin, Paris, London, and Leiden to study fossils, visit colleagues, and try to secure a professional position for Von Koenigswald. Everywhere Von Koenigswald goes, he talks up the controversy about *P.e.* and the Modjokerto child's skull. It is the only find of great significance that he has made, the only way to demonstrate the importance of his work. He hopes the Modjokerto skull will earn him an academic appointment, and the more people talk about it, the better.

Von Koenigswald delays calling on Dubois at his Haarlem home until February 17, 1937, though this is one of the most important of his planned visits. Von Koenigswald later describes the visit:

> He was stated to be ill and unable to see anyone; but, when I gave
> my name I was allowed in since I came fresh from Java. Dubois was
> sitting quietly in his living-room, a big, broad-shouldered, imposing
> man with a stereotyped, almost embarrassed smile round his
> mouth. When I cautiously made my request to be allowed to see his
> original finds (which had been deposited for some years in [the]
> Leiden Museum, where they were safer than at Haarlem), I received
> permission only after he had assured himself in an open telephone

conversation with his assistant that I had not already tried to force
my way into the sacred halls at Leiden behind his back.

The following day, in Leiden, the double safe was opened for me,
and I was allowed to take the finds themselves in my hands. The
fragments were dark brown in color, weighty, and heavily fossilized.
The smooth, round skull-cap was deeply corroded by acid ground-
water.... The convolutions of the brain were clearly imprinted on
the inside of the skull-cap. Holding the hollow, fragile object in one's
hand—during casting it is always filled—one is particularly con-
scious of its fragmentary nature....

That morning in Leiden was decisive for me in many respects. I
was in great difficulties at the time.... But that morning it became
clear to me that I must return to Java.

This meeting might have been an opportunity to establish a more friendly relationship, but Dubois admits no correct opinion on matters Javanese except his own. To succeed, Von Koenigswald must overturn Dubois' work on human evolution in Java; he is determined to bring the old man down. Though Dubois is nearly eighty years old, and weary from a lifetime of conflict, he will never allow his scientific work to be supplanted without a fight. He is an aging bear, but not yet a dead one.

Youth and energy favor Von Koenigswald, as does the fact that he has new fossils, the ultimate currency of paleoanthropology. Nonetheless, Von Koenigswald is fearful of the damage that Dubois' fierce opposition may do to his reputation. The showdown will come at a symposium on early man to be held at the Academy of Natural Sciences in Philadelphia on March 17–20, 1937. As a relative unknown, Von Koenigswald has been invited at the request of the sponsor of the symposium, John Merriam, who wants to look over this young friend of Teilhard's. Von Koenigswald prepares with great care.

Dubois is, of course, also invited to the conference. However, the transatlantic trip is long, his health is poor, and he has attended enough scientific conferences to last a lifetime, he thinks, never realizing the advantage he is ceding to Von Koenigswald. Dubois prefers to stay at home, with his books and the familiar comforts of De Bedelaar, though he sends a manuscript as a contribution for the conference proceedings. Oppenoorth likewise stays in Holland. Thus it transpires that Von Koenigswald is the only expert on Indies fossils to attend. He can present the situation, explain his ideas, and secure his reputation without challenges from other authorities.

On the opening day of the conference, Von Koenigswald speaks first. He begins his talk innocuously.

> In 1890, Professor Eugène Dubois discovered in Trinil, in Central Java, the remains of the famous Pithecanthropus, also called Java Man. Since then, Java has become of special interest to scientists working on the problems of fossil man. This short address will give only a review of the latest results on the stratigraphy of the Pleistocene of Java, and its relations to early man.
>
> Professor Dubois was the first to start excavations for fossil mammals in Java (1889–91).

That having been said, Von Koenigswald moves swiftly to attack Dubois' work.

> All the remains which he collected belong, in his opinion, to one and the same stratigraphic zone, to which he gave the name "Kendeng or Trinil zone," and of "Pleistocene" age, according to his first publications. He later changed his opinion and called this fauna "Pliocene." In 1909–10 a German expedition under the leadership of Mrs. Selenka undertook new excavations in Trinil. They confirmed the Pleistocene age of Trinil. The fauna found by this expedition was, however, not as rich as that listed by Dubois.

Painting his own work as far more meticulous, Von Koenigswald reveals a new stratigraphic sequence of *seven* superimposed faunal zones, each recognizable by its typical constellation of mammalian species, where Dubois only saw one. The Trinil zone, near the middle of the sequence, is the one in which the Modjokerto child's skull was found. Von Koenigswald glosses over the confusion about the place and manner of finding the Modjokerto skull, although these are key issues if this skull is to be the basis of a revision of Dubois' stratigraphy.

> The skull . . . is perfectly fossilized, and we are certain that it was found in situ, because the bone is so thin that it would have been destroyed by any movement or rewashing. . . . [I]t was found in a stratum older than that near Trinil, where Pithecanthropus was found.
>
> We also use the name Trinil zone, but in a sense different from that of Dubois. Our Trinil fauna is exactly the same as that

> *described by the Selenka expedition, for the animals of the Dubois*
> *list, which are missing here, belong really to an older level, namely*
> *the Djetis zone which Dubois did not recognize.*

Von Koenigswald then offers an abbreviated, unflattering, and rather inaccurate review of Dubois' assessment of *P.e.*

> *The first suggestion Dubois made after the find of the skull-cap was,*
> *that it belonged to a kind of chimpanzee* ("Anthropopithecus").
> *Later, when he found the first femur, he chose the name*
> Pithecanthropus erectus, *which he regarded as a primitive human.*
> *He changed his mind a few years ago and now considers it as*
> *belonging to a giant gibbon. . . . But since* Sinanthropus *was found*
> *in China, which is quite definitely to be considered as a human*
> *being, and closely related to* Pithecanthropus, *we are sure about the*
> *hominid character of the latter.*

Few present at the symposium remember Dubois as the strong, brilliant young paleontologist of forty years ago, the man who returned from Java with stunning fossils that turned ideas about human evolution upside-down. All they see in their mind's eyes is the picture Von Koenigswald paints of the elderly Dubois: old-fashioned, dictatorial, unscientific.

In closing, Von Koenigswald suggests that the *Pithecanthropus* skullcap is simply that of a primitive human, but *less* primitive than the Chinese *Sinanthropus*. In fact, the *Pithecanthropus* skullcap may even be nothing but a female of the same type of human as the Ngandong Neanderthalers, while the femur derives from another creature entirely.

Dubois is not there to object in person, nor is his paper ever read in its entirety to the conference. His abstract is read aloud on the last day, along with the titles of the papers of other absentees, but by then the damage has been done. Von Koenigswald is in the ascendancy and Dubois has been largely dismissed as an out-of-date scholar—a lucky physician, really. Much the same fate awaits Oppenoorth, whose cogent arguments against Von Koenigswald's grouping of the Ngandong skulls with Neanderthals go unheard.

Merriam and the Carnegie Institution of Washington promise Von Koenigswald financial support, so Von Koenigswald writes to his chief collector "in my best Malay," enclosing a check to finance the purchase of more fossils at Sangiran.

CHAPTER 56 THE LETTER

Dubois' perspective on these events is very different. When Von Koenigswald comes to Haarlem, Dubois makes a special trip from De Bedelaar, and he does not feel he is being uncooperative or suspicious. Indeed, when Von Koenigswald visits in the early spring of 1937, Dubois is in an unusually mellow frame of mind. On February 11, he received an unexpected letter from his old, dear friend, Adam Prentice.

Prentice is the best man Dubois has ever known, the one person who has never failed him, never betrayed him, never worked against him. That business—that silly suspicion—about Anna and Prentice is completely discredited, long forgotten. The light of memory bathes Prentice and those days in Java in a golden glow of youth, of opportunity, of resonant companionship and deep understanding. To hear from Prentice again is like being transported back to those days. Dubois is lost in his fond memories of those times for weeks to come. Though he has always been suspicious of others' ambitions, Dubois is for a time blind to Von Koenigswald's dark intent.

Prentice's letter comes to him through the hand of a fellow countryman, C. van den Koppel.

February 9, 1937

By sending you the enclosed letter from Mr. Adam Prentice in Kediri, I fulfill a promise given to Mr. Prentice, and that I should have fulfilled earlier. Mr. Prentice asked me to bring the letter to you personally, but although I am almost 1/2 year in the country, I have not found the opportunity to do so, and I send it now by post to avoid further delay.

As you can see the letter is written almost a year ago. I met Mr. P. at the end of 1935, while traveling from Java to the west coast of Australia, and during this journey of 14 days I became acquainted with him, so that we, notwithstanding the great difference in age, became good friends. Actually he escaped the Indian tax for a year, by being in Scotland and British India, and, because the year was not yet completely expended, he made a trip to and from Singapore to West Australia to use up the rest of the time. He told me a lot about you, and asked me to say hello to you when I returned to Holland.

Later he wrote to me, however, that he preferred to send you a letter, and sent me this one. Because I traveled via New Zealand, the Philippines, China and Japan, and America, I did not receive his letter until my arrival in Holland in August 1936.

I offer you my apologies, that I did not see to it that the letter reached its destination sooner.

I believe that Mr. Prentice will appreciate it very much when you write him back. He speaks Dutch very well. During the journey I made with him, he was still very healthy, a stately upright figure, who does not look his age. His address is: Mr. Adam Prentice, Kediri, Java.

For weeks, Dubois does not think of conferences, or scientific papers, or academic rivals. He does not think of P.e. He only thinks of himself young and strong, and of the true companion of his Java days, Adam Prentice, and of his letter.

Kediri   February 7, 1936

An echo of the Past!
"Dost thou recall?"

My dear doctor,
You will hardly expect a letter from <u>me</u>! It is long, so very long since last we saw each other. . . .
Dost thou recall from the quietness of your peaceful study in the home-land—the days now long, long flown which we passed together in the peaceful atmosphere of dear old Mr. Boyd's Koffeeland Mringin,—the good old man's dwelling Ngrodjo, Willisea the block house he put up for you at Jonojang, my own quiet abode at remote Tempoersarie?
Do you remember the many pleasant meetings we had at Ngrodjo when the old gentleman & I listened with so much interest to your enlightening & informative conversation? Indeed we learned <u>much</u> from you, and our minds ever reverted with satisfaction to the many agreeable meetings we three had together. Do you recall our excursion to Trinil the scene of your labors (where the famous Pithecanthropus erectus *was found), when contrary to your wont you regaled us at dinner in the evening with a <u>bottle of wine</u> saying it "aided digestion." Do you remember our bathing the next day in the river, our pleasant walk in the afternoon to the station along the country road where a snake swallowed a frog and you at once ran to the rescue forcing the snake to disgorge the frog which, still quite alive, first looked <u>to the right</u> & <u>to the left,</u> and then lightheartedly plunged into the stream by the roadside? Do you remember the beautiful flowers at the station which we looked at while waiting for the train? One had a delicate blue tint and you said that <u>was well nigh</u> your <u>favorite</u> color!*

Do you remember the long walks you & I had through the widespread coffee gardens at Mringin? Do you remember the Nekkie, the big man at Toeloeng Agoeng; the worldly Regent who got a decoration from Batavia but sordidly thought <u>a sum of money instead</u> would have been something more useful?

Do you remember the two corporals of the engineers who looked after your team of convicts, at the excavation work? Their mode of life ever amused you—terribly grand, living like kings, at the <u>beginning</u> of each month when money was plentiful, and ever on <u>very</u> short rations towards the <u>end</u> of the month when the money was <u>all spent!</u> Through your favorable report they got promoted in time to the rank of Sergeant. Then you photographed them & noticed how they were maneuvering to bring full into the picture their arm shewing the new sergeant's <u>stripes!</u> . . . And Mr. Mulder, P. T. Sanvraar, & Mr. Turner, controller, at Toeloeng Agoeng. Do you remember our age—you, Mr. Mulder & I—was 34 years. Ah, yes, the golden days of youth! Perhaps we had our troubles, too, but we had <u>youth, health, home,</u> and <u>length of days</u> before us! Dost thou recall?

As oft as I look back, the recollection of that happy time is a green spot in my memory, and will endure as long as life lasts!

Good old Mr. Boyd died in 1902 at Kediri under Van Buren's care from <u>cancer of the throat,</u> aged 74, & was buried at Toeloeng Agoeng. We were all present, and the Asst. Resident, Regent & etc & etc attended also. It may be the kind old man smoked too heavily?

While he still lived we often spoke of you after your departure from Java, very, very, often, & always with esteem & affection. Yes, we both loved you, and never could forget you! Like a sun that had come into our orbit you brought us light and happiness—it was just a chance in life never likely to recur, for <u>when</u> does it happen that a man of learning ever comes to live on a coffee estate <u>for any length of time</u>? . . .

Dubois is now an old man lost in time. The fire in his heart that once propelled him up mountain slopes and down riverbanks is banked and barely smoldering, the light of his intellect that once burned his path through the academic centers of Europe is dimmed. The memories bring a glow to his once-sharp Garuda eyes, eyes that could see the truth in a glance and that are now faded and rimmed with wrinkles of soft, almost translucent skin. He muses for many months before he composes an answer to his friend. He must write from the heart, not a simple letter of news or fact, but one of love and remembrance. Oh, yes, he remembers Prentice.

CHAPTER 57 PRETENDER TO THE THRONE

Dubois is saved from immediate apoplexy but not from harm by his ignorance of Von Koenigswald's actions. Von Koenigswald sends letters full of flattery and praise to the older man, never hinting at the harsh criticisms he has leveled at Dubois.

July 12, 1937

Now I am again back in Java, after a very stimulating study journey, and like to thank you heartily once more for the collegial and friendly reception you have given me. Even when I cannot share your scientific opinion on all points, I wish to assure you, esteemed Professor, that I feel for you the greatest personal respect. . . .

Further I have an additional message which you, honored Professor, will find more interesting than anyone else: I have discovered a mandible fragment of Pithecanthropus! *That piece was found by my collector, who searches for stone tools and fossils on my behalf in Middle Java. It comes certainly from the Trinil layer. . . .*

In fact, the mandible is in the very first basket of fossils from Sangiran that Von Koenigswald examines upon his return. Dubois is in poor health when he receives Von Koenigswald's letter, but he responds quickly to the exciting news.

August 26, 1937

It gave me a great deal of pleasure to learn that you have had a rewarding and fruitful study journey in America and China—and especially that your return to Java was met with such a find as a mandible fragment of Pithecanthropus! *From your indication of its geological placement and your short description, I cannot doubt that this is really another piece of the mandible of* Pithecanthropus. *The ramus, the large molars, the small canine . . . completely agree with that which I found on the fragment from Kedoeng Broebus and also the premolar from Trinil. Your discovery is for me really a great pleasure. I look forward greatly to receiving the photographs which you kindly promise to send me.*

Yet something in Von Koenigswald's letter disturbs Dubois. His intuition alerts him to some intangible problem between them. Perhaps he should offer to be a sort of elder brother to the younger man, guiding him through his scientific work. Von Koenigswald has a very different relationship in mind: Dubois must recant his ancient ideas and clear the way for the young. The next day, Dubois writes to Von Koenigswald again:

This is a continuation of my letter of yesterday. . . . Because of an unexpected disturbance it was written hastily and remained incomplete. I omitted, namely, to say something about the difference in our scientific opinions "on several points" that you mentioned and to express the hope, which I have long cherished, that we will come to agree, perhaps soon. Allow me to mention something of my own scientific evolution. For many years, I was hindered by my teaching duties (which are now carried out by four men). Under these circumstances my ideas about bioevolution, especially with regard to Pithecanthropus, *could develop or come to expression only slowly. Actually these ideas started to deviate from the current opinion in 1895 (the year of my return to Europe), because at the time I was firmly convinced that the "Trinil femur" is not completely human. Facts that I have established during my cephalization research (since 1897), and by the new biology, which considers each organism to be an indivisible whole or entity . . . have made me realize that* Pithecanthropus *cannot be reckoned to be among the hominids but belongs between the hominids and the most generalized anthropoids, the gibbons. As late as 1923, I did not yet see this. Descent must certainly be accepted from the origin of humans, but not gradual development. Now I hope wholeheartedly that your new discovery will lead to agreement between us about evolution and that we will work together, which in my earnest conviction will be beneficial to science.*

It is not a likely outcome.

Von Koenigswald replies:

September 3, 1937

My heartfelt thanks for the friendly lines you wrote me and the trust that you expressed in me and my identification. I believe that I have meanwhile succeeded in proving that the mandible fragment belongs to Pithecanthropus: *a few days ago, from the same layer and totally unexpectedly, there came to light also a skullcap, which completely agrees with your find from Trinil!*

. . . So much for today. The skull is not yet completely prepared, so I cannot yet measure my find, but I wanted to inform you personally of it at once, knowing how much it interests you. I really hope that the new find may contribute more clearly to the recognition of the peculiar nature and the systematic position of your Pithecanthropus.

Von Koenigswald first heard of the new skullcap when his collector Atmowidjojo wrote saying that they had not found additional orang-utan fossils, but enclosed a part of a human skull. Von Koenigswald left for Sangiran by the night train, anxious to be on the scene and to collect additional pieces of this skull.

Arriving at Sangiran, Von Koenigswald showed the piece around, promising ten cents for every additional piece of skull and half a cent or a cent for a tooth. This strategy backfired.

We had to keep the price so low because we were compelled to pay cash for every find; for when a Javanese has found three teeth he just won't collect any more until these three teeth have been sold. Consequently we were forced to buy an enormous mass of broken and worthless dental remains and throw them away in Bandung—if we had left them at Sangiran they would have been offered to us for sale again and again! In spite of the low price, we used to pay several hundred guilders a month for fossils.

Cautiously we began to hunt through the hill-side foot by foot, and soon the first fragments of skull did really come to light. Unfortunately, they were extremely small: too late I realized that my opportunist brown friends were breaking up the largest pieces behind my back, in order to get a bigger bonus. I had the good luck to find part of the frontal bone with the eyebrow ridge myself. We hunted on into the afternoon and found in all forty fragments. It was already perfectly clear that we had discovered a new Pithecanthropus *skull. . . .*

In 1937, Von Koenigswald (in pith helmet at right, holding skull fragment) and his workers find many pieces of a skull of Pithecanthropus *at Sangiran.*

Now, the region round the ear is decisive in answering the question, Man or ape? . . . This find, therefore, proved at last that Pithecanthropus *was human.*

After the skull had been reconstituted I immediately sent a preliminary photograph to old Dubois. I thought he would share my joy that the problem had finally been solved, even hoped he would declare that his first impression of Pithecanthropus *had been right after all. I was very much mistaken, however.*

This account, written twenty years after the fact, is not wholly accurate, for Dubois' response to Von Koenigswald's letter is generous.

September 14, 1937

You will understand how great my joy is about this skull find, which, with the mandible you have lately discovered, I am convinced will certainly contribute to the clearer recognition of the peculiar nature and systematic position of Pithecanthropus. *. . . I am very desirous to be allowed to learn more of your most important finds. . . .*

Von Koenigswald replies:

October 3, 1937

The preparation of the Pithecanthropus *skull is now nearly finished. It is no longer possible to doubt that one really deals here with a primitive* hominid, *one which stands at an even lower step than* Sinanthropus. *I will, when there is a chance, report further to you in more detail.*

The next month, Von Koenigswald writes,

November 12, 1937

The reconstruction of the new Pithecanthropus *skull is now complete. I was astonished at how closely the profile of the frontal region resembled your skullcap from Trinil. . . .*

Following your suggestion, I have compared the skull once again accurately with a skull of a gibbon, and unfortunately must confess that I cannot confirm your opinion. The hominid character of the new skull can no longer be doubted, for it shows in addition to primitive characters, the following:

—the ear is positioned <u>under</u> the root of the cheekbone,

—the jaw joint has a marked articular tubercle,

which traits monkeys do not possess and which appear only in humans.

The affinity of Pithecanthropus *with* Sinanthropus *appears even more clearly in the ... new skull.*

Von Koenigswald prepares a paper on the new jaw, calling it *Pithecanthropus,* which is published in the *Proceedings Koninklijke Akademie van Wetenschappen* (Proceedings of the Royal Academy of Science) on November 27, 1937. Dubois cannot accept this identification, he warns Von Koenigswald, so he prepares an opposing paper, which he will present at the Royal Academy meeting on January 29, 1938.

January 12, 1938

Now that I have looked closely at the photographic illustration you published and have studied attentively your description in the Proceedings Koninklijke Akademie van Wetenschappen *of November 27, 1937, I must agree completely with you, that this is the mandible of a human, and in my opinion highly probably of a* Homo soloensis.

With this important discovery, I wholeheartedly wish you luck. I regret, however, that you have given the owner of this jaw the name Pithecanthropus.

Had you sent me the photograph, which you promised me in your letter of July 12, and which I asked for urgently in my letter of October 20 ... there would be no need for me to offer a communication at the next meeting of the Academy, in which I will demonstrate the inaccuracy of the name you have assigned this jaw.

Von Koenigswald preempts Dubois' January presentation with yet another find. Shortly before the Royal Academy meeting is to take place, Von Koenigswald announces the new skullcap to a meeting of the Natuurwetenschappelijke Raad van Nederlandsch-Indië (Natural Science Council of the Dutch East Indies). As is widely reported in the press, he proclaims this specimen to be so similar to the Trinil skullcap that the human identity of *Pithecanthropus* is indisputable. Dubois is disheartened and infuriated. Try as he may, he cannot seem to publish or speak fast enough to keep up with Von Koenigswald's furious stream of announcements and publications. As Dubois reaches his eightieth birthday on January 28, 1938, he is unhappy. Even a cheerful letter from his beloved daughter, Eugénie, and telegrams and cards from dozens of well-wishers around the world (including Von Koenigswald himself) cannot console him. No matter what anyone says, Dubois senses he is being overtaken and thrust aside by a younger man. He hates the sensation.

He lectures on Von Koenigswald's jaw to the Royal Academy on January 29 in an attempt to reestablish the correctness of his own opinions. He quotes extensively from Von Koenigswald's paper on the jaw, contrasting the younger man's views with his own. He points out every minute anatomical detail in which Von Koenigswald's new jaw differs from *P.e.* The area of the jaw where the digastric muscle attaches, the chin region, and the form of the premolar tooth are "entirely different." Throughout the article, he emphasizes how gibbonlike the Kedoeng Broebus jaw is and how humanlike Von Koenigswald's jaw from Sangiran is. Dubois also fervently denies the charge Von Koenigswald made at the Philadelphia conference, that Dubois now considers *P.e.* to be a giant gibbon.

> *I never imagined* Pithecanthropus *as a "giant* Hylobates,*" only as a giant descendent of a "generalized" form, which had inherited from its ancestor the "gibbonlike appearance," but had . . . doubled [its] cephalization. . . . Probably, an essential change of diet caused the canine teeth to reduce and the fore part of the mandible and maxilla to shorten and resemble [the] human appearance.*

His words make little impact on the listeners, for at that very meeting, Von Koenigswald's description of his new skullcap is read. Dubois sounds like a jealous old man, carping about details, while Von Koenigswald is obviously the up-and-coming expert on Javan paleontology.

On March 26, 1938, Dubois battles back again at a Royal Academy meeting, complaining about Von Koenigswald's reconstruction of the new skullcap. He bases his remarks on Von Koenigswald's own words (in letters to Dubois, dated September 3 and November 12, 1937, an article about the find in the *Illustrated London News* of December 11, 1937, and Von Koenigswald's communication to the Royal Academy in January 1938), and on Von Koenigswald's photographs of the specimen in various stages of preparation. In particular, Dubois draws attention to three photographs of the new skullcap. Two of these, from the *Illustrated London News,* show the specimen in pieces, as it was recovered, and at an early stage of preparation. The third is one Von Koenigswald sent to Dubois in November, which shows the "preparated" skullcap. All three photographs reveal the "great thickness of the skullcap," which distinguishes this new skull from *P.e.* "For the general thickness of the cranial bones of *Pithecanthropus* is only moderate, as is clearly visible and, indirectly, measurable on the exact photographs, natural size, in my description of 1924."

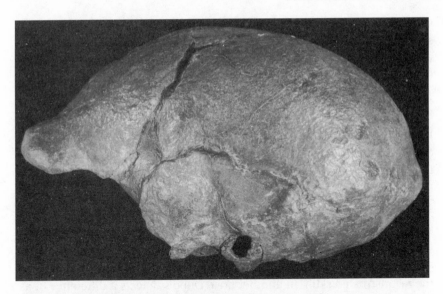

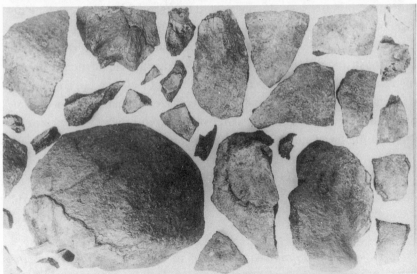

Dubois charges that Von Koenigswald makes serious errors in reconstructing the skull of Sangiran II (top) from the many fragments that were found (bottom).

Then Dubois homes in on the reconstruction of the skullcap. He denies the great similarity in shape of the midline profiles of *P.e.* and the new skullcap (a similarity of which Von Koenigswald has written), because the resemblance is false, caused by the "too abundant substitution" of plasticine for the "bone substance" that is "deficient" at the front of the new skull. The extra plasticine eliminates the "distinct and broad furrow" that originally separated the brow ridge from the forehead in the new skullcap, making the reconstituted skull resemble *P.e.* in lacking this furrow. However, a similar furrow is characteristic of more humanlike forms such as *Sinanthropus* and *Javanthropus*, Dubois observes. In short, Von Koenigswald's reconstruction has artificially altered the shape of the new skull, producing a false resemblance to *Pithecanthropus* and minimizing features that ally it with *Sinanthropus* or *Javanthropus*. "The real fossil," Dubois remarks acidly, "got lost in the reconstruction."

There is more. Dubois compares the size of the pieces of the fragmentary skullcap with that of the "prepared" skullcap, as shown in Von Koenigswald's own photographs. Enlarged to natural size, these images reveal startling changes in the fossil. Dubois finds, for example, that the height of the skull above the ear hole is 112.5 mm on the fragments but only 102.5 mm on the reconstructed skull. In another area, Dubois detects a difference of as much as 18 mm between the pieces and the reconstructed whole. The net effect of these inaccuracies and distortions, Dubois concludes, is to transform the shape of the skull "artificially, from that of an immature individual into that of a quite adult one." He shows Von Koenigswald no mercy: "Even if this reconstructed skull were not obviously artificial, to a large extent, its importance would not be overwhelming, considering that this is not the only human skullcap which resembles that of *Pithecanthropus*. This was the case also with the first *Sinanthropus* skull . . . [found by] PEI and DAVIDSON BLACK. . . . *Sinanthropus* in every case is distinguished from *Pithecanthropus* by . . . significant characters of the brain."

Dubois' charge is a serious one, provoking a furious response from Von Koenigswald.

May 5, 1938

Our difference of opinion comes first of all from the fact that you see in Pithecanthropus *an ape. As I wrote you already, very honored Professor, the new find does not confirm your opinion. . . .*

Unfortunately, I have to remind you, as I said in my original letter, that my letters to you, as well as the photographs, are <u>only</u> *meant for* <u>your personal information</u> *and* <u>not for publication.</u>

Dubois replies on June 3.

I may remind you that on November 12, 1937, you <u>literally</u> *wrote to me: "To give you directly an impression of the new find, I send for your personal information a photograph that I made just a few days ago."*

The same photograph, given by you to Professor Weinert, was published already in Die Umschau *of January 23, 1938. . . . The photograph in your "A new* Pithecanthropus *skull," oriented only slightly differently, was announced by you on January 29 and appeared in print towards the end of February.*

The implication is clear: the photographs had already been published, by both Weinert and Von Koenigswald, when Dubois included them in his paper. Von Koenigswald writes to the Royal Academy, arguing that Dubois has accused him of fakery without evidence, since photographs taken at different angles may distort linear measurements. Although Von Koenigswald's reply is not published, Dubois is asked to apologize. He concedes that he does not think Von Koenigswald *deliberately* altered the form of the skull.

In 1938, Dubois (second from left) tries in vain to persuade Franz Weidenreich (left) that the Sinanthropus *fossils now in his charge are distinctly different from both* Pithecanthropus *and Von Koenigswald's new finds. (Third from left: H. Boschma; right, L. D. Brongersma.)*

The correspondence between the men is ended, but their fierce struggle is not. In April, Von Koenigswald is visited again in Bandung by Teilhard de Chardin, now head of the Laboratoire Palaeontologique of the Sorbonne in Paris. Teilhard is among the first to see Von Koenigswald's newest find from Sangiran, another fragmentary skull that will come to be known as *Pithecanthropus* II. On his side, Dubois tries to enlist Franz Weidenreich, the distinguished German anatomist, who now supervises research and the excavation of *Sinanthropus* remains at Chou Kou Tien. He invites Weidenreich to Haarlem on August 8, 1938, although "unfortunately I cannot then show you the fossil remains of *Pithecanthropus.*" Weidenreich and Dubois enjoy their meeting, each recognizing the superb anatomist in the other. After the visit, Dubois writes to Weidenreich,

August 9, 1938

I cannot let this day end without sending a heartfelt greeting to you in Basel, with sincere thanks for what I . . . know now, for certain, that we both work in harmony on the greatest scientific question that there is for us humans. Then, I am now completely convinced of it: that Sinanthropus *is the most primitive human, from which descended all later humans.*

Pithecanthropus is, to him, the ape-man ancestor of *Sinanthropus*. A few weeks later, after further cordial correspondence with Weidenreich, Dubois writes happily to Eugénie:

August 23, 1938

The meeting with Professor Weidenreich, in which he made it evident that he inclines toward my view in many points, could make a union between us possible if he had not tied himself to the position that Pithecanthropus = Sinanthropus, *like that. Also therefore he would not be willing to disown Von Koenigswald, although Weidenreich has already deviated from him concerning some published points.*

Dubois feels he perhaps has found a powerful ally against Von Koenigswald. In September, he writes again to Eugénie: "Weidenreich wrote that he is willing with great delight to work together. . . . The opponents have become partisans, fighting on the same side. That makes me hopeful."

Though alliances are being forged in the scientific world, the political world turns ever darker and more divisive. Adolf Hitler controls Germany, where Jews such as Franz Weidenreich are no longer welcome however great their learning. Everyone in Europe, Dubois included, keeps a wary eye on the military actions of their aggressive German neighbors.

CHAPTER 58 OLD FRIENDS

Dubois pauses in his research to write several long overdue letters to Prentice. For the first time in many years, he feels at ease. As in the old days, it is a relief to spell out to Prentice all his concerns and thoughts, his trials and his battles. Prentice, loyal as ever, replies late in November.

Kediri   November 28, 1938

My dear old Friend,

Your letters of 15th and 19th August reached me safely and I was ever so pleased to know you still remembered me kindly notwithstanding the passing of so many long years since we sojourned at Mringin & enjoyed daily converse, often too in the company of our dear friend and wellwisher good old Mr. Boyd, but for whom we might never have met.

Very much has happened since those days yet my thoughts oft-times revert to you, and I ever hoped all would be well with you & yours through the years that intervened.

Yes, in quiet moments the past returns fondly to mind:—

"Oft in the stilly night
When other thoughts have found me,
Fond memory brings the light
Of other days around me."

As one advances in years one oft looks back on sweet days that are gone & on the sunny friendships of radiant youth. When you resided in Java we were in the bloom of early manhood physically fit for any task we set our- selves. We knew no pain or ache and were able to walk for hours without fatigue & could climb mountains! I ascended the Ardjoeno above Lalie Djuvo once. You climbed it too, & wrote your name on the rock beside my name. Can you recollect? It was long long ago.

Do you remember the visit to Trinil when Mr. Boyd & I came to pass a day with you at the place you made <u>famous;</u> of our bathing next day in the river; and our long pleasant walk in the afternoon (the three of us) from Trinil to the railway station? It was such a pleasant leisurely walk that afternoon through the peaceful countryside engaged in agreeable and interesting conversation all the way. We arrived at the railway in ample time. Outside the halte [station] there were flowers growing. The blossoms were of a delicate blue, & you told me that was a favorite color of yours. The recollection of our visit to Trinil & the happiness you made us feel when there can never be forgotten. As often as I see the name Trinil mentioned I think of you and your labors, & of the happy day spent with

you there. You had been some days busy at Trinil when, by appointment,
Mr. Boyd & I called over, eager to see the place so often mentioned. And it
was pleasant the three of us returning together in a happy frame of mind
to Toeloeng Agoeng, no one left behind solitary & alone. Ah those delight-
ful days!

Good Mr. Boyd passed away in 1902 of cancer of the throat aged 74
years. His two younger sons Alex & Robert, both engaged in coffee, died
some years ago aged 60 & 70. The eldest son William Boyd my oldest
remaining friend in Java after being long sick died 6 weeks ago over 81
years of age. His memory had gone, yet revived at intervals. Sometimes he
didn't know even his own children & asked who they were, yet he always
knew ME when I called. His decline arose largely I fear from the worry of
straitened circumstances. When too old to work in coffee he became
Secretary of the Kediri Planters Association. When his memory failed later
on, the Association gave him a small pension of f75 a month, not much,
but better than nothing at all in these hard times. The world crisis that
struck Java in 1930, & is still with us, half ruined Java, & Wm. Boyd like
ALL OF US suffered heavily! All values are 50 to 80% down here today.
Hardly anyone can make any money now. People—Europeans & Native
alike—are just able to live by practicing economy. What money Wm. Boyd
made—it wasn't much—went into houses, & house property here is down
75% & 80% in value. This no doubt preyed upon my friend's mind for he
had a mortgage on his house property & sale today would hardly cover the
mortgage, while leaving nothing to his heirs. Well William Boyd was the
last of my old planting friends. That generation has now all but passed
away!

We are all getting older. You are now in your 81st year, and like Newton
still deeply engaged in scientific work. I am very glad indeed of THAT, for I
know full well that to YOU study and research are labors of love without
which life would lose its chief interest! I earnestly hope you will yet long be
spared in good health to continue your researches and that you will see
your views adopted by all whose opinion you value!

The luminous dissertation in your last letter to me re Pithecanthropus *I*
have read and reread with deep interest, & must read & reread it again.
Assuredly you have the gift of clear exposition! And as I read your convinc-
ing statement of the points at issue I could not help thinking of old Mr.
Boyd & the pleasure he always felt when you dropped in upon him of an
afternoon or evening at Ngrodjo. He delighted in your discourse & appreci-

ated so much the clear way in which you would explain questions he asked concerning matters on which he yearned for more light. Your stay at Mringin was most agreeable to him, & ever afterwards he spoke of you in terms of sympathy & sincere good will. We felt that a dear friend had left us and we might perchance meet him no more, but he would never fade from our mind & heart!

Returning to your disquisition I see that there are three groups among the learned—

(a) those who consider P.e. to be a missing link

(b) those who consider him a man ape

(c) those who regard him as the most primitive of all types of man.

You hold the first view (group a) still, just as you did before. I am glad of that. I remember at Mringin you gave me as your opinion that P.e. was really a missing link, there would be a series, not one link only (yet no other has been found). Of course I am speaking from memory here, for I find I have preserved nothing in writing.

Well, not long ago, it might be in January of this year, I read in a newspaper that you had altered your view, & now considered P.e. to be a . . . GIBBON. Somehow I felt sad on reading that, but felt: —Well, if my friend has really ultimately changed his view it will only have been after yet deeper study on the subject. For the seeker after truth must ever keep an open mind, willing to suppress all predilections and weigh honestly any further evidence adduced even should it lead him to modifying beliefs long & fondly held. I knew YOUR mind would ever be open to the truth & give new facts & evidence an impartial hearing.

By post I send you the newspaper article in question where I read of your "change" of view. It is the Soerabaja Handelsblad (I think of January this year but unfortunately the date is NOT printed above). It is written by G. H. R. von Koenigswald. Thinking of you, I put the paper aside for reference. . . .

In your own long long strenuous study of the subject endless questions and suggestions will have presented themselves to your mind & each been sifted & weighed as it arose.

When the learned are at variance it would be a mere presumption for an outsider like ME to speak positively. But looking at the question broadly I think the great majority of those who reflect on the matter will accept YOUR views. For, as far as scientific training & mental endowment are concerned, you are as fully qualified as any man living to arrive at the

truth of the matter. And with qualifications equal to the most learned,
YOU have made a profounder, more continuous study of the subject from
month to month & year to year during half a century. So you have DEEPER
knowledge & a FIRMER GRASP of all the intricacies of the case, and are
consequently MORE ABLE to speak WITH AUTHORITY than any man.
Were it for a court of law to judge, your evidence would win the case. But it
is here no question for law courts but for the considered verdict of the sci-
entist after due study & full & searching examination.

The German mission which came to Java in 1907 went to Trinil &
worked there but found nothing. They averred however that at the time
Pithecanthropus erectus *lived in Java, MAN had already appeared ELSE-*
WHERE on the globe. Here too I am speaking from MEMORY. Their asser-
tion gives rise to much thought but I won't go farther in the matter as my
letter is long enough already & I would not weary you with the musings of
an outsider.

I also send a newspaper 23 Aug 1938 with a startling article entitled:—
"Revolution in Science."

It is too deep for ME! It is also written in such an iconoclastic spirit that
I would rather form no opinion at all but wait for the learned to deal with
the matter. Well, I must now close with every good wish from your old
friend who remains

Very Sincerely Yours,

Adam Prentice

P.S. I am sending my letter by the ordinary mail so that it allows of the
newspapers reaching you AT THE SAME TIME, & not leave you in any
doubt as to what the newspaper article said about changed views (which
are not changed at all). If I sent my letter by air mail & the newspapers by
steamer mail you would be puzzled & be filled with uncertainty FOR
WEEKS before the newspapers came to hand.

The letter from Prentice brings Dubois peace at the closing of the year.
The world is in turmoil, the scientific world is being turned upside-down by
young hotheads like Von Koenigswald who do not know of what they write,
but Prentice is still Prentice: loyal, supportive, sound, and true. Dubois
passes a cold but contented Christmas at De Bedelaar, his mind filled with
thoughts of warmer climes and golden friendships. It is the last communication between the two men, Prentice and Dubois, who have meant so
much to each other through the years.

CHAPTER 59 THE FINAL CONFLICT

Dubois' hoped-for alliance with Weidenreich against Von Koenigswald does not materialize. Weidenreich leaves Dubois to meet with Von Koenigswald in Java and study his new finds. Weidenreich is swayed by the young man's marvelous fossils, his charm, and his knowledge. Besides, Dubois' denial that any of the new fossils from Java or China are the same as *P.e.* is simply irrational. Before leaving, Weidenreich invites Von Koenigswald and his wife to bring the Sangiran fossils to Peking, so that the Chinese and Javan materials may be compared side by side. The two Germans, each far from home, become frequent collaborators and coauthors. The planned visit to Peking, described later by Von Koenigswald, occurs early in 1939.

> *We laid out our finds on the large table in Weidenreich's modern laboratory: on the one side the Chinese, on the other the Javanese skulls. The former were bright yellow and not nearly so strongly fossilized as our Javanese material; this is no doubt partly owing to the fact that they were much better protected in their cave than the* Pithecanthropus *finds, which had been embedded in sandstone and tufa. Every detail of the originals was compared: in every respect they showed a considerable degree of correspondence. . . . The two fossil men are undoubtedly closely allied, and Davidson Black's original conjecture that* Sinanthropus *and* Pithecanthropus *are related forms—against which Dubois threw the whole weight of his authority—was fully confirmed by our detailed comparison.*

Convinced that they have resolved the matter at last, Von Koenigswald and Weidenreich write an article for *Nature* which they call "The Relationship between *Pithecanthropus* and *Sinanthropus*." Their conclusion is straightforward and clear.

> *Considered from the general point of view of human evolution,* Homo sapiens *and* Sinanthropus, *the two representatives of the Prehominid stage, are related to each other in the same way as two different races of present mankind, which may also display certain variations in the degree of their advancement.*

But they cannot convince Dubois, although scientific opinion is now in their favor and remains so.

For Dubois, only *P.e.* is a transitional ape-man; only *P.e.* occupies that precious ancestral position in human evolution. He will never relinquish this view. Weary, saddened, and frightened by recent political events, Dubois works slowly on his response to Von Koenigswald and Weidenreich's paper. Remembering his own times of trial, Dubois decides to take up the weapon of public pressure against Von Koenigswald as it was once used against him. He tries to organize an international protest from the scientific community to the Royal Academy of Science, asking that the new Javan fossils be sent to Holland where they may be accessible for study. It is a failure. Some of his friends and colleagues send letters urging the government to send the fossils back to Europe. Others decline to write on the grounds that Von Koenigswald is making quick work of publishing the remains and ought to be allowed to complete his studies, as a reward for his energy in finding them.

By the time Von Koenigswald and Weidenreich's paper equating *Pithecanthropus* with *Sinanthropus* is published, other matters have assumed greater importance than controversies over human evolution. On September 1, 1939, Germany invades Poland, and the world is at war. Dubois remains at De Bedelaar, deep in the country and far away from major military targets and cities. Soon Anna, her sister, and some of her sister's children also seek refuge

there, for they fear correctly that it will not be long before Germany invades the Netherlands, too. At least in the country there is simple food and some rural safety, away from the horrors of bombs and tanks and troops. Dubois works slowly on yet another attack on Von Koenigswald; Anna and her sister try to stock up on food and provisions to get them through the winter, with the help of a girl from the village. Dubois is still hale for a man his age, but there is a great deal of wood to be chopped for fuel and cooking and he must seek

In his last years, Dubois lives at De Bedelaar, where Anna and her sister also seek refuge after the Germans' invasion of the Netherlands. This photograph was taken December 19, 1939.

help from younger neighbors. He still walks through his grounds and swims when the air is warm enough, but he is not the vigorous man he once was. Just before his eighty-second birthday, he writes to Eugénie, who will yet herself flee to De Bedelaar as the war progresses:

January 3, 1940

I am here, indeed, ailing, working in this much too confined space, now I still have so much to do and must do. For all that, I cannot let what I have achieved AS BY A WONDER come to nothing, by the blindness (and worse) of SOME others. Unfortunately difficulties put in the way from that quarter become continually bigger. It also becomes clearer to me, the longer this goes on, that from that quarter it is not about the TRUTH, but about the fight for what they see as hostile for mankind, for on that side they think completely wrongly about me. . . . I have explained to them that my researches have delivered certain evidence for the animal descent and definite monkeylike descent of humans, not by gradual transformation (as according to Darwinism) but by two big leaps, leaps that have taken place everywhere in the animal kingdom—which the religious believers can and will consider to be creation. My opponents do not understand that, or rather refuse to understand it, and ascribe to me other motives than the search after the objective truth.

In March, Dubois' paper against Von Koenigswald—the first of three parts—is read before the Royal Academy of Science. It states Dubois' objections clearly.

The child skull of Modjokerto, the mandible and the skull of Sangiran, all of them undoubtedly human, I was convinced in my own mind were morphologically distinct from Pithecanthropus *and belonged at the same time to the proto-Australian fossil man of Java, well known under the name of* Homo soloensis. *. . .*

Nevertheless, this "Pithecanthropus IV" skull was formerly, at the time that it was complete, uncrushed, and unweathered—viz., during life—as it appears to me, an exact Wadjak-man, or Solo-man, skull. . . .

Needless to repeat that Sinanthropus pekinensis *is another member of the same proto-Australian group. . . .*

Dubois cannot help but see the parallels between his scientific struggles and world politics. There is strife on all sides: political, scientific, and personal. So strong is his faith in science that he hopes a correct and accurate view of the unity of man's ancestry, and the primitiveness that has been left behind, might do much to help. He writes again to Eugénie of these thoughts in April 1940.

April 3, 1940

The circumstances are now such that nobody knows what the present situation will become, but we all feel that this, in many ways, will become SOMETHING BETTER than the recent past, because ALL nations will be compelled to reconsider themselves, and us all, in our former conceptions. Many already feel some favorable effects from that reconsideration, concerning the merits of acquiescence and intention to do their best.

On May 10, 1940, German troops invade the Netherlands. Three days later, the Dutch government flees to exile in London and the Netherlands become occupied territory. Events are so grim and shocking that scholarly issues seem insignificant. In June, the second part of Dubois' paper is presented, in which he speaks of the way ideas about the evolutionary process distort assessments of the fossils themselves.

Now twenty-two years after Stewart Arthur Smith's publication [on the fossil human skulls of Talgai, Queensland], we observe, in a strikingly similar case, Von Koenigswald and Weidenreich led astray to the same mistake, by the same belief in evolution from Ape to Man through gradual transformation of parts of the body. This belief, in matters of evolution, results from almost exclusive morphological consideration of the organisms. Hence the apparent ... inconsiderate [ill-considered] supposition of those investigators, that the large, bestial canines of the Apes are mainly organs of defense, and became gradually smaller and less apelike in the course of Man's phylogenetic development.... Such a gradual transformation from Ape to Man is hardly conceivable....

Apparently, Von Koenigswald, in his opinion about the nature of his finds ... [relies] more on (insufficient) stratigraphical data than on unprejudiced examination of the fossil remains themselves. No wonder that his implicit belief in human evolution from Ape to Man, by gradual transformation of parts of the body, would lead him astray about the true nature of his finds, and wrongly guide the hand which made the restorations....

No one really listens to Dubois anymore. International congresses attended by all the scientists of Europe and America are a thing of the past.

Weidenreich flees China for America, getting out shortly before the Japanese invasion, but he is unable to save the *Sinanthropus* fossils. They are seized by Japanese troops and disappear forever, leaving only the casts, photographs, x-rays, and measurements Weidenreich has compiled. Communications with Java are so difficult after the Japanese invade that it is years before anyone knows whether Von Koenigswald is alive. He survives internment in a prisoner-of-war camp and eventually joins Weidenreich in New York, at the American Museum of Natural History. The Sangiran and Modjokerto fossils were safely hidden by neutral Swedish and Swiss friends, while Von Koenigswald's indomitable wife, Luitgarde, kept the one she deemed most important in her apron pocket throughout the war.

Events in the Netherlands are also grim. Month by month, week by week, Dutch Jews are deprived of all ordinary rights. The abundant food produced by the rich Dutch countryside is seized and shipped to the German troops, leaving little for the populace. Resistance to the German occupation grows, covertly, as German treatment of the Dutch worsens.

Dubois tries to reassure Eugénie, but the news is never good.

October 4, 1940
Also in these surroundings it is not at all restful, although De Bedelaar remains untouched. We saw MANY Very lights, for months, and many bombs exploded in the environs. For cooking and heat we mainly have wood. Food is not plentiful, but sufficient. Certainly there are "indeed worse things than to suffer from cold and to be hungry" and the "worse things" are bearable through moral strength. The world will, IN THE END, for mankind in general, become better. What is nature, the sun, the plants and animals, but an encouragement—and the more one can study nature, the better one can appreciate her magnitude and beauty. Then, one understands even the present world as a natural phenomenon, too, and perceives all horror as such.

He writes again in a few weeks.

October 28, 1940
Now I suffer from cold, in my study (second story) with the small base-burner, which stood in times gone by in the reception room, bell floor, in Haarlem. . . . I have still so much urgent work to do, that becomes clearer to me every day from what is published around me . . . because on these subjects (especially anthropological ones), the ideological principles of the present world war rest.

When the German patrols come through De Bedelaar, they see only an aged man, his elderly wife, and her sister. No one ever guesses that this nondescript old man shaped the course of modern paleoanthropology. Indeed, once Eugénie is at De Bedelaar, the apparent reality is even more deceptive. With the courage and independence that she has inherited from her father, Eugénie—seemingly an ordinary, middle-aged woman—sets up an escape line for downed Allied fliers. She houses them in the cellars of De Bedelaar and hides their weapons under the apples stored for the winter.

In November, the last of Dubois' trilogy of protests against Von Koenigswald is read at the meeting of the Royal Academy. Attendance is very poor, as is expected. Even now, Dubois cannot shake his bone-deep certainty of the importance of *P.e.* for understanding both the process of evolution and the origin of the human species. He cannot abandon *P.e.*, though she is imprisoned where he cannot reach her, at the Natural History Museum in Leiden. Still, he cannot let her be slandered or forgotten. *P.e.* has been his life, his great discovery, his tribute to Science and Truth, in pursuit of which he has expended his life and energies. He must persevere.

Dubois' arguments are familiar by now, having been reiterated and elaborated in many venues. *Pithecanthropus* is the only true missing link; *Sinanthropus* and *Homo soloensis* and the new Sangiran fossils are merely early humans. Dubois' closing words are poignant:

> *It is most regrettable, that for the interpretation of the important discoveries of human fossils in China and Java, WEIDENREICH, VON KOENIGSWALD, and WEINERT were thus guided by preconceived opinions, and consequently did not contribute to (on the contrary they impeded) the advance of knowledge of man's place in nature, what is commonly called human phylogenetic evolution. Real advance appears to depend on obtaining material data in an unbiased way. . . .*

There, Dubois thinks as he pens the last lines, I have said it. I have stated my credo, my opinions, and it remains for the world to judge who is right. But it is I who will be remembered as the man who found the missing link, and it is I who have defended and understood her better than anyone.

As the bitterly cold winter closes in, Dubois spends more and more time thinking of days gone by, living in his mind in a warm place, where the sun shines so brightly and the mountains are cloaked in shadowy green forests. Prentice is always by his side, young and handsome and fit. He remembers

*Dubois dies December 16, 1940, and is buried in an unconsecrated corner of the ceme-
tery at Venlo, under a special tombstone.*

sitting with Prentice at the top of Ardjoena, tired from the climb and yet too
young and healthy to really notice the exertion. There they sat, back to back,
looking at the world spread at their feet. There they carved their names in
the rock, side by side to show they were companions, not one the leader and
the other the follower. There their names rest still, hardly softened by the
ravages of time or the hurtling downpour of tropical monsoon, he supposes. That is something which will last: that, and *Pe.*

He supposes Prentice is dead by now, and he feels he probably will not
survive the winter himself. He aches with the cold and the deprivation and
the misery of the German occupation, for every freedom or right that the
Dutch once took for granted is now denied them. Science is still there, true
and brave and difficult, but he has no heart to pursue her any longer. He is

very tired: tired of trying so hard, tired of fighting, tired of being alone and misunderstood, tired of seeing the hellish downward spiral the world has taken. To continue to hope and believe in the face of all this takes more courage than he can muster on most days.

On December 16, 1940, Dubois dies in his bed of a heart attack. He is buried in the cemetery at nearby Venlo. His grave lies in an unconsecrated corner, for Eugénie knows he would not rest easy among the graves of fervent Catholics and Protestants. Besides, it would be an abomination to deny his convictions now. She orders a special tombstone.

Prof. Dr. Eug. Dubois
28-1-1858   16-12-1940

Above these words is a carving of a skull and crossbones, the universal symbol of poison. In this case, the bones are recognizably those of *P.e.*

EPILOGUE: AN UNDERESTIMATED MAN

In the years that have passed since his death, Eugène Dubois has not been forgotten in scientific circles. His discovery of *Pithecanthropus erectus*, the first specimen of a species that anthropologists now call *Homo erectus*, would be sufficient to ensure his fame, even without the extraordinary story of his deliberate decision to search for the missing link and his courage in doing so.

Despite the significance of his finds and an excellent appraisal of his scientific work by Theunissen, Dubois has been an underestimated man. Few anthropologists—and fewer still among the general public—have credited him with focusing the attention of the scientific community on human origins. The record clearly shows that his research and his perseverance made human evolution a controversial and exciting topic of general interest. The importance of his monograph is largely overlooked, yet it showed the missing link to be a transitional form between humans and apes, rather than a primitive human race. Perhaps the brevity of his monograph worked against Dubois, for later scholars produced monographs that dwarfed his in length and thoroughness. Surely, too, the harsh criticism meted out to Dubois—the very criticism that established the standard format for such monographs—slighted its true quality. It must not be forgotten that Dubois, sitting in a remote area of Java, single-handedly created the conceptual framework now used in the analysis of all fossil hominid remains. It was groundbreaking work.

So, too, was his research on body size and brain size ratios, work that was the forerunner of the enormous field of allometric studies in biology. Dubois' contributions to cephalization studies were strongly criticized for their lack of statistics in the 1930s and 1940s, and, sadly, fell into disrepute for decades. Not until his innovative work was rediscovered and defended by scholars such as Harry Jerison, in his 1973 book *Evolution of the Brain and Intelligence,* did the study of the allometry of the brain become once again a central issue in evolutionary biology.

Tragically, Dubois' strong personality and irascible disposition harmed his reputation. Whatever Dubois did, it was with such focused intensity that he drew criticism as a lightning rod draws electricity. He challenged, argued, insisted, persevered; he stretched the minds of his friends and enemies alike; he demanded of those around him an ever-higher standard of thinking and performance. No one ever forgot Dubois; no one who ever knew him was the same afterward. Though inadvertently, this paranoid, brilliant, and stubborn man truly cast his own fate.

When and how Dubois' steadfast friend Adam Prentice died is unknown. His death is not recorded in the archives at the Centraal Bureau voor Genealogie. I could not find his grave in Java, nor was his son Gerardus Prentice-MacLennan listed in the Malang, Solo, or Jakarta phone books. If Prentice was still alive when the Japanese invaded Java in 1942, he surely died during the occupation, when many Europeans were interned and even those at liberty suffered harsh conditions. However, in 1973, Adam Boyd— the son of William Boyd (the Old Warrior's son) and of Prentice's daughter by his nyai—chanced to see the photograph of Robert Boyd and the tiger cub at Mringin in a book on human evolution. Because the same photo had hung in his father's office in Java, Boyd recognized it. Eventually he contacted Jean M. Dubois, Eugène's grandson, who gave Boyd a duplicate photo and put him in touch with Eugénie and Carel Hooijer. The correspondence among these descendants of Prentice and Dubois provides all I know of the fate of Adam Prentice.

Dubois' last battle, against Von Koenigswald, was particularly unfortunate. His position was scientifically indefensible, for Von Koenigswald's finds were surely the same species as the Trinil fossils (as were Davidson Black's *Sinanthropus* fossils from China). Worse yet, the feud with Von Koenigswald was politically unwise. Even in his lifetime, Dubois was portrayed as a lunatic who hid the fossils away; in his latter years, he was accused of having recanted and assigned his precious *P.e.* to some species of fossil gibbon. Neither of these is true; he was neither a madman nor a fool. He was simply a man of singular vision, a man who longed for but did not need affirmation of his ideas in order to act. When he died he left behind many enemies, younger scientists who were all too ready to believe the worst of him. His story became notorious, told and mistold many times, until it reached mythic proportions. It is time now for the truth, and I have told it.

NOTES

Nearly all primary sources can be found in the Dubois Archives at the Naturalis, Leiden, the Netherlands. Letters held in the Dubois Archives are referred to by sender, recipient, and date. In this context, "Dubois" refers to Eugène Dubois only; other members of his family (Anna, Victor, Eugénie) are referred to by their first names only. The exceptions are Jean M. Dubois, the living grandson of Eugène, and Jean M. F. Dubois, the elder son of Eugène and the father of Jean M. Dubois. Other major sources are Dubois' pocket agendas (a sort of daily calendar), his journals, diaries, and notes; newspaper clippings he saved on scientific issues; drafts of manuscripts and lectures; and various drafts of brief autobiographies, which are also kept in the Archives. I relied as well on the diaries of J. J. A. Bernsen, O. F. M., Dubois' assistant from 1930 to 1932, in which many conversations with Dubois are recorded apparently verbatim. His diary entries are cited as "Bernsen diary," followed by the date of the entry. They are quoted by permission of the Dutch Franciscans.

I have also used unpublished manuscripts and information about Dubois compiled by the late L. D. Brongersma, Dubois' last assistant, stored in the Dubois Archives in Leiden.

ABBREVIATIONS

B.R.P.   Bloys van Treslong Prins, P. C. van. *Bronnenpublikaties van de Indische Genealogische Vereniging, vol. 5: Grafschriften van Europeanen in Nederlandsche-Indie.* 's Gravenhage, 1993: E. J. C. Boutong de Katzmann.

B.S.   *B.S. Name-List/Address-Book of the Dutch Indies 1822–1923*

G.H.G.   Bloys van Treslong Prins, P. C. van. *Genealogische en Heraldische Gedenkwaardigheden betreffende Europeanen op Java,* vols. 1–4. 1934–1939. Batavia: Drukkerij Albrecht.

I.N.N.   Anonymous, "Uitgegeven de Indische Genealogische Vereniging." *De Indische Navorscher,* vols. 1–100, 1988–1999. 's Gravenhage.

J.A.I.   *Journal of the Anthropological Institute of Ireland and Great-Britain*

N.T.   *Natuurkundig Tijdschrift voor Nederlandsch-Indië*

P.K.A.W.   *Proceeding Koninklijke Akademie van Wetenschappen*

T.A.G.   *Tijdschrift van het Koninklijk Nederlandsch Aardrijkskundig Genootschap*

Verslag   *Verslag van de gewone vergaderingen der wis- en natuurkundige Afdeeling der Koninklijke Akademie van Wetenschappen te Amsterdam.* After 1925, this journal was continued as *Verslag van de gewone vergaderingen der Afdeeling natuurkunde, Koninklijke Akademie van Wetenschappen te Amsterdam*

V.M.   *Verslag van het Mijnwezen. Extra bijvoegel der Javasche courant.*

CHAPTER 1 AN ECHO OF THE PAST

1   "My dear doctor": letter, Prentice to Dubois, February 7, 1938.

5   He had a beautiful nyai: letter, Adam Boyd to Jean M. Dubois, November 5, 1973; letter, C. R. Hooijer to Jean M. Dubois, December 11, 1974.

CHAPTER 2 THE BEGINNING

9   He was born on January 28: Basic information about Dubois' life is summarized in Dubois' autobiographical notes in the Dubois Archives; Anonymous, "Professor Dr. Eug. Dubois," *Vierde-Blad Avondblad,* January 27, 1933; Anonymous, "Prof. Dubois blikt over zijn leven terug," *De Telegraaf,* January 27, 1938; L. D. Brongersma, "Eugène

Dubois," *Natuurhistorisch maandblad* 1973, vol. 62, pp. 107–109; L. D. Brongersma, *"Pithecanthropus*: Echt en 'Pseudo'," *Verslag* 1982, vol. 91, pp. 34–36; L. D. Brongersma, "Professor dr. Eug. Dubois," *Nieuwe Rotterdamsche Courant*, January 28, 1938; P. Tesch and L. D. Brongersma, "Eugène Dubois," *Geologie en Mijnbouw*, 1941, vol. 3, no. 2, pp. 29–33; B. Theunissen, *Eugène Dubois and the Ape-man from Java*, 1989, Dordrecht: Kluwer Academic Publishers; B. Theunissen and J. de Vos, "Eugène Dubois, ondekker can de rechtopgaande aapmens," *Natuurhistorisch maandblad*, 1982, vol. 71, nos. 6–7, pp. 107–114; C. D. W. Vrijland, "Nóg herinnering aan Professor Dubois," *Teylers Magazijn*, 1993, voorjaar, p. 14; A. van der Welff, "Herinneringen een merkwaardige geleerde: Eugène Dubois," *Teylers Magazijn*, 1992, najaar, pp. 6–7.

15 There, sitting on that bench: Description based on author's visit to the house in Eijsden and information from the current owners, Mr. and Mrs. Hovens.

17 "do apes have churches?": Karl Vogt's lecture and this specific question are described in Anonymous, "Prof. Dubois blikt," *De Telegraaf*, 1938; Dubois' autobiographical notes in the Dubois Archives; and Brongersma, "Eugène Dubois," 1973.

19 By the end of his first year: Bernsen diary, September 25, 1931; letter, Dubois to Boeke, March 3, 1932.

19 "I always knew": Bernsen diary, February 13, 1931, last two sentences of quotation added.

20 Roermond is a good place: Dubois' autobiographical notes; Anonymous, "Prof. Dubois blikt," *De Telegraaf*, 1938.

20 "As a consequence": E. Haeckel, *The History of Creation*, 1892, fourth English edition, London: Kegan Paul, Trench, Trubner & Co., pp. 7–9. Dubois' copy was the German version, *Natürliche Schöpfungsgeschichte*, 1868, Berlin.

CHAPTER 3 THE GAME

22 "The Ape-like men": Haeckel, *History*, 1892, p. 398.

22 "Those processes of development": ibid., pp. 405–406.

22 Haeckel even predicts: ibid., p. 406.

23 Marie—his favorite sister: That Marie became an Ursuline sister is documented in Brongersma, "Eugène Dubois," *Natuurhistorisch maandblad*, 1973. The impact of her decision upon Dubois is unknown.

CHAPTER 4 AMBITION

27 It is Friday: Bernsen diary, November 19, 1931, and February 13, 1931.

29 Now that he can organize things: ibid., November 19, 1930.

30 Dubois would have preferred: ibid., December 17, 1930.

30 Why had Fürbringer: ibid., December 17, 1930, and January 29, 1931.

31 If he becomes the man who understands the larynx: Dubois' autobiographical notes.

CHAPTER 5 LIGHTNING ROD

31 Dubois calls on Mia Cuypers: Bernsen diary, November 19, 1930; also Theunissen, *Eugène Dubois*, 1989, p. 48.

35 the Honorable Mr. F. G. Taen-Err-Toung: Mia Cuypers' love affair with Taen-Err-Toung and marriage to him was a notorious scandal later used by her cousin, Lodewijk Van Deyssel, as the basis for his 1892 novel, *Blank en geel* (*White and Yellow*) according to Theunissen, *Eugène Dubois*, 1989, p. 48.

38 "I have been told": Bernsen diary, November 19, 1931.

CHAPTER 6 LOVE AND CONFLICT

41   Anna Geertruida Lojenga: The meeting and courtship of Anna and Dubois are described in Jean M. F. Dubois, "Trinil: A Biography of Prof. Dr. Eugène Dubois the Discoverer of *Pithecanthropus erectus*," 1957, unpublished ms. in Dubois Archives, pp. 20ff, and Bernsen diary, September 25, 1931. Engagement announcement and wedding invitation in the Dubois Archives, Naturalis.

42   Anna's parents insist: J. M. F. Dubois, "Trinil," 1957, pp. 20ff.

45   Was he trying to claim credit: Bernsen diary, January 29, 1931.

46   intellectually, she stands not so high: Bernsen diary, May 4, 1931.

46   out of romantic chivalrousness: ibid.

46   They are wed: The dates of the marriage and of the Dubois children's births are recorded in the family tree and genealogical quarters compiled by Dubois and stored in the Dubois Archives, along with the wedding invitation.

47   Teaching is for Dubois a nightmare: Theunissen, *Eugène Dubois*, 1989, pp. 27, 49, based on a personal communication from one of Dubois' students, F. H. van der Marel.

47   "So the question arises": letter, Fürbringer to Dubois, October 2, 1886.

47   a colleague, Max Weber: F. Pieters and J. de Visser, "The scientific career of the zoologist Max Wilhelm Carl Weber (1852–1937)," *Bijdragen tot de Dierkunde*, 1993, vol. 62, no. 4, pp. 193–214.

CHAPTER 7 TURNING POINT

54   Max Lohest, . . . and . . . Marcel de Puydt: Anonymous, "Prof. Dubois blikt," *De Telegraaf*, 1938.

56   Johannes Fuhlrott [and] Hermann Schaaffhausen: H. Schaaffhausen, "On the crania of the most ancient races of man," *Natural History Review*, April 1, 1861, no. 2, pp. 155–80. Tr. by George Busk from H. Schaaffhausen, "Zur Kentniss der ältesten Rassenschädel," *Archiv für Anatomie, Physiologie und wissenschaftliche Medicin*, 1858, vol. 2, pp. 453–88.

56   a powerful and formidable opponent: An account of the debates over the original Neanderthal fossil is given in E. Trinkaus and P. Shipman, *The Neandertals*, 1992, New York: Alfred A. Knopf.

56   Virchow had once boasted: quoted in S. Nuland, *Doctors: The Biography of Medicine*, 1988, New York: Vintage Books, p. 336.

56   "The supposition of Virchow": newspaper clipping in Dubois' notebook marked "Nota Palaeontolog. etc."; no indication of source.

57   These new finds: Anonymous, "Prof. Dubois blikt," *De Telegraaf*, 1938.

58   In 1877, when Darwin went to Cambridge: H. Litchfield, ed., *Emma Darwin: A Century of Family Letters*, 1915, London: Murray, vol. 2, p. 230.

CHAPTER 8 TO FIND THE MISSING LINK

59   He begins to work out his reasoning: E. Dubois, "Over de wenschelijkheid van een onderzoek naar de diluviale fauna van Ned. Indië in het bijzonder van Sumatra," *N.T.*, 1888, vol. 48, pp. 48–165; also manuscript versions of Dubois' final lecture at the University of Amsterdam in the Dubois Archives.

60   After perusing some reports from . . . Richard Lydekker: R. Lydekker, "Notices of Siwalik Mammals," *Records of the Geological Survey of India*, 1879, vol. 12, pp. 33–52; R. Lydekker, "Siwalik Mammalia," Memoirs of the Geol. Survey of India, *Palaeontologica India*, 1886, s. 10, *Indian Tertiary and Post-Tertiary Vertebrata*, vol. 4, pt. 1, pp. 1–18.

61 Finally his ideas begin: Dubois' arguments are loosely summarized from Dubois, "Over de wenschelijkheid," 1888.

62 some . . . crucial reports about the Indies fossils: K. Martin, "Ueberreste vorweltlicher Proboscidier auf Java und Banka," *Sammlungen des Geologischen Reichs-Museums in Leiden*, 1884–1889, vol. 4, pp. 1–24; K. Martin, "Fossile Säugethierreste von Java und Japan," *Sammlungen des Geologischen Reichs-Museums in Leiden*, 1884–1889, vol. 4, pp. 25–69.

62 Alfred Russel Wallace's great ideas: A. Wallace, *The Geographical Distribution of Animals*, 1876. London: Macmillan.

65 "Don't throw away": Dubois' autobiographical notes, Dubois Archives, Naturalis.

66 "For all that" [and subsequent exchanges]: Bernsen diary, December 10, 1930.

CHAPTER 9 LOGISTICS

67 "that crazy book of Darwin's": Anonymous, "Prof. Dubois blikt," *De Telegraaf*, 1938.

69 "After thinking more about your plans": letter, Weber to Dubois, undated, probably 1887.

69 "he writes inquiring": This can be deduced from letter, De Referendaris, chief of Military Affairs Colonial Department to Dubois, August 4, 1887. Copies of the contracts of enlistment are preserved in the Dubois Archives.

72 Dubois' father, adamantly opposed: Bernsen diary, March, 1931.

75 "I advise against this": ibid., December 10, 1931.

75 He offers to edit: letters, Fürbringer to Dubois, October 15, 1887, and October 26, 1887.

76 the steamship S.S. *Prinses Amalia*: dates of departure and arrival from letter, Dubois to Stoomvaartmaatschappij, Nederlands, March 3, 1931.

CHAPTER 10 PADANG

84 "Captain Hendrik Krull": A note in the Dubois Archives mentions that H. Krull of the Dutch Indies Army collected Dubois and his family upon arrival in Padang. His Christian name was confirmed by searching the enlistment records of the Koninklijk Nederlandsch-Indisch Leger, the Royal Dutch East Indies Army.

85 a totok, a newly arrived European: During the late nineteenth and early twentieth century, colonists in the Indies had an elaborate system of classifying people according to their ethnic origin. I have used the terms of that era as I understand them. "Pures" was the common term for those of full European ancestry. "Totoks" were Pures who had only recently come to the Indies; the term implied a certain naïveté about Indies ways. "Indos," now considered offensive in some circles, was used to describe those with a mixture of European and Indonesian ancestry. "Natives" were those of unadulterated Indonesian ancestry; manual laborers were routinely referred to as "coolies." "Indische" refers to families that were longtime residents of the Indies; such families very often intermarried with Indonesians.

90 "'walking dictionary'": For a recent consideration of interracial sexual attitudes and behavior in the colonial Indies, see E. Locher-Schloten, "So Close and Yet So Far," pp. 131–153, and P. Pattynama, "Secrets and Danger," pp. 84–107, both in J. Clancy-Smith and F. Gouda, eds., *Domesticating the Empire*, 1998, Charlottesville: University Press of Virginia. For a turn-of-the-century perspective, see L. Couperus, *The Hidden Force*, 1990, Amherst: University of Massachusetts Press (originally *De Stille Kracht*, 1900); E. Breton de Nijs, *Faded Portraits*, 1982, Amherst: University of Massachusetts Press (originally *Vergeelde portretten uit een Indisch familiealbum*, 1954); and Multatulti (Edward Douwes Dekker), *Max Havelaar*, 1982, Amherst: University of Massachusetts Press (originally 1860, Amsterdam: Van Oorschot).

93 Dubois offers to lecture: Dubois' lecture notes, Dubois Archives, Naturalis.

95 "It is obvious that scholars": Dubois, "Over de wenschelijkheid," 1888, p. 165.

CHAPTER 11 PAJAKOMBO

95 Kroesen is a thoughtful man: letter, Kroesen to Dubois, September 8, 1888.

96 Anna gives birth: J. M. F. Dubois, "Trinil," 1957, pp. 35–36.

97 In September, another physician arrives: ibid., pp. 31ff; Bernsen diary, February 13, 1931; L. D. Brongersma, "The Vicissitudes of the Dubois Collection," unpublished manuscript, n.d., Dubois Archives, Naturalis, p. 5.

CHAPTER 12 FOSSILS

101 "We are really going to look forward": letter, Kroesen to Dubois, September 8, 1888.

104 a provisional report of his finds: E. Dubois, "Voorloopig verslag over palaeontologi-schen asporingen in grotten bij Pajakombo (Padangsche Bovenlanden)." Unpublished report in Dubois Archives, Naturalis.

105 "REGISTER OF RESOLUTIONS": copy of document (in Dutch) in Dubois Archives.

CHAPTER 13 GARUDA

107 They come to a large cave: J. M. F. Dubois, "Trinil," 1957, p. 33ff; Bernsen diary, February 13, 1931.

CHAPTER 14 FEVERS AND SPELLS

113 A slight noise: J. M. F. Dubois, "Trinil," 1957, p. 34; Bernsen diary, February 13, 1931.

116 "Everything here has gone against me": letter, Dubois to Jentink, October 17, 1888.

117 Anthonie De Winter and Gerardus Kriele: dates of their starting to work for Dubois are recorded in his daily agenda for 1889; first names confirmed by searching enlistment records of the Koninklijk Nederlandsch-Indisch Leger, the Royal Dutch East Indies Army.

CHAPTER 15 TO JAVA

118 He writes for advice: letter, Dubois to Verbeek, October 17, 1889; letters, Verbeek to Dubois, November 22, 1888, March 8, 1889, and March 9, 1889.

118 "I received your letter": letter, Sluiter to Dubois, December 21, 1888; B. D. van Rietschoten, "Uit een schrijven van den heer Van Rietschoten te Blitar," N.T. 1889, vol. 48, pp. 346–47, based on a letter, Van Rietschoten to Sluiter, October 13, 1888.

119 "In every respect": letter, Dubois to Sluiter, no date, 1889, basis of E. Dubois, "Uit een schrijven van den heer Dubois te Pajakombo naar aanlideing van den aan dien heer toegezonden schedel, door den heer Van Rietschoten in zijn marmergroeven in het Kedirische opgegraven," N.T. 1890, vol. 49, pp. 209–210.

121 Permission is duly granted: copy of resolution in Dubois Archives, Naturalis.

123 the house formerly occupied by the Dutch Assistant Resident: J. M. F. Dubois, "Trinil," 1957, pp. 36–38; photographs in the Dubois archives; details of similar houses from descriptions and photographs of Indies houses in Breton de Nijs, *Faded Portraits*, 1982, and R. Wassing and R. Wassing-Visser, *Adoeh, Indië! Het best van Hein Buitenweg*, 1992, Atrium.

126 a marabou stork: J. M. F. Dubois, "Trinil," 1957, p. 38.

CHAPTER 16 JAVA FOSSILS

128   As he explains to Groeneveldt: E. Dubois, monthly report for November 1890, unpublished manuscript in Dubois Archives, Naturalis.

130   Soon the fossils usurp: photograph, Dubois Archives, Naturalis.

131   "Amidst the remains": E. Dubois, "Palaeontologische onderzoekingen op Java," *V. M.* 4th Quarterly Report, 1890, p. 14.

CHAPTER 17 COOLIES

132   "Will you also be so kind": letter, De Winter to Anna, June 28, 1890.

133   "All the forced laborers": letter, De Winter to Dubois, July 11, 1890.

133   "Now I have here": letter, Kriele to Dubois, July 11, 1890.

133   "So it is also very hard": letter, De Winter to Dubois, n.d. but between September 29, 1890, and December 30, 1890.

133   on January 16, 1891, Victor Marie Dubois: birth recorded in *B.S.*

133   Sluiter replies cheerfully: letter, Sluiter to Dubois, December 29, 1890.

133   "I really would like": letter, Sluiter to Dubois, January 6, 1891.

134   "Why are you so disconsolate": letter, Sluiter to Dubois, January 23, 1891.

134   De Winter, working in the north: letter, De Winter to Dubois, March 29, 1891.

135   Even Dubois' own workers: Bernsen diary, March 18, 1931.

137   In contrast, Kriele and De Winter: J. M. F. Dubois, "Trinil," 1957, p. 48ff.

CHAPTER 18 DISCOVERIES AT TRINIL

139   Dubois' uncanny ability [and much of paragraph]: Bernsen diary, February 13, 1931.

141   "The most important find": E. Dubois, "Palaeontologische onderzoekingen op Java," *V. M.* 3rd Quarterly Report, 1891, pp. 13–14.

143   "Near the place on the left bank": E. Dubois, "Palaeontologische onderzoekingen op Java," *V. M.* 4th Quarterly Report, 1891, p. 13.

143   "Besides, the most important fact": ibid.

143   "all that we possessed": Dubois, "Palaeontologische onderzoekingen op Java," *V. M.* 4th Quarterly Report, 1891, pp. 14–15.

144   "The creature": letter, Dubois to Kroesen, December 30, 1891.

CHAPTER 19 GATHERING RESOURCES

145   Early in 1892: letters from Weber to Dubois, April 4, 1892, and October 21, 1892.

145   The older man is Robert Boyd: Information on Robert Boyd and his family, their marriages, and their children is derived from *B.S., B.R.P., G.H.G., I.N.N.,* and the dossier "Boyd" at the Centraal Bureau voor Genealogy, The Hague.

147   Their wedding . . . January 16, 1890: *B.S.*

148   By the evening of the twenty-first: The birth of the child is recorded in *B.S.*

149   "Jane de Clonie MacLennan": The death of Jane de Clonie MacLennan-Prentice and her epitaph are recorded in *G.H.G.* The birthdate of Gerard Alexander Prentice MacLennan is recorded in the same source and mentioned in a letter, Prentice to Dubois, July 13, 1893.

152   Bishop Wilberforce being bested in debate: L. Huxley, ed., *Life and Letters of Thomas Huxley*, 1900, New York: Appleton, p. 199; see also F. Darwin, ed., *Life and Letters of Charles Darwin*, 1887, London: John Murray, pp. 114–16.

CHAPTER 20 FRIENDSHIP

157   "they romp with a . . . tiger cub": letter, Adam Boyd to Jean M. Dubois, November 23, 1974.

CHAPTER 21 TRINIL

159   "A few cool windy days": E. Dubois, notes, July 28, 1892, Dubois Archives, Naturalis.
160   "De Winter told me": letter, Kriele to Dubois, September 7, 1892.
163   small tin soap dish: item preserved in Dubois Archives, Naturalis.

CHAPTER 22 THE BIRTH OF *PITHECANTHROPUS*

165   "The most important find": E. Dubois, "Palaeontologische onderzoekingen op Java," *V. M.* 3rd Quarterly Report, November 28, 1892, p. 10.
166   "This being was in no way equipped": ibid., pp. 12–13.
166   "Because of this find": ibid., pp. 11, 14.
167   "In calculating the relative volume": letter, Dubois to Groeneveldt, December 4, 1892.
167   "I have made a terrible mistake": There is no documentary evidence of this letter whatsoever.
168   "I received your letter": letter, Dubois to Weber, December 19, 1892.
170   In the very act of writing: letter, Dubois to Groeneveldt, December 28, 1892, complete with "A" overwritten by "P."
171   "I have, Your Excellency": letter, Dubois to Groeneveldt, December 28, 1892.

CHAPTER 23 1893

174   its editor and publisher, P. A. Daum: Information about Daum from E. M. Beekman, "Introduction," 1987, pp. 1–2, in P. A. Daum, *Ups and Downs of Life in the Indies*, tr. by Elsje Qualm Sturtevant and Donald W. Sturtevant. Amherst: University of Massachusetts Press. (Originally published as P. A. Daum, *Ups en downs in het Indische leven*, 1890, Batavia: feuilleton.)
176   "IN PURSUANCE": Anonymous ("Homo Erectus"), "Palaeontologische onderzoekingen op Java," *Bataviaasch Nieuwsblad*, February 6, 1893, no. 57. Almost certainly, this piece was written by P. A. Daum, the editor of the newspaper.
180   "rather hastily": J. A. C. A. Timmerman, "Belangrijke palaeontologische vondstein op Java," *T.A.G.*, 1893, s. 2, vol. 10, pp. 310–12, 312.

CHAPTER 24 DISASTER

182   "Ngrodjo, Saturday morning": letter, Prentice to Dubois, n.d. but certainly early May of 1893, as Dubois starts for Lalie Djuvo on May 4 according to his daily agenda.
185   And there he carves his name: This incident is mentioned in a letter, Prentice to Dubois, November 28, 1938. The date of their climb up Gunung Ardjoena is recorded in Dubois' 1893 daily agenda, as is the date of their departure from Lalie Djuvo.

CHAPTER 25 LETTERS FROM A FRIEND

186   "Mringin   July 9, 1893": letter, Prentice to Dubois, July 9, 1893.
187   "Tempoersarie   July 13, 1893": letter, Prentice to Dubois, July 13, 1893; a few Dutch words translated into English for clarity.
190   "Tempoersarie   July 16, 1893": letter, Prentice to Dubois, July 16, 1893. Prentice refers to Dubois' site as Ngawi, the nearest sizable town, but I have substituted "Trinil" for clarity.
191   "Shall we take a photograph": letter, Prentice to Dubois, February 7, 1938.

CHAPTER 26 AFTERMATH

196 "Anna abortus": Dubois' daily agenda has a handwritten entry that says simply "Anna abortus," crossed out and then rewritten, on August 30, 1893. This is the only documentation of this event.

196 He practices science: Bernsen diary, October (n.d., but before October 7) 1931.

CHAPTER 27 PERSEVERANCE

198 "Trinil   August 8, 1893": letter, Kriele to Dubois, August 8, 1893.

198 "The Resident from Solo": letter, De Winter to Dubois, August 25, 1893.

198 "Regarding the free people": letter, De Winter to Dubois, September 3, 1893. "Free people" refers to workers who are not forced laborers.

199 Last year, in 1892: E. Dubois, "Nashrift op 'De Klimaten der Voorwereld en de Geschiedenis der Zon,'" *N.T.*, 1892, vol. 51, pt. 1, pp. 93–100.

199 "Geboegan, Oenarang   September 27, 1893": letter, Prentice to Dubois, September 27, 1893. "Trinil" substituted for "Ngawi" for clarity.

200 He decides to erect a monument: The monument still stands at the site museum at Trinil in East Java.

CHAPTER 28 THE MONOGRAPH

202 "It would be foolish to doubt": E. Dubois, *Pithecanthropus erectus: Eine menschen-aehnliche Uebergangsform aus Java*, 1894, Batavia: Landsdrukkerij, p. 2.

203 Adjusting his calculations for *P.e.* accordingly: ibid., pp. 10–11.

204 In his monograph, Dubois remarks: ibid., p. 6.

205 Too, the general gracility: ibid., pp. 10–13.

205 Prentice has called at Toeloeng Agoeng: letter, Prentice to Dubois, October 31, 1893.

206 Boyd's daughter, Anna Grace: The marriage of Anna Grace Boyd to M. G. de Witte, an employee of a sugar company, on October 11, 1893, is recorded in *B.S.*

206 "Oenarang   October 31, 1893": letter, Prentice to Dubois, October 31, 1893.

CHAPTER 29 WRITING UP

207 *P.e.* must have walked like a man: Dubois, *Pithecanthropus erectus*, 1894, p. 23.

208 "The points suffice": ibid.

208 "pillar, girder, and siphon": ibid., pp. 26–27.

208 *P.e.* stood perhaps 1.7 meters: ibid.

209 "An Anthropopithekos has become": ibid., p. 31.

210 Evolution, Dubois becomes convinced: ibid., p. 37.

210 "forerunner of the gibbon *Hylobates*": ibid., pp. 37–39.

210 "Although already quite advanced": ibid., p. 31.

CHAPTER 30 SEPARATION AND LOSS

211 "Do not listen to pontianaks": This incident is based on a similar one in Couperus, *Hidden Force*, 1985, p. 176ff.

214 "the djongas, Nassi": Photograph of Nassi, labeled by Dubois, in Dubois Archives.

CHAPTER 31 INTERMISSION

219 "Geboegan, Oenarang   June 30, 1894": letter, Prentice to Dubois, June 30, 1894.

221 "Mringin   October 20, 1894": letter, Prentice to Dubois, October 20, 1894. "Trinil" substituted for "Ngawi."

222   "Toeloeng Agoeng   November 3, 1894": letter, Prentice to Dubois, November 3, 1894.

222   "Toeloeng Agoeng   November 7, 1894": letter, Prentice to Dubois, November 7, 1894.

CHAPTER 32 TO INDIA

223   "Mringin   December 21, 1894": letter, Prentice to Dubois, December 21, 1894. Also letters, Prentice to Galloway and Prentice to Lyon, December 21, 1894.

224   "*December 29, 1894*": Dubois India journal, dates given.

226   "*January 10, 1895*": ibid., date given.

CHAPTER 33 CALCUTTA

228   "*January 22, 1895*": ibid., date given.

230   "*January 23, 1895*. Alcock is": ibid., date given.

230   "*January 23, 1895*. Yesterday I studied in the museum": ibid.

231   "The significant name": Anonymous, "Notes," *Nature*, 1895, January 3, vol. 51, p. 230.

231   "*January 27, 1895*". Dubois India journal, date given.

231   a group of natives: Ethnographic exhibits and showcases displaying "exotic" indigenous peoples were common from 1870 to about 1930. For an analysis, see R. Corbey, "Ethnographic Showcases," *Cultural Anthropology*, 1993, vol. 8, no. 3, pp. 338–69.

231   "*February 2, 1895*": ibid., date given.

232   "*February 3, 1895*": ibid., date given.

233   He is far better off: Dates, times, and stations for Dubois' outward journey in India given in C. L. Griesbach's written advice, 1895, in the Dubois Archives, Naturalis.

237   "Indian Museum   February 9, 1895": letter, Alcock to Dubois, February 9, 1895. Phrases in German translated into English for clarity.

237   "Monday. Call on Deputy Commissioner": Griesbach, advice, 1895.

239   "For horse get about": ibid.

240   "The Punjab Government has been informed": ibid.

240   "Let me know when you": ibid.

241   Dubois writes a suitably officious letter: letter, Dubois to Collector and Magistrate of Saharapur, February 13, 1895.

243   "Sir, Please explain": There is no direct evidence of the contents or intended recipient of the telegram, although the existence of a telegram that Alcock refuses to send is apparent from a letter, Alcock to Dubois, February 17, 1895. Dubois' later accusation that Prentice had an affair with Anna while Dubois was in India is shown by Prentice's denial in a letter, Prentice to Dubois, June 23, 1895.

CHAPTER 34 SIRMOOR STATE

244   He has set up a school: The existence of this school is mentioned in an undated letter, M. M. Carliton to Sukh Chain Sinha, sent to Dubois by Sukh Chain Sinha on April 3, 1895.

245   "*younger brothers*": The metaphor of the Dutch ruler in the East Indies as the elder brother of the native regent or prince was a common one, reflected in an official directive from the Dutch government in 1820. This policy figures prominently in such Dutch colonial literature as Multatuli, *Max Havelaar*, 1982, and Couperus, *Hidden Force*, 1985.

246   he receives a letter from the Collector: letter and purwana, Collector and Magistrate of Saharanpur to Dubois, February 15, 1895.

246   Letters are written and notes filed: telegrams, Griesbach to Punjab Government, February 18, 1895.

246   "Indian Museum   February 17, 1895": letter, Alcock to Dubois, February 17, 1895.

CHAPTER 35 SIWALIK ADVENTURES

249   His itinerary sounds like: Brongersma, notes on India trip, Dubois Archives, Naturalis.

249   Dubois sends another telegram . . . A reassuring reply: telegram, Dubois to Maharajah of Sirmoor, February 25, 1895; telegram, Maharajah of Sirmoor to Dubois, February 26, 1895.

249   "Obstruction by Rajah Nahan": telegram from Foreign Department to Government of the Punjab Revenue and Agricultural Departments, March 1, 1895.

250   "Indian Museum   March 3, 1895": letter, Alcock to Dubois, March 3, 1895.

250   a copy of an article: P. Matschie, "Noch einmal *Pithecanthropus erectus*," *Naturwissenschaftliche Wochenschrift*, 1895, vol. 10, pp. 122–23.

251   Dubois has already written Gamble: The existence of this letter is mentioned in a letter, Gamble to Dubois, March 15, 1895.

251   He sends yet another letter; His letter of March 4: letter, Dubois to Maharajah of Sirmoor, March 4, 1895; letter, Maharajah of Sirmoor to Dubois, March 4, 1895.

251   Dubois replies immediately: second letter, Dubois to Maharajah of Sirmoor, March 4, 1895.

251   By letter the next morning: letter, Maharajah of Sirmoor to Dubois, March 5, 1895.

251   The guide is Sukh Chain Sinha: letter, Sukh Chain Sinha to Dubois, April 3, 1895, and included letters of reference.

252   "some 5 fossil bones": letter, Maharajah of Sirmoor to Dubois, March 10, 1895.

253   "Toeloeng Agoeng   March 7, 1895": letter, Prentice to Dubois, March 7, 1895.

254   "The case of my missing link": Brongersma, notes taken on a letter, Dubois to Anna, March 16, 1895. Original letter not in archives.

254   "I hope you will succeed": letter, Gamble to Dubois, March 15, 1895.

254   "Did you read Lydekker's critique": letter, Griesbach to Dubois, March 18, 1895.

255   "Griesbach writes about the critique": Brongersma, notes on letter, Dubois to Anna, March 28, 1895. Original letter not in archives.

255   he sends a desperate telegram: telegram, Dubois to Griesbach, March 29, 1895.

CHAPTER 36 LEAVING INDIA

256   "Review of Dubois' *Pithecanthropus erectus*": All quotations from the review come from R. Lydekker, "Review of Dubois' *Pithecanthropus erectus: Eine menschenaehnliche Uebergangsform aus Java*," *Nature*, 1895, vol. 51, p. 291.

258   It contains a begging letter: letter, Sukh Chain Sinha to Dubois, April 3, 1895.

259   "It is only justice": O. Marsh, "On the *Pithecanthropus erectus*, from the Tertiary of Java," *American Journal of Science*, 1896, s. 4, vol. 1, pp. 475–82.

259   "Marsh sent me a reprint": Brongersma, notes on letter, Dubois to Anna, April 9, 1895. Original letter not in archives.

260   "April 14, 1895": Brongersma, notes on letter, Dubois to Anna, April 14, 1895. Original letter not in archives.

CHAPTER 37 TOELOENG AGOENG

265   "You have been seducing my wife": That Dubois made this accusation is plain from a letter, Prentice to Dubois, June 23, 1895.

CHAPTER 38 DEPARTURE

268   The Dutch anthropologist: H. ten Kate, "Review of Dubois' *Pithecanthropus erectus, eine menschenaehnliche Uebergangsform aus Java*," *Nederlandsch Koloniaal Centralblad*, 1894–1895, vol. 1, pp. 127–29.

268   And Rudolf Martin: R. Martin, "Kritische Bedenken gegen den *Pithecanthropus erectus* Dubois," *Globus*, 1895, vol. 67, pp. 213–17.

268   The next is Daniel Cunningham: D. J. Cunningham, "Dr. Dubois' So-called Missing Link," *Nature*, 1894–95, vol. 51, pp. 428–29; D. J. Cunningham, "A Paper on *Pithecanthropus erectus*, the Man-Like Transitional Form of Dr. Eug. Dubois," *Journal of Anatomy and Physiology*, 1895, vol. 29, n.s. 8, pp. xviii–xix.

268–269   "By a series of easy"; "The fossil cranium": Cunningham, "Dr. Dubois' So-called Missing Link," 1894–95, pp. 428–29.

269   accompanied by Janet Boyd: She accompanied the Dubois family on this voyage "to keep some order among the 3 young children" and then entered school in the Netherlands, according to a letter from C. R. Hooijer to Jean M. Dubois, December 11, 1974. Hooijer describes Janet as a "half-sister" to Adam Boyd; Adam Boyd was the son of William Boyd and Samila, who was in turn the daughter of Adam Prentice and his nyai. Janet is probably one of the two children, "Errol and Janet," of William Boyd and his first Javanese wife, mentioned in a letter, Adam Boyd to Jean M. Dubois, November 5; Janet's birth is not recorded in the *B.S.* (not an uncommon circumstance for a female child of mixed race at the time).

270   "Mringin June 23, 1895": letter, Prentice to Dubois, June 23, 1895.

CHAPTER 39 EUROPE

271   a ferocious storm: The storm and the exchange between Dubois and Anna are described in J. M. F. Dubois, "Trinil," 1957, p. 190ff.

274   "Look at that skull!": This incident is described in ibid., p. 191.

275   "It is the ape-man!": The homecoming is remembered in a letter, Ant Spitzen to Brongersma, September 9, 1941.

276   "But, boy, what use is it?": Bernsen diary, March (n.d.) 1931.

CHAPTER 40 THE BATTLEFIELD

278   the Ministry for the Colonies: Brongersma, "Vicissitudes," p. 24.

278   Third International Congress of Zoology: Program preserved in the Dubois Archives, Naturalis.

279   Virchow is his fiercest opponent: See P. Shipman, *The Evolution of Racism*, 1994, New York: Simon & Schuster, for a recounting of the battles between Virchow and Haeckel over evolution.

279   "Here the fantasy": R. Virchow, "Commentary on Krause's discussion of Dubois' *Pithecanthropus erectus: eine menschenaehnliche Uebergangsform aus Java*," *Zeitschrift für Ethnologie*, 1895, vol. 27, pp. 81–88; also, Virchow, "Die Frage von dem *Pithecanthropus erectus*," *Zeitschrift für Ethnologie*, 1895, vol. 27, pp. 435–42.

280   Now, for the first time: E. Dubois, "*Pithecanthropus erectus*, eine menschen-aehnliche Uebergangsform," *Compte-rendu des séances du Troisième Congrès International de Zoologie, Leyde, 11–16 septembre 1895*, 1896, Leiden, pp. 251–71.

284   the American Marsh comes forward: O. C. Marsh, "A Commentary on Dubois' '*Pithecanthropus erectus*, eine menschenaehnliche Uebergangsform.'" *Compte-rendu des séances du Troisième Congrès International de Zoologie, Leyde, 11–16 septembre 1895*, 1896, Leiden, p. 272.

284   Although some believe: See J. F. van Bemmelen, "Het Leidsche internationale zoologencongres: Dubois's aapmensch voor de vierschaar der wetenschaap," *Java-bode*, November 1895, vol. 16.

284   van Bemmelen's findings: J. M. van Bemmelen, "Der Gehalt an Fluorcalcium eines fossilen Elephantenknochen aus der Tertiärzeit," *Zeitschrift für anorganische Chemie*, 1897, vol. 15, pp. 84–122.

284   "September 17, 1895/Professor Virchow": postcard, Jentink to Dubois, September 17, 1895.

285   "September 17, 1895/Judging from": letter, Dubois to Jentink, September 17, 1895.

286   the abandoned suitcase: J. M. F. Dubois, "Trinil," 1957, p. 78ff; also A. Hrdlička, *The Skeletal Remains of Early Man*, 1930, Washington, D.C.: Smithsonian Miscellaneous Collections, vol. 83, p. 38.

287   Sir William Turner: W. Turner, "On M. Dubois' Description of Remains Recently Found in Java, Named by Him *Pithecanthropus erectus*. With Remarks on So-called Transitional Forms between Apes and Man," *Journal of Anatomy and Physiology*, 1896, vol. 29, n.s. 9, pp. 424–45.

288   "Of course, not being a special anatomist": letter, Munro to Dubois, December 23, 1896.

288   "the most important hitherto recorded": W. Turner, "Discussion of Dubois' 'On *Pithecanthropus erectus:* A Transitional Form between Man and the Ape,'" *J.A.I.*, 1896, vol. 25, p. 250ff. The discussion followed Dubois' presentation to the Royal Dublin Society on November 20, 1895.

288   "After the erect position": letter, Munro to Dubois, November 26, 1895.

288   "lowest human cranium": D. J. Cunningham, "A paper on *Pithecanthropus erectus*, the Man-like Transitional Form of Dr. Eug. Dubois," *Journal of Anatomy and Physiology*, 1895, vol. 29, n.s. 9, pp. 18–19.

289   "Professors Sir W. Turner": Quotations in this paragraph come from E. Dubois, "On *Pithecanthropus erectus:* A Transitional Form between Man and Apes," *J.A.I.*, 1896, vol. 25, pp. 240–55, 244.

289   "It is unfortunate": W. H. Flower, "Discussion of Dubois' 'On *Pithecanthropus erectus*, A Transitional Form between Man and the Ape,'" *J.A.I.*, 1896, vol. 25, p. 248.

289   "The opportunity which Dr. Dubois": Turner, "Discussion," 1896, pp. 249–51.

290   "I have studied": J. Garson, "Discussion of Dubois' 'On *Pithecanthropus erectus*, A Transitional Form between Man and the Apes,'" *J.A.I.*, 1896, vol. 25, pp. 251–52.

291   "The femur is extremely": ibid.

291   "What strikes me most forcibly": J. A. Thomson, "Discussion of Dubois' 'On *Pithecanthropus erectus*, A Transitional Form between Man and the Apes,'" *J.A.I.*, 1896, vol. 25, pp. 253–54.

291   "As to whether or no": ibid., p. 254.

292   "The chief question": A. Keith, "Discussion of Dubois' 'On *Pithecanthropus erectus*, A Transitional Form between Man and the Ape,'" *J.A.I.*, 1896, vol. 25, p. 253.

292   "To my mind": ibid.

293   "Saturday November 23, 1895": O'Mulligan, "Bones of Contention," *The Evening Telegraph* (Dublin), November 23, 1895.

295   "According to all the rules of classification": R. Virchow, "Commentary on Dubois' *Pithecanthropus erectus*, betrachtet als eine wirkliche Uebergangsform und als Stammform des Menschen," *Zeitschrift für Ethnologie*, 1895, vol. 27, pp. 744–47, 744.

295   the most satisfactory bit: J. M. F. Dubois, "Trinil," 1957, p. 81.

295   "I am well aware": E. Dubois, "*Pithecanthropus erectus*, betrachtet als eine wirkliche Uebergangsform und als Stammform des Menschen," *Zeitschrift für Ethnologie*, 1895, vol. 27, pp. 723–38, 737.

296   Wilhelm Branco and William Dames: Theunissen, *Eugène Dubois*, 1989, p. 109.

CHAPTER 41 MORE SKIRMISHES

297   In February: letter, Dubois to Haeckel, February 22, 1896.

297   "May 12, 1896": letter, Schwalbe to Dubois, May 12, 1896.

298   "the Prix Broca": certificate in Dubois Archives, Naturalis.

299   On January 8, 1897: Brongersma, "Vicissitudes," p. 19.

299   On February 5: letter, Place to Dubois, February 5, 1897.

300   "Among the students": letter, Schrijnen to Dubois, February 10, 1897.

300   "February 18, 1897": letter, Dubois to Place, February 18, 1897.

300   But if Seydel is rejected: letter, Place to Dubois, February 19, 1897.

301   "Dr. Victor Dubois, Venlo": telegram, Dubois to Victor, February 20, 1897.

301   "I believe": letter, Victor to Dubois, February 21, 1898.

302   "February 22, 1898": letter, Dubois to Victor, February 22, 1898.

304   "Toeloeng Agoeng   15 April 1896": letter, Prentice to Dubois, April 15, 1896.

CHAPTER 42 USING HIS BRAINS

305   the Fourth International Congress of Zoology: program preserved in the Dubois Archives, Naturalis.

306   "The next question": E. Haeckel, "On our present knowledge of the origin of Man," *Annual Report of the Board of Regents of the Smithsonian Institution for the Year Ending June 30, 1898,* 1899, pp. 461–80, 468–70. Translation of a discourse given at the Fourth International Congress of Zoologists at Cambridge, England, August 26, 1898.

308   "To this momentous interpretation": ibid., p. 471.

308   "Virchow further asserted": ibid., pp. 471–72.

311   "It is actually": E. Dubois, "De verhouding van het gewicht der hersenen tot de grootte van het lichaam bij de zoogdieren," *P.K.A.W.,* 1897, s. 2, vol. 5, no. 10, pp. 1–41, 10.

313   "From all these considerations": E. Dubois, "Remarks upon the Brain-cast of *Pithecanthropus,*" *Proceedings of the Fourth International Congress of Zoology, Cambridge, 22–27 August 1898,* 1899, London, pp. 78–96, 96.

315   "IVth International Congress of Zoology": copy of resolution in Dubois Archives, Naturalis.

CHAPTER 43 BETRAYAL AND RESURRECTION

317   "December 20, 1897": letter, Schwalbe to Dubois, December 20, 1897.

318   "to contribute to the important question": G. Schwalbe, "Ziele und Wege einer vergleichenden physischen Anthropologie," *Zeitschrift für Morphologie und Anthropologie,* 1899, vol. 1, pp. 1–15, 2.

318   "The apparently wide chasm": ibid., p. 5.

318   "In this way, paleontology": ibid., p. 15.

319   Although in 1899 Dubois is offered: Brongersma, "Vicissitudes," p. 19; Theunissen, *Eugène Dubois,* 1989, p. 3.

319   more than eighty publications: Theunissen, *Eugène Dubois,* 1989, pp. 80, 122.

321   "You'll do very nicely": J. M. F. Dubois, "Trinil," 1957, p. 194.

321   "*not* my face!": ibid., p. 195.

322   "That is my father!": ibid., p. 196.

CHAPTER 44 FAMILY

323   In company, she knows how: Bernsen diary, May 4, 1931.

323   Eugénie does not worry him: author's interview with Victor E. Dubois and Nelleke Hooijer.

323   He establishes a firm routine: J. M. F. Dubois, "Trinil," 1957, p. 207.

325   Jean and Victor do not do well: Bernsen diary, March, 2, 1931.

325   When Jean and Victor are ready: ibid., December 12, 1930.

326   "Did you save the skull?": J. M. F. Dubois, "Trinil," 1957, p. 203.

CHAPTER 45 THE NEW CENTURY

327   Now that I have found: Bernsen diary, March 2, 1931.

328   There is a move afoot: Brongersma, "Vicissitudes," p. 20.

328   He leads the Ministry of the Colonies: ibid.

328   "April 30, 1903": postcard, Lorié to Dubois, April 30, 1903.

329   "May 4, 1903": letter, Lorié to Dubois, May 4, 1903.

329   recalled to active duty: Brongersma, "Vicissitudes," p. 20.

329   Emil Selenka, a German zoologist: Theunissen, *Eugène Dubois*, 1989, p. 118.

330   "The excavations which at present": E. Dubois, "Eenige van Nederlandschen kant verkregen uitkomst met betrekking tot de kennis der Kendeng-fauna (fauna van Trinil)," *T.A.G.*, 1907, s. 2, vol. 24, pp. 449–58, 449.

330   "In short, I consider": ibid., tr. and quoted by E. Dubois, "The fossil human remains discovered in Java by Dr. G. H. R. von Koenigswald and attributed by him to *Pithecanthropus erectus*, in reality remains of *Homo sapiens soloensis* (conclusion)," *P.K.A.W.*, 1940, vol. 43, pp. 1268–75, 1271.

331   "April 25, 1907/*Bataviaasch Nieuwsblad*": Anonymous, interview with Mme Selenka, *Bataviaasch Nieuwsblad*, April 25, 1907.

332   "From the two teeth": E. Dubois, "Das geologische Alter der Kendeng- oder Trinil-fauna," *T.A.G.*, 1908, s. 2, vol. 25, pt. 6, pp. 1235–70, 1252.

332   On February 9, 1909: M. Selenka, "Die fossile Zähne von Trinil," *T.A.G.*, 1909, s. 2, vol. 26, p. 398. Following Selenka's article are contributions from Schlosser, pp. 398–99, Walkhoff, p. 399, and Dubois, pp. 400–401.

333   Dragutin Gorjanović-Kramberger: D. Gorjanović-Kramberger, "Der diluviale Mensch von Krapina in Kroatien: ein Beitrag zur Paläoanthropologie," 1906, in O. Walkhoff, ed., *Studien über die Entwicklungsmechanik des Primatenskeletts*, Kreidel, Wiesbaden, pp. 59–277.

333   a massive monograph on the La Chapelle-aux-Saints Neanderthal: M. Boule, "L'homme fossile de La Chapelle-aux-Saints," *Annales de Paléontologie*, 1911–1913, Paris.

334   Boule's work is underpinned: See also discussion of Boule's perspective in Trinkaus and Shipman, *Neandertals*, 1992.

334   he lifted the corner: Paraphrased from P. J. van der Feen and W. S. S. van Bentham-Jutting, "Antje Schreuder, Amsterdam, 15 november 1887–Amsterdam, 2 februari 1952," *Geologie en mijnbouw*, 1952, vol. 14, pp. 121–25, 122.

CHAPTER 46 DIVERSIONS

335   In August, work starts: M. Gijsbers, pers. comm. to P. Storm.

335   a sort of prehistoric nature park: Leakey and Slikkerveer, *Man-ape*, 1993, p. 155.

336   "December 17, 1910": letter, Dubois to Sluiter, December 17, 1910.

337   "Man, witness of the Flood"; "rare relic"; "melancholy skeleton": A. C. Haddon, *History of Anthropology*, 1910, London: Watts & Co., p. 70.

337   "fade into relative insignificance": ibid., p. 76.

337   "Dubois published his account": Haddon, *History*, 1910, pp. 77–78.

337   "'I believe that in *Pithecanthropus erectus*'": Haddon, ibid., p. 77, was quoting W. H. L. Duckworth, *Morphology and Anthropology*, 1904, p. 520.

338   In 1911, Dubois' old ally: information about Keith from A. Keith, *An Autobiography*, 1950, London: Watts.

338   "now Professor of Geology": A. Keith, *Ancient Types of Man*, 1911, New York and London: Harper & Bros., p. 131.

338   "Went out to Java": ibid.

339   "Whatever the exact date may be": ibid., p. 134.

340   "Here we find a creature": W. H. L. Duckworth, *Prehistoric Man*, 1912, Cambridge, Eng.: Cambridge University Press, pp. 2–3.

340   Calling their new find: A thorough account of the Piltdown affair can be found in F. Spencer, *Piltdown, A Scientific Forgery*, and F. Spencer, *The Piltdown Papers*, both 1990, both New York: Oxford University Press.

342   In fact, Fraipont fears: F. Spencer, *Aleš Hrdlička, M.D., 1869–1943: A Chronicle of the Life and Work of an American Physical Anthropologist*, 1979, Ph.D. dissertation, University of Michigan. Ann Arbor: University Microfilms, p. 425, citing letter, Hrdlička to Holmes, June 22, 1912.

342   "all this talk of descent": ibid., p. 418, citing letter, Hrdlička to Strickler, June 22, 1912.

345   "The anthropologists of the world": A general account of this incident is given in Spencer, *Hrdlička*, 1979; visiting card preserved in Dubois Archives, Naturalis.

346   "On account of the peculiar circumstances": A. Hrdlička, *The Most Ancient Skeletal Remains of Man*, 1914, Annual Report, Smithsonian Institution, p. 498.

346   he was unlucky: Bernsen diary, March 2, 1931, with minor paraphrasing for clarity.

CHAPTER 47 TRAGEDY

347   "In species of Vertebrates": E. Dubois, "On the Relation between the Quantity of Brain and the Size of the Body in Vertebrates," *P.K.A.W.*, 1914, vol. 16, pp. 647–68, p. 655.

349   Eugénie . . . announces her intention: author's interview with Victor E. Dubois and Nelleke Hooijer; Dubois family genealogy, Dubois Archives.

350   the Talgai skull was famous: A. Keith, *Antiquity of Man*, 1925, London: Williams and Northgate, pp. 440–41, 448.

351   "The differences may nearly all be attributed": E. Dubois, "The Proto-Australian Fossil Man of Wadjak, Java," *P.K.A.W.*, 1922, vol. 23, pp. 1013–51, 1030.

352   they meet two American girls: author's interview with Jean M. Dubois.

352   billing himself as a photographer: Poster for a lecture by Jean M. F. Dubois is reproduced in Leakey and Slikkerveer, *Man-Ape*, 1993, p. 145.

352   In 1922, at age thirty-one: obituary notices for Jean M. F. Dubois preserved in Dubois Archives, Naturalis.

352   Anna's heart and mind: Bernsen diary, May 4, 1931.

353   "I have seen how you look at Claartje": Antje Schreuder's account of the story of Dubois' affair with Claartje is recorded in ibid., April 23, 1931.

CHAPTER 48 DANGEROUS TIMES

355   "scattered and scanty materials": H. F. Osborn, *Men of the Old Stone Age*, 1915, New York: Scribners, p. 81.

355   "Asia is the mother": H. F. Osborn, foreword in R. C. Andrews, *On the Trail of Ancient Man*, 1926, New York: G. P. Putnam's Sons, p. vii.

355   "It is possible": Osborn, *Men of the Old Stone Age*, 1915, p. 511.

355   Andrews has modern vehicles: Expedition details from Andrews, *On the Trail*, 1926, pp. 3–23.

356   "August 1, 1922": letter, Osborn to Dubois, August 1, 1922.

356   McGregor also wanted: letter, McGregor to Dubois, June 2, 1921.

357   "As Professor Lorentz": telegram, Dubois to Osborn, n.d.

357   "December 2, 1922": letter, Osborn to Bolk, December 2, 1922.

358   Bolk writes sternly to Dubois: letter, Bolk to Dubois, January 3, 1923.

358   Bolk clumps stiffly: The amputation of Bolk's right leg in 1918 is attested to by Dr. Robert Baljit, personal communication to author.

360   "I know he made a lot of noise": Bernsen diary, March 2, 1931.

361   "First: Is the Academy competent": Report of the meeting of the Koninklijke Akademie van Wetenschappen, Dubois Archives, Naturalis.

362   "The unwinding of this tangled affair": ibid.

362 "I ask the Board to clarify": letter, Dubois to Bolk, February 14, 1923, language slightly modified for clarity.

362 He is able to write: letter, Dubois to Bolk, June 29, 1923.

363 "a creature that stood at the threshold"; "If science was to construct": Hrdlička, lecture given at the American University in 1920, quoted in Spencer, *Aleš Hrdlička*, 1979, p. 503.

363 "We found Professor Dubois": A. Hrdlička, *The Skeletal Remains of Early Man*, 1930, Washington, D.C.: Smithsonian Miscellaneous Collections 83.

364 "I am awaiting with great impatience": letter, Osborn to Dubois, April 2, 1924.

365 "I am very glad indeed": letter, Osborn to Bolk, February 28, 1925.

365 "You may make all the use": letter, Dubois to Osborn, March 24, 1925.

366 an Honorary Fellow: certificate preserved in the Dubois Archives, Naturalis.

CHAPTER 49 A NEW SKULL

366 a new and complete skull: letter, Heberlein to Dubois, November 7, 1926; E. Dubois, "The So-called New *Pithecanthropus* Skull," *P.K.A.W.*, 1927, vol. 30, no. 1. Read at the meeting of December 2, 1926.

367 "Accept my grateful homage": telegram, Dubois to Heberlein, October 2, 1926.

367 "The newspaper notices": letter, Osborn to Dubois, October 6, 1926.

367 "According to the ANETA telegram": copy of press release in Dubois Archives dated October 7, 1926.

368 Heberlein telegraphs Dubois: telegram, Heberlein to Dubois, October 16, 1926.

368 "Thanks can you inform me": telegram, Dubois to Heberlein, November 2, 1926.

368 "I found the *Pithecanthropus* skullcap": letter, Heberlein to Dubois, November 7, 1926.

369 "But for me": letter, Dubois to Weber, December 7, 1926.

371 "What a *sad* story that is": letter, Weber to Dubois, December 7, 1926.

371 "It is not even a skull": E. Dubois, "*Pithecanthropus erectus*. De 'nieuwe *Pithecanthropus* vondst,'" *Het Algemeen Handelsblad*, December 8, 1926.

371 Dubois cables Smith: telegram, Dubois to Grafton Elliot Smith, December 8, 1926; letter, Smith to Dubois, December 8, 1926.

372 "Since I first heard": letter, Dubois to Heberlein, December 10, 1926.

372 "On a photograph": Dubois, "The So-called . . . Skull," 1927, p. 137.

373 Mijsberg and Bijlmer in Java: This incident is described in Brongersma, "Vicissitudes," pp. 41–42.

CHAPTER 50 RUMORS AND ISOLATION

373 "Their discoverer, Dr. Dubois": F. W. MacBride, *Evolution*, 1928, London: Sixpenny Library Series, p. 75.

374 On January 28, 1928: Numerous letters of congratulations and a book signed by many of the attendees at the celebration are preserved in the Dubois Archives, Naturalis.

375 "February 11, 1928": letter, F. A. B. J. Vroom, S. J., to *The Tablet*, February 11, 1928.

376 "Wanted a servant-girl": Bernsen diary, April 23, 1931.

376 It is a debacle: ibid.

377 "It seems": ibid.

CHAPTER 51 BRAIN WORK

378 "absence of evidence of continuous development": E. Dubois, "The law of the necessary phylogenetic perfection of the psychoencephalon," *P.K.A.W.*, 1928, vol. 31, pt. 3, p. 304.

378 "Clearly paleontology bears evidence": ibid., pp. 304–314, 304–305.

380 "Besides, in the case of the psychoencephalon": ibid., pp. 313–14.

380 That he has done too little: Bernsen diary, January 14, 1931, March 2, 1931, and November 12, 1931.

381 "striking confirmation"; "New Tertiary man"; "one more link": quotes and account from J. Lanpo and H. Weiwen, *The Story of Peking Man from Archaeology to Mystery*, 1990, Beijing: Foreign Languages Press, pp. 26–27.

381 a new name: *Sinanthropus pekinensis:* D. Black, "On a Lower Molar from the Chou Kou Tien Deposits," *Paleontologica Sinica*, 1927, ser. C, vol. 7, fasc. I.

381 On December 2: Account and telegram from Lanpo and Weiwen, *The Story*, 1990, p. 65.

382 "A new chapter": G. Balchior, "The Peking Man—An Undamaged Skull," *Manchester Guardian*, December 30, 1929.

382 "February 11, 1930": letter, Dubois to Black, February 11, 1930.

383 "April 3, 1930": letter, Black to Dubois, April 3, 1930.

383 "Boule pointed out long ago": ibid.

384 "Since my last report before the Society": D. Black, "Interim Report on the Skull of *Sinanthropus*," March 29, 1930, Annual Meeting of the Geological Society of China; preserved in Dubois Archive.

385 "June 22, 1930": letter, Dubois to Black, June 22, 1930.

386 "March 28, 1933": letter, Dubois to Black, March 28, 1933.

CHAPTER 52 THE DILIGENT ASSISTANT

387 Their choice is a curious one: Brongersma, "Vicissitudes," p. 23ff.

387 "The conversation comes around": Bernsen diary, November 10, 1930.

388 "Dubois realizes that the study": ibid., November 12, 1930.

388 "I am very nervous": ibid., November 23, 1930.

389 "I myself speak": ibid., December 13, 1930.

390 "Miss Dr. Schreuder from Amsterdam": ibid., April 23, 1931.

390 "He suffers from nerves": ibid., April 24, 1931.

390 "He is always afraid": ibid.

391 "I have said": ibid., June 10, 1931; last sentence of paragraph added.

391 One day Dubois admits: ibid., February 19, 1932.

391 "Yet I have never written": ibid., January 19, 1931.

391 "I believe": ibid., May 29, 1931. Sentences beginning "Ah, you raise," "You ask," and "I'll tell" added.

391 "I never had the time": ibid., January 14, 1931.

392 "The difficulties that I have experienced": ibid., December 10, 1930.

392 "Also *Pithecanthropus* caused me a lot of trouble": ibid.

392 "What a misery I have had": ibid., March 2, 1931.

392 "I get the impression": ibid., April 29, 1931.

393 "Every day I study them": ibid., May 29, 1931.

393 "You know, I admire Black": ibid.

394 "Black gives a pleasant solution": ibid.

394 "the more I look at the smaller skull": ibid., June 8, 1931.

395 Sometimes Dubois speaks of giving Bernsen: Brongersma, "Vicissitudes," p. 24.

395 "The more I look at the pictures": Bernsen diary, June 10, 1931.

396 "Although it certainly resembles": ibid., June 15, 1931.

396 "June 15, 1931": letter, Dubois to Black, June 15, 1931.

396 One day, he commits: Brongersma, "Vicissitudes," p. 24.

397 "I have not published enough": Bernsen diary, March 2, 1932.

397 "After finding the truth": ibid.

397 "Only after 1923": ibid., with minor additions for clarity.

397 Bernsen can remain quiet: incident and dialogue, with minor additions for clarity, from ibid., March 3, 1932.

399 "No word of regret": ibid., March 5, 1932.

400 the burgomaster informs him: incident and dialogue from ibid., March 9, 1932.

400 Even Van der Klaauw: ibid., March 12, 1932.

401 "That is easy for you to say": ibid., March 16, 1932.

401 "And shall you walk": ibid.

401 "There is something hostile in you": ibid., May 12, 1932.

401 "Now I see that you are hard": ibid.

402 "I have always felt": ibid.

402 Working systematically: ibid., June 2, 1932.

403 "After a great deal of inspection": paraphrased slightly from ibid.

403 "After a close comparison": ibid., June 1, 1932.

404 "This is a solemn moment": ibid.

404 "these pieces must certainly have come": ibid.

404 "In the box": ibid.

405 "Dubois will surely find it unpleasant": ibid., June 4, 1932.

405 he dies of internal hemorrhaging: Brongersma, "Vicissitudes," p. 27.

405 "Forty years ago": E. Dubois, "The distinct organization of *Pithecanthropus* of which the femur bears evidence, now confirmed from other individuals of the described species," *P.K.A.W.*, 1932, vol. 35, no. 6, pp. 716–22, 716–18. Some editing to improve clarity.

408 "The morphological evidence acquired": ibid., pp. 721–22.

408 Bernsen is replaced: Brongersma, "Vicissitudes," p. 31.

CHAPTER 53 NEW SKULLS FROM JAVA

408 In August 1931, the Geological Survey of Java: The account of the discovery comes from G. H. R. von Koenigswald, *Meeting Prehistoric Man*, 1956, London: Thames & Hudson, p. 65ff.

409 After the three finds in 1931: Dates from F. Weidenreich, "Morphology of Solo Man," *Anthropological Papers of the American Museum of Natural History, N.Y.*, 1951, vol. 43, part 3, p. 217.

410 "provisional description": W. F. F. Oppenoorth, "*Homo (Javanthropus) soloensis*, een pleistocene mensch van Java," *Wetenschappelijke mededelingen Dienst van den Mijnbrouw in Nederlandsch-Indië*, 1932, vol. 20, pp. 49–63.

410 "May 31, 1932": letter, Oppenoorth to Dubois, May 11, 1932.

411 Dubois publishes a brief letter: E. Dubois, "Early man in Java," *Nature*, 1932, vol. 130, p. 20.

411 "one identical type": ibid.

411 "June 15, 1932": letter, Dubois to Oppenoorth, June 15, 1932.

411 Oppenoorth describes Ngandong IV and V: W. F. F. Oppenoorth, "De vondst van palaeolitische menschelijke schedels op Java," *De Mijningingenieur*, 1932, June.

412 Oppenoorth proposed a geographic separation: ibid., p. 114.

412 Dubois' next task: E. Dubois, "The Shape and the Size of the Brain in *Sinanthropus* and *Pithecanthropus*," *P.K.A.W.*, 1933, vol. 36, no. 4, pp. 415–23.

412 "There is obviously little difference": ibid., p. 419.

413 "Such a volume": ibid., pp. 422–23.

414 "oblong and narrow": ibid.

415 "It is therefore possible": E. Dubois, "On the Gibbonlike Appearance of *Pithecanthropus erectus*," *P.K.A.W.*, 1935, vol. 38, pp. 578–85, 580, 585.

CHAPTER 54 A WORTHY OPPONENT

416 "Our excavation site": For quotations and a general account of Von Koenigswald's experiences in Java, see Von Koenigswald, *Meeting Prehistoric*, 1956, p. 74.

417 all nonessential government employees: M. C. Ricklefs, *A History of Modern Indonesia Since c. 1300*, 1993, Stanford: Stanford University Press, p. 186ff.

418 When Teilhard arrives: Information about Teilhard de Chardin's visit comes from P. V. T. Tobias, "Life and Work of Professor Dr. G. H. R. von Koenigswald," in *Auf den Spuren des Pithecanthropus*, 1984, Frankfurt: Waldemar Kramer, pp. 25–95; letter, Teilhard de Chardin to Mlle. Teilhard-Chamdon, January 24, 1936, quoted in P. Teilhard de Chardin, *Letters from a Traveller*, 1962, London: Collins, p. 380.

418 "If *Pithecanthropus* is to be found anywhere": quoted in Von Koenigswald, *Meeting Prehistoric*, 1956, p. 92, and Tobias, "Life and Work," 1984, p. 39.

418 "During the stratigraphic survey": G. H. R. von Koenigswald, "Een nieuwe *Pithecanthropus* ondekt," *Het Algemeen Indische Dagblad*, March 28, 1936.

419 "On April 15 I got": E. Dubois, "Nieuwe *Pithecanthropus* ondekt?" *Algemeen Handelsblad*, April 18, 1936; E. Dubois, "Een nieuwe *Pithecanthropus* ondekt?" *Het Vaderland*, April 18, 1936.

420 "It is inconceivable": G. H. R. von Koenigswald, "*Pithecanthropus erectus*, Antwoord dr. Von Koenigswald," *Algemeen Handelsblad*, May 7, 1936.

421 Unfortunately, Andojo is by then elderly: J. de Vos, "*Homo modjokertensis*—vind-plaats, ouderdom en fauna," *Cranium*, 1994, vol. 11, no. 2, pp. 103–107.

421 a series of articles: E. Dubois, "Fossil humans and *Pithecanthropus*. The youngest find in Java no *Pithecanthropus*," *Nieuwe Rotterdamsche Courant*, August 2, 1936.

421 "Concerning the new find": Anonymous, "*Homo modjokertensis*: Lecture of Dr. Von Koenigswald," *Handelsblad*, August 29, 1936.

421 "a POSSIBILITY": G. H. R. von Koenigswald, "Erste Mitteilung über einen fossilen Hominiden aus dem Altpleistocän Ostjavas," *P.K.A.W.*, 1936, vol. 39, no. 8, pp. 1000–1009. Communicated at the meeting of September 26, 1936.

421 a desire to be courteous: Von Koenigswald, *Meeting Prehistoric*, 1956, p. 82.

421 a sign of uncertainty: letter, Oppenoorth to Dubois, October 28, 1936.

CHAPTER 55 TO THE BATTLEFRONT

422 "glorious disdain"; "almost inconceivable influence": Anonymous, "Dr. Van Stein Callenfels," *Handelsblad*, October 22, 1936.

422 Tuan Setan: Anonymous, *Handelsblad*, July 24, 1936.

422 Stein declares Von Koenigswald's: Anonymous, "Prehistoric Congress in Oslo, lecture of Dr. Van Stein Callenfels," *De Telegraaf*, August 8, 1936.

422 "In former times": Information and quotations below on this lecture taken from L. D. Brongersma, report on lecture of Van Stein Callenfels, Delft, October 28, 1936, and Anonymous, "Praehistorische vondsten op Java; Voordracht Dr. Van Stein Callenfels," *Handelsblad*, October 29, 1936.

423 "The Chief of Mining": Brongersma, report, 1936.

424 They sail on the S.S. *Baloeran*: Anonymous, "Dr. Von Königswald, Studiereis naar Europa," *Handelsblad*, November 7, 1936; Von Koenigswald, *Meeting Prehistoric*, 1956, p. 33.

424 Von Koenigswald delays: Dubois gives February 17, 1937, as the date of this visit in E. Dubois, "On the fossil human skull recently described and attributed to *Pithecanthropus erectus* by G. H. R. von Koenigswald," *P.K.A.W.*, 1938, vol. 41, pp. 380–86, though Von Koenigswald, *Meeting Prehistoric*, 1956, pp. 32–33, suggests the visit occurred in October 1936. The October date is contradicted by the report in the newspaper *Handelsblad* (see previous note) of the departure of the Von Koenigswalds from the Indies on November 18, 1936, so I have accepted Dubois' date.

424   "He was stated to be ill": Von Koenigswald, *Meeting Prehistoric*, 1956, pp. 32–33.

425   Nonetheless, Von Koenigswald is fearful: In a letter, Von Koenigswald to Smit, May 6, 1982, written more than forty years after Dubois' death, Von Koenigswald still remembers his deep concern over Dubois' opposition.

426   "In 1890, Professor Eugène Dubois": G. H. R. von Koenigswald, "A review of the stratigraphy of Java and its relations to early man," in G. G. MacCurdy, ed., *Early Man*, 1937, Philadelphia: J. B. Lippincott & Co., pp. 23–32, 23.

426   "All the remains": ibid., pp. 25–26, 27.

426   "The skull . . . is perfectly fossilized": ibid., p. 28.

427   "The first suggestion": ibid.

427   His abstract is read aloud: E. Dubois, "Early Man in Java and *Pithecanthropus erectus*," in G. G. MacCurdy, ed., *Early Man*, 1937, pp. 315–22, 315.

427   "in my best Malay": Von Koenigswald, *Meeting Prehistoric*, 1956, p. 93.

CHAPTER 56 THE LETTER

428   "February 9, 1937": letter, Van den Koppel to Dubois, February 9, 1937.

429   "Kediri   February 7, 1936": letter, Prentice to Dubois, February 7, 1936.

CHAPTER 57 PRETENDER TO THE THRONE

431   "July 12, 1937": letter, Von Koenigswald to Dubois, July 12, 1937.

431   "August 26, 1937": letter, Dubois to Von Koenigswald, August 26, 1937.

432   "This is a continuation": letter, Dubois to Von Koenigswald, August 27, 1937.

432   "September 3, 1937": letter, Von Koenigswald to Dubois, September 3, 1937.

433   "We had to keep the price so low": Von Koenigswald, *Meeting Prehistoric*, 1956, pp. 97–98.

434   "September 14, 1937": letter, Dubois to Von Koenigswald, September 14, 1937.

434   "October 3, 1937": letter, Von Koenigswald to Dubois, October 3, 1937.

434   "November 12, 1937": letter, Von Koenigswald to Dubois, November 12, 1937.

435   Von Koenigswald prepares a paper: G. H. R. von Koenigswald, "Ein Unterkieferfragment des *Pithecanthropus* aus den Trinilschichten Mittenjavas," *P.K.A.W.*, 1937, vol. 40, pp. 883–93.

435   "January 12, 1938": letter, Dubois to Von Koenigswald, January 12, 1938.

435   As is widely reported: Anonymous, "De Fossile Mensch van Java; De Weg der Evolutie; De beroemdste is nog altijd de *Pithecanthropus*," *Handelsblad*, January 18, 1938.

435   Even a cheerful letter: letter, Eugénie to Dubois, January 27, 1938.

436   "entirely different": E. Dubois, "The mandible recently described by G. H. R. von Koenigswald, compared with the mandible of *Pithecanthropus erectus* described in 1924 by Eug. Dubois," *P.K.A.W.*, 1938, vol. 41, pp. 139–47, 146–47.

436   "I never imagined": ibid., pp. 145–46.

436   an article about the find: G. H. R. von Koenigswald, "*Pithecanthropus* received into human family," *Illustrated London News*, December 11, 1937; G. H. R. von Koenigswald, "Ein neuer *Pithecanthropus*-Schädel," *P.K.A.W.*, 1938, vol. 41, pp. 185–92.

436   "preparated": E. Dubois, "On the fossil human skull recently described and attributed to *Pithecanthropus erectus* by G. H. R. von Koenigswald," *P.K.A.W.*, 1938, vol. 41, pp. 380–86, 380.

436   "great thickness of the skullcap": The quotations in this paragraph are from Dubois, "On the fossil human skull," 1938, pp. 382–83.

438   "The real fossil got lost": ibid., p. 384.

438   "artificially, from that of an immature individual": ibid., p. 385.

438   "Even if this reconstructed skull": ibid.

438 "May 5, 1938": letter, Von Koenigswald to Dubois, May 5, 1938.

438 "I may remind you": letter, Dubois to Von Koenigswald, June 3, 1938.

439 Von Koenigswald writes: Von Koenigswald, *Meeting Prehistoric*, 1956, p. 99.

440 "unfortunately I cannot": letter, Dubois to Weidenreich, July 23, 1938.

440 "August 9, 1938": letter, Dubois to Weidenreich, August 9, 1938.

440 "August 23, 1938": letter, Dubois to Eugénie, August 23, 1938.

440 "Weidenreich wrote": letter, Dubois to Eugénie, September 2, 1938.

CHAPTER 58 OLD FRIENDS

441 "Kediri   November 28, 1938": letter, Prentice to Dubois, November 28, 1938. Some words translated into English for clarity.

CHAPTER 59 THE FINAL CONFLICT

445 "We laid out our finds": Von Koenigswald, *Meeting Prehistoric*, 1956, p. 101.

445 "Considered from the general": G. H. R. von Koenigswald and F. Weidenreich, "The relationship between *Pithecanthropus* and *Sinanthropus*," *Nature*, 1939, vol. 144, pp. 926–29, 928.

447 "January 3, 1940": letter, Dubois to Eugénie, January 3, 1940.

447 "The child skull": E. Dubois, "The fossil human remains discovered in Java by Dr. G. H. R. von Koenigswald and attributed by him to *Pithecanthropus erectus*, in reality remains of *Homo wadjakensis* (syn. *Homo soloensis*)," *P.K.A.W.*, 1940, vol. 43, pp. 494–96, 494. Read at the meeting of March 30, 1940.

448 "April 3, 1940": letter, Dubois to Eugénie, April 3, 1940.

448 "Now twenty-two years": E. Dubois, "The fossil human remains discovered in Java by G. H. R. von Koenigswald and attributed by him to *Pithecanthropus erectus*, in reality remains of *Homo soloensis*. Continuation." *P.K.A.W.*, 1940, vol. 43, pp. 842–52, 843–44.

449 They are seized: The best accounts of the loss of the *Sinanthropus* fossils are given in H. Shapiro, *Peking Man*, 1974, New York: Simon & Schuster, and J. Lanpo and H. Weiwen, *The Story of Peking Man from Archaeology to Mystery*, 1990. A more lively but possibly less accurate recounting is C. Janus and W. Brashler, *The Search for Peking Man*, 1975, New York: Macmillan.

449 Von Koenigswald's indomitable wife: Tobias, "Life and Work," 1984, pp. 60–63.

449 "October 4, 1940": letter, Dubois to Eugénie, October 4, 1940.

449 "October 28, 1940": letter, Dubois to Eugénie, October 28, 1940.

450 an escape line for downed Allied fliers: author's interview with Jean M. Dubois, Victor E. Dubois, and Nelleke Hooijer.

450 "It is most regrettable": E. Dubois, "The fossil human remains discovered in Java by G. H. R. von Koenigswald and attributed by him to *Pithecanthropus erectus*, in reality remains of *Homo sapiens soloensis*. Conclusion," *P.K.A.W.*, 1940, vol. 43, pp. 1268–75, 1275. Punctuation changed slightly for clarity.

452 "Prof. Dr. Eug. Dubois": Dubois' tombstone, as described, can be seen in the cemetery at Venlo.

EPILOGUE

452 an excellent appraisal: Theunissen, *Eugène Dubois*, 1989.

453 allometry of the brain: H. Jerison, *Evolution of the Brain and Intelligence*, 1973, New York: Academic Press.

GLOSSARY

Note: Spellings and usages are nineteenth- and early twentieth-century, not modern.

adik (Indies)—boy or younger brother; form of address for younger native man

ado (Indies)—an expression of disbelief and astonishment

ajo (Indies)—an exclamation of encouragement, like "Let's go!" or "Okay!"

alang-alang (Indies)—species of tall, coarse, wild grass with sharp edges

anak mas (Indies)—literally, a golden child; a native child adopted by a European family

Ardjoena (Indies)—Prince Ardjoena, a heroic figure from the Hindu Ramayana myths; also, the name of a volcano in Java

babu (Indies)—nursemaid or ladies' maid

barang-barang (Indies)—carried items or luggage

bhisti (Raj)—native water-bearer

bok'n (Indies)—no, not

cepat (Indies)—hurry

chit (Raj)—small official receipt

chuprassi (Raj)—office servant or messenger

delman (Indies)—type of horse-drawn carriage

desa (Indies)—village

diam (Indies)—quiet; be quiet

djati (Indies)—teak

djongas (Indies)—houseboy or butler; the head male servant of the household

dokar (Indies)—small horse-drawn cart

dukun (Indies)—medicine man or traditional healer

durian (Indies)—a fruit with an extremely pungent odor

fabriek (Dutch)—a plantation or manufacturing plant

gamelan (Indies)—traditional Javanese or Balinese orchestra consisting of gongs, xylophones, and other percussion instruments

Garuda (Indies)—eagle who carries the god Vishnu on his back

gunung (Indies)—mountain

hati-hati (Indies)—take care; beware

head-jaksa (Indies)—native magistrate

Indische (Indies)—a European or part-European who has lived a long time in the Indies; related to the attitudes, opinions, or fashions typical of this group

Indo (Indies)—a person of mixed European-Indies ancestry; now considered a derogatory usage *ja* (Dutch, Indies)—yes

kampong (Indies)—village or neighborhood

kassian (Indies)—common expression of sympathy, meaning "Take pity" or "It's a pity!"

kebaya (Indies)—loose, hip-length overblouse trimmed in lace, originally Chinese but worn by many European women in the Indies

khitmagar (Raj)—native foreman, "head boy," or overseer

kokkie (Indies)—cook

maidan (Raj)—large parade ground or public space

mandi (Indies)—bathing room. Water for bathing is stored in a huge pottery jug and ladled over the body.

mandur (Indies)—foreman

mejjuffrouw (Dutch)—miss; respectful term of address for an unmarried well-to-do woman

melati (Indies)—jasmine

mevrouw (Dutch)—mrs. or madame; respectful term of address for a well-to-do married woman

mijnheer (Dutch)—sir; respectful term of address for a well-to-do man

mungkin, mungkin tidak (Indies)—maybe, maybe not

nasi goreng (Indies)—the common fried rice dish of the Indies

ngrodjo (Indies)—headquarters; also, the name given to Boyd's house

njonja (Indies)—madame; a respectful term of address for a married European woman

nyai (Indies)—native mistress of a European; concubine

obat (Indies)—medicine

orang belanda (Indies)—person of Dutch ancestry; generally, a European

orang djager (Indies)—hunter

panjang kursi (Indies)—long chair or chaise, often made of rattan with wooden leg rests that swivel out from the arms

patjol (Indies)—short-handled hoe used by Indies native farmers

pontianak (Indies)—female spirit or demon

poodle-faker (Raj)—womanizer or ladies' man

prahu (Indies)—native boat or outrigger

Pure (Indies)—person of European ancestry

purwana (Raj)—permit to work in a region, in the form of an open letter to the populace asking that assistance be rendered as needed to the holder of the purwana

rijstafel (Indies)—literally, "rice table," a dish consisting of rice accompanied by many small dishes of spicy stews and condiments

roro (Indies)—princess or maiden of noble birth

sadhu (Raj)—holy man

Sahib (Raj)—sir or lord; similar to Indies "tuan," a respectful form of address used by natives speaking to a European male

sarong (Indies)—piece of cloth wrapped like an ankle-length skirt and worn by both sexes; the pattern on the sarong may indicate place of origin within the Indies, and status

sedaka (Indies)—ceremonial offering

sinjo (Indies)—boy, of either European or Eurasian parentage

songket (Indies)—an intricately patterned cloth with gold or silver threads, made in Sumatra

susa (Indies)—fuss, nuisance, or bother

syce (Raj and Indies)—native groom, one in charge of the horses

tehsildar (Raj)—local tax collector

tempo doeloe (Indies)—times gone by; an expression of nostalgia for the colonial past in the Indies

tiffin (Raj)—luncheon; the midday meal

tjempaka (Indies)—type of magnolia

toeloeng (Indies)—Help!

totok (Indies)—European recently arrived in the Indies; newcomer

tuan (Indies)—lord or master; respectful form of address used by natives speaking to male Europeans

tukang kebun (Indies)—gardener

tunggu (Indies)—Wait!

tutup (Indies)—literally "closed"; slang for a common garment of the Indies, a lightweight, high-collared man's jacket that buttons up the front

V.O.C. (Dutch)—Vereenigde Oost-Indische Compagnie, the Dutch East Indies Company, which controlled trade in the Indies from the late 1500s until the early 1800s

wallah (Raj)—man; when hyphenated, slang for a specialist in some particular trade or task, as in a "punkah-wallah," who pulls a string that moves a large fan or punkah

waringin (Indies)—banyan tree

warung (Indies)—small native shop or stall

BIBLIOGRAPHY

The first section of the bibliography is a listing of the publications of Eugène Dubois, drawn from an unpublished bibliography in the Dubois Archives that was compiled in 1979 by L. D. Brongersma, with additions or corrections based on Theunissen, 1989. Entries in this part are organized by year of publication. I do not list as publications the printing of a title only, prior to publication of the text, nor do I consider reprints of articles with separate pagination to be publications. The second section gives literature cited in the notes or text, which is organized alphabetically by first author. All publications by Dubois have been omitted in the second section. Abbreviations are as for the notes.

I. PUBLICATIONS OF EUGÈNE DUBOIS

1884

"Over Anatomie in hare betrekking tot de Beeldende Kunst. I." *Maandag gewijd aan de belangen van het teekenonderwijs en kunstnijverheid in Nederland,* no. 1, June 1, 1884, pp. 2–4.

1885

"Over Anatomie in hare betrekking tot de Beeldende Kunst. II." *Maandag gewijd aan de belangen van het teekenonderwijs en kunstnijverheid in Nederland,* no. 8, January 1, 1885, pp. 65–69.

1886

"Ueber den Larynx." In Max Weber, *Studien über Säugerthiere, ein Beitrag zur Frage nach dem Ursprung der Cetaceen.* Jena: Gustav Fischer, pp. 88–111.

"Zur Morphologie des Larynx." *Anatomische Anzeiger,* vol. 1, no. 7, pp. 176–86; no. 9, pp. 225–31.

1888

"Over de wenschelijkheid van een onderzoek naar de diluviale fauna van den Nederlandsch Indië, in het bijzonder van Sumatra." *N.T.,* vol. 48, pp. 148–65.

1889

"Uittreksel van een schrijven van het Heer Dubois te Pajacombo naar aanleiding van den aan dien Heer toegezonden schedel, door den Heer van Rietschoten in zijn marmergroeven in het Kedirische opgegraven." *N.T.,* vol. 49, pp. 209–10.

1890

"Beschriving van een bloeienden *Amorphophallus titanum,* Beccari, aangetroffen te Boea bij de grot der Batang Pangian, den 24sten November 1889." *Teysmannia,* vol. 1, pp. 89–91.

1891

"*Anoa santeng* Dubois." In F. A. Jentink, "On *Lepus netscheri* Schlegel, *Felis megalotis* Müller and *Anoa santeng* Dubois." *Notes Leyden Museum,* vol. 13, pp. 217–22.

"De Klimaten der Voorwereld en de Geschiedenis der Zon." *N.T.,* vol. 51, pt. 1, pp. 37–92.

"Palaeontologische onderzoekingen op Java." *V.M.,* 3rd quarter 1890, 1891, no. 9, pp. 12–15.

"Palaeontologische onderzoekingen op Java." *V.M.,* 4th quarter 1890, 1891, no. 41, pp. 14–18.

"Voorloopig Bericht omtrent het Onderzoek naar de Pleistocene en Tertiare Vertebraten-Fauna van Sumatra en Java, gedurdende het jaar 1890." *N.T.*, vol. 51, pt. 1, pp. 93–100.

"Palaeontologische onderzoekingen op Java." *V.M.*, 1st quarter 1891, 1891, no. 58, pp. 12–13.

"Palaeontologische onderzoekingen op Java." *V.M.*, 2nd quarter 1891, 1891, no. 93, pp. 11–12.

1892

"Palaeontologische onderzoekingen op Java." *V.M.*, 3rd quarter 1891, 1892, no. 3, pp. 12–14.

"Palaeontologische onderzoekingen op Java." *V.M.*, 4th quarter 1891, 1892, no. 36, pp. 12–15.

"Palaeontologische onderzoekingen op Java." *V.M.*, 2nd quarter 1892, 1892, no. 87, pp. 14–18.

Naschrift op "De Klimaten der Voorwereld en de Geschiedenis der Zon." *N.T.*, vol. 51, ser. 8, vol. 12, pp. 270–74.

1893

"Palaeontologische onderzoekingen op Java." *V.M.*, 3rd quarter 1892, 1893, no. 10, pp. 10–14.

"Palaeontologische onderzoekingen op Java." *V.M.*, 4th quarter 1892, 1893, no. 36, pp. 11–12.

"Die Klimate der geologischen Vergangenheit und ihre Beziehung zur Entwicklungsgeschichte der Sonne." Nijmegen: H. C. A. Thieme and Leipzig: Max Spohr.

1894

"Palaeontologische onderzoekingen op Java." *V.M.*, 3rd quarter of 1893, 1894, no. 2, pp. 15–17.

"Palaeontologische onderzoekingen op Java." *V.M.*, 4th quarter of 1893, 1894, no. 81, pp. 12–15.

Pithecanthropus erectus. Eine menschenaehnliche Uebergangsform aus Java. Batavia: Landsdrukkerij.

1895

"*Pithecanthropus erectus*. Eine menschenaehnliche Uebergangsform aus Java." *Jahrboek van het Mijnwezen Nederlandsch–Oost-Indië*, vol. 24, pp. 1–77.

The Climates of the Geological Past and Their Relation to the Evolution of the Sun. London: Swan Sonnenschine & Co.

1896

"The Place of 'Pithecanthropus' in the Genealogical Tree." *Nature,* vol. 53, no. 1368, pp. 245, 247.

"*Pithecanthropus erectus*, eine Stammform des Menschen." *Anatomische Anzeiger,* vol. 12, no. 1, pp. 1–22.

"*Pithecanthropus erectus*, eine menschenaehnliche Uebergangsform." *Compte-rendu des séances du Troisième Congrès International de Zoologie, Leyde, 11–16 septembre 1895,* pp. 251–71.

"*Pithecanthropus erectus*, betrachtet als eine wirkliche Uebergangsform und als Stammform des Menschen." *Verhandlungen der Berliner Gesellschaft für Anthropologie, Ethnologie und Urgeschichte,* ausserordentl. Sitzung vom 14. Dezember 1895, pp. 723–38. (Reprinted in *Zeitschrift für Ethnologie,* vol. 27, pp. 723–38.)

"Résumé d'une communication de M. le Dr. Eug. Dubois sur le *Pithécanthropus erectus* du pliocène de Java." *Bulletin de la Société Belge Géologique, Paléontologique et Hydrologique*, vol. 9, Procès-Verbaux, pp. 151–60.

"Näheres über den *Pithecanthropus erectus* als menschenähnliche Uebergangsform." *Internationale Monatschrift Anatomie und Physiologie*, vol. 13, pt. 1, pp. 1–26.

"On *Pithecanthropus erectus:* a Transitional Form between Man and the Apes." *Transactions of the Royal Dublin Society*, ser. 2, vol. 6, pp. 1–18.

"On *Pithecanthropus erectus:* a Transitional Form between Man and the Apes (Abstract)." *J.A.I.*, no. 96, February, pp. 240–48.

"De thans bekende soorten van fossilen Menschapen." *Tijdschrift der Nederlandsche Dierkundige Verslagen*, ser. 2, vol. V, pt. 2, pp. 70–74.

"On the occurrence of *Crocodilus porosus* far above the tideway in a Sumatran river." *Notes Leyden Museum*, vol. 18, p. 134.

"Le '*Pithécanthropus erectus*' et l'origine de l'homme." *Bulletin de la Société d'Anthropologie de Paris*, ser. 4, vol. 7, pt. 5, pp. 460–67.

1897

"Ueber drei ausgestorbene Menschaffen." *Neues Jahrbuch für Mineralogie, Geologie und Paleontologie*, vol. 1, pp. 83–104.

"De Verhouding van het Gewicht der Hersenen tot de Grootte van het Lichaam bij de Zoogdieren." *P.K.A.W.*, Amsterdam, 2de Sectie, vol. 5, no. 10, pp. 1–41.

"Sur le rapport du poids de l'encéphale avec la grandeur du corps chez les mammifères." *Bulletin de la Société d'Anthropologie de Paris*, sér. 4, vol. 8, pt. 4, pp. 337–76.

1898

"Ueber die Abhängigkeit des Hirngewichtes von der Körpergrösse bei den Säugethieren." *Archiv für Anthropologie*, vol. 25, pts. 1–2, pp. 1–28.

"Ueber die Abhängigkeit des Hirngewichtes von der Körpergrösse beim Menschen." *Archiv für Anthropologie*, vol. 25, pt. 4, pp. 423–41.

1899

"The Brain-cast of *Pithecanthropus erectus*." *Proceedings of the Fourth International Congress of Zoology, Cambridge, 22–27 August 1898*, pp. 78–95.

"Over den Kringloop der Stof op Aarde." Rede uitgesproken bij de aanvaardin van het ambt van buitengewoon hoogleraar aan de l'Universiteit van Amsterdam, den 20sten Februari. Leiden: E. J. Brill.

1900

"Données justificatives sur l'essai de reconstruction plastique du *Pithécanthropus erectus*." Leaflet distributed at the International Exhibition in Paris.

"Données justificatives sur l'essai de reconstruction plastique du *Pithécanthropus erectus*." *Petrus Camper, Nederlandsche bijdragen tot de anatomie*, vol. 1, pt. 2.

"*Pithecanthropus erectus:* A Form from the Ancestral Stock of Mankind." *Report of the Smithsonian Institution for 1898*, pp. 445–49.

"Over den Ouderdom der Aarde." *T.A.G.*, ser. 2, vol. 17, pp. 697–734.

"De groote van den kringloop der koolzure kalk en de ouderdom der aarde. I." *Verslag*, vol. 9, pp. 12–28.

"De groote van den kringloop der koolzure kalk en de ouderdom der aarde. II." *Verslag*, vol. 9, pp. 99–115.

"The Amount of the Circulation of the Carbonate of Lime and the Age of the Earth. I."
P.K.A.W., vol. 3, pp. 43–62.

"The Amount of the Circulation of the Carbonate of Lime and the Age of the Earth. II."
P.K.A.W., vol. 3, pp. 116–30.

1901

"Paradoxe klimatische toestanden in het Palaeozoïsche tijdvak, beschouwd in verband
met den vroeheren aard der zonnestraling." *Handelingen 8ste Nederlandsch Natuur-
en Geneeskundig Congres, Rotterdam, 11–14 April 1901*, pp. 311–26.

"Zur systematischen Stellung der ausgestorbenen Menschaffen." *Zoologische Anzeiger*,
vol. 24, no. 652, pp. 556–60.

"Les causes probables du phénomène paléoglaciaire permo-carboniférien dans les
basses latitudes." *Archives Musée Teyler*, ser. 2, vol. 7, pt. 4, pp. 311–60.

"Notes et Corrections." Ibid., pp. 361–62.

1902

"Over den toevoer van natrium en chloor door de rivieren aan de zee." *Verslag*, vol. 10,
pp. 493–504.

"On the Supply of Sodium and Chlorine by the Rivers to the Sea." *P.K.A.W.*, vol. 4, pt. 7,
pp. 388–99.

"De geologische samenstelling en de wijze van otstaan van den Hondsrug in Drenthe."
Verslag, vol. 11, pp. 43–50.

"De geologische samenstelling en de wijze van otstaan van den Hondsrug in Drenthe."
Verslag, vol. 11, pp. 150–52.

"The geological structure of the Hondsrug in Drenthe and the origins of that ridge."
P.K.A.W., vol. 5, pt. 2, pp. 93–101.

Idem, "Second communication." Ibid., pp. 101–103.

"Staring en het Steenkolenvraagstuk in Zuid-Limburg." *T.A.G.*, ser. 2, vol. 19, pp. 869–70.

"Les causes probables du phénomène paléoglaciaire permo-carboniférien dans les
basses latitudes (deuxième étude)." *Archives Musée Teyler*, ser. II, vol. VIII, pp. 73–91.

"Notes sur les conditions locales dans lesquelles se sont formés les dépôts paléoglaciaires
permo-carbonifériens dans l'Afrique australe, l'Inde et l'Australie." *Archives Musée
Teyler*, ser. 2, vol. 8, pt. 1, pp. 157–63.

"La structure géologique et l'origine du Hondrug dans la Province Drenthe." *Archives
Néerlandaises des Sciences Exactes et naturelles*, ser. 2, vol. 7, pt. 4/5, pp. 484–96.

1903

"De geologische gesteldheid van onzen bodem en drinkwater voor Amsterdam."
Algemeen Handelsblad, February 3, 1903, Avonblad, 2de blad.

"De drinkwaterquaestie." *Algemeen Handelsblad*, February 9, 1903, Ochtendblad.

"Nog eens het drinkwater." *Algemeen Handelsblad*, February 11, 1903, Ochtendblad.

"De Water-quaestie." *Algemeen Handelsblad*, February 12, 1903, Ochtendblad, 1ste blad.

"De Watervraag." *Algemeen Handelsblad*, February 12, 1903, Avonblad, 2de blad, p. 5.

"De Waterquaestie." *Algemeen Handelsblad*, February 13, 1903, Ochtendblad, 1ste blad,
p. 2.

"Toch overvloed van zoetwater uit hen duin te halen." *Algemeen Handelsblad*, March 9,
1903, Avonblad.

"Diep gelegen keinenleem van een jongeren ijstijd in den bodem van Noord-Holland."
Verslag, vol. 12, pp. 17–22.

"Feiten ter opsporing van de bewegingsrichting en der oorsprong van het grondwater onzer zeeprovinciën." *Verslag*, vol. 12, pp. 187–212.

"Over de herkomst van het zoete water in den ondergrond van eenige minder diepe polders." *Verslag*, vol. 12, pp. 593–603.

"Deep Boulder-Clay Beds of a latter glacial Period in North-Holland." *P.K.A.W.*, vol. 6, pt. 4, pp. 340–45.

1904

"Over de herkomst van eenige chemische bestanddeelen van het grondwater in ons laagland." *Pharmaceutisch Weekblad*, vol. 41, no. 3, pp. 46–51.

"Niet-biologische vorming van limoniet." *Pharmaceutisch Weekblad*, vol. 41, no. 7, pp. 137–39.

"Facts leading to trace out the motion and the origin of the underground water in our sea-provinces." *P.K.A.W.*, vol. 6, pt. 10, pp. 738–60.

"Richting en uitgangspunt der diluviale ijsbeweging over ons land." *Verslag*, vol. 13, pp. 44–45.

"Klei van Tegelen, Teyler lezing." Newspaper clipping, November 23.

"On the direction and the starting point of diluvial ice motion over the Netherlands." *P.K.A.W.*, vol. 7, pt. 1, pp. 40–44.

"On the Origin of the Fresh-water in the Subsoil of a few shallow Polders." *P.K.A.W.*, vol. 7, pt. 1, pp. 53–63.

"Over een equivalent van het Cromer Forest-Bed, in Nederland." *Verslag*, vol. 13, pp. 243–51.

"On an Equivalent of the Cromer Forest-Bed in the Netherlands." *P.K.A.W.*, vol. 7, pt. 3, pp. 214–22.

"Corrigenda en Addenda bij de mededeeling van den Heer Eugène Dubois 'Over een equivalent van het Cromer Forest-Bed in Nederland.'" *Verslag*, vol. 13, pp. 453–54.

"Corrigenda et Addenda to the paper 'On an Equivalent of the Cromer Forest-Bed in the Netherlands.'" *P.K.A.W.*, vol. 7, pt. 5, pp. 382–83.

"Etudes sur les eaux souterraines des Pays Bas. I. L'eau douce de sous-sol des Dunes et des Polders." *Archives Musée Teyler*, ser. 2, vol. 9, pt. 1, pp. 1–96.

1905

"Sur un équivalent du Forest-Bed de Cromer dans les Pays-Bas (Traduction avec une note additionelle par M. O. van Ertborn." *Bulletin de la Société Belge Géologique, Paléontologique et Hydrologique*, vol. 18, pp. 240–52.

"Note sur un espèce de Cerf d'Age Icénien (Pliocène Supérieur). *Cervus falconeri* Dawk., trouvée dans les argiles de la Campin." *Bulletin de la Société Belge Géologique, Paléontologique et Hydrologique*, vol. 10, Mémoires, pp. 121–24.

"De geographische en geologische beteekenis van den Hondsrug en het onderzoek der zwersteenen in ons noordsch diluvium." *Verslag*, vol. 14, pp. 360–68.

"The geographical and geological signification of the Hondsrug, and the examination of the erractics in the Northern Diluvium of Holland." *Verslag*, vol. 8, pt. 5, pp. 427–36.

"L'âge de l'argile de Tégelen et les espèces de cervidés qu'elle contient." *Archives Musée Teyler*, ser. 2, vol. 9, pt. 4, pp. 605–15.

"De voorzienig van Amsterdam met drinkwater uit de duinen." *Nederlandsch Tijdschrift van Geneeskundig*, 2de helft, no. 20, pp. 1346–65.

"Bestaat er gevaar voor verzouting eener goed aangelegde prise d'eau in de duinen?" *Nederlandsch Tijdschrift van Geneeskundig*, ade helft, no. 25, pp. 1707–1708.

"L'âge des différentes assizes englobées dans la série du 'Forest-Bed' ou le Cromerian." *Archives Musée Teyler,* ser. 2, vol. 10, pt. 1, pp. 59–74.

"L'âge des différentes assizes englobées dans la série du 'Forest-Bed' ou le Cromerian." *Société Belge Géologique, Paléontologique et Hydrologique,* no further information.

1906

"Ueber Facettengeschiebe im niederländischen Diluvium." *Centraalblad für Mineralogie, Geologie, Palaeontologie,* 1906, pt. 1, p. 15.

"Etudes sur les eaux souterraines des Pays-Bas. II. L'eau salée peut-elle envahir une prise d'eau dans les dunes?" *Archives Musée Teyler,* ser. 2, vol. 10, pt. 2, pp. 75–84.

"La Pluralité des Périodes Glaciaires dans les dépôts pleistocènes et pliocènes des Pays-Bas. I." *Archives Musée Teyler,* ser. 2, vol. 10, pt. 2, pp. 163–79.

1907

"Note sur une novelle espèce de cerf des argiles de la Campine *Cervus ertbornii,* n. sp." *Taxandria, Gedenkschriften der Geschieden Oudheidkundigen Kring der Kempen, Turnhout,* vol. 4, pp. 80–74.

"Eenige van Nederlandschen kant verkregen uitkomsten met betrekking tot de kennis der Kendeng-fauna (Fauna van Trinil)." *T.A.G.,* ser. 2, vol. 24, pt. 3, pp. 449–58.

"Sur quelle échelle s'accomplit le phénomène du transport atmosphérique de sel marin?" *Archives Musée Teyler,* ser. 2, vol. 10, pt. 4, pp. 461–71.

1908

"Heeft Roodharigheid de Beteekenis van atavistiche Varieteit?" *Nederlandsch Tijdschrift van Geneeskundig,* 1ste helft, no. 7, pp. 553–56.

"Het zootgehalte van zeewater." *Oprechte Haarlemsche Courant,* Stadseditie, May 8, 1908.

"On the Correlation of the Black and the Orange Coloured Pigments and its Bearing on the Interpretation of Redhairedness." *Man,* vol. viii, pt. 6, art. no. 46, pp. 87–89.

"Das geologische Alter der Kendeng-Oder Trinil-Fauna." *T.A.G.,* ser. 2, vol. 25, pt. 6, pp. 1235–70.

"Een Duitsch opstel in een Nederlandsch wetenschappelijk tijdschrift." *Nieuwe Rotterdamsche Courant,* Zaterdag 15 December 1908, 2de Blad A.

1909

"Over een veeljarige schommeling van den grondwaterstand in de Hollandsche duinen." *Verslag,* vol. 17, pp. 782–90.

"On a long-period Variation in the Height of the Groundwater in the Dunes of Holland." *P.K.A.W.,* vol. 11, pt. 3, pp. 674–81.

"Fossiele tand van Sondé." *Nieuwe Rotterdamsche Courant,* March 25, 1909, Ochtenblad B.

"Die Fossilie Zähne von Trinil." *T.A.G.,* ser. 2, vol. 26, pt. 3, pp. 399–401. Response to a note by Margarethe Selenka.

"Een 'Raadsel,' dat geen Raadsel is." *Album der Natuur,* no indication of volume, p. 283.

"Een en ander over Geologie en Hydrologie onzer duinen." *Verslag van den Elfde Algemeene Vergadering van de Vereeniging voor Waterleidingsbelangen in Nederland, Nijmegen 17–18 September 1909,* pp. 81–98, 100, 101.

"Over het Ontstaan van de Vlakten in het Duin." *T.A.G.,* ser. 2, vol. 26, pt. 6, pp. 896–910.

"De Prise d'eau der Haarlemsche Waterleiding." *Rapport uitgebracht aan Burgemeester en Wethouders van Bloemendaal,* 53 pp.

1910

"Geen 'artiesisch' drinkwater voor Amsterdam." *Der Amsterdammer,* no. 1700, January 23, 1910, p. 2.

"Over Duinvalleien, den vorm der Nederlandsche Kustlijn en het ontstaan van laagveen in verband met bodembewegingen." *T.A.G.,* ser. 2, vol. 27, pt. 3, pp. 395–402.

"De Haarlemsche Waterlediing. Over Dr. Pareau's Twijfel aan de aanzienlijke daling van het grondwaterpeil door de Haarlemsche Waterleiding nabij hare prise d'eau teweeggebracht en nog wat." *Haarlem's Dagblad,* February 28, 1910, 2de blad.

1911

"De Beteekenis der palaeontologische gegevens voor de ouderdomsgepaling der Klei van Tegelen." *T.A.G.,* ser. 2, vol. 28, pt. 2, pp. 234–46.

"De Hollandsche Duinen, Grondwater en Bodemdaling." *T.A.G.,* ser. 2, vol. 28, pt. 3, pp. 395–413.

"Over den Vorm van het Grondwatervlak in het Duin." *T.A.G.,* ser. 2, vol. 28, pt. 6, pp. 895–902.

"Berekening van een grondwaterstroom." *De Ingenieur,* vol. 26, no. 50, p. 1074.

1912

"Over de Plaats van *Pithecanthropus* in het Zöologisch Systeem." *Archives Musée Teyler,* ser. 3, vol. 1, pp. 142–49.

1913

"De betrekking tusschen hersenmassa en lichaamsgrootte bij de gewervelde dieren." *Verslag,* vol. 22, pp. 593–614.

1914

"On the Relation between the Quantity of Brain and the Size of the Body in Vertebrates." *P.K.A.W.,* vol. 16, pt. 2, nos. 1–5, pp. 647–68.

"Die gesetzmässige Beziehung von Gehirnmasse zu Körpergrösse bei den Wirbeltieren." *Zeitschrift für Morphologie und Anthropologie,* vol. 18, pp. 323–50.

"Evenwichten in den Kringloop der Stof op de Aarde." *Archives Musée Teyler,* ser. 3, vol. 2, pp. 61–75.

"Van Oude tot Nieuwe Levenswerelden." *Archives Musée Teyler,* ser. 3, vol. 2, pp. 103–21.

"Het Liedsche Duinwater. Ein Hydrologische Studie." Printed for the Water Company at Leiden, pp. 1–31.

1915

"De natuurlijke grens van Nederland beschouwd in verband met de daling van den bodem." *Handelsblad 15de Natuur- en Geneeskundig Congres, Amsterdam 8–10 April, 1915,* pp. 435–43.

1916

"Hollands Duin als natuurlijke Zeewering en de tijd." *T.A.G.,* ser. 2, vol. 33, pt. 3a, pp. 395–415.

1917

"De Maat der Hersenen." *Archives Musée Teyler,* ser. 3, vol. 3, pp. 282–94.

"Hoe ontstonden de vennen bij Oisterwijk." *Verslag Algemeen Vergaderingen 1913–1917*, pp. 81–91.

1918

"De Betrekking der hoeveelheden van de Hersenen, het neuron en zijn deelen tot de lichaamsgrootte." *Verslag*, vol. 26, pp. 1416–25.

"On the Relation between the Quantities of the Brain, the Neurone and its Parts, and the Size of the Body." *P.K.A.W.*, vol. 20, pts. 9/10, pp. 1328–37.

1919

"De beteekenis der grootte van het neuron en zijn deelen." *Verslag*, vol. 27, pt. 4, pp. 503–20.

"The Significance of the Size of the Neuron and its Parts." *P.K.A.W.*, vol. 21, pt. 5, pp. 711–29.

"Vergelijking van het hersengewicht, in functie van het lichaamsgewicht, tusschen de twee seksen." *Verslag*, vol. 27, pp. 713–32.

"Comparison of the Brain Weight in Function of the Body Weight, between the Two Sexes." *P.K.A.W.*, vol. 21, pts. 6/7, pp. 850–69.

"Over het ontstaan en de Geologische Geschiedenis van Vennen, Venen en Zeeduinen." *Archives Musée Teyler*, ser. 3, vol. 4, pp. 266–93.

1920

"De Hoeveelheidsbetrekking van de Hersenen tot het Lichaam der Gewervelde Diersooten en hare Beteekenis." *Vakblad voor Biologie*, vol. 1, pt. 6, pp. 83–89.

"De hoeveelheidsbetrekking van het zenuwstelsel bepaald door het mechanisme van het neuron." *Verslag*, vol. 28, pt. 6, pp. 623–38.

"The Quantitative Relations of the Nervous System determined by the Mechanism of the Neurone." *P.K.A.W.*, vol. 22, pts. 7/8, pp. 665–80.

"De proto-Australische fossiele Mensch van Wadjak, Java, I." *Verslag*, vol. 29, pt. 1, pp. 88–105.

1921

"De proto-Australische fossiele Mensch van Wadjak, Java, II." *Verslag*, vol. 29, pt. 6, pp. 866–87.

"De beteekenis der groote schedelcapaciteit van *Homo neandertalensis*." *Verslag*, vol. 29, pt. 7, pp. 1021–22.

(With Molengraaf, G. A. F.) "Preadvies over de Vraag van den Minister van Arbeid waaraan de aanwezigheid van artesisch grondwater in de duingronden te danken is." *Algemeen Handelsblad*, November 10, 1921, avondblad. (*Algemeen Handelsblad*, October 20, 1921, Ochtendblad; reprinted in *Verslag*, vol. 30, pts. 4/5, pp. 208–13.)

1922

"Over den schedelvorm van *Homo neandertalensis* en van *Pithecanthropus erectus*, bepaald door mechanisme factoren." *Verslag*, vol. 30, pt. 7, pp. 391–411.

"The Proto-Australian Fossil Man of Wadjak, Java." *P.K.A.W.*, vol. 23, pt. 7, pp. 1031–51.

"On the Significance of the Large Cranial Capacity of *Homo neandertalensis*." *P.K.A.W.*, vol. 23, pt. 8, pp. 1271–88.

"On the Motion of Ground Water in Frost and Thawing Weather." *P.K.A.W.*, vol. 24, pts. 6/7, pp. 313–32.

"On the Cranial Form of *Homo neandertalensis* and *Pithecanthropus erectus*, Determined by Mechanical Factors." *P.K.A.W.*, vol. 24, pts. 6/7, pp. 307–32.

"Phylogenetische en Ontogenetische toeneming van het volumen der hersenen bij de Gewervelde Dieren." *Verslag*, vol. 31, pt. 6, pp. 307–32.

"Hat sich das Gehirn beim Haushund, im Vergleich mit Wildhundarten, vergrössert oder verkleinert?" *Bijdragen t.d. Dierkunde*, pt. 22, pp. 315–20.

1923

"Phylogenetic and Ontogenetic Increase of the Volume of the Brain in Vertebrata." *P.K.A.W.*, vol. 25, pts. 7/8, pp. 230–55.

"Limburg's bodem als getuige can klimaatsveranderingen." *Handelingen 19de Nederlandsch Natuur- en Geneeskundig Congres, Maastricht, 5–7 April 1923,* pp. 50–68.

"De voornamste eigenschappen van den schedel en de hersenen van *Pithecanthropus erectus*." *Algemeen Handelsblad*, May 27, 1923, Ochtendblad.

"Over de onderkaak en het gebit van *Pithecanthropus erectus*." *Algemeen Handelsblad*, November 25, 1923, Ochtendblad.

1924

"Over de voornamste eigenschappen van den schedel en de hersenen, de onderkaak en het gebit van *Pithecanthropus erectus*." *Verslag*, vol. 32, pt. 9, pp. 135–48.

"On the Principal Characters of the Cranium and the Brain, the Mandible and the Teeth of *Pithecanthropus erectus*," *P.K.A.W.*, vol. 27, pts. 3/4, pp. 265–78.

"Over de hersenhoeveelheid van gespecialiseerde zoogdiergeslachten." *Verslag*, vol. 33, pt. 4, pp. 319–26.

"On the Brain Quantity of Specialized Genera of Mammals." *P.K.A.W.*, vol. 27, pts. 5/6, pp. 430–37.

"Figures of the Calvarium and Endocranial Case, a Fragment of the Mandible and Three Teeth of *Pithecanthropus erectus*." *P.K.A.W.*, vol. 27, pts. 5/6, pp. 459–64.

1925

"Geologische aanvulling van de mededeeling an den heer Bolk: over het bestaan van een langhoofdig Gorillaras." *Verslag*, vol. 34, pt. 3, p. 285.

"Het palaeothermale probleem en de evoutie der Zon." *Algemeen Handelsblad*, March 7, 1925, Avonblad.

"Het palaeothermale probleem in het licht can de reus- en dwegtheorie der stellaire evolutie." *Algemeen Handelsblad*, May 31, 1925, Ochtendblad. Also published in *Verslag*, vol. 34, pt. 6, pp. 539–56.

"The Paleothermal Problem in Light of the Giant and Dwarf Theory of Stellar Evolution." *P.K.A.W.*, vol. 28, pt. 6, pp. 587–604.

"Over de Plaats van den *Pithecanthropus* onder de Primaten en zijn genealogische Betrekkingen, naar de Beschouwingen van Boule." *Vakblad voor Biologen*, vol. 7, no. 1, pp. 1–5.

1926

"Het palaeothermale probleem, poolverplaatsing en de evolutie der zon." *Verslag Geologische Sectie Geologie-Mijnbouw Generale Nederlandsch Kolonie*, vol. 3, pt. 5, pp. 123–31.

"Over de voornamste onderscheidende eigenschappen van het femur van *Pithecanthropus erectus*." *Verslag*, vol. 35, pt. 3, pp. 443–55.

"On the Principal Characters of the Femur of *Pithecanthropus erectus*." *P.K.A.W.*, vol. 29, pt. 5, pp. 730–43.

"Over brokken van glaciaal geslepen en geschramde rotsbodems als zwerfsteenen in ons Diluvium." *Verslag Geologische Sectie Geologie-Mijnbouw Generale Nederlandsch Kolonie*, vol. 3, pt. 5, pp. 144–45.

"*Manis palaejavanica*, het reuzenschubdier der Kendeng-fauna." *Algemeen Handelsblad*, October 31, 1926, Ochtendblad.

"*Manis palaejavanica*, het reuzenschubdier der Kendeng-fauna." *Verslag*, vol. 35, pt. 8, pp. 949–58.

"*Pithecanthropus erectus*. De 'nieuwe *Pithecanthropus*-vondst.'" *Nieuwe Rotterdamsche Courant*, December 8, 1926, Avonblad.

"De zoogenamde '*Pithecanthropus*'-vondst." *Nieuwe Rotterdamsche Courant*, December 8, 1926, Avonblad C.

"De zoogenamde nieuwe *Pithecanthropus*-vondst." *Algemeen Handelsblad*, December 18/19, 1926, Avon/Ochtenblad.

"*Manis palaejavanica*, the Giant Pangolin of the Kendeng Fauna." *P.K.A.W.*, vol. 29, pt. 9, pp. 1233–43.

"Figures of the femur of *Pithecanthropus erectus*." *P.K.A.W.*, vol. 29, pt. 9, pp. 1275–77.

1927

"De zoogenamde nieuwe *Pithecanthropus*-vondst." *Verslag*, vol. 36, pt. 1, pp. 62–66.

"The So-Called New *Pithecanthropus* Skull." *P.K.A.W.*, vol. 30, pt. 1, pp. 134–37.

"Über die Hauptmerkmale des Femur von *Pithecanthropus erectus*." *Anthropologische Anzeiger*, vol. 4, pt. 2, pp. 131–46.

"Opmerkingen over eenige uitkomsten der Palaeontologie met betrekking tot de evolutie der dierenwereld." *Handelingen 21ste Nederlandsche Natuur- en Geneeskundig Congres, Amsterdam 19–21 April 1927*, pp. 150–51.

(With L. M. R. Rutten.) "Verslag omtrent het verzoek van het Natuurhistorisch Genootschap in Limburg, gericht tot den Minister can Onderwijs, Kunsten, en Wetenschappen, om een jaarlijksch subsidie groot f. 1000.– etc." *Verslag*, vol. 36, pt. 9, pp. 252–59.

"De *Pithecanthropus*." *Haarlem's Dagblad*, September 30, 1927.

1928

"De Wet der noodwendige phylogenetische volmaking van het psychoencephalon." *Verslag*, vol. 37, pt. 3, pp. 252–59.

"The Law of the Necessary Phylogenetic Perfection of the Psychoencephalon." *P.K.A.W.*, vol. 31, pt. 3, pp. 304–14.

1930

"De Aapmensch." Catalogus van de Afdeeling Wetenschap in het Nederlandsch Paviljoen der Internationale Tentoonstelling te Luik. Leiden: S. C. van Doesburgh, pp. 118–20, 122.

"L'homme singe." Catalogue de la Section des Sciences dans le Pavillon Néerlandais de l'Exposition Internationale à Liège. Leiden: S. C. van Doesburgh, pp. 119–121, 123.

"Die phylogenetische Grosshirnzunahme autonome Vervollkommnung der animalen Funktionen." *Biologica Generalis*, vol. 6, pt. 2, pp. 247–92.

"Corrections and additions to 'Die phylogenetische Grosshirnzunahme . . .'" *Biologica Generalis*, vol. 6, unknown page.

"Phyloblastese het beginsel en de grondvoorwaarde der aanpassingsbetrekkingen."
Algemeen Handelsblad, September 28, 1930.

1931

"Over de afneming van het zoetwaterlichaam in Holland's duinen door menschelijk
ingrijpen." *Handelingen 23ste Nederlandsch Natuur- en Geneeskundig Congres, Delft
7–9 April, 1931*, pp. 251–53.

1932

"The distinct organization of *Pithecanthropus* of which the femur bears evidence, now
confirmed from other individuals of the described species." *P.K.A.W.*, vol. 35, pt. 6,
pp. 716–22.

"Praehistorische Vondsten." *Algemeen Handelsblad*, May 12, 1932, Ochtendblad.

"Schakelvorm tusschen Mensch en Dier." *Algemeen Handelsblad*, June 10, 1932, Avonblad.

"Early Man in Java." *Nature*, vol. 130, no. 3270, p. 20.

"De afzonderlijke organisatie van *Pithecanthropus*, waarvan het femur getuigt, thans
bevestigd door andere individuen van de beschreven soort." *Verslag*, vol. 41, pt. 6,
pp. 76–77.

1933

"De schijnbare en de werkelijke cephalisatie van den Australischen inboorling."
Algemeen Handelsblad, January 29, 1933, Ochtendblad.

"De schijnbare en de werkelijke cephalisatie van den Australischen inboorling." *Verslag*,
vol. 42, pt. 1, p. 10.

"The seeming and real cephalization of the Australian aborigine." *P.K.A.W.*, vol. 36, pt. 2,
pp. 2–13.

"Corrigendum to 'The seeming and real cephalization of the Australian aborigine.'"
P.K.A.W., vol. 36, pt. 2, p. 240.

"De voorm der Hersenen bij *Sinanthropus* en bij *Pithecanthropus*." *Algemeen
Handelsblad*, March 26, 1933, Ochtendblad.

"De voorm en de groottee der hersenen bij *Sinanthropus* en bij *Pithecanthropus*."
Verslag, vol. 42, pt. 3, pp. 40–42.

"Corrigenda." *Verslag*, vol. 42, pt. 4, p. 53.

"The Shape and Size of the Brain in *Sinanthropus* and in *Pithecanthropus*." *P.K.A.W.*,
vol. 36, pt. 4, pp. 415–23.

"Repliek." *T.A.G.*, vol. 50, p. 785.

1934

"New Evidence of the Distinct Organization of *Pithecanthropus*." *P.K.A.W.*, vol. 37, pt. 3,
pp. 139–45.

"Bespreking van 'The Age of *Pithecanthropus*' door L. J. C. van Es." *Mensch en
Maatschappij*, vol. 10, pt. 3, 6 pp.

"Phylogenetic Cerebral Growth." *Congrès International de Sciences Anthropologiques et
Ethnologiques*, comptes rendus de la première session, Londres 1934, pp. 71–75.

"Über die Gleichheit der Cephalisationsstufe aller Menschengruppen bei verschiedenem
spezifischem Gehirnvolumen." *Biologica Generalis*, vol. 10, pt. 1, pp. 185–93.

1935

"On the Gibbon-like appearance of *Pithecanthropus erectus.*" *P.K.A.W.*, vol. 38, pt. 6, pp. 578–85.

"*Pithecanthropus erectus* als organisme." *Handelingen 25ste Nederlandsch Natuur- en Geneeskundig Congres, Leiden 23–25 April 1935,* pp. 147–59.

"The sixth (fifth new) femur of *Pithecanthropus erectus.*" *P.K.A.W.*, vol. 38, pt. 8, pp. 850–53.

1936

"Nieuwe *Pithecanthropus* ontdekt?" *Algemeen Handelsblad,* April 18, 1936, Avonblad, 5de blad, p. 10.

"Een nieuwe *Pithecanthropus* ontdekt?" *Het Vaderland,* April 18, 1936, Avonblad 13.

"Strijd om een Schedel." *De Telegraaf,* April 19, 1936, 5de blad, p. 11.

"Brief aan Prof. Dr. J. P. Kleiweg de Zwaan, Voorzitter van het Nederlandsch Nationaal Bureau voor Anthropologie, te Amsterdam." *Mensch en Maatschappij,* vol. 12, pt. 4, pp. 303–304.

"Over fossiele Menschen en den *Pithecanthropus.*" *Nieuwe Rotterdamsche Courant,* July 26, 1936, Ochtendblad 13.

"Fossiele Menschen en *Pithecanthropus.* De jongste vondst op Java geen *Pithecanthropus.* II." *Nieuwe Rotterdamsche Courant,* August 2, 1936, Ochtendblad 13, p. 2.

"Fossiele Menschen en *Pithecanthropus. Pithecanthropus* geen mensch. III." *Nieuwe Rotterdamsche Courant,* August 12, 1936, Avonblad E, p. 2.

"Fossiele Menschen en *Pithecanthropus.* Rectificatie." *Nieuwe Rotterdamsche Courant,* August 22, 1936, Avonblad 13, p. 3.

"Fossiele Menschen en *Pithecanthropus.*" *Nieuwe Rotterdamsche Courant,* August 1936. (Reprint of four publications immediately above with corrections and an additional figure.)

"Racial Identity of *Homo soloensis* Oppenoorth (including *Homo modjokertensis* Von Koenigswald) and *Sinanthropus pekinensis* Davidson Black." *P.K.A.W.*, vol. 39, pt. 10, pp. 1180–85.

1937

"On the Fossil Human Skulls recently discovered in Java and *Pithecanthropus erectus.*" *Man,* vol. 37, art. no. 1, pp. 1–7.

"The osteone arrangement of the thigh-bone compacta of Man identical with that first found of *Pithecanthropus.*" *P.K.A.W.*, vol. 40, pt. 10, pp. 864–70.

"Early Man in Java and *Pithecanthropus erectus,*" in G. G. MacCurdy, ed., *Early Man.* Philadelphia: J. B. Lippincott Co., pp. 315–22.

1938

"Prof. Dr. Eug. Dubois over den *Pithecanthropus erectus* (verslag van een lezing, gemaakt door Dr. E. M. Krutzer)." *Natuurhistorisch maandblad,* vol. 27, pt. 9, pp. 92–95.

"The mandible recently described and attributed to the *Pithecanthropus* by G. H. R. von Koenigswald, compared with the mandible of *Pithecanthropus erectus* described in 1924 by Eug. Dubois." *P.K.A.W.*, vol. 41, pt. 2, pp. 139–47.

"On the fossil human skull recently described and attributed to *Pithecanthropus erectus* by G. H. R. von Koenigswald." *P.K.A.W.*, vol. 41, pt. 4, pp. 380–86.

"Over de spronggrootte van den phylogenetischen en ontogenetischen Hersengroei."
   Vakblad voor Biologen, vol. 20, pt. 3, pp. 37–46.

1939

"Explanatory Statement concerning the communication of Eug. Dubois in Proceedings
   K.N.A.v.W., vol. 41, no. 4, 1938; On the fossil human skull recently described and
   attributed to *Pithecanthropus erectus* by G. H. R. von Koenigswald." *P.K.A.W.,* vol. 42,
   pt. 1, p. 53.
"On insignificance of cranial vault-height in phylogenetic braingrowth." *P.K.A.W.,* vol. 42,
   pt. 2, pp. 125–26.

1940

"The fossil human remains discovered in Java by Dr. G. H. R. von Koenigswald and attributed by him to *Pithecanthropus erectus,* in reality remains of *Homo wadjakensis*
   (syn. *Homo soloensis).*" *P.K.A.W.,* vol. 43, pt. 4, pp. 494–96.
"The fossil human remains discovered in Java by Dr. G. H. R. von Koenigswald and attributed by him to *Pithecanthropus erectus,* in reality remains of *Homo soloensis.*"
   P.K.A.W., vol. 43, pt. 5, p. 653. (This is an announcement of a change of title of the
   publication immediately above.)
"The fossil human remains discovered in Java by Dr. G. H. R. von Koenigswald and attributed by him to *Pithecanthropus erectus,* in reality remains of *Homo soloensis.*
   Continuation." *P.K.A.W.,* vol. 43, pt. 7, pp. 842–51.
Idem., *P.K.A.W.,* vol. 43, pt. 9, p. 1142. (Corrigenda to above.)
"The fossil human remains discovered in Java by Dr. G. H. R. von Koenigswald and attributed by him to *Pithecanthropus erectus,* in reality remains of *Homo sapiens soloen-
   sis.* Conclusion." *P.K.A.W.,* vol. 43, no. 10, pp. 1268–75.

II. LITERATURE CITED

Andrews, R. C. *On the Trail of Ancient Man.* 1926. New York: G. P. Putnam's Sons.
Anonymous. *B.S. Name-List/Address-Book of the Dutch Indies 1922–1923.*
Anonymous ("Homo Erectus"). "Palaeontologische onderzoekingen op Java."
   Bataviaasch Nieuwsblad, February 6, 1893, no. 57.
Anonymous. "Notes." *Nature,* 1895, January 3, vol. 51, p. 230.
Anonymous. "Interview with Madame Selenka." *Bataviaasch Nieuwsblad,* April 25, 1907.
Anonymous. "Report of views of Dr. Van Stein Callenfels." *Handelsblad,* July 24, 1936.
Anonymous. "Prehistoric Congress in Oslo, lecture of Dr. Van Stein Callenfels." *De
   Telegraaf,* August 8, 1936.
Anonymous. "*Homo modjokertensis:* Lecture of Dr. Von Koenigswald." *Handelsblad,*
   August 29, 1936.
Anonymous. "Lecture of Dr. Van Stein Callenfels." *Handelsblad,* October 22, 1936.
Anonymous. "Praehistorische vondsten op Java; Voordracht Dr. Van Stein Callenfels."
   Handelsblad, October 29, 1936.
Anonymous. "Dr. Von Königswald, Studiereis naar Europa." *Handelsblad,* November 7,
   1936.
Anonymous. "De Fossile Mensch van Java; De Weg der Evolutie; De beroemdste is nog
   altijd de *Pithecanthropus.*" *Handelsblad,* January 18, 1938.
Anonymous. "Prof. Dubois blikt over zijn leven terug." *De Telegraaf,* January 27, 1938.
Anonymous. "Uitgegeven de Indische Geneaologische Vereniging." *I.N.N.,* 1988–1997.

Balchior, G. "The Peking Man—An Undamaged Skull." *Manchester Guardian,* December 30, 1929.

Beekman, E. M. "Introduction." In P. A. Daum, *Ups and Downs of Life in the Indies.* 1987, Amherst: University of Massachusetts Press, pp. 1–49.

Bemmelen, J. F. van. "Het Leidsche internationale zoologencongres: Dubois' aapmensch voor de vierschaar der wetenschaap." *Java-bode,* November 1895, vol. 16.

Bemmelen, J. M. van. "Der Gehalt an Fluorcalcium eines fossilen Elephantenknochen aus der Tertiärzeit." *Zeitschrift für anorganische Chemie,* 1897, vol. 15, pp. 84–122.

Black, D. "On a Lower Molar from the Chou Kou Tien Deposits." *Paleontologica Sinica,* 1927, ser. C, vol. 7, fasc. I.

———. "Interim Report on the Skull of *Sinanthropus.*" *Annual Meeting of the Geological Society of China,* March 29, 1930.

Bloys van Treslong Prins, P. C. *G.H.G.,* 1934–1939. Batavia: Drukkerij Albrecht.

———. "Grafschiften van Europeanen in Nederlandsch-Indië: naar de originele aantekeningen." *B.R.P.* 1993, 5. 's Gravenhage: J. C. Boutong de Katzmann.

Boule, M. "L'homme fossile de La Chapelle-aux-Saints." *Annales de Paléontologie,* 1911–1913, Paris.

Brongersma, L. D. "Professor Dr. Eug. Dubois." *Nieuwe Rotterdamsche Courant,* 1938, January 28.

———. "Eugène Dubois." *Natuurhistorisch maandblad,* 1973, vol. 62, pp. 107–109.

———. *Pithecanthropus:* Echt en 'Pseudo.'" *Verslag,* 1982, vol. 91, pp. 34–36.

———. "The Vicissitudes of the Dubois Collection." Unpublished manuscript, n.d., Dubois Archives, Naturalis.

Corbey, R. "Ethnographic Showcases, 1870–1930." *Cultural Anthropology,* 1993, vol. 8, no. 3, pp. 338–96.

Couperus, L. *The Hidden Force.* 1985. Translated by A. Teixeira de Mattos. Amherst: University of Massachusetts Press. Originally published as *Die Stille Kracht.* 1900. Amsterdam: L. J. Veen.

Cunningham, D. J. "Dr. Dubois' So-called Missing Link." *Nature,* 1894–95, vol. 51, pp. 428–29.

———. "A Paper on *Pithecanthropus erectus,* the Man-like Transitional Form of Dr. Eug. Dubois." *Journal of Anatomy and Physiology,* 1895, vol. 29, n.s. 8, pp. 18–19.

Daum, P. A. *Ups and Downs of Life in the Indies.* 1987. Amherst: University of Massachusetts Press. Translated by Elsje and Donald Sturtevant. Originally published as *Ups en downs in het Indische leven.* 1890. Batavia: Feuilleton.

Dekker, E. D. (pseudonym Multatuli). *Max Havelaar.* 1982. Amherst: University of Massachusetts Press. Translated by Roy Edwards. Originally published 1860. Amsterdam: Van Oorschot.

Deyssel, L. van. *Blank en Geel.* 1979. Amsterdam. (Originally published 1892.)

Dubois, J. M. F. "Trinil: A Biography of Prof. Dr. Eugène Dubois, the Discoverer of *Pithecanthropus erectus.*" 1957. Unpublished ms. in Dubois Archives.

Duckworth, W. H. L. *Prehistoric Man.* 1912. Cambridge, Eng.: Cambridge University Press.

Feen, P. J. van der, and W. S. S. van Bentham-Jutting. "Antje Schreuder, Amsterdam, 15 november 1887–Amsterdam, 2 februari 1952." *Geologie en mijnbouw,* 1952, vol. 14, pp. 121–25.

Flower, W. H. "Discussion of Dubois' 'On *Pithecanthropus erectus,* A Transitional Form between Man and the Apes.'" *J.A.I.,* 1896, vol. 25, p. 248.

Garson, J. "Discussion of Dubois' 'On *Pithecanthropus erectus,* A Transitional Form between Man and the Apes.'" *J.A.I.,* 1896, vol. 25, pp. 251–52.

Gorjanović-Kramberger, D. "Der diluviale Mensch von Krapina in Kroatien: ein Beitrag zur Paläoanthropologie," 1906. In O. Walkoff, ed., *Studien über die Entwicklungsmechanik des Primatenskeletts.* Kreidel, Wiesbaden, pp. 59–277.

Haddon, A. C. *History of Anthropology.* 1910. London: Watts and Co.

Haeckel, E. *The History of Creation,* 4th English ed. 1892. London: Kegan Paul, Trench, Trubner & Co, Ltd.

———. "On our present knowledge of the origin of Man." *Annual Report of the Board of Regents of the Smithsonian Institution for the year ending June 30, 1898.* 1899. Pp. 461–80. Translation of a discourse given at the Fourth International Congress of Zoologists at Cambridge, England, August 26, 1898.

Hrdlička, A. "The Most Ancient Skeletal Remains of Man." *Annual Report of the Smithsonian Institution,* 1914, pp. 491–552.

———. *The Skeletal Remains of Early Man.* 1930. Washington, D.C.: Smithsonian Miscellaneous Collections, vol. 83.

Janus, C., and W. Brashler. *The Search for Peking Man.* 1975. New York: Macmillan.

Jerison, H. *Evolution of the Brain and Intelligence.* 1973. New York: Academic Press.

Kate, H. ten. "Review of Dubois' *Pithecanthropus erectus,* eine menschenaehnliche Uebergangsform aus Java." *Nederlandsch koloniaal centraalblad,* 1894–95, vol. 1, pp. 127–29.

Keith, A. "Discussion of Dubois' 'On *Pithecanthropus erectus,* A Transitional Form between Man and the Ape.'" *J.A.I.,* 1896, vol. 25, p. 253.

———. *Ancient Types of Man.* 1911. New York and London: Harper & Bros.

———. *Antiquity of Man.* 1925. London: Williams and Northgate.

———. *An Autobiography.* 1950. London: Watts.

Koenigswald, G. H. R. von. "*Pithecanthropus erectus,* Antwoord Dr. Von Koenigswald." *Algemeen Handelsblad,* May 7, 1936.

———. "Een nieuwe *Pithecanthropus* ondekt." *Algemeen Indische Dagblad,* March 28, 1936.

———. "Erste Mitteilung über einen fossilen Hominiden aus dem Altpleistocän Ostjavas." *P.K.A.W.,* 1936, vol. 39, no. 8, pp. 1000–1009. Communicated at the meeting of September 26, 1936.

———. "Ein Unterkieferfragment des *Pithecanthropus* aus den Trinilschichten Mittenjavas." *P.K.A.W.,* 1937, vol. 40, pp. 883–93.

———. "A review of the stratigraphy of Java and its relations to early man." In G. G. MacCurdy, ed., *Early Man.* 1937. Philadelphia: J. B. Lippincott and Co., pp. 23–32.

———. "*Pithecanthropus* received into human family." *Illustrated London News,* December 11, 1937.

———. "Ein neuer *Pithecanthropus*-Schädel." *P.K.A.W.,* 1938, vol. 41, pp. 185–92.

———. *Meeting Prehistoric Man.* 1956. London: Thames & Hudson.

———, and F. Weidenreich. "The relationship between *Pithecanthropus* and *Sinanthropus.*" *Nature,* 1939, vol. 144, pp. 926–29.

Lanpo, J., and H. Weiwen. *The Story of Peking Man from Archaeology to Mystery.* Translated by Yin Zhinqui. 1990. Beijing: Foreign Languages Press.

Leakey, R. E., and J. Slikkerveer. *Man-ape, Ape-man.* 1993. Leiden: Netherlands Foundation for Kenya Wildlife Service.

Litchfield, H., ed. *Emma Darwin: A Century of Family Letters.* 1915. London: Murray.

Locher-Schloten, E. "So Close and Yet So Far." In J. Clancy-Smith and F. Gouda, eds., *Domesticating the Empire.* 1998. Charlottesville: University Press of Virginia, pp. 131–53.

Lydekker, R. "Notices of Siwalik Mammals." *Records of the Geological Survey of India,* 1879, vol. 12, pp. 33–52.

———. "Siwalik Mammalia." Memoirs of the Geological Survey of India, *Palaeontologica India,* 1886, s. 10, *Indian Tertiary and Post-Tertiary Vertebrata,* vol. 4, pt. 1, pp. 1–18.

———. "Review of Dubois' *Pithecanthropus erectus.* Eine menschenaehnliche Uebergangsform aus Java." *Nature,* 1895, vol. 51, p. 291.

MacBride, F. W. *Evolution.* 1928. London: Sixpenny Library Series.

Marsh, O. "On the *Pithecanthropus erectus,* from the Tertiary of Java." *American Journal of Science,* 1895, s. 4, vol. 1, pp. 475–82.

———. "A Commentary on Dubois' '*Pithecanthropus erectus,* eine menschenaehnliche Uebergangsform.'" *Compte-rendu des séances du Troisième Congrès International de Zoologie, Leyde, 11–16 September 1895.* 1896. Leiden. P. 272.

Martin, K. "Ueberreste vorweltlicher Proboscidier auf Java und Banka." *Sammlungen des Geologischen Reichs-Museums in Leiden,* 1884–89, vol. 4, pp. 1–24.

———. "Fossile Säugethierreste von Java und Japan." *Sammlungen des Geologischen Reichs-Museums in Leiden,* 1884–89, vol. 4, pp. 25–69.

———. "Kritische Bedenken gegen den *Pithecanthropus erectus* Dubois." *Globus,* 1895, vol. 67, pp. 213–17.

Matschie, P. "Noch einmal *Pithecanthropus erectus.*" *Naturwissenschaftliche Wochenschrift,* 1895, vol. 10, pp. 122–23.

Nuland, S. *Doctors: The Biography of Medicine.* 1988. New York: Vintage Books.

O'Mulligan (unknown first name). "Bones of Contention." *The Evening Telegraph* (Dublin), November 23, 1895.

Oppenoorth, W. F. F. "*Homo (Javanthropus) soloensis.* Een pleistocene mensch van Java." *Wetenschappelijke mededelingen Dienst van den Mijnbrouw in Nederlandsch-Indië,* 1932, vol. 20, pp. 49–63.

———. "De vondst van palaeolitische menschelijke schedels op Java." *De Mijningingenieur,* June 1932.

Osborn, H. F. *Men of the Old Stone Age.* 1915. New York: Scribners.

———. Foreword. In R. C. Andrews, *On the Trail of Ancient Man.* 1926. New York: G. P. Putnam's Sons.

Pattynama, P. "Secrets and Danger." In J. Clancy-Smith and F. Gouda, eds., *Domesticating the Empire.* 1998. Charlottesville: University Press of Virginia, pp. 84–107.

Pieters, F., and J. de Visser. "The scientific career of the zoologist Max Wilhelm Carl Weber (1852–1937)." *Bijdragen tot de Dierkunde,* 1993, vol. 62, no. 4, pp. 193–214.

Ricklefs, M. C. *A History of Modern Indonesia Since c. 1300.* 1993. Stanford: Stanford University Press.

Rietschoten, B. D. van. "Uit een schrijven van den heer Van Rietschoten te Blitar." *N.T.,* 1889, vol. 48, pp. 346–47.

Schaaffhausen, H. "On the crania of the most ancient races of man." *Natural History Review,* April 1, 1861, no. 2, pp. 155–80. Translated by G. Busk from H. Schaaffhausen, "Zur Kenntnis der ältesten Rassenschädel," *Archiv für Anatomie, Physiologie und wissenschaftliche Medicin,* 1858, vol. 2, pp. 453–88.

Schwalbe, G. "Ziele und Wege einer vergleichenden physischen Anthropologie." *Zeitschrift für Morphologie und Anthropologie,* 1899, vol. 1, pp. 1–15.

Selenka, M. "Die fossile Zähne von Trinil." *T.A.G.,* 1909, s. 2, vol. 26, pp. 398–99. (Includes comments from Schlosser, Walkoff, and Dubois.)

Shapiro, H. *Peking Man.* 1974. New York: Simon & Schuster.

Shipman, P. *The Evolution of Racism.* 1994. New York: Simon & Schuster.

Spencer, F. *Piltdown, A Scientific Forgery.* 1990. New York: Oxford University Press.

———. *The Piltdown Papers.* 1990. New York: Oxford University Press.

———. *Aleš Hrdlička, M.D., 1869–1943: A Chronicle of the Life and Work of an American Physical Anthropologist.* 1979. Ph.D. dissertation, University of Michigan. Ann Arbor: University Microfilms.

Teilhard de Chardin, P. *Letters from a Traveller.* 1962. London: Collins.

Tesch, P., and L. D. Brongersma. "Eugène Dubois." *Geologie en Mijnbouw,* 1941, vol. 3, no. 2, pp. 29–33.

Theunissen, B. *Eugène Dubois and the Ape-man from Java.* 1989. Dordrecht: Kluwer Academic Publishers.

———, and J. de Vos. "Eugène Dubois, ondekker can de rechtopgaande aapmens." *Natuurhistorisch maandblad,* 1982, vol. 71, nos. 6–7, pp. 107–114.

Thomson, J. A. "Discussion of Dubois' 'On *Pithecanthropus erectus,* A Transitional Form between Man and the Apes.'" *J.A.I.,* 1896, vol. 25, pp. 253–54.

Timmerman, J. A. C. A. "Belangrijke palaeontologische vondstein op Java." *T.A.G.,* 1893, s. 2, vol. 10, pp. 310–12.

Tobias, P. V. T. "Life and Work of Professor Dr. G. H. R. von Koenigswald." In *Auf den Spuren des Pithecanthropus.* 1984. Frankfurt: Waldemar Kramer, pp. 25–95.

Trinkaus, E., and P. Shipman. *The Neandertals.* 1992. New York: Knopf.

Turner, W. "On M. Dubois' Description of Remains Recently Found in Java, Named by Him *Pithecanthropus erectus.* With Remarks on So-called Transitional Forms between Apes and Man." *Journal of Anatomy and Physiology,* 1896, vol. 29, n.s. 9, pp. 424–45.

———. "Discussion of Dubois' 'On *Pithecanthropus erectus,* A Transitional Form between Man and the Ape.'" *J.A.I.,* 1896, vol. 25, p. 249.

Virchow, R. "Commentary on Krause's discussion of Dubois' '*Pithecanthropus erectus, eine menschenaehnliche Uebergangsform aus Java.*'" *Zeitschrift für Ethnologie,* 1895, vol. 27, pp. 81–88.

———. "Die Frage von dem *Pithecanthropus erectus.*" *Zeitschrift für Ethnologie,* 1895, vol. 27, pp. 435–42.

———. "Commentary on Dubois' '*Pithecanthropus erectus,* betrachtet als eine wirkliche Uebergangsform und als Stammform des Menschen.'" *Zeitschrift für Ethnologie,* 1895, vol. 27, pp. 744–47.

Vos, J. de. "*Homo modjokertensis*—vindplaats, ouderdom en fauna." *Cranium,* 1994, vol. 11, no. 2, pp. 103–107.

Wallace, A. R. *The Geographical Distribution of Animals.* 1876. London: Macmillan.

Wassing, R., and R. Wassing-Visser. *Adoeh, Indië!* 1992. Atrium: The Hague.

Weidenreich, F. "Morphology of Solo Man." *Anthropological Papers of the American Museum of Natural History, N.Y.,* 1951, vol. 43, pt. 3.

ACKNOWLEDGMENTS

Some ten or so years ago, John de Vos of the Naturalis asked if I would be interested in "doing something" with the underutilized Dubois Archives, of which he had become the curator. The twinkle in his eye should have warned me that he was inviting me to acquire an obsession, but even had I understood—and I did not—I would not have passed up the opportunity. John has been endlessly patient, knowledgeable, and helpful. At an early stage of this work, John put me in touch with a potential assistant, Paul Storm, then a graduate student studying the Wadjak fossils. Over the years, Paul earned his own Ph.D. and then worked part-time for me translating nearly all of the Dutch and German documents used in this work and tirelessly researching obscure bits of information in response to my queries. Paul has been an invaluable colleague, and this book would not have happened without his efforts. Additional translations from the Dutch were made by Marcel van Tuinen, to whom I am grateful. Ans Molenkamp and Caroline Pepermans of Naturalis were wonderfully helpful with illustrations, some of which were printed from the original glass negatives by Hans de Herder of Nationaal Fotografie Restoratie Atelier. My first forays into the Dubois Archives were funded by a grant to me from the Wenner-Gren Foundation.

Marijke Gijsbers kindly provided information about De Bedelaar. Living descendants of Eugène and Victor Dubois—Jean M. Dubois in California, and Victor Dubois, Nelleke Hooijer, and A. Hooijer-Ruben in Holland—trusted me with family documents and photographs, told me family stories, and gave me permission to quote from Jean M. F. Dubois' unpublished book, *Trinil.* Joost P. M. van Heijst, Secretary of the Dutch Franciscans, gave permission for the quotations from Bernsen's diary. Christine Hertler kindly shared information with me, as did Bert Theunissen. The Hovens family showed me through Dubois' childhood home in Eijsden.

The adventure of my first trip to Java would not have been nearly so much fun without my traveling companion and friend Sally McCutcheon. Our drivers, Anto Sanyoto (kindly recommended to me by Heri Harjono) and Dodif Hendro Susilo, helped me find places where Dubois had lived or worked and showed me traces of Dutch colonial life in Java. The amazingly efficient and charming Yodha Susanti, of the Hotel Tugu in Malang, understood my unusual quest and arranged many things for me, including a visit to a coffee plantation in Wlingi which serves as my model for Mringin. In Jakarta, Veronica Grasman and Cynthia Mackie willingly gave time to help a friend of a friend. The people of Java themselves were astonishingly kind to a blond, blue-eyed foreigner who came looking for the past in their country.

The BBC paid my way to Java for my second visit, and the film crew—Charlie Smith, Philip Martin, Alicky Lockhart, Steve Robinson, Mike Carling, Fraser Barber, and, from the Discovery Channel, Alex Mittendorf—made for a marvelous, productive, and laughter-filled trip. Without them, I would never have been paddled down the Solo River in a prahu or sprained my ankle and covered myself in mud slipping down the banks of the Solo.

At home, the mainstay of my life, my first reader, and my best friend is my husband, Alan Walker, who only rarely complained about my obsession with a man dead nearly sixty years. Alan provided several astonishing insights into Dubois' character. My cat, Amelia, assisted in much of the typing, and my horse, Wigston Magna, did his best to give me perspective. My editorial colleagues at Penn State—Barbara Kennedy, Nancy Marie Brown, and Gigi Marino—read parts of the manuscript, giggled over my Indonesian adventures, and generally served as stalwart friends. Cheryl Glenn, Bill Earnshaw, Chris Dean, and Claire Van Vliet encouraged me to take on this strange and unusual book. Claire, bookmaker and printer extraordinaire, even offered to design it for me. Marla Caplan gave me the courage to go to Java, without which I could not have understood Dubois' life. At Simon & Schuster, Denise Roy had faith in my ability to write such a book and provided intelligent, gentle editing and occasional cheerleading.

Thank you all, so much. It has been great fun.

Picture Credits

Naturalis, the Nationaal Natuurhistorisch Museum (copyright Nationaal Natuurhistorisch Museum, Leiden, the Netherlands): pages 12, 31, 41, 76, 109, 117, 120, 127, 129, 131, 138, 140, 141, 142, 146, 157, 162, 175, 191 (left), 191 (right), 201, 204, 213, 218, 229, 252, 276, 280, 296, 298, 312, 314, 316, 321, 323, 328, 370, 374, 384, 387, 407, 414 (left), 414 (right), 439, 451; A. Hooijer-Ruben and Nelleke Hooijer: pages 71, 446; Jean M. Dubois: page 157; by permission of the Société d'Anthropologie de Paris: page 281; Anatomische les: G. J. van Rooij/Universiteitmuseum De Agnietenkapel, Amsterdam: pages 44, 48, 358; Natur-Museum Senckenberg: pages 417, 433, 437; Smithsonian Institution Archives, Record Unit 9521, Oral History interviews with Thomas Dale Stewart 1975, 1986 (negative number 4816): page 342; Neg./Transparency number 36106, courtesy the Library, American Museum of Natural History: page 354; Förderverein Ernst-Hackel-Hause V.: page 306; La Famille Lohest: page 55. Pages 10, 61, 81, 234 drawn by Jeff Mathison expressly for this publication.

INDEX

Page numbers in *italics* refer to illustrations.

Aborigines, 167, 350–51, 410, 411
Academy of Natural Sciences
    (Philadelphia, Pa.), 425–27, 436
Adjutant (Dubois' pet stork), 126–27, *127*,
    227
Alcock, A., 229–30, *229*, 233–35, 237, 242,
    246–47, 250, 256–59
allometric studies, 453
American Museum of Natural History,
    354–66, *354*, 449
Anatomischer Anzeiger, 43–46
Ancient Types of Man (Keith), 338–39
Andaman Islanders, 167, 231–32
Andojo (native collector), 418, 420, 421
Andrews, Roy Chapman, 355–56
ANETA, 366, 367, 371
animals, extinct, *see* fauna, ancient
antelopes, 104, 128, 153–54
anthropology:
    evolutionary theory in, *see* evolution
    historical accounts of, 336–40
    "measuring school" of, 281–82
    see also paleontology
anthropometry, 337
Anthropopithecus erectus, 166, 170, 178,
    296
Anthropopithecus javanensis, 144, 160,
    166
Anthropopithecus sivalensis:
    ancient fauna associated with, 62, 63,
        64, 65, 103, 131, 209, 233, 252, 259,
        355
    Dubois' excavations for, 173, 187, 216,
        228, 229, 230, 231, 233–55, 256, 259,
        260, 263, 299
    Dubois' views on, 128, 144, 173, 177,
        209, 210, 257, 259
    location of, 60, 64, 65, 103, 128, 242,
        247
    Lydekker's excavations for, 60, 62, 103,
        128, 139, 143, 144, 166, 170, 173,
        177, 187, 230, 233, 247, 252, 289,
        305
    phylogenetic position of, 60, 144, 209,
        257
    Pithecanthropus erectus compared
        with, 128, 139, 141–44, 166, 170,
        173, 177, 209, 210, 257

apes, 56, 60, 103, 104, 119, 143, 152, 173,
    187, 203–4, *204*, 209, 210, 284, 355,
    405, 415
    see also chimpanzees; gibbons; goril-
        las; orang–utans
Asian zoogeographic region, *61*, 63–64
Atma (native collector), 416–17
Atmowidjojo (native collector), 432–33
Australia:
    Aborigines of, 167, 350–51, 410, 411
    fossil record of, 350–51, 448
    hominid development in, 336, 411
Australian zoogeographic region, *61*,
    63–64
babu, 83–87, 126, 148–49, 193–94, 211,
    212, 262, 273–74

Bali, 63
Batavia, East Indies, 121–23, 125, 134, 216,
    261, 267, 270
Bataviaasch Nieuwsblad, 174–80, 202, 251,
    331–32
Bemuller, Johannes, 295
Bengawan Solo river, 130–32, *138,* 139,
    144, 202, 218, 317, 368, 409
Bernsen, J. J. A., 386–408, *387*
    death of, 405, 406
    diary of, 387–90, 399
    as Dubois' assistant, 386–408
    Dubois' conflicts with, 396–402
    second femur discovered by, 402–8,
        407
Bijlmur, H. J. T., 371, 373
Bischoff, Theodor, 166–67, 203–4
Black, Davidson, 380–86, 393–96, 410, 412,
    413, 417, 424, 438, 453
Boekit Ngalau Sariboe mountain range,
    105, 107–8, 116
Bolk, Louis, 299–300, 303, 357–62, *358,*
    364–65, 372–73
Boschma, H., *439*
Boule, Marcellin, 333–34, 341, 355, 383–84,
    394
Boyd, Adam, 453
Boyd, Alexander, 156, 442
Boyd, Anna Grace Penelope, 146, 156, 206
Boyd, Elena, 156
Boyd, Erroll, 156

Boyd, Grace, 146, 156
Boyd, Janet, 156, 269, 271, 273, 305, 465n
Boyd, Robert:
    background of, 145–46
    coffee plantation of, see Mringin coffee
      plantation
    death of, 3, 430, 442
    Dubois' first meeting with, 145–46,
      150–53
    Dubois' friendship with, 2, 3, 145–46,
      150–59, 185, 189–92, 197, 220, 221,
      222, 270, 305, 338, 441, 442–43
    Dubois' monograph as viewed by, 220,
      221, 222
    Dubois' quest supported by, 152–59,
      161–65, 167, 172–74, 181, 304,
      442–43
    natural history as interest of, 151, 153
    as "Old Warrior," 146, 154, 192, 222, 304
    personality of, 154
    photographs of, 146, 157
    physical appearance of, 146
    Prentice's friendship with, 147, 149–50,
      199, 253, 304
    at Trinil site, 189–92, 199–200, 217, 221,
      268, 429, 441–42
Boyd, Robert (son), 156, 442
Boyd, William, 156, 442, 453
brain, evolution of, 341, 378–80, 453
Branco, Wilhelm, 296, 303
breccia, 419
British Association for the Advancement
    of Science, 288, 349–50
Brongersma, L. D., 408, 422, 439
buffalo, 103, 104, 139, 409
Burger, H., 374

"c" (cephalization coefficient), 311
calvarium, 289, 291–92, 421
Cambridge University, 58
camera lucida, 161, 181, 186
caput humeri, 369–73, 370
Carnegie Institution of Washington, 418,
    427
cat family, 131, 380
cave bears, 56
cell division, 378–80
Centraal Bureau voor Genealogie, 453
Central Asiatic Expedition, 355–56
cephalization, 310–11, 313, 348–49,
    414–15, 453
cerebrum, 380
chimpanzees, 59–60, 62, 142–45, 161–72,

187, 203–7, 204, 303, 307, 427
Siwalik, see Anthropopithecus sivalensis
China:
    fossil remains in, 380–86
    hominid evolution in, see
      Sinanthropus pekinensis
    Japanese invasion of, 449
Chou Kou Tien (Dragon Bone Hill) site,
    381, 382–83, 385, 393–96, 440
Claartje (housemaid), 353, 390
Clara (Anna Dubois' friend), 90–91
climates, ancient, 199, 200, 207
Climates of the Geological Past and Their
    Relation to the Evolution of the Sun,
    The (Dubois), 199, 200, 207
Commission for Natural Sciences, 61–62
Committee for the Promotion of Research
    in the Natural Sciences in the
    Dutch Colonies, 104–5, 173, 174
condyles, 164, 207
Cope, Matthew, 259, 318
cranial vault, 385
Cunningham, Daniel J., 268–69, 288, 289,
    294, 305
Cuypers, Mia, 31–40, 41, 42, 60
Cuypers, Petrus, 31–32, 36, 37, 39

Dames, William, 296, 303
Darwin, Charles, 25, 30, 58
    evolutionary theory of, 9, 18–19, 20,
      59–60, 62, 67, 68, 152–53, 166, 176,
      177, 178, 186, 211, 220, 257, 269,
      279, 306, 378–80, 391, 447
Daum, P. A., 174–80, 251
Dawson, Charles, 340, 417
De Bedelaar, 1–9, 11, 325, 335–36, 353,
    425, 428, 444, 446–47, 446, 449–52
De Bosse, Anna Antoinette, 104
deer, 103, 139, 321
Dehra Doon, India, 235, 247, 249
De Jong, De Josselin, 48
De Mortillet, Gabriel, 287, 298, 303
Department of Education, Religion, and
    Industry, 104, 105–6, 126, 132, 150,
    165
De Preangerbode, 419
De Puydt, Marcel, 54–55
Descent of Man, The (Darwin), 59–60
De Terra, Helmut, 417
De Try, Bastin, 28
De Vries, Hugo, 25, 75, 98, 278, 303
De Winter, Anthonie, 117, 117, 121, 128–38
    Trinil site supervised by, 141–42, 156,

159, 160, 174, 182, 189–91, *191*, 197–98, 200, 208, 218, 267, 303, 315, 330, 420, 423, 430
De Witte, M. G., 206
digastric muscle, 436
dinosaurs, 259
Djetis zone, 427
djongas, 84, 86, 87, 126, *213*, 214
Dryopithecus fontani, 143
Dubois, Anna Geertruida Lojenga:
    at De Bedelaar, 446, *446*
    domestic servants of, 83–87, 126, 267, 273–74
    Dubois' correspondence with, 242–43, 246–49, 250, 254, 255, 258, 260, 261, 262
    Dubois' engagement to, 41–42, 43, 44, 46, 53
    Dubois' Indian trip and, 228, 242–43, 246–49, 250, 254, 255, 258, 260, 261–67
    Dubois' marriage to, 1, 5, 9, 46–47, 49–54, 130, 133, 158, 182, 196–97, 205–6, 211–15, 218, 242–43, 255, 261–67, 271, 272–73, 298–99, 322, 323, 335–36, 352–54, 376, 390
    Dubois' monograph as viewed by, 220
    Dubois' quest supported by, 69–74, 100–101, 109–10, 120–21, 158, 168, 169–70
    in Eijsden, 275–77, *276*
    in The Hague, 277–78, 299
    health of, 77, 192–96
    Hrdlička's meeting with, 342–43
    isolation of, 111–12, 114, 122, 192–93
    mental stability of, 192–96, 211–15, 216
    as mother and wife, 49–54, 70–71, 100–101, 130, 133, 154, 155–58, 171–72, 192–96, 211–15, 242–43, 269–70, 272–74, 326, 349, 352–53
    in Padang, 80–96, *81*
    in Pajakombo, *81,* 96, 97, 100–101, 109–12, 120–21, 122
    personality of, 41–42, 46, 47, 51–54, 123, 298–99
    photographs of, *41, 71, 76, 276*
    physical appearance of, 41, 47, 90
    Prentice's relationship with, 205–6, 222, 243, 246–49, 250, 252–53, 260, 261–67, 270–71, 299, 428
    return to Holland by, 271–77
    social life of, 122–23, 145, 206
    stillborn child of, 192–96, 197, 200, 205,

211–15, 218, 223, 242–43, 269–70, 271, 319
    in Toeloeng Agoeng, *81,* 120–21, 124–27, 171–72, 192–96, 211–15, 242–43, 261–67, 269–70
    transoceanic voyages of, 76–80, *76,* 267, 271–73
Dubois, Anna Jeannette, 192–96, 197, 200, 205, 211–15, 218, 223, 242–43, 269–70, 271, 319
Dubois, Jeannette Gérardine, 11–13, *12*
Dubois, Jean Joseph Balthasar, 11, *12,* 13, 15–19, 23, 69, 72–73, 108–10, 132, 152, 182–85, 196, 268, 322, 325
Dubois, Jean M., 453
Dubois, Jean Marie François:
    Dubois' relationship with, 321–26, 353
    in Java, 121–22, 127, 158, 168, 171, 182, 248, 263, 322–23
    as model for *Pithecanthropus erectus* statue, 321–22, *321*
    in Netherlands, 273–74, 275, *276,* 322–23, *323*
    return to East Indies by, 325–26, 351
    in Sumatra, 96–97, 101
    in U.S., 351–52, 353
Dubois, Marie Antoinette, 11–12, *12,* 23–25, 196, 342
Dubois, Marie Catherine Floriberta Agnes Roebroeck (Trinette), 11, *12,* 13, 72, 108–10, 182, 196, 274, 275–77, *276,* 301, 302, 322
Dubois, Marie Eugène François Thomas:
    abrasive personality of, 11, 28–30, 41, 47, 52–54, 70–71, 83, 102–3, 106, 113–15, 135–38, 205–6, 246–53, 261–67, 270–71, 295, 324–25, 374–77, 380, 389–90, 453
    academic career of, 30–31, 42, 43, 46, 66–69, 72–76, 152, 180–81, 297, 299–303, 319, 320, 325, 328, 335, 374, 375, 380, 391–92, 432
    academic position sought by, 297, 299–303
    in Ambala, *234,* 236–43, 255
    ambition of, 19–20, 22–23, 29–30, 31, 57–58, 64, 93, 102–3, 106, 113–15, 135–38, 152–59, 161–65, 167, 172–74, 181, 298–99, 303–5, 327, 442–44
    in Berlin, 294–96, *296*
    birth of, 9
    in Calcutta, 227–33, 253, 255–60

Dubois, Marie Eugène François Thomas
(cont.)
Catholic background of, 11, 19, 23–29,
32, 42, 159, 342, 373–76, 387, 388,
391, 392, 397, 398, 401, 452
cave collapse avoided by, 113–15, 171
childhood of, 9–21
death of, *451*, 452
De Bedelaar estate of, 1–9, 11, 325,
335–36, 353, 425, 428, 444, 446–47,
446, 449–52
destiny of, 6, 30, 138–39
determination of, 101–3, 107–8,
112–13, 137–38
in Dublin, 288–94
in Edinburgh, 287–88
education of, 13, 18, 19–23, 25, 29,
30–31, 69
eightieth birthday of, 435
in Eijsden, 9–21, *11*, 73, 126, 274,
275–77
extramarital dalliances of, 353, 376–78,
390
family background of, 11–13, 28
as father and husband, 52–54, 74, 133,
192–96, 218, 261–67, 272–73, 274,
301–3, 322–26, 352–54
finances of, 69–70, 105–6, 126, 132,
299, 301–3, 320, 328, 329, 345
"Garuda eyes" of, 102–3, 106, 113, 137,
430
general health of, 77, 80, 110–13, 130,
216
gravesite of, *451*, 452
in The Hague, 277–78, 299
honorary doctorate of (University of
Amsterdam), 297, 299
as Honorary Fellow of the American
Museum of Natural History, 365,
366
Indian journal of, 223, 224–33
isolation of, 7, 99, 134, 158–59,
196–202, 376–78, 391
in Jena, 297
at Lalie Djuvo, 182–85, 200, 217, 441,
450–51
languages spoken by, 77, 186–87, 227,
282
leadership of, 102–3, 106, 113–15,
135–38
as lecturer in anatomy, 42, 43, 46, 47,
54, 69, 74–76
legacy of, 452–54

library of, 79–80, 87, 133–34, 172, 173,
180
in Liège, 285
malaria contracted by, 110–13, 115,
116, 117, 152, 159–60, 171, 205–7,
216, 254, 299, 315, 320, 329, 424
in Marseilles, 267, 273
medicine studied by, 23, 25, 29, 30–31,
69
as military surgeon, 69–70, 74, 77,
88–89, 92–93, 96, 97–99, 106, 111,
150, 172, 187, 216, 278, 329, 338
Order of the Knights of the
Netherlands Lion awarded to, 374
in Padang, 80–96, *81*
in Pajakombo, *81*, 96, 97–121, 122, 134
in Paris, 273–74, 285–87
personal correspondence of, 108–10,
132, 133–34, 167, 168; *see also individual correspondents*
photographs of, *ii, 11, 31, 71, 76, 314,
328, 439, 446*
physical appearance of, 11, 47, 53, 77,
82, 83, 102–3, 123, 227, 282
portrait of, 374, *374*
press coverage of, 174–80, 202, 251,
293–94, 331–32, 345, 360, 363, 366,
367–68, 371–73, 419–21
Prix Broca awarded to, 298–99
as Professor Extraordinarius of
Crystallography, Mineralogy,
Geology, and Palaeontology, 319,
320, 328, 342–43, 374, 375, 380,
391–92
as prosecutor in anatomy, 30–31, 42,
43, 48, 54
religion rejected by, 19, 20–21, 23–29,
46, 387, 389, 391, 392, 452
resignation of, 74–76
retirement of, 374, 380
return to Holland by, 219, 271–77
seventieth birthday of, 374–75, *374*
social life of, 29, 31–42, 122–23, 145,
206
suspiciousness of, 205–6, 217, 222, 243,
246–49, 252–53, 254, 255, 258, 260,
261–67, 270–71, 299, 316–20,
341–46, 390–91, 395, 424–25
in tiger's lair, 107–8, 114, 171
in Toeloeng Agoeng, *81*, 120–21,
124–27, *127, 129*, 130, 153, 158, 160,
171–72, 173, *175*, 205–7, 261–67,
269–71

trans-Indian railway journey of,
     233–37, *234*, 255
transoceanic voyages of, 76–80, *76*,
     267, 271–73
wooden suitcase of, 267, 271, 272–73,
     275–77, 286–87
Dubois, Marie Eugène François Thomas,
     as paleontologist:
anatomical research of, 30–31, 42–48,
     54, 69, 74–76, 99–100, 144, 153–54,
     164, 187, 254, 258, 281–82, 295, 300,
     302, 310–15, 317, 356
assistants of, 359, 374, 376–77,
     386–408, 424–25
Betche-aux-Rotches discovery as influ-
     ence on, 54–58, *55*, 65
bibliography of, 479–91
false leads and setbacks overcome by,
     107–8, 112–17, 120, 128–29
field trips of (Java), 2–3, 6–8, 128–38,
     153–54, 267; *see also* Trinil site
field trips of (Sumatra), 93–121
government support for, 66–68, 95–96,
     98, 104–6, 126, 132, 278, 338–39,
     345, 361, 375
Haeckel's *History* as influence on,
     20–22, 57–58, 152
initial disbelief and discouragement
     encountered by, 64–69, 72–76,
     108–10, 132
intellectual acumen of, 25–26, 29, 47,
     53, 58, 79, 113, 123, 138, 150–53,
     158, 195, 254–60, 389–90, 447–48,
     451–52
notebooks of, x, 56, 62, 79, 99, 103–4,
     159, 218, 267, 286
official correspondence of, 116–20,
     121, 130, 144, 167, 170–71, 411
official reports of, 104, 105, 128, 130–32,
     139–41, 143, 165–71, 179, 180, 203
publications of, 43–49, 59, 94–95, 98,
     105, 199, 200, 207, 216, 327–28,
     347–51, 366, 378–80, 453
publications of, on *Pithecanthropus
     erectus*, 257, 278, 310–15, 328–30,
     331, 333, 351, 362, 364, 366, 373,
     375, 391, 397, 412–15, 419–21, 435,
     446–48, 452–53; *see also* mono-
     graph on *Pithecanthropus erectus*
     (Dubois)
reputation of, 64, 70, 73, 97, 104–10,
     122, 132, 137–38, 171, 174, 182,
     201–2, 220, 232, 254, 268, 271, 294,

298–302, 352, 354–66, 371–76, 397,
     404, 425–27, 436, 441, 452–54
scientific training of, 13, 18, 19–23, 25,
     29, 30–31, 69
species identified by, 104, 143–44,
     170–71, 210–11, 366, 394, 452
Vogt's lectures as influence on, 13–19
Dubois, Marie Eugénie:
birth of, 49–54, 77
Dubois' relationship with, 447, 449, 452
in Java, 121–22, 127, 158, 168, 171, 248,
     263, 322–23
marriage of, 349, 453
in Netherlands, 273–74, 275, *276*, 322,
     323, 435
in Sumatra, 60, 64, 69, 84, 86, 87, 88, 90,
     91–92, 101
Dubois, Victor, 11, 26–30, 255, 300, 301–3
Dubois, Victor Marie:
death of, 352–54
Dubois' relationship with, 322–26,
     352–54
in Java, 133, 158, 168, 171, 182, 263,
     322–23
in Netherlands, 273–74, 275, *276*,
     322–23, *323*
return to East Indies by, 325–26, 351
in U.S., 351–52
Dubois Collection:
analysis and sorting of, *129*, 130,
     153–54, 158, 160–65, 173–74, *175*,
     179–80, 197, 216, 278, 304–5, 315,
     346–47
Bernsen's organization of, 386–408
cataloguing of, 297, 304–5, 327–30,
     346–47, 374, 380, 386–408
Dubois as curator of, 299, 327–31,
     354–66, 386–408, 423–24
Dubois' sequestering of, 179–80, 317,
     319, 327–30, 341–47, 349, 354–66,
     373–76, 393–402, 424–25, 453–54
Hrdlička's attempted inspection of,
     341–46
outside examination of, 284–85,
     289–90, 341–46, 363–64, 424–25
Royal Academy report on (1932), 405–8
storage of, 220, 297, 304–5, 327–30,
     341–47, 386–88, 450
Von Koenigswald's inspection of,
     424–25
see also Trinil site
Dubois' Law, 347–49
Duckworth, W. H. L., 303, *314*, 337, 339–40

dukun, *213,* 214–15
Dutch East India Company, 214
Dutch East Indies, *see* East Indies
Dutch Guiana, 326
Dutch Indies Archaeological Service, 422
Duyfjes, J., 420

East Indies:
    ancient fauna of, 62, 63, 65, 93–94,
        99–104, 115–21, 128, 130–31, 139,
        153–54, 198, 202, 208, 218, 229,
        256–57, 281, 284, 290–91, 330, 409,
        415, 420, 422–23, 426–27
    British rule in, 95, 145
    diseases in, 64, 67, 74, 97, 110–13, 115,
        116, 117, 120; *see also* malaria
    Dubois' arrival in, 69–93
    Dubois' departure from, 267–71
    Dubois' residence in, 60, 64, 69–271,
        338–39
    as Dutch colony, 32, 36, 60, 64, 70, 74,
        81, 122–23, 138, 145, 228, 232,
        244–45
    education in, 70, 244–45
    fossil record of, 61–67, 93–94, 99–104,
        105, 107, 110, 115–16, 118–20, 121,
        128–32, 160, 209; *see also* Dubois
        Collection
    "Indische" in, 5, 83, 90–91, 102, 125,
        353, 458*n*
    "Indos" in, 90, 102, 458*n*
    Japanese invasion of, 453
    Kediri province of, 1, 120, 124
    legal system of, 134–36
    living conditions in, 64, 67, 74, 80–93
    maps of, *61, 67, 81*
    natives of, 81–83, 85–86, 119, 134–36,
        458*n*
    "Pures" in, 90, 91–92, 458*n*
    rainy season in, 85, 88, 118, 120, 144,
        159, 165
    superstitions in, 211–15
    "Totoks" in, 458*n*
    tribes of, 94
    see also Java; Sumatra
Ecole d'Anthropologie, 285, 287, 298–99
Edwards, A. Milne, 279
Elbert, Johannes, 330
elephants, 100, 103, 130–31, 139, 251–52,
        252, 254, 348–49, 356, 369–73, *370*
Elephas, 130–31
embryology, 21, 31, 43–46, 57, 378
Eoanthropus dawsoni, 340–41, 344, 417

Escher, Prof., 390, 399
Evans, John, 305
evolution:
    biological aspect of, 21, 31, 43–46, 57,
        60, 209–10, 378–80, 432, 453
    of brain, 341, 378–80, 453
    cellular mechanism for, 378–80, 453
    Darwin's theory of, 9, 18–19, 20, 59–60,
        62, 67, 68, 152–53, 166, 176, 177,
        178, 186, 211, 220, 257, 269, 279,
        306, 378–80, 391, 447
    embryology and, 21, 31, 43–46, 57, 378
    fossil record of, 9, 54–58, *55,* 60, 61, 65,
        128, 210, 306, 318, 333–34, 336–38,
        340–41, 350–51, 378–79
    geographical distribution in, *61,* 62–64
    geological evidence on, 56, 58, 62
    Haeckel's views on, 20–22, 25, 30,
        57–58, 60, 68, 75, 152, 170, 179, 210,
        257, 279, 297, 303, 305–10
    histories of, 336–40, 354–55, 373–76
    humanity as defined by, 14–17
    "missing link" in, 14–17, 21–22, 55–60,
        61, *61,* 67, 70, 152–53, 174–80, 187,
        204, 209–11, 220, 268–69, 293–96,
        317–19, 336–40, 354–55, 366,
        378–86, 393–96
    natural selection in, 22, 62, 380, 391
    Neanderthal development in, 54–56,
        128, 204, 209, 333–34, 383–84
    per saltum, 378–80
    phylogenetic lineages in, 210–11,
        295–96, *296,* 313, 318, 319, 378–80,
        383, 412, 415, 448
    popular misconceptions about, 13–19,
        67, 68, 152–53, 174–80
    rate of, 209–10, 378–80, 432
    religious opposition to, 152, 309
    scientific opposition to, 56, 57, 176,
        255, 269, 279, 297, 305–10, 309
    Virchow's views on, 56, 57, 204, 257,
        297, 309, 317, 318, 413
    Vogt's lectures on, 13–19
Evolution (MacBride), 373–76
Evolution of the Brain and Intelligence
        (Jerison), 453
Exposition Universelle (1900), 320–22,
        321, 327, 356
eye, diameter of, 348

fauna, ancient:
    of East Indies, 62, 63, 65, 93–94,
        99–104, 115–21, 128, 130–31, 139,

153–54, 198, 202, 208, 218, 219,
     256–57, 281, 284, 290–91, 330, 409,
     415, 420, 422–23, 426–27
  of India, 62, 63, 64, 65, 103, 131, 209,
     233, 252, 259, 355
  of Java, 62, 63, 65, 128, 130–31, 202,
     256–57, 281, 330, 409, 415, 420,
     422–23, 426–27
  of Sumatra, 93–94, 99–104, 115–21
  see also specific animals
femurs, Pithecanthropus erectus:
  casts of, 320, 346, 356–57, 362–66
  comparative analysis of, 161, 166, 167,
     169, 177–78, 268, 279, 280–81,
     284–86, 289–92, 310, 313, 317, 340,
     346, 364, 404, 427, 432
  discovery of, 160–65, 352
  in Dubois' monograph, 171, 173, 186,
     202, 204–5, 207–9, 231, 250, 254,
     269, 281, 285, 289–90
  exostosis of, 161, 164, 205, 208, 297–98,
     308–9, 316
  illustrations of, 161, 162, 181, 186, 407
  second example of, 402–8, 407
  upright posture implied by, 161, 166,
     167, 177–78, 205, 207–9, 281, 288,
     294, 295, 307, 339, 405–6, 408
Finn, Frank, 229, 229
flogging, 134–36, 137
Flower, W. H., 279, 289, 305
fluorine, 284
foramen magnum, 295, 416
Fort Van Den Bosch, 126
Fort William, 227–28
fossils:
  in Australia, 350–51, 448
  in China, 380–86
  condition of, 99–100, 134
  in East Indies, 61–67, 93–94, 99–104,
     105, 107, 110, 115–16, 118–20, 121,
     128–32, 160, 209; see also Dubois
     Collection
  evolutionary record of, 9, 54–58, 55, 60,
     61, 65, 128, 210, 306, 318, 333–34,
     336–38, 340–41, 350–51, 378–79
  in India, 60, 61, 62, 65, 128, 187, 355
  in Java, 61, 62, 65, 105, 118–20, 121,
     128–32, 209
  from Lida Adjer cave, 100–104, 105,
     115–16, 119, 120, 121
  on "missing link," 22, 64, 65–66, 119
  Neanderthal, 9, 54–56, 55, 65, 119, 177,
     204–5, 333–34, 338, 351, 383, 393

  of Sinanthropus pekinensis, 412–15,
     414, 419, 449
  in Sumatra, 62–64, 93–94, 99–104, 105,
     107, 110, 115–16, 119, 120, 121, 160
  from Trinil site, see Trinil site
Fraipont, Julien, 54, 173, 342
Fuhlrott, Johannes, 55–56
Fürbringer, Max, 25, 278, 299, 301, 303, 310
  Dubois' troubled relationship with, 30,
     42, 44–49, 54, 56, 57, 58, 68–69,
     72–76, 152, 311, 338
  photograph of, 44

Galloway (Prentice's friend), 223–24, 226,
     261
Gamble (Alcock's friend), 235, 237, 247–48,
     251, 254
Garson, George, 289, 290–91
Garuda (eagle-eyed god), 102–3, 106, 113,
     137, 430
Gaudry, Albert, 334
gaur (wild cattle), 209
geckos, 87–88, 89
Geographical Distribution of Animals, The
     (Wallace), 62–63
Geographical Society of China, 385
Geological Museum (Leiden), 328
Geological Society of London, 340
Geological Survey of India, 231, 243
Geological Survey of Java, 408–9, 410, 415,
     417, 418
gibbons, 60, 103
  Pithecanthropus erectus compared
     with, 143, 144–45, 160, 168, 172, 187,
     203, 204, 208, 210, 281, 284, 291,
     295, 296, 307, 308, 318, 395, 396,
     414–15, 419, 427, 434, 436, 443, 454
gill cartilage, 43–46
Giring-Giring, East Indies, 109
glaciers, 366
Globus, 268
gorillas, 59–60, 119, 143, 166, 187, 204,
     307, 351
Gorjanović-Kramberger, Dragutin, 333
Grand Hotel (Calcutta), 228–29
Grand Trunk Road, 233
Great Chain of Being, 58
Griesbach, C. L., 231, 233, 237–40, 241,
     242, 246, 249–50, 254, 255, 256, 259
Groeneveldt, Willem, 104, 121, 126, 128,
     130, 132, 139, 143, 165–66, 167, 170,
     171, 303
Gunung Ardjoena mountain, 184–85, 200,

Gunung Ardjoena mountain *(cont.)*
217, 441, 450–51
gyri, 295

Haddon, Alfred, 336–37
Haeckel, Ernst:
    Dubois' relationship with, 297, 305–10,
        341, 391
    evolutionary views of, 20–22, 25, 30,
        57–58, 60, 68, 75, 152, 170, 179, 210,
        257, 279, 297, 303, 305–10
    Pithecanthropus erectus lecture of,
        305–10
    portrait of, *306*
    Virchow vs., 57, 297, 305–10, 317
Haripur, India, 252, 254–55
Heberlein, C. E. J., *Pithecanthropus erectus*
    "skull" found by, 366–73, *370*, 411
Het Algemeen Indische Dagblad, 418
Het Algemeen Indische Handelsblad, 345,
    360, 367, 371, 418–19, 422
High Javanese language, 78
hippopotamus, 131
History of Creation (Haeckel), 20–22,
    57–58, 68, 152
Hitler, Adolf, 440
Holocene age, 423
Hominidae, 210, 381
Homo diluvii testis, 336–37
Homo erectus, 452
Homo heidelbergensis, 338, 412
Homo javanensis, 339
Homo modjokertensis, 416–27
    child's skull of (Sangiran Skull I),
        416–27, *417*, 449
    Dubois' views on, 415–27
    Neanderthal man compared with, 420,
        427
Homo primigenius, 319, 333
Homo sapiens:
    articulate speech of, 22, 31, 43–46, 57
    evolutionary definition of, 14–17
    origins of, 59–60, 62–64, 180–81, 267,
        296, 354, 355, 381, 383, 412, 415
    phylogenetic position of, *296*, 383
    skull of, 168, 169, 172, 204, *204*, 291
    upright posture of, 22, 58, 60, 166,
        207–9
Homo sapiens neandertalensis, *see*
    Neanderthal man
Homo soloensis, 408–16
    casts of, 410
    cranial capacity of, 410, 411

Dubois' views on, 409–15
full designation of, 410
Homo wadjakensis compared with,
    410, 411, 412, 447
Neanderthal man compared with, 410,
    411, 412, 415
Ngandong site of, 408–12
phylogenetic position of, 412, 415
Pithecanthropus erectus compared
    with, 409–16
Sinanthropus pekinensis compared
    with, 399, 410, 412
Skull I of, 409, 410, 412
Skull II of, 409, 410
Skull III of, 409, 410
Skull IV of, 409, 411, 415–16, 447
Skull V of, 409, 411
Skull VI of, 409
Skull VII of, 409
Skull VIII of, 409, 416
Skull IX of, 409
Skull X of, 409
Skull XI of, 409, 416
Homo spec. indet., 132
Homo wadjakensis, 118–20
    Dubois' views on, 118–20, 123, 133,
        303, 336, 350–51, 410, 412, 447
    Homo soloensis compared with, 410,
        411, 412, 447
    phylogenetic position of, 350–51, 410,
        419, 420
    Pithecanthropus erectus compared
        with, 131, 350–51, 410, 412, 447
    skulls of, 118–29, *120*, 123, 128, 336, 356
Hooijer, Carel, 349, 453
Hooijer, Marie Eugénie Dubois, *see*
    Dubois, Marie Eugénie
horses, 56
Hrdlička, Aleš, 341–46, *342*, 357, 360,
    363–64, 394
Hubrecht, A. A. W., 295
Huxley, Thomas, 20, 22, 25, 30, 55–56, 59,
    152, 204
hyenas, 131
Hylobates genus, *see* gibbons

ice age, *see* Pleistocene age
Illustrated London News, 367–68, 436
India:
    ancient fauna of, 62, 63, 64, 65, 103,
        131, 209, 233, 252, 259, 355
    British control of, 60, 64, 227–28,
        232–33, 245, 260

Dubois' visit to, 173, 187, 216, 219, 220, 223–61, *229, 234, 252,* 263, 299
    fossil record of, 60, 61, 62, 65, 128, 187, 355
    Siwalik Hills discoveries in, *see Anthropopithecus sivalensis*
Indian Museum, 228, 229–31, 243, 260
Industrial Arts Museum, 322
Institut International d'Anthropologie, 386
International Congress of Zoology (1895), 278–85, *280,* 306–7, 308
International Congress of Zoology (1898), 305–16, *312, 314,* 338, 339, 373–74

Janse, Jacobus, *48*
Java:
    ancient fauna of, 62, 63, 65, 128, 130–31, 202, 256–57, 281, 330, 409, 415, 420, 422–23, 426–27
    British control of, 95
    caves of, 128–29, 132, 138
    coffee plantations in, 144–47, 155–59, 161–65, 188–89, 192, 197, 206, 219, 220, 253, 305, 326, 417, 442; *see also* Mringin coffee plantation
    coolie labor in, 8, 128–38, 159, 160, 174, 190, 192, 198, 218, 254, 303, 315, 355, 423–24, 430, 458*n*
    Dubois' explorations of, 2–3, 6–8, 128–38, 153–54, 267; *see also* Trinil site
    Dutch engineers in, 121, 128–38; *see also* Trinil site, engineers used for
    fossil record of, 61, 62, 65, 105, 118–20, 121, 128–32, 209
    Kediri province of, 145, 150, 176
    maps of, *61,* 62, 118, 120, *140, 141*
    Ngandong site in, 408–12
    Ngawi region of, 139
    riverbeds of, 128–32, 138–39, *138,* 159–60, 165
    Sangiran site in, 416–27, 432–40, *433, 437,* 447–48, 449, 450, 453
    topography of, 129, 236
    Trinil site in, *see* Trinil site
Java man, *see Pithecanthropus erectus*
Jentink, F. A., 116–17, 165, 284
Jerison, Harry, 453
Junghuhn, Franz, 61–62, 69, 118, 120, 122

Kalimantan, 63
Kedoeng Broebus site, 128, 131–32, *131,* 133, 202–3, 362, 394–95

Kedoeng Loeboe site, 128
Keith, Arthur, 289, 292–93, *292,* 305, *314,* 338–39, 341
Kendeng Hills, 128, 408–9
khitmagar, 238–39, 242, 249, 251
Klaatsch, Herman, 295
kokkie, 84, 86, 92, 126, 127, 191, 195
Krause, Wilhelm, 279, 295
Kriele, Gerardus, 117, *117,* 121, 128–38
    Trinil site supervised by, 141–42, 156, 159, 160, 174, 182, 189–91, *191,* 197–98, 200, 208, 218, 267, 303, 315, *316,* 330, 408, 420, 423, 430
Kroesen, R. C., 95–96, 98, 100, 104, 121, 144, 165, 303
Krull, Hendrik, 84–85

"L" (length of body), 348
Lamarck, Jean-Baptiste de Monet de, 166, 177
Lankester, E. Ray, 305
larynx:
    as developed from fourth branchial arch, 43–46
    differentiation of, 22, 31, 43–49, 58
    Dubois' research on, 43–49, 56, 57, 72–73, 75–76
    of whales, 48–49
lemurs, 22
Lida Adjer cave, 100–104, 105, 115–16, 119, 120, 121
ligamentum teres, 164
Lilik (ladies' maid), 87, 97, 126
Limburg region, 9–11, *10,* 13, 16, 26, 42
Linnaean Society, 62
Lohest, Max, 54–55, 173
Lombok Strait, 63–64
Lorentz, Hendrik Antoon, 357
Lorié, Jan, 328–29
Lubbock, John, 289, 305
Lydekker, Richard, 60, 62, 103, 128, 139, 143, 144, 166, 170, 173, 177, 187, 230, 233, 247, 252, 289, 305
Lyon (Prentice's friend), 223–24, 226, 261

MacBride, D. W., 373–76
McGregor, J. H., 356, 364
MacLennan, Doortje, 148
MacLennan, Isabella Boyd, 147, 156
MacLennan, Theo, 147, 156
Maidan (Calcutta), 227–28
Major, John Forsythe, 305
Malang, East Indies, 149, 151

malaria, 67, 97, 110–13, 115, 116, 117, 120,
    152, 159–60, 171, 205–7, 216, 254,
    299, 315, 320, 329, 424
Malay language, 77–78, 83–84, 86, 92, 369
    glossary of, 476–78
mammals:
    body weight vs. brain size of, 310–15,
        328, 340, 347–49, 351, 364, 366, 404,
        414–15, 453
    marsupial vs. placental, 63–64
    metabolism of, 311
    nervous system of, 310–11
mammoth, 56
Manchester Guardian, 382
mandibles, Pithecanthropus erectus:
    comparative analysis of, 131–32,
        202–3, 362, 394–95
    discovery of, 131–32, 431–32
    illustration of, 131
    from Kedoeng Broebus site, 128,
        131–32, 131, 133, 202–3, 362,
        394–95
    Von Koenigswald's example of, 431–32,
        435, 436, 447
mandur, 101, 102, 108
Manis palaejavanica, 366
Manouvrier, Léonce-Pierre, 281–82, 281,
    285–87, 289, 298, 303, 310, 334, 341,
    342
Man's Place in Nature (Huxley), 20, 55–56
Mardall, Captain, 239, 241–42
Marsh, Othniel, 258–60, 279, 284, 289, 303,
    309, 315
Martin, Karl, 62, 65, 105, 109–10, 328–29
Martin, Rudolf, 268, 289, 294–95
Marx, Karl, 391
Matschie, Paul, 250–51, 254–55, 289
Max, Gabriel, 306
Men of the Old Stone Age (Osborn), 354–55
Merriam, John C., 418, 425, 427
Mijnbouwkundige te Delft, 422–24
Mijningingenieur, 411
Mijsberg, W. A., 371, 373, 411
Minangkabau tribe, 94
Ministry of the Colonies (Netherlands),
    328–29, 345
Miocene age, 209
"missing link":
    concept of, 58
    Dubois' search for, 21–23, 30, 57–58,
        64–76, 100, 104–6, 109–10, 116–17,
        122, 128, 131, 139, 152–67, 172–74,
        180–81, 218, 232, 275–77, 304,
        442–43

Dubois' theories on, 59–66, 67, 68, 93,
        94–95, 150–53
    evolutionary position of, 14–17, 21–22,
        55–60, 61, 61, 67, 70, 152–53,
        174–80, 187, 204, 209–11, 220,
        268–69, 293–96, 317–19, 336–40,
        354–55, 366, 378–86, 393–96
    fossil record on, 22, 64, 65–66, 119
    geographical location of, 59–64
    Pithecanthropus erectus as, 57–58, 144,
        159–65, 179, 180–81, 187, 209–11,
        220, 222, 254, 256–60, 275–77, 278,
        283–84, 293–94, 305–10, 312–14,
        327, 340, 408, 414, 422, 427, 445–48,
        450, 452
    research on, 59–64, 79–80
Modjokerto skull, see Homo modjokerten-
    sis
molars, Pithecanthropus erectus:
    casts of, 320, 346, 356–57, 362–66
    comparative analysis of, 143, 160–61,
        163, 165, 169, 176–78, 279, 297, 307,
        332, 346, 404, 436
    discovery of, 139–41, 142, 163, 283,
        314, 352
    in Dubois' monograph, 186, 202, 207,
        210, 231, 250, 268, 283, 288
    illustration of, 161, 162, 181, 186
    left pre–, 314, 332, 346, 436
    second, 283, 288, 307, 332, 346
    third, 139–41, 142, 143, 160–61, 162,
        163, 165, 169, 176–78, 186, 202, 207,
        210, 231, 250, 268, 279, 283, 288,
        297, 307, 332, 346, 352, 404, 407
monkeys, 22, 56, 128, 348, 394, 434
monograph on Pithecanthropus erectus
    (Dubois), 277–96
    Boyd's views on, 220, 221, 222
    Dubois' defense of, 277–96, 302,
        318–19, 334, 452
    femur described in, 171, 173, 186, 202,
        204–5, 207–9, 231, 250, 254, 269,
        281, 285, 289–90
    German title of, 480
    illustrations in, 161, 162, 181, 186, 204
    molar described in, 186, 202, 207, 210,
        231, 250, 268, 283, 288
    Prentice's views on, 220, 303–5, 327
    publication of, 181, 187, 188, 199, 216,
        218, 219, 220–22
    reviews on, 230–31, 250–51, 254–58,
        268–69
    skull described in, 280–82, 283, 285,
        288, 289, 290, 291–96, 318

"*thöricht*" used in, 202, 268, 279
writing of, 171–74, 180, 181, 185,
  186–87, 188, 196, 198–99, 202–11
Most Ancient Skeletal Remains of Man, The
  (Hrdlicka), 346
Mount Willis, 120, 123, 146
Mringin coffee plantation, 2, 3–4, 146–47,
  155–59, *157*, 161–65, 188, 189, 197,
  199, 205, 206, 220–23, 266–70, 274,
  304, 429, 441–43
Munro, Robert, 288
Musée National d'Histoire Naturelle,
  333–34

Nahan, Maharajah of, 241, 249–50, 251,
  252, *252*, 254
Nassi (djongas), *213*, 214
natural selection, 22, 62, 380, 391
Nature, 230–31, 232, 250, 254, 256, 268,
  411, 445
Naturwissenschaftliche Wochenschrift,
  250–51
*Natuurkundig Tijdschrift voor
  Nederlandsch-Indië*, 94–95, 199
Natuurwetenschappelijke Raad van
  Nederlandsch-Indië, 435
Neanderthal man:
  at Betche-aux-Rotches (Spy, Belgium),
    55–58, *55*, 173, 268, 283, 285, 309,
    342
  Boule's monograph on, 333–34, 383–84
  evolutionary development of, 54–56,
    128, 204, 209, 333–34, 383–84
  femur of, 204–5
  fossil remains of, 9, 54–56, *55*, 65, 119,
    177, 204–5, 333–34, 338, 351, 383,
    393
  on Gibraltar, 55, 65
  Homo modjokertensis compared with,
    420, 427
  Homo soloensis compared with, 410,
    411, 412, 415
  at Krapina, 333
  at La Chapelle-aux-Saints, 333–34, 383
  at La Quina, 333
  at Le Moustier, 333
  in Neander Valley, 55, 65
  Pithecanthropus erectus compared
    with, 57, 59, 131, 143, 204–5, 257,
    258–59, 268, 283, 285, 288, 289, 290,
    309, 318, 334, 339, 341, 355, 383,
    393–94, 420, 427
  Sinanthropus pekinensis compared
    with, 382–86, 393–94, 413

skulls of, 55–56, *55*, 119, 204, 351, 383,
  393
Neander Valley (Germany), 55, 65
Nederlandsch Koloniaal Centraalblad, 268
nerves, motor vs. sensory, 311
Netherlands Society for Nature
  Conservancy, 335
neurons, 378–80
Ngrodjo (Boyd's house), *see* Mringin coffee
  plantation
Nightingale, Florence, 33
njonja, 85
"Nota Paleontologica" (Dubois), 56
nyai, 5, 90–91, 146, 156, 157–58, 377, 453

"Oak ape," 143
occipital torus, 205
Oerder, Frans David, 374, *374*
"O'Mulligan," 293–94
"On the Gibbonlike Appearance of
  Pithecanthropus erectus" (Dubois),
  414–15, 419
Oppenoorth, W. F. F., 330, 409–12, 415, 425,
  427
optic nerve, 348
orang-utans, 60, 103, 143, 160, 166, 187,
  207, 307, 340, 432–33
Origin of Species, The (Darwin), 9, 20, 62,
  68
Osborn, Henry Fairfield, 303, 354–66, *354*,
  367, 371–72
Oudemans, Johannes, *48*

"P" (surface area of body), 311
paleontology, 58, 59, 318, 333–34, 356,
  378–79, 450
"paleothermal problem," 366
pangolins, 366
patjol, 135, 160
Pavillon des Indes Néerlandaises, 320–22,
  327
Pei, W. C., 381, 438
Peking man, *see Sinanthropus pekinensis*
"Peking Man—An Undamaged Skull, The,"
  382
Peking Union Medical College, 381
Pettit, August, 287, 289, 298
phylogenetic lineages, 210–11, 295–96,
  296, 313, 318, 319, 378–80, 383, 412,
  415, 448
pigs, wild, 112, 128, 131, 153
Piltdown man, 340–41, 344, 417
Pithecanthropi, 22
Pithecanthropidae, 210

"*Pithecanthropus* Child's Skull," 418–19

Pithecanthropus erectus:

    Anthropopithecus sivalensis compared with, 128, 139, 141–44, 166, 170, 173, 177, 209, 210, 257

    articulate speech of, 57, 58, 170

    body weight vs. brain size of, 310–15, 340, 404, 414–15

    casts of, 295, 316–17, 320, 346, 356–57, 362–66, 412–15, *414,* 425

    competitive claims on, 173–74

    Cunningham's views on, 268–69, 288, 294

    discovery of, 2, 6, 7, 131–32, 138–45, 159–65, 180–81, 187

    Dubois' analysis of, 139–45, 165–74, 180–81, 187, 202–11, 231, 250–51, 277–96, 302, 318–19, 334, 452–54

    Dubois' defense of, 230–32, 334, 452

    Dubois' monograph on, *see* monograph on *Pithecanthropus erectus* (Dubois)

    Dubois' official report on, 165–71, 179, 180, 203

    Dubois' publications on (non-monographic), 257, 278, 310–15, 333, 351, 362, 364, 366, 373, 375, 391, 397, 412–15, 419–21, 435, 446–48, 452–53

    Dubois' silence on, 327–33

    Dubois' tombstone representation of, *451,* 452

    femurs of, *see* femurs, *Pithecanthropus erectus*

    fossilization of, 144, 203, 280–81, 284, 295, 312, *312,* 332–33, 366–68

    Garson's views on, 290–91

    gender of, 165, 205

    gibbonlike appearance of, 143, 144–45, 160, 168, 172, 187, 203, *204,* 208, 210, 281, 284, 291, 295, 296, 307, 308, 318, 395, 396, 414–15, 419, 427, 434, 436, 443, 454

    Haeckel's views on, 305–10

    hominid character of, 131–32, 166, 279, 288, 289–90, 293, 395, 427, 434–35, 445–48, 452

    as *Homo erectus,* 452

    Homo modjokertensis compared with, 416–27

    Homo soloensis compared with, 409–16

    Homo wadjakensis compared with, 131, 350–51, 410, 412, 447

    illustrations of, *x, xii, 131, 142,* 161,

    162, 174, 181, 186, *204,* 216, *298, 306, 312, 321,* 346, 366, *407, 414*

    Krause's views on, 279, 295

    Lydekker's views on, 254–58, 259

    mandibles of, *see* mandibles, *Pithecanthropus erectus*

    Marsh's views on, 258–60, 284

    Martin's views on, 268

    Matschie's views on, 250–51, 254–55, 259

    as "missing link," 57–58, 144, 159–65, 179, 180–81, 187, 209–11, 220, 222, 254, 256–60, 275–77, 278, 283–84, 293–94, 305–10, 312–14, 327, 340, 408, 414, 422, 427, 445–48, 450, 452

    molars of, *see* molars, *Pithecanthropus erectus*

    morphology of, 202–11, 414–15, 447

    name of, 144, 166, 170–71, 172, 231, 339, 405

    Nature announcement on, 230–31, 232, 250

    Neanderthal man compared with, 57, 59, 131, 143, 204–5, 257, 258–59, 268, 283, 285, 288, 289, 290, 309, 318, 334, 339, 341, 355, 383, 393–94, 420, 427

    as new species, 210–11, 394, 452

    phylogenetic position of, 210–11, 295–96, *296,* 313, 318, 319, 378–80, 383, 412, 415, 448

    reconstructions of, 287, *298,* 394–95, 434, 436–39, *437*

    Schwalbe's article on, 316–20, 333, 341

    scientific opposition to, 180–81, 220, 250–51, 254–60, 268–69, 277–98, 299, 305–16, 319–20, 327, 334, 337, 354, 392, 404, 422–24, 452–53

    scientific support for, 258–60, 279, *281,* 284, 285–87, 289, 292–93, 296–97, 303, 305–10, 337, 338–40

    Sinanthropus pekinensis compared with, 380–86, *384,* 393–96, 412–15, *414,* 419, 427, 435, 438, 440, 445–48, 450, 453

    single individual represented by, 202, 203, 207, 250, 257, 268, 269, 279, 291–92, 296, 304, 402–8, *407,* 427

    skulls of, *see* skulls, *Pithecanthropus erectus*

    statue of ("Piet"), 320–22, *321,* 327, 356

    Ten Kate's views on, 268

    Thomson's views on, 291–92

as transitional form, 163–68, 173, 187, 203–11, 254, 268–69, 281–82, 283, 288–96, 304–13, 318–22, 337, 340, 347, 363, 422, 432

Turner's views on, 287–90

upright posture of, 58, 161, 166, 167, 177–78, 205, 207–9, 281, 288, 294, 295, 307, 339, 405–6, 408

Virchow's views on, *278*, 279, 282–85, 293, 294, 295, 296, 305–10, 318, 404, 414

Von Koenigswald's views on, 415–28, *417*, 431–40, 443–50, 453

Pithecanthropus erectus. Eine menschenaehnliche Uebergangsform aus Java (Dubois), *see* monograph on *Pithecanthropus erectus* (Dubois)

Place, Thomas, 25, 30, *278*, 299–301, 303, 389

plasticine, 437–38

Pleistocene age, 56, 62, 64, 100, 103, 131, 143, 166, 202, 209, 210, 281, 284, 350, 356, 381, 420, 423, 426

Pliocene age, 131, 143, 209, 281, 290, 293, 306, 335, 339, 340, 423, 426

Pliopithecus antiquus, 143

point bars, 129–30, 139

Pollak, Dr., 97–99, 111, 112, 115, 134

porcupines, 100, 128

Prehistoric Man (Duckworth), 339–40

"Preliminary Note on additional *Sinanthropus* Material Discovered in Chou Kou Tien during 1928" (Black), 382

Prentice, Adam:
    Anna Dubois' relationship with, 205–6, 222, 243, 246–49, 250, 252–53, 260, 261–67, 270–71, 299, 428
    background of, 146–50
    Boyd's friendship with, 147, 149–50, 199, 253, 304
    coffee plantation owned by, 3, 220, 253, 305
    death of, 451, 453
    Dubois' correspondence with, 1–9, 182–83, 186–89, 197, 199–200, 206–7, 219–24, 252–53, 260, 261, 262, 270–71, 303–5, 428–30, 441–44
    Dubois' final meeting with, 261–67
    Dubois' first meeting with, 145, 150–53
    Dubois' friendship with, 1–7, 145, 150–59, 182–92, 197, 199–200, 205–7, 216–24, 243, 246–49, 252–53,

261–68, 270–71, 303–5, 327, 338, 428–30, 441–44, 450–51

    Dubois' Indian trip and, 223–24, 226, 261
    Dubois' monograph as viewed by, 220, 303–5, 327
    Dubois' quest supported by, 152–59, 161–65, 167, 172–74, 191, 303–5, 327, 442–44
    Dubois' telegram to, 243, 246–47, 250, 463*n*
    as estate manager, 146–47, 149–50, 155–59, 161–65, 188–89, 192, 197, 206, 219, 220
    natural history as interest of, 151, 163
    personality of, 145, 147, 154–55, 253
    physical appearance of, 145, 146–47, 217
    Tempoersarie house of, 147–48, 149
    at Trinil site, 189–92, 199–200, 217, 221, 268, 429, 441–42
    as widower, 148–51, 155, 158, 171–72

Prentice, Gerard (Gerardus Prentice-MacLennan), 148–49, 150, 151, 158, 186, 188, 206–7, 216, 221, 305, 453

Prentice, Jane de Clonie MacLennan, 147–50, 158

Primates, 22

Prix Hollandais, 386, 412

Proceedings Koninklijke Akademie van Wetenschappen, 414–15, 419

Prothylobates, 210

psychoencephalon, 380

Pycraft, William, 305

Quaternary period, 294

Quatorzième Conférence Annuelle Transformiste de la Société d'Anthropologie, 287

Raffles, Stamford, 145

reindeer, 56

"Relationship between *Pithecanthropus* and *Sinanthropus*, The" (Von Koenigswald and Weidenreich), 445–48

Renan, Joseph-Ernest, 2

R. F. Damon & Co., 365–66

rhinoceros, 56, 100, 103, 104, 128, 131

rickets, 56

Rijksmuseum, 31–32

Rockefeller Foundation, 381

Royal Academy of Science, 350, 357–63, 367, 372–73, 397, 405–8, 412, 419,

Royal Academy of Science, *(cont.)*
435, 436–39, 446, 447, 450
Royal Dublin Society, 268–69, 288–94
Royal Dutch East Indies Army, 61, 69–70
Royal Society, 288, 305, 386
Royal Zoological Society Natura Artis
Magistra, 48

Sahul Shelf, 63–64
Saleh, Raden, 61, 62, 105, 118, 128, 328, 416
Samira, Raden Roro, 146, 156
Sanders, M., 408
Sanikem (nursemaid), 87, 88, 97, 101
Santos skull, 309
Schaaffhausen, Hermann, 56
Schreuder, Antje, 359, 374, 376–77,
389–91, 399
Schrijnen, Pierre, 300–301, 303
Schwalbe, Gustav, 297–98, 316–20, 333,
341, 383
Sedgwick, Adam, 305
Selenka, Emil, 329–30
Selenka, Margarethe Lenore, 330–33, 369,
409, 410, 426–27
Semarang, East Indies, 128, 216–17
sensory-motor units, 348–49
Seydel, Otto, 299–303
Shah Jahan, 255
Shakespeare, William, 51
"Shape and Size of the Brain in
Sinanthropus and
Pithecanthropus" (Dubois), 412–15,
419
siamangs, 204
Sikhs, 236, 241
Simiidae, 210
Simons Artz Emporium, 79
Sinanthropus pekinensis, 380–86
casts of, 412–15, *414*, 449
Chou Kou Tien (Dragon Bone Hill) site
of, 381, 382–83, 385, 393–96, 440
cranial capacity of, 412–15, *414*, 419
discovery of, 380–86, 440
fossil remains of, 412–15, *414*, 419, 449
hominid character of, 412–15, 440,
445–48
Homo soloensis compared with, 399,
410, 412
Neanderthal man compared with,
382–86, 393–94, 413
photographs of, *384*, 413–14, *414*, 449
phylogenetic position of, 415
Pithecanthropus erectus compared

with, 380–86, *384*, 393–96, 412–15,
414, 419, 427, 435, 438, 440, 445–48,
450, 453
Singapore, 224, 225–26, 261
Sinha, Sukh Chain, 251, 254, 258
Sirmoor State, *234*, 241, 243–54, *252*
Siwalik Hills fossils, *see Anthropopithecus
sivalensis*
Skeletal Remains of Early Man, The
(Hrdlicka), 363–64
skulls, *Pithecanthropus erectus*:
calvarium of, 289, 291–92, 421
casts of, 295, 316–17, 320, 346, 356–57,
362–66, 412–15, *414*, 425
comparative analysis of, 144–45,
160–72, 176–78, 202–7, 210, 211,
231, 250, 254, 257, 268–69, 283–90,
297, 303, 307–18, 332, 338, 346, 385,
393–96, 404–15, 425, 427, 435
cranial capacity of, 166–67, 203–4, *204*,
210, 268–69, 281–82, 288, 291,
292–93, 307, 308, 310–15, 339, 340,
363, 364, 412–15, *414*, 419, 438
discovery of, 141–45, 352
in Dubois' monograph, 280–82, 283,
285, 288, 289, 290, 291–96, 318
Dubois' reconstruction of, 287, *298*,
394–95, 434
foramen magnum of, 295, 416
gyri and sulci of, 295
Heberlein's "skull" as, 366–73, *370*, 411
Homo soloensis skulls vs., 408–16, 447
illustrations of, *142, 204, 298, 306, 370,
407, 414, 433*, 434, 435, 436–39, *437*
matrix in, 144, 203, 280–81, 284, 295,
312, *312*, 332–33, 366–68
Sangiran Skull I (*Homo
modjokertensis*) vs., 416–27, *417*,
449
Sangiran Skull II vs., 432–40, *433, 437*,
447–48, 449, 450, 453
temporal fossa of, 309
Sluiter, C. P., 118–19, 122, 133–34, 303, 336
Smith, Grafton Elliot, 305, *314*, 341, 350,
367, 371–72, 381, 386, 394
Smith, Stewart Arthur, 350, 448
Smithsonian Institution, 341, 363
Smith Woodward, Arthur, 305, 340, 341
Snell, Otto, 310, 311
social Darwinism, 391
Société Belge de Géologie, 285
Sollas, William, 303
Spice Islands, 122

S.S. *Prinses Amalia,* 76–80, *76*
State HBS, 19–23
Stegodon, 130–31, 139, 369–73, *370*
stellar evolution, 366
Straké, Louis, *48*
Suez Canal, 79, 80
Sumatra:
    ancient fauna of, 93–94, 99–104,
        115–21
    caves of, 94, 98–108, 112–21
    coolie labor in, 95–117, *117,* 458*n*
    Dubois' explorations of, 93–121
    Dutch engineers in, 105, 106, 116, 117,
        117, 121
    fossil record of, 62–64, 93–94, 99–104,
        105, 107, 110, 115–16, 119, 120, 121,
        160
    Lida Adjer cave in, 100–104, 105,
        115–16, 119, 120, 121
    maps of, 81, 105, 118
Sunda Shelf, 63–64
"survival of the fittest," 22, 62, 380, 391
susa, 369
Swan Sonnenschine & Co., 199
syce, 126, 239

Tablet, 374–75
Taen-Err-Toung, F. G., 35–40, 42, 135
Taj Mahal, 255
Talgai skull, 350–51, 448
Tanjung Priok, East Indies, 121–22
tapirs, 103, 104
Tasmanians, 350
Tegelen Clays, 335, 387, 396–97
Teilhard de Chardin, Pierre, 417–18, 440
temporal fossa, 309
Ten Kate, Herman, 268
Ter Haar, C., 408–9, 415–16
Tertiary period, 22, 293, 294, 355, 381
Teyler Museum, 299, 336, 363, 375–76
    Dubois Collection at, *see* Dubois
        Collection
Tez (Dubois' pony), 243, 246, 249
Thompson, D'Arcy, 305
Thomson, John Arthur, 289, 291–92
thyroid cartilage, 43–46
tigers, 100, 103, 107–8, 112, 114, 117, 171
Timmerman, J. A. C. A., 180–81
tools, stone, 55, 208, 334
Trinil site, 138–45, 159–65
    ancient fauna of, 139–41, 153–54, 198,
        202, 209, 218, 229, 284, 290–91,
        333–34, 369–73, 395, 396–97

Congress of Zoology's resolution on,
        315–16
    coolies used for, 8, 159, 160, 174, 190,
        192, 198, 218, 254, 303, 315, 355,
        423–24, 430
    Dubois' collection from, *see* Dubois
        Collection
    Dubois' excavations of, *xii,* 2, 8, 138–45,
        138, 140, 141, 142, 156, 159–65, *162,*
        173, 174, 182, 189–92, *191,* 197–202,
        208, 218, *218,* 254, 267, 280–81, *280,*
        284, 303, 304–5, 315, *316,* 330, 355,
        408, 420, 422–24, 430
    Dutch government's excavation of,
        314, 315–16, 329, 330
    engineers used for, 8, 138–42, 156, 159,
        160, 174, 182, 189–91, *191,* 197–98,
        200, 208, 218, 267, 303, 315, *316,*
        330, 355, 408, 420, 423–24, 430
    find-spots for, 118, 128–29, 138–39, *280*
    geographical location of, *81,* 138–39,
        138, 201, 202, 409–10
    geological strata of, 159, 160, 164, 165,
        202, 209, 218, 280–81, *280,* 284,
        288–91, 319–20, 331, 332–33, 339,
        368, 369, 418–19, 421, 422, 426, 431
    Heberlein's "skull" from, 366–73, *370*
    maps of, *vi,* 130, *140, 141,* 218
    Modjokerto site compared with, 419–27
    Prentice's visit to, 189–92, 199–200,
        217, 221, 268, 429, 441–42
    records of, 142, 159, 218, 267
    Selenka's excavations of, 329–33, 369,
        409, 410, 426–27
    site marker for, 200–202, *201,* 218
    standpoints of, *141, 316*
Troglodytes sivalensis, see
        Anthropopithecus sivalensis
tukang kebun, 84, 85, 86
Turner, William, 287–90, 305
Tyndall, John, 219

University of Amsterdam, 25–31, 42–43,
        48, 54, 74–76, 152, 297, 299–303,
        319, 320, 328, 335, 342–43, 374
University of Jena, 297
University of Leiden, 386, 399, 424–25
University of Strasbourg, 297
University of Utrecht, 42, 48
Ursuline Sisters, 24–25

Van Bemmelen, J. M., 284
Van den Koppel, C., 4, 428–29

Van der Leeuw, P. J., 188, 189, 199, 200,
    219
Van der Steen (Dubois' assistant), 403, 405
Van der Waals, J. D., 25
Vanity Fair, 278
Van Rietschoten, B. D., 118–19, *120,* 128
Van Stein Callenfels, Pieter Vincent,
    422–24
Verbeek, Rogier D. M., 118, 119–20, 174
Verneau, René, 287, 298
*Verslagen der Kon. Akademie voor
    Wetenschappen,* 375
Virchow, Rudolf:
    evolution as viewed by, 56, 57, 204,
        257, 297, 309, 317, 318, 413
    Haeckel vs., 57, 297, 305–10, 317
    Pithecanthropus erectus as viewed by,
        278, 279, 282–85, 293, 294, 295, 296,
        305–10, 318, 404, 414
Vogt, Karl, 13–19, 21, 30
Von Koenigswald, G. H. R.:
    Dubois' correspondence with, 419,
        431–35, 438–40
    Dubois' meetings with, 424–25, 428
    Dubois vs., 415–17, *417,* 428, 431–40,
        443, 444–50, 453
    financial support for, 418, 427
    native collectors used by, 416–17, 418,
        420, 421, 432–33
    Ngandong excavations of, 408–12
    photograph of, *433*

Pithecanthropus erectus as viewed by,
    415–28, *417,* 431–40, 443–50, 453
Pithecanthropus erectus mandible
    fragment found by, 431–32, 435,
    436, 447
Sangiran excavations of, 416–27,
    432–40, *433, 437,* 447–48, 449, 450,
    453
Stein's support for, 422–24
Teilhard de Chardin and, 417–18, 440
Von Koenigswald, Luitgarde, 418, 449

Wadjak skulls, *see Homo wadjakensis*
Wallace, Alfred Russel, 62–63, 267
Wallace's line, *61,* 62–63
Warsina, Embok Maas, 146
Weber, Max, 47–48, *48,* 64–66, 69, 75, 98,
    104–5, 109–10, 145, 165, 168, 203,
    303, 338, 367, 369–71
Weidenreich, Franz, 394, *439,* 440, 445–49,
    450
Weinert, Hans, 394, 439, 450
Went, Friedrich, *48*
whales, 48–49
Wilberforce, Samuel, 152
Willem (Anna Dubois' admirer), 123
World War I, 349, 350
World War II, 440, 446–52

*Zeitschrift für Morphologie und
    Anthropologie,* 317–18, 319